Evolution, Literature, and Film

Evolution, Literature, and Film

A READER

EDITED BY

BRIAN BOYD, JOSEPH CARROLL,

AND JONATHAN GOTTSCHALL

Columbia
University
Press
New York

Columbia University Press
Publishers Since 1893
New York   Chichester, West Sussex

Library of Congress Cataloging-in-Publication Data
Evolution, literature, and film : a reader / edited by Brian Boyd, Joseph Carroll,
and Jonathan Gottschall.
  p. cm.
Includes bibliographical references and index.
ISBN 978-0-231-15018-7 (cloth : alk. paper)
ISBN 978-0-231-15019-4 (pbk : alk. paper)
1. Literature and science. 2. Darwin, Charles, 1809–1882—Influence. 3. Evolution in
literature. 4. Evolution in motion pictures. 5. Literature—History and criticism—Theory,
etc. 6. Motion pictures—History. 7. Motion pictures—Philosophy. I. Boyd, Brian, 1952–
II. Carroll, Joseph, 1949– III. Gottschall, Jonathan.
PN55.E86   2010
809'.9336—dc22   2009052788

Contents

Humankind

Part III. Literature, Film, and Evolution: Theory

Part IV. Interpretations

Part V. Literature as Laboratory

Editors' Note

Most of the journal articles and excerpts from books included in *Evolution, Literature, and Film* have been abridged, and their various forms of citation have been standardized. In condensing the articles, the editors have aimed at shaping coherent essays and have not indicated omissions through ellipses. Sources for all the reprinted essays are indicated in the credit note on the first page of each contribution and are also included in the bibliography. Essays that were originally published as journal articles or as book chapters in edited collections can usually be located in the bibliography by their titles in this volume.

Evolution, Literature, and Film

Introduction

Brian Boyd, Joseph Carroll,
and Jonathan Gottschall

Literary and Film Studies Now: Death or Rebirth?

Adding to a long litany of similar laments, William Deresiewicz recently declared that "the real story of academic literary criticism today is that the profession is, however slowly, dying." We need not sit passively by, watching as the patient succumbs. Instead, we should take seriously an observation by the philosopher Kwame Anthony Appiah: "In the humanities . . . we are always engaged in illuminating the present by drawing on the past; it is the only way to make a future worth hoping for" (*Experiments* 1–2). By drawing on the deep past of our evolutionary history, studies in literature and film can ensure their survival and enrich their—and our—common human future.

For at least a decade, many scholars in literature have felt that their discipline is in crisis. They have been demoralized by falling enrollments and funding, by eroding prestige within and beyond academia, and by a sense of intellectual repetition and exhaustion (see chapter 36). Poststructuralism swept through departments of literature and film in the late 1970s and early 1980s, but its once fresh questions have hardened into habit or dogma. Women's studies, queer theory, ethnic literatures, cultural studies, postcolonialism, and ecocriticism opened up new subject areas, but they have been thoroughly explored through now-familiar research modes. Science studies, under the aegis of Foucauldian discourse theory, have offered another new field but suffer from the same malaise that afflicts poststructuralism in general. The Sokal hoax, in 1996, was only the most spectacular exposé of mortal weaknesses at the heart of this school (Sokal and Bricmont, *Fashionable Nonsense*).

Many scholars working under the influence of "New Historicism" or "cultural studies" now claim that they are "post-theory" because they focus not on theories but on "empirical" historical data, especially data gleaned from archives. In reality, the archivalists have not left poststructuralist theory behind but have only internalized it. The categories they use derive chiefly from Foucauldian traditions: versions of Marxism and Freudianism filtered through deconstructive epistemology and cast in a programmatically oppositional mode. Simply eschewing explicit theoretical commitment does not invest findings with empirical validity or insulate them from theoretical critique. It has certainly done little to reverse the declining status of the humanities within the larger world of knowledge.

While the established forms of study in literature and cinema have been drifting into disarray, the evolutionary analysis of human nature has been maturing. It has increasingly placed due emphasis on cooperation, culture, intelligence, and imagination. For more than three decades, many observers inside and outside academia have been fascinated to see how evolutionary studies can illuminate human lives, feelings, thoughts, and behavior. Over the past fifteen years or so, evolutionary study in literature and film has emerged as a distinct movement. In the past few years, it has gained rapidly in visibility and impact, with many articles and books, and with much attention from the popular and scholarly press, from *Nature* to the *New York Times*.

Evolutionary thinking has had revolutionary effects across the human sciences, but it has most dramatically transformed psychology. Those who think that psychological research limits itself to rats running mazes or pigeons pecking levers should think again. Since it began to ask the evolutionary question *why*—for what benefits—do our minds work as they do, psychology has begun to grapple with much at the heart of literature and film: our core emotions like love, fear, sorrow, and happiness; our social and moral emotions like generosity, trust, fairness, and indignation; our core relations like parent and child, partner and partner, friends, allies, and enemies; Theory of Mind, our capacity to understand other minds; and metarepresentation, our capacity to understand representations *as* representations. Cognitive, developmental, and evolutionary psychology track empathy in humans and other animals, and our attunement to the emotions of others. Memory researchers study the relationship between our remembered pasts and imagined futures, or between our memory for character traits and our memory for the past. Evolutionary, developmental, and cognitive psychology together show why these and other capacities arise, where they emerge from, when they develop in individuals, and how they operate. How could literature and film *not* seize on all this with excitement?

Although many theorists in literary and film studies have reacted with alarm to claims that biology might help shape culture, acknowledging the reality of human evolution presents no serious dangers and offers immense opportunities. An evolutionary perspective allows us to see ourselves both in the widest angle and with the most precise focus, as individuals solving particular problems within specific contexts, physical and social, using the cognitive equipment—including a predilection for culture—acquired through natural selection. Evolutionary anthropology, biology, economics, and psychology offer an integrated explanatory framework and a host of new insights that we can apply to literature and film.

Evolutionary theorists of the arts aim not just to offer one more "school" or "approach" to fit within the grab bag of current theories, but to alter the paradigm within which studies in art and culture are conducted. We have rallied to Edward O. Wilson's cry for "consilience" among all the branches of learning. We envision an integrated body of knowledge extending from theories of subatomic particles to theories of the arts. Within this consilient worldview, evolutionary biology is the pivotal discipline uniting the hard sciences with the human sciences and the arts. Evolutionists seek to investigate how evolution has shaped human bodies, minds, and behavior; how culture has emerged out of nature; and how culture has equipped us to modify our behavior.

Virtually all evolutionary theorists of the arts formulate biocultural ideas. That is, we believe that works of art are shaped by our evolved human nature, by culture, and by individual experience. We therefore distinguish ourselves from "cultural constructivists," who effectively attribute exclusive shaping power to culture. We give close attention to "human universals" or cross-cultural regularities that derive from regularities in human nature, but we also recognize the uniquely intense human capacity for culture. We welcome thick descriptions of local context but argue that a true understanding of culture must be rooted in the biological characteristics from which all human cultures grow. Adopting an evolutionary perspective enables us to build theories of literature and film not from near the end of the story but from the start, from the ground up. By building in this way, we can ask altogether new questions and return to older questions with sharper eyes and surer hands.

By insisting on the separateness of humanistic subjects and modes of inquiry, many in the humanities have deprived themselves of the resources discovered in other fields. They have also rendered their own research irrelevant to the interests that most actively engage the minds of the larger educated world. It need not be this way.

What Dangers Does Evolutionary Theory Pose?

The social sciences and humanities have long resisted evolutionary biology, assuming that an evolutionary perspective can only minimize culture, glorify ruthless self-seeking, and deny human uniqueness, diversity, and purpose. In fact, evolution lets us understand the full measure of human uniqueness *and* human continuity with other life, the power of cooperation, culture, diversity, and the emergence of purpose. Let us quickly allay misgivings about supposed dangers in applying evolution to human minds and behavior.

- *Ignoring human differences.* An evolutionary view of human nature does not ignore or deny the enormous cultural differences among peoples, but makes it possible to *explain* cultural differences in a way that insisting humans are completely "culturally constructed" cannot. (Constructed out of what, in any case?)
- *Biological or genetic determinism.* All modern evolutionists recognize that *phenotypes* (the observable properties of organisms) are not determined solely by *genotypes* (the genetic recipes in DNA). Behavior is always *co*-determined by the interaction of genes and environments. Environments shape, constrain, and elicit the behaviors of organisms. Failing to account for complex interactions between genes and environments is, in fact, profoundly *un*biological.
- *Nature versus culture.* Biology is not an alternative to society or culture. Sociality occurs only within living species, and culture exists only within the social and therefore the biological realm. Moreover, culture is far from being uniquely human. Culture—the nongenetic transmission of behavior, including local customs and even fashions—has been discovered over the past few decades in many social species, among birds as well as mammals.
- *The natural as right.* To argue that biology provides a base for human life does not mean that it must impose a model for human morality. A particular origin need not predetermine a particular end. Biology, in any case, offers a vast array of different models, and culture is itself a part of biology with the power to generate a cascade of new possibilities.
- *Genetic selfishness.* Genes are "selfish" in the sense that they prosper according to what benefits them in successive reproductive rounds, but most genes benefit from the health of a whole organism or even from the success of a whole group of individuals. Richard Dawkins points out that he could just as aptly have called his famous first book not *The Selfish Gene* but *The Cooperative Gene* (a title adopted for a more recent book by the biologist Mark Ridley) (*Ancestor's Tale* 158). Evolutionary psychology and evolutionary economics concern themselves with the complex mix of cooperation and competition in

social life. Indeed, they have placed far *more* emphasis on generosity, trust, and fairness than *non*evolutionary psychology or economics ever had.

- *The supposedly fixed and unchangeable character of human nature.* Evolution *is* change, and species have evolved multiple ways to respond more sensitively to rapid change, including sexual reproduction, nervous systems, flexible intelligence, social learning, and culture. Biologist David Sloan Wilson observes that because of the unique importance of culture in humans, "we have not escaped evolution, as so commonly assumed. We experience evolution in hyperdrive" (Foreword 16).

- *Nature and power.* Some fear that "Nature, red in tooth and claw," would valorize power and class. In fact, evolutionary anthropology and biology increasingly stress that a major difference between humans and other mammals is that humans have found ways to control the urge for dominance. They collaborate to resist being dominated and thus unleash the power of human cooperation (Boehm). A key concern of evolutionary psychology has been to explain not the redness of our teeth and nails, but why we are, as primatologist Frans de Waal puts it, so "good natured" (*Good Natured*).

- *Nature versus human singularity.* Evolution can explain both our substantial continuity with and our substantial differences from other forms of life. Life has passed through a number of major transitions, each involving new forms of cooperation at one level that enable radically new and more complex possibilities at a higher level. Many biologists now see the cooperation that makes human culture possible as the latest major transition in evolution, as dramatic as the transitions from single-celled to multicellular organisms, or from individual organisms to societies (Wilson and Wilson).

What Would We Lose by Adopting an Evolutionary Perspective on Literature and Film?

Traditional humanists approach literature and film from the perspective of cultivated common sense. The scope of their research extends from scholarship on intellectual and artistic traditions to subtle analyses of tone, theme, style, and form in particular works. At its best, this kind of study embodies in its own perspective the qualities we look for in literary prose and cinematic art: elegance, power, humor, wit, wisdom, and passion. The great scholars and critics have assimilated and articulated rich cultural traditions. In constructing and expressing their worldviews, they draw on works of imagination and thus exemplify the value and importance that can attach to the study of artistic culture.

Are any of these values and goals incompatible with an evolutionary perspective on literature and film? We see no reason they should be. Modern science now enters ever more directly into the investigation of what most concerns students of literature and film: human motives and behavior, perception, emotions, cognition, personality, and social dynamics. Integrating the new human sciences with humanist literary cultivation presents not only an immense challenge but also an immense opportunity.

A rapprochement with science need not diminish the creative contribution of the individual literary scholar. The idea that science must mean sterile impersonality is merely a prejudice. In reality, as the physiologist Robert Root-Bernstein explains, the forms of imagination "manifested in styles of scientific creativity" are "just as unique as those of any artist" (53). Einstein famously affirmed that "imagination is more important than knowledge." If creativity is essential to the best work in both the humanities and the sciences, surely it is essential also to work that integrates these domains.

For the past several decades, studies in literature and film have taken a "theoretical" turn. Without abandoning the ideal of cultivating minds through art, poststructuralist scholars and critics have often rejected the perspective of cultivated common sense. In its place they have looked for theories that open up deeper explanations of the forces shaping human experience and products of the human imagination. Semiotics and deconstructive linguistic philosophy stressed the central role language plays in human consciousness. Freudian psychoanalysis opened up psychosexual symbolism emerging from the most intimate family relationships and the phases of childhood development. Gender theory foregrounded the power conflicts built into human sexual relations. Marxist social theory has inquired into the way works of imagination articulate socioeconomic conditions.

Is an evolutionary perspective unable to deal effectively with issues that arise out of linguistic philosophy, depth psychology, gender theory, and socioeconomic theory? No. Evolutionary human science embraces cognitive neuroscience and cognitive linguistics. Evolutionary psychology concentrates heavily on the often-conflicted relations in the core reproductive relations of families—mothers, fathers, and children—and offers new and penetrating insights into gendered social roles. Evolutionary social theory identifies affiliation and dominance as elemental forces in human social interaction. The way those forces ramify into the complexities of specific social economies provides a rich field of exploration for scholars and scientists.

Culture is part of human nature, and all the forces we have been describing—linguistic, psychological, and social—manifest themselves in imaginative culture. An adequate understanding of human nature and the products of the human imagination will require the combined and collective work of biologists, psychologists, social theorists, and scholars in the humanities. We need lose nothing of the best that has been thought and said. We need only add to it.

Frequently Asked Questions

Over the years, those who approach art from evolutionary perspectives have met recurrent criticisms, questions, and complaints. Here are some of the most common, along with our responses.

- *Isn't an evolutionary approach to literature reductive?*

"Reductive" is a pejorative term signifying that an approach oversimplifies the complex and rigidifies the flexible. "Reductiveness" is a fault, but "reduction" is essential to causal explanation, which necessarily involves reduction to more fundamental principles. Successful reduction connects basic causal principles with the particular features of a specific object. For instance, an explanation of heredity that reduces it to genes and DNA can unify our understanding of diverse phenomena and at the same time bring to light undreamed-of diversity and intricacy. Similarly, evolutionary explanations in the arts can identify elemental causal forces and at the same time open out into a powerfully expansive understanding of particular works or artistic features.

- *There have been many failed attempts to examine literature and film from professedly scientific foundations. Why is this different?*

Structuralism, Marxist economics, and Freudian psychoanalysis all claimed a scientific basis for studying literature and film. None of these claims survived prolonged empirical scrutiny. The theory of evolution by means of natural selection, in contrast, has been firmly established as the most fertile and robust theory in the life sciences. Evolutionary psychological and social science has emerged as a cumulative, continually productive research program only in the past forty years. That program provides the necessary context for evolutionary studies in the arts.

- *How would an evolutionary approach to literature and film avoid the dogmatism and guruism that have characterized much recent literary theory?*

The theory of evolution by means of natural selection is not a cult or dogma but a scientific research program that took decades to establish on a firm empirical footing. It has to face the continual challenge of consistency with established and emerging data from multiple disciplines. Evidence, not assertions or the authority of particular advocates, provides its criterion for validity.

- *Frank Lentricchia claims that if he knows a literary critic's theory, he knows in advance how the critic will interpret particular works (64). How do you keep evolutionary approaches from sterile apriorism, if you assume from the outset that all people are ultimately motivated by survival and reproduction?*

Survival and reproduction are "ultimate" principles that have shaped the characteristics of human nature over evolutionary time. In any individual artist, these characteristics operate through "proximate" mechanisms that are often only tenuously connected to survival and reproduction. Individuals vary widely in their genetic makeup and in the experiences—personal and cultural—that shape their motives. Humans fabricate imaginative worlds that can detach them from reproduction and survival—driving them, for instance, into celibacy or into sacrificing their lives for an abstract cause. Each artistic work challenges us to draw connections between the elemental components of human nature and the particular features of a unique imagined world.

- *Aren't the claims in part II just-so stories? Isn't this whole approach underpinned by mere storytelling?*

Evolutionary hypotheses about the adaptive or nonadaptive role of the arts are not final products to be accepted or rejected according to their narrative or rhetorical force. They must be evaluated on the basis of their explanatory and predictive power. The process of testing alternative predictions should—and already has begun to—yield unexpected new information about art and its causes and effects, just as do evolutionary accounts of other familiar but biologically puzzling activities like sleep, dreaming, and play.

- *Isn't it a "category error" to study questions about art in terms of the sciences?*

It would certainly be a category error to confuse art itself with science. Art and science have different origins, purposes, effects, and criteria for success. But we can *study* art as we can human psychology. While we would be wrong to confuse an actual human being with a psychological theory, we can reasonably study both human beings and their cultural products using concepts from the human sciences.

- *Doesn't analyzing literature and film require us to invoke concepts and methods different from those in biology and the human sciences?*

Like any other subject area, literature has "emergent" concepts that are peculiar and appropriate to its own subject. Literary and film scholars necessarily employ categories such as narrative structure, genre, point of view, verse, mise-en-scène, camera angles, cuts, tone, style, and thematic structure. All such categories engage attention, imagination, cognition, and emotion in ways that can be most deeply explored in terms of an evolved and culturally developed human nature.

- *If the paradigm of literary and film study were to shift toward evolution, would everyone have to start reading from the "gene's-eye view"?*

Modern evolutionary analysis requires explanation at multiple levels: at the level of the gene, the organism, the group (pairs, families, alliances), the population, and the species. Recent cultural and literary studies have stressed group-level (class, community, nation) effects, underplaying other levels. The most complete forms of explanation in the arts would connect high-level phenomena like artistic meaning and effect with causal forces at multiple levels.

- *If scholars in the humanities adopt evolutionary perspectives, will there be any room left for imaginative performance—for the kind of work that reveals the depth and richness of the critic's own mind?*

The sciences have developed methods like experiment and quantitative analysis to *overcome* the biases of individuals, cultures, and species. Artists, by contrast, have developed methods to *appeal* to the biases of individuals, cultures, and our species. Critics of literature and film aim at uncovering and explaining facts, but they also respond emotionally and aesthetically to the imaginative qualities in art. Good critics read widely and intensively and write with the boldness and flair that comes from long and deep immersion in great writing. Good criticism requires not only responsive individual experience, but also the objectivity and impartiality that characterize the scientific ethos.

- *How do evolutionary approaches to literature and film relate to cognitive approaches?*

Cognition occurs only within evolution. First-generation cognitive psychology derived from early artificial intelligence, from the kinds of information processing involved in computer programming. Second- and third-generation cognitive psychology has incorporated a greater recognition of evolved minds, especially of the prelinguistic bases of cognition, the emotions, and the connections among senses, thoughts, and feelings. Evolutionary and cognitive approaches are only different facets of a common process of inquiry.

Plan of the Volume

EVOLUTION AND HUMAN NATURE

Historical Overview Part I opens with "Evolutionary Psychology," David M. Buss's overview of the historical development of evolutionary biology and evolutionary psychology. The excerpts in the two subsections that follow were selected for their classic status, their representative character, and their power to illuminate central themes in evolution and human nature. The evolutionary ideas delineated here provide the historical and theoretical background for the inquiries into imaginative culture in the rest of the book.

The Theory of Evolution The most significant scientific theories bring the largest range of phenomena within the smallest compass of causal explanation. Judged by this criterion of significance, Charles Darwin's theory of adaptation by means of natural selection is one of the most successful efforts at explanation in the history of science, ranking with the theories of Copernicus, Newton, and Einstein.

In the selections from *On the Origin of Species* that open the first subsection, Darwin delineates the core elements in his theory of evolution. Not only has Darwin's theory stood the test of time, but it required three-quarters of a century before science caught up with his prescience and confirmed in detail the validity of his explanation for evolutionary change. Within just a few years after the publication of *On the Origin of Species* in 1859, most scientists had accepted the idea of "evolution," or, in Darwin's terms, "descent with modification." That is, they had acknowledged that all currently existing life-forms had descended from previous forms and changed over the course of geologic time. What remained in doubt was the mechanism of change: the theory of "adaptation by means of natural selection." These elements of Darwin's theory were not fully established as the unifying principle of evolutionary biology until the 1930s, in the integrative movement now known as the "Modern Synthesis."

Darwin himself knew nothing about the specific mechanism of inheritance through which characteristics descend from parents to offspring. When he published *On the Origin of Species*, genetics as a science did not exist, and it did not begin to achieve scientific maturity until the early twentieth century. DNA, the molecule that carries genetic information, was not understood until the middle of the twentieth century. In "The Digital River," Richard Dawkins formulates the theory of natural selection in modern genetic terms.

Humankind Darwin not only founded the modern concept of evolution but also made the first major contribution to evolutionary psychology in *The Descent of Man, and Selection in Relation to Sex*, key passages of which open the second subsection. Natural selection had to wait decades to be fully confirmed by subsequent scientific research. Darwin's account of the evolved and adapted character of human nature had to wait still longer. In the first decades of the twentieth century, most social science resolutely turned away from evolutionary conceptions, partly as a reaction to "social Darwinism," which in fact owed more to the laissez-faire philosopher Herbert Spencer than to Darwin. Until the last quarter of the twentieth century, most social scientists rejected any appeal to evolutionary biology. The final chapter of Edward O. Wilson's *Sociobiology*, "Man: From Sociobiology to Sociology," thus marked a pivotal event in the transition to modern Darwinian social science. We include excerpts from that chapter and from Wilson's retrospective and forward-looking commentary, "Sociobiology at Century's End," in *Sociobiology*'s twenty-fifth–anniversary edition.

The evolutionary understanding of human nature has often concentrated on the "species-typical" or "universal" characteristics of human nature. Anthropologist Donald E. Brown's synthetic portrait of "The Universal People" offers an engaging introduction to the characteristics shared by humans in all known cultures. In "Evolution and Explanation," a concise theoretical and historical overview of modern psychology, cognitive psychologist Steven Pinker reveals the explanatory power that evolutionary theory brings to the social sciences. And, finally, biologist David Sloan Wilson argues in "Evolutionary Social Constructivism" that culture reshapes human motives, minds and environments and that "cultural constructivism" should, accordingly, be included in an evolutionary understanding of human nature.

THE RIDDLE OF ART

Homo sapiens is a strangely artistic ape. Most humans spend much of their time outside work and sleep lost in landscapes of make-believe. Across the whole breadth of human history, across the wide mosaic of world cultures, there has never been a society in which people have *not* devoted much of their time to seeing, creating, and hearing fictions—from folktales to film, from theater to television. Nor has there ever been a culture in which people have not devoted precious resources—in time, skill, and material—to producing songs, music, and dances; pictures and carvings; and ornaments, decorations, and designs.

Other animals have adaptations for cooperation in social groups with specialized functions and status hierarchies. Other animals engage in play, produce technology, and share information. Humans alone produce imaginative artifacts designed to provide aesthetic pleasures, evoke subjective sensations, express emotions, depict nature or human experience, or delineate through symbols the salient features of their experience. Dispositions for producing and consuming art constitute uniquely human, species-typical characteristics, and the arts offer rich insights into the human mind. No understanding of human nature that leaves out the arts could possibly give an adequate account of its subject.

Art poses an evolutionary riddle: Why are we storytelling apes? Why do we compulsively concoct and consume fictional stories, stories we know to be untrue? Why do we spend (waste?) so much time telling tales and shaping objects for aesthetic effects—time that could be devoted to activities that produce obvious biological benefits: securing resources, courting mates, or caring for offspring and other kin?

The selections in part II show some of the range of positions on this contentious and complex issue. In "Art and Adaption," Steven Pinker argues that aesthetic aptitudes arose as side effects of other adaptively significant mental powers. In "Arts of Seduction," Geoffrey Miller offers the most influential version of the idea that the arts serve the purposes of sexual display. Edward O. Wilson ("The Arts and Their Interpretation"), Ellen Dissanayake ("Art and Intimacy"), John Tooby and Leda Cosmides ("Does Beauty Build Adapted Minds?"), and Denis Dutton ("The Uses of Fiction") provide overlapping versions of arguments that the arts fulfill definite adaptive functions. Brian Boyd, in part IV, also addresses the adaptiveness of art in tandem with a close reading of a single avant-garde work. In part V, Joseph Carroll, Jonathan Gottschall, John Johnson, and Daniel Kruger offer empirical evidence for one of story's adaptive functions.

LITERATURE, FILM AND EVOLUTION: THEORY

Evolutionary and cognitive approaches to the arts have been strongly at odds with recently prevailing assumptions in the humanities. There have been two core disagreements: evolutionists claim (1) that we humans share with one another a great deal of our mental and motivational makeup, that differences across cultures and ideologies are dwarfed by similarities; and (2) that our minds have been shaped to provide us with mostly valid information about our world, rather than to operate by arbitrary convention.

Clashes between bioculturalism and cultural constructivism have sometimes been sharp. For instance, surveying and lamenting the decline of literary studies, critic Louis Menand appealed for renovation but explicitly rejected just one possible form of renovation, the one we regard as most promising: the integration of the humanities with evolutionary biology. "Consilience," he claimed, "is a bargain with the devil" ("Dangers" 14). Menand's article prompted Brian Boyd's rejoinder: "Getting It All Wrong," which opens part III. While making a sweeping comparison of the key positions in postmodern and evolutionary cultural theory, Boyd delineates the false conceptions that inform Menand's resistance to consilience. In "Imagining Human Nature," Joseph Carroll, Jonathan Gottschall, John Johnson, and Daniel Kruger provide an overview of the literary theory of the twentieth century and contrast it with the emerging evolutionary paradigm. In "Two Worlds," writing from outside literature, Edward Slingerland, a scholar of early Chinese thought, argues that a biologically grounded cognitive science can reconcile the humanities and the sciences. In "Consilient Literary Interpretation," Marcus Nordlund incisively formulates a naturalistic basis for a literary hermeneutics alert to the unpredictable complexities of specific interpretive problems. Refuting attempts to present Shakespeare and his contemporaries as deniers of shared human nature, Robin Headlam Wells ("Humanism and Human Nature in the Renaissance") shows that they had an explicit belief in human nature and a desire to illuminate it.

In "The Reality of Illusion," reviewing the history of film theory, Joseph Anderson suggests that the accessibility of film depends on its appeal to evolved human modes of "ecological" perception—that is, our perception of the environment that we negotiate as mobile creatures. Murray Smith, in "Darwin and the Directors," shows how evolution forces us to take the emotions seriously; deepens the analysis of film's expression of emotions in posture, gesture, and speech; and alters the weighting of editing and image in film theory. Beginning with what his American students can understand of an undubbed, unsubtitled Taiwanese martial arts movie—a great deal, it turns out—David Bordwell proposes a naturalistic approach to film response that explains both the considerable convergence and the partial divergence in our responses to art in "What Snakes, Eagles, and Rhesus Macaques Can Teach Us."

INTERPRETATIONS

Despite first the "theoretical" and then the "archival" turn in much recent scholarship, readings of individual works still dominate the activity of criticism. To vindicate its paradigmatic claims, evolutionary study in literature and film must also demonstrate that it can give compelling interpretive accounts of particular works. Part IV showcases evolutionary readings that display both explanatory power and sensitivity to particular meanings and effects.

If evolutionary approaches work for any art—if evolution has indeed shaped all artists and audiences—they should work for all arts. The readings in part IV offer test cases chiefly in one main form of art—stories. The readings move chronologically from Homeric epic to avant-garde comics. Although many selections focus on English literature, others discuss ancient or modern European literature or cinema from Africa, the Middle East, and Asia, as well as America and Europe. Modes include prose, drama, verse, film, and comics. Genres range from epic, comedy, tragedy, realistic novel, and story, to science fiction, film, and serially composed comic. While most readings focus on a single work, others consider multiple examples as they concentrate on genre, like Daniel Nettle discussing tragedy and comedy in "The Wheel of Fire and the Mating Game," or on technique, like David Bordwell discussing shot-reverse shot composition in "Convention, Construction, and Cinematic Vision."

Some critics stress what might seem like naturally evolutionary topics: anxieties about paternity certainty in Judith P. Saunders's essay "Paternal Confidence in Zora Neale Hurston's 'The Gilded Six-Bits,'" for instance, or mating strategies in Homer's *Iliad* in Jonathan Gottschall's "Homeric Women: Reimagining the Fitness Landscape." While Saunders focuses on a human universal, Gottschall attempts to explain the peculiar ecological and reproductive pressures that promoted conflict in the society Homer depicts. For those with no prior acquaintance with evolutionary approaches to literature, other contributions might be more surprising: William Flesch's "Vindication and Vindictiveness," on altruistic punishment in *Oliver Twist*; Joseph Carroll's "The Cuckoo's History," an analysis of *Wuthering Heights* in terms of human life-history theory; and Brian Boyd's discussion, in "Art and Evolution," of dominance and counter-dominance in Art Spiegelman's contribution to *The Narrative Corpse*.

Many of the critics address the question of human nature and its relative persistence across time and circumstance. In "New Science, Old Myth," Michelle Scalise Sugiyama reads *Oedipus Rex* in the light of evolutionary evidence. Marcus Nordlund considers local legal and cultural conditions that modify male sexual jealousy in early-seventeenth-century Europe ("Jealousy in *Othello*"), but he argues strongly against the idea of Othello's jealousy as a predominantly local construct. In "Human Nature, Utopia and Dystopia," Brett Cooke shows how Yevgeny Zamyatin, in his science-fiction dystopia, *We*, critiques Soviet Communism's attempts to ignore and re-engineer human nature. Others set Freudian against evolutionary accounts of human nature. In "Wordsworth, Psychoanalysis and the 'Discipline of Love,'" drawing on the developmental theories of John Bowlby and Daniel Stern, Nancy Easterlin revises feminist and psychoanalytic accounts of mother–infant bonding in Wordsworth's *Prelude*.

While evolutionary criticism has a deep interest in human nature, it should not neglect local or individual differences or the effects of artistry. Indeed, just as life itself is a problem-solving process, an evolutionary and cognitive approach to the arts should focus on the partly unique and partly shared problem–solution landscape within which individual artists work. Essays that place particular emphasis on form and individual artistry include those by Nordlund on Shakespeare, Easterlin on Wordsworth, Joseph Anderson on Orson Welles ("Character in *Citizen Kane*"), Bordwell on shot–reverse shot composition, and Boyd on Spiegelman.

LITERATURE AS LABORATORY

Most evolutionary literary scholars believe that mainstream literary studies could do a much better job of generating reliable and cumulative knowledge. We all concur in seeking better sources of knowledge in the biologically grounded human sciences, but some of us have taken a further step: using empirical methods to study subjects in the humanities.

In taking this step, we distance ourselves sharply from two salient features of postmodernist literary theory. We reject both the radical skepticism that denies the possibility of ever gaining reasonably objective knowledge, and the practice, so regrettably common in literary study, of looking for evidence only or mainly in *support* of our hypotheses, especially those that we deem politically desirable.

Science has established many ways to avoid being misled by selective evidence, confirmation bias, and a priori thinking. We believe that many of these methods can be adapted to questions about art. We cannot do without reading, thinking, and writing well, but we want to extend the scholarly tool kit by adding scientific methods, which can test, tease apart, and extend intuitions about literature.

Literature offers an unparalleled source of thick descriptions of imagined human behavior and actual human preferences over thousands of years— evidence that can be used to inspire, enrich, or challenge hypotheses about human nature. In part V, we include examples of humanities scholars and scientists flowing back and forth across the divide between the two cultures of the arts and the sciences. Their goal is not to promote a conquest of humanistic culture by scientific culture. Rather, they are seeking to establish a Third Culture on the fertile ground between the humanities and the sciences. We include selections suggesting that (1) stories are a valuable and much-neglected resource in the scientific study of mind and behavior, and (2) humanities scholars can draw on the methods, as well as the theories and findings, of the sciences.

Jonathan Gottschall's "Literature, Science, and a New Humanities" opens part V—and his book of that name—by diagnosing the widespread sense of malaise within literary studies and proposing, as his prescription for recovery, that we add the objective methods of science to standard subjective methodologies in order to create a more durable and more rapidly cumulative literary knowledge. Catherine Salmon and Donald Symons's "Slash Fiction and Human Mating Psychology" investigates the first-time responses to slash fiction of female romance readers and argues that, despite its male–male lovers, slash fiction appeals to mainstream female mating psychology. In "Cultural Variation Is Part of Human Nature," Michelle Scalise Sugiyama reconsiders anthropologist Laura Bohannan's classic study of the unexpected responses of the Tiv people of West Africa to her retelling of *Hamlet*. Scalise Sugiyama concludes not that Bohannan's report demonstrates deep-seated human differences but that cultural variation and universal cognitive design are complementary rather than contradictory. Literary scholars Joseph Carroll and Jonathan Gottschall teamed with psychologists John Johnson and Daniel Kruger to amass and analyze over 500 readers' responses to an online questionnaire about the personality and romantic and social ambitions of 2,000 characters in 200 nineteenth-century British novels. Among many findings in "Paleolithic Politics in British Novels of the Nineteenth Century," the authors suggest that the demonization of antagonists for their single-minded pursuit of social dominance matches the counter-dominance strategies evident in hunter-gatherer societies, which "stigmatize and suppress status-seeking in potentially dominant individuals."

The essays in this volume offer a chance to reshape the landscape of scholarship. The contributions from evolutionary biology, the human sciences, and the humanities provide complementary and convergent perspectives on a single grand subject: human nature in the works of the human imagination. Despite productive disagreements, the authors represented here embody a set of shared attitudes and beliefs. We all believe that evolutionary theory promises the deepest, widest, and most reliable knowledge about human-kind and all its works. Still more broadly, we believe in science; we believe that the world can be known in a reasonably objective way, that knowledge in any field can accumulate, and that knowledge in diverse fields can be causally connected. We are eager to extend scientific knowledge into the fields of the humanities and to make use of the tools available from both the humanities and the sciences. From the humanities, we incorporate the skills of historical scholarship and close reading. We respect the sensitivity critics display in responding to the artistry of literature and film, and we are committed to preserving and celebrating the rich heritage of the humanities. From evolutionary biology and the human sciences, we incorporate a knowledge of human universals and of the cognitive mechanisms underpinning art. We recognize that the building blocks of human nature combine in different ways and under different conditions to form unique structures in any given culture. We also recognize the rich genetic and experiential differences between individuals. Whether or not we adopt empirical, quantitative methods, we share a profound respect for the spirit of impersonal, disinterested inquiry. We recognize that a passionate responsiveness to the arts is natural to scholarship in the humanities, but we are also determined to have access to the impersonal, objective scrutiny of science.

If the impulses behind this anthology were to become active across the humanities, this would constitute an epistemic revolution expanding the scope of both science and the humanities. In the short term, it would open the products of the human imagination to the human sciences and the methods and results of the human sciences to the humanities. In the long term, it would enable humanists to join with scientists in contributing to the continual development of more reliable and durable knowledge. The revolutionary impulses implicit in this volume are thus fundamentally creative and constructive. We have no illusions that our formulations are fixed and final, but we have felt the excitement of making new discoveries and look forward confidently to more. We have all been inspired by the exhilarating sense that we are joining together in an intellectual adventure of great scope. We invite you to join us.

Part I

Evolution and Human Nature

HISTORICAL OVERVIEW

1

Evolutionary Psychology

THE NEW SCIENCE OF THE MIND

David M. Buss

Evolution Before Darwin

Evolution refers to change over time in organic (living) structure. Change in life-forms was postulated by scientists to have occurred long before Darwin published his classic 1859 book, *On the Origin of Species* (Glass, Temekin, and Straus; C. L. Harris).

Jean Pierre Antoine de Monet de Lamarck (1744–1829) was one of the first scientists to use the word *biologie*, thus recognizing the study of life as a distinct science. Lamarck believed in two major causes of species change: first, a natural tendency for each species to progress toward a higher form and, second, the inheritance of acquired characteristics. Lamarck said that animals must struggle to survive and that this struggle causes their nerves to secrete a fluid that enlarges the organs involved in the struggle. Giraffes evolved long necks, he thought, through their attempts to eat from higher and higher leaves. Lamarck believed that the neck changes that came about from these strivings were passed down to succeeding generations of giraffes, hence the phrase "the inheritance of acquired characteristics." Another theory of change in life-forms was developed by Baron Georges Léopold Chrétien Frédérick Dagobert Cuvier (1769–1832). Cuvier proposed a theory called *catastrophism*, according to which species are extinguished periodically by sudden catastrophes, such as meteorites, and then replaced by different species.

Biologists before Darwin also noticed the bewildering variety of species, some with astonishing structural similarities. Humans, chimpanzees, and orangutans, for example, all have exactly five digits on each hand and foot. The wings of birds are similar to the flippers of seals, perhaps suggesting that one was modified from the other (Daly and Wilson, *Sex*). Comparisons among these species suggested that life was not static, as some scientists and theologians had argued. Further evidence suggesting change over time also came from the fossil record. Bones from older geological strata were not the same as bones from more recent geological strata. These bones would not be different, scientists reasoned, unless there had been a change in organic structure over time.

Another source of evidence came from comparing the embryological development of different species (Mayr, *Growth*). Biologists noticed that such development was strikingly similar in species that otherwise seemed very different from one another. An unusual loop-like pattern of arteries close to the bronchial slits characterizes the embryos of mammals, birds, and frogs. This evidence suggested, perhaps, that these species might have come from the same ancestors many years ago. All these pieces of evidence, present before 1859, suggested that life was not fixed or unchanging. The biologists who believed that organic structure changed over time called themselves evolutionists.

Another key observation had been made by various evolutionists before Darwin: Many species possess characteristics that seem to have a purpose. The porcupine's quills help it fend off predators. The turtle's shell helps to protect its tender organs from the hostile forces of nature. The beaks of many birds are designed to aid in cracking nuts. This apparent functionality, so seemingly abundant in nature, also required an explanation.

Missing from the evolutionists' accounts before Darwin, however, was a theory to explain how change might take place over time and how such seemingly purposeful structures like the giraffe's long neck and the porcupine's sharp quills could have come about. A causal mechanism or process to explain these biological phenomena was needed. Charles Darwin provided the theory of just such a mechanism.

Darwin's Theory of Natural Selection

Darwin's task was more difficult than it might at first appear. He wanted not only to explain why change takes place over time in life-forms, but also to account for the particular ways it proceeds. He wanted to determine how new species emerge (hence the title of his book *On the Origin of Species*), as well as how others vanish. Darwin wanted to explain why the component parts of animals—the long necks of giraffes, the wings of birds, the trunks of elephants—existed in those particular forms. And he wanted to explain the apparent purposive quality of those forms, or why they seem to function to help organisms accomplish specific tasks.

The answers to these puzzles can be traced to a voyage Darwin took after graduating from Cambridge University. He traveled the world as a naturalist on a ship, the *Beagle*, for a five-year period, from 1831 to 1836. During this voyage he collected dozens of samples of birds and other animals from the Galápagos Islands in the Pacific Ocean. On returning from his voyage he discovered that the Galápagos finches, which he had presumed were all of the same species, actually varied so much that they constituted different species. Indeed, each island in the Galápagos had a distinct species of finch. Darwin determined that these different finches had a common ancestor but had diverged from each other because of the local ecological conditions on each island. This geographic variation was likely pivotal to Darwin's conclusion that species are not immutable but can change over time.

What could account for why species change? This was the next challenge. Darwin struggled with several different theories of the origins of change, but rejected all of them because they failed to explain a critical fact: the existence of adaptations. Darwin wanted to account for change, of course, but perhaps even more important he wanted to account for why organisms appeared so well designed for their local environments.

> It was . . . evident that [these others theories] could [not] account for the innumerable cases in which organisms of every kind are beautifully adapted to their habits of life—for instance, a woodpecker or tree-frog to climb trees, or a seed for dispersal by hooks and plumes. I had always been much struck by such adaptations, and until these could be explained it seemed to me almost useless to endeavour to prove by indirect evidence that species have been modified. (Darwin, from his autobiography; cited in Ridley, *Origins* 9)

Darwin unearthed a key to the puzzle of adaptations in Thomas Malthus's *An Essay on the Principle of Population* (published in 1798), which introduced Darwin to the notion that organisms exist in numbers far greater than can survive and reproduce. The result must be a "struggle for existence," in which favorable variations tend to be preserved and unfavorable ones tend to die out. When this process is repeated generation after generation, the end result is the formation of a new species.

More formally, Darwin's answer to all these puzzles of life was the theory of *natural selection* and its three essential ingredients: *variation, inheritance,* and *selection.* (The theory of natural selection was discovered independently by Darwin and Alfred Russel Wallace [1858]; Darwin and Wallace co-presented the theory at a meeting of the Linnaean Society.) First, organisms vary in all sorts of ways, such as in wing length, trunk strength, bone mass, cell structure, fighting ability, defensive ability, and social cunning. Variation is essential for the process of evolution to operate—it provides the "raw materials" for evolution.

Second, only some of these variations are inherited—that is, passed down reliably from parents to their offspring, which then pass them on to their offspring down through the generations. Other variations, such as a wing deformity caused by an environmental accident, are not inherited by offspring. Only those variations that are inherited play a role in the evolutionary process.

The third critical ingredient of Darwin's theory is selection. Organisms with some heritable variants leave more offspring *because* those attributes help with the tasks of *survival* or *reproduction.* In an environment in which the primary food source might be nut-bearing trees or bushes, some finches with a particular shape of beak, for example, might be better able to crack nuts and get at their meat than would finches with other shapes of beaks. More finches who have beaks better shaped for nut cracking survive than those with beaks poorly shaped for nut cracking and thereby can contribute to the next generation.

An organism can survive for many years, however, and still not pass on its inherited qualities to future generations. To pass its inherited qualities to future generations it must reproduce. Thus, *differential reproductive success,* brought about by the possession of heritable variants that increase or decrease an individual's chances of surviving and reproducing, is the "bottom line" of evolution by natural selection. Differential reproductive success or failure is defined by reproductive success relative to others. The characteristics of organisms who reproduce more than others, therefore, get passed down to future generations at a relatively greater frequency. Because survival is usually necessary for reproduction, it took on a critical role in Darwin's theory of natural selection.

Darwin's Theory of Sexual Selection

Darwin had a wonderful scientific habit of noticing facts that seemed inconsistent with his theories. He observed several that seemed to contradict his theory of natural selection, also called "survival selection." First he noticed weird structures that seemed to have absolutely nothing to do with survival; the brilliant plumage of peacocks was a prime example. How could this strange luminescent structure possibly have evolved? The plumage is obviously metabolically costly to the peacock. Furthermore, it seems like an open invitation to predators, suggesting fast food. Darwin became so obsessed with this apparent anomaly that he once commented, "The sight of a feather in a peacock's tail, whenever I gaze at it makes me sick!" (quoted in Cronin, *Ant* 113). Darwin also observed that in some species, the sexes differed dramatically in size and structure. Why would the sexes differ so much, Darwin pondered, when both have essentially the same problems of survival, such as eating, fending off predators, and combating diseases?

Darwin's answer to these apparent embarrassments to the theory of natural selection was to devise what he believed to be a second evolutionary theory: the theory of *sexual selection*. In contrast to the theory of natural selection, which focused on adaptations that have arisen as a consequence of successful survival, the theory of sexual selection focused on adaptations that arose as a consequence of successful mating. Darwin envisioned two primary means by which sexual selection could operate. The first is *intrasexual competition*—competition between members of one sex, the outcomes of which contributed to mating access to the other sex. The prototype of intrasexual competition is two stags locking horns in combat. The victor gains sexual access to a female either directly or through controlling territory or resources desired by the female. The loser typically fails to mate. Whatever qualities lead to success in the same-sex contests, such as greater size, strength, or athletic ability, will be passed on to the next generation by virtue of the mating success of the victors. Qualities that are linked with losing fail to get passed on. So evolution—change over time—can occur simply as a consequence of intrasexual competition.

The second means by which sexual selection could operate is *intersexual selection*, or preferential mate choice. If members of one sex have some consensus about the qualities that are desired in members of the opposite sex, then individuals of the opposite sex who possess those qualities will be preferentially chosen as mates. Those who lack the desired qualities fail to get mates. In this case, evolutionary change occurs simply because the qualities that are desired in a mate increase in frequency with the passing of each generation. If females prefer to mate with males who give them nuptial gifts, for example, then males with qualities that lead to success in acquiring nuptial gifts will increase in frequency over time. Darwin called the process of intersexual selection *female choice* because he observed that throughout the animal world, females of many species were discriminating or choosy about whom they mated with. However, it is clear that *both* sexes engage in preferential mate choice and *both* sexes compete with members of their own sex for access to desirable members of the opposite sex.

Darwin's theory of sexual selection succeeded in explaining the anomalies that had given him nightmares. The peacock's tail, for example, evolved because of the process of intersexual selection: Peahens prefer to mate with males who have the most brilliant and luminescent plumage. Males are often larger than females in species in which males engage in physical combat with other males for sexual access to females—the process of intrasexual competition.

Although Darwin believed that natural selection and sexual selection were two separate processes, it is now known that they stem from the same fundamental process: differential reproductive success by virtue of heritable differences in design. Nonetheless, some biologists believe that it is useful to distinguish between natural selection and sexual selection. The distinction crisply highlights the importance of two classes of adaptations: those that evolved because of the survival advantage they gave organisms (e.g., tastes for sugar and fat help guide us to eat certain foods that lead to survival; fear of snakes helps us to avoid getting poisonous bites) and those that evolved because of the mating advantage they gave organisms (e.g., greater fighting ability in the sex that competes).

The Role of Natural and Sexual Selection in Evolutionary Theory

Darwin's theories of natural and sexual selection are relatively simple to describe, but many sources of confusion surround them even to this day. This section clarifies some important aspects of selection and its place in understanding evolution.

First, natural and sexual selection are not the only causes of evolutionary change. Some changes, for example, can occur because of a process called *genetic drift*, which is defined as random changes in the genetic makeup of a population. Random changes come about through several processes, including *mutation* (a random hereditary change in the DNA), founder effects, and genetic bottlenecks. Random changes can arise through a *founder effect*, which occurs when a small portion of a population establishes a new colony and the founders of the new colony are not entirely genetically representative of the original population. Imagine, for example, that the 200 colonizers who migrate to a new island happen by chance to include an unusually large number of redheads. As the population on the island grows, say, to 2,000 people, it will contain a larger proportion of redheads than did the original population from which the colonizers came. Thus founder effects can produce evolutionary change—in this example, an increase in genes coding for red hair. A similar random change can occur through *genetic bottlenecks*, which happen when a population shrinks, perhaps owing to a random catastrophe such as an earthquake. The survivors of the random catastrophe carry only a subset of the genes of the original population. In sum, although natural selection is the *primary* cause of evolutionary change and the only known cause of adaptations, it is not the only cause of evolutionary change. Genetic drift—through mutations, founder effects, and genetic bottlenecks—can also produce change in the genetic makeup of a population.

Second, evolution by natural selection is not forward-looking and is not "intentional." The giraffe does not spy the juicy leaves stirring high in the tree and "evolve" a longer neck. Rather, those giraffes that, owing to an inherited variant, happen to have longer necks have an advantage over other giraffes in getting to those leaves. Hence they have a greater chance of surviving and thus of passing on their slightly longer necks to their offspring (recent work suggests that the long neck of giraffes may serve other functions, such as success in same-sex combat). Natural selection merely acts on those variants that happen to exist. Evolution is not intentional and cannot look into the future and foresee distant needs.

Another critical feature of selection is that it is *gradual*, at least when evaluated relative to the human life span. The short-necked ancestors of giraffes did not evolve long necks overnight or even over the course of a few generations. It has taken dozens, hundreds, thousands, and in some cases millions of generations for the process of selection to gradually shape the organic mechanisms we see today. Of course, some changes occur extremely slowly, others more rapidly. And there can be long periods of no change, followed by a relatively sudden change, a phenomenon known as *punctuated equilibrium* (Gould and Eldredge). But even these "rapid" changes occur in tiny increments each generation and take hundreds or thousands of generations to occur.

Darwin's theory of natural selection offered a powerful explanation for many baffling aspects of life, especially the origin of new species (although Darwin failed to recognize the full importance of geographic isolation as a precursor to natural selection in the formation of new species [Cronin, *Ant*]). It accounted for the modification of organic structures over time. It also accounted for the apparent purposive quality of the component parts of those structures—that is, that they seem "designed" to serve particular functions linked with survival and reproduction.

Perhaps most astonishing to some (but appalling to others), in 1859 natural selection united all species into one grand tree of descent in one bold stroke. For the first time in recorded history, each species was viewed as being connected with all other species through a common ancestry. Human beings and chimpanzees, for example, share more than 98 percent of each other's DNA and shared a common ancestor perhaps 6 million years ago (Wrangham and Peterson). Even more startling is the finding that many human genes turn out to have counterpart genes in a transparent worm called *Caenorhabditis elegans*. They are highly similar in chemical structure, suggesting that humans and this worm evolved from a distant common ancestor (Wade, "Dainty Worm"). In short, Darwin's theory made it possible to locate humans in the grand tree of life, showing their place in nature and their links with all other living creatures.

Darwin's theory of natural selection created a storm of controversy. Lady Ashley, a contemporary of Darwin, remarked on hearing his theory that human beings descended from apes: "Let's hope it's not true; but if it is true, let's hope that it does not become widely known." In a famous debate at Oxford University, Bishop [Samuel] Wilberforce bitingly asked his rival debater Thomas Huxley whether the "ape" from which Huxley descended was on his grandmother's or his grandfather's side.

Even biologists at the time were highly skeptical of Darwin's theory of natural selection. One objection was that Darwinian evolution lacked a coherent theory of inheritance. Darwin himself preferred a "blending" theory of inheritance, in which offspring are mixtures of their parents, much like pink paint is a mixture of red and white paint. This theory of inheritance is now known to be wrong, as we will see later in the discussion of the work of Gregor Mendel, so early critics were correct in the objection that the theory of natural selection lacked a solid theory of heredity.

Another objection was that some biologists could not imagine how the early stages of the evolution of an adaptation could be useful to an organism. How could a partial wing help a bird, if a partial wing is insufficient for flight? How could a partial eye help a reptile, if a partial eye is insufficient for sight? Darwin's theory of natural selection requires that each and every step in the gradual evolution of an adaptation be advantageous in the currency of reproduction. Thus, partial wings and eyes must yield an adaptive advantage, even before they evolve into fully developed wings and eyes. For now it is sufficient to note that partial forms can indeed offer adaptive advantages; partial wings, for example, can keep a bird warm and aid in mobility for catching prey or avoiding predators, even if they don't afford full flight. This objection to Darwin's theory is therefore surmountable (Dawkins, *Blind Watchmaker*). Further, it is important to stress that just because biologists or other scientists have difficulty imagining certain forms of evolution, such as how a partial wing might be useful, that is not a good argument against such forms having evolved. This "argument from ignorance," or as Dawkins calls it, "the argument from personal incredulity"

(*Extended Phenotype*) is not good science, however intuitively compelling it might sound.

A third objection came from religious creationists, many of whom viewed species as immutable (unchanging) and created by a deity rather than by the gradual process of evolution by selection. Furthermore, Darwin's theory implied that the emergence of humans and other species was "blind," resulting from the slow, unplanned, cumulative process of selection. This contrasted with the view that creationists held of humans (and other species) as part of God's grand plan or intentional design. Darwin had anticipated this reaction, and apparently delayed the publication of his theory in part because he was worried about upsetting his wife, Emma, who was deeply religious.

The controversy continues to this day. Although Darwin's theory of evolution, with some important modifications, is the unifying and nearly universally accepted theory within the biological sciences, its application to humans, which Darwin clearly envisioned, still meets with vigorous resistance. But humans are not exempt from the evolutionary process, despite our profound resistance to being analyzed through the same lens used to analyze other species. We finally have the conceptual tools to complete Darwin's revolution and forge an evolutionary psychology of the human species.

Evolutionary psychology is able to take advantage of key theoretical insights and scientific discoveries that were not known in Darwin's day. The first among these is the physical basis of inheritance—the gene.

The Modern Synthesis: Genes and Particulate Inheritance

When Darwin published *On the Origin of Species* he did not know the nature of the mechanism by which "inheritance" occurred. Indeed, as mentioned previously, the dominant thinking at the time was that inheritance constituted a sort of "blending" of the two parents, whereby offspring would be an intermediate between them. A short and a tall parent, for example, would produce a child of intermediate height, according to the blending theory. This theory is now known to be wrong.

An Austrian monk named Gregor Mendel showed why it did not work. Inheritance was "particulate," he argued, and not blended. That is, the qualities of the parents are not blended with each other, but rather are passed on intact to their offspring in distinct packets called *genes*. Furthermore, parents must be born with the genes they pass on; they cannot be acquired by experience.

Unfortunately for the advancement of science, Mendel's discovery that inheritance is particulate, which he demonstrated by crossbreeding different strains of pea plants, remained unknown to most of the scientific community for some thirty years. Mendel had sent Darwin copies of his papers, but either they remained unread or their significance was not recognized.

A *gene* is defined as the smallest discrete unit that is inherited by offspring intact, without being broken up or blended—this was Mendel's critical insight. *Genotypes*, in contrast, refer to the entire collection of genes within an individual. Genotypes, unlike genes, are not passed down to offspring intact. Rather, in sexually reproducing species such as our own, genotypes are broken up with each generation. Thus, each of us inherits a random half of genes from our mother's genotype and a random half from our father's genotype. The specific half of the genes we inherit from each parent, however, is identical to half of those possessed by that parent because they get transmitted as a discrete bundle, without modification.

The unification of Darwin's theory of evolution by natural selection with the discovery of particulate gene inheritance culminated in a movement in the 1930s and 1940s called the "Modern Synthesis" (Dobzhansky; Huxley; Mayr, *Systematics*; Simpson). The Modern Synthesis discarded a number of misconceptions in biology, including Lamarck's theory of the inheritance of acquired characteristics and the blending theory of inheritance. It emphatically confirmed the importance of Darwin's theory of natural selection, but put it on a firmer footing with a well-articulated understanding of the nature of inheritance.

The Ethology Movement

To some people, evolution is most clearly envisioned when it applies to physical structures. We can easily see how a turtle's shell is an adaptation for protection and a bird's wings an adaptation for flight. We recognize similarities between ourselves and chimpanzees, and so most people find it relatively easy to believe that human beings and chimps have a common ancestry. The paleontological record of skulls, although incomplete, shows enough evidence of physical evolution that most concede that change has taken place over time. The evolution of behavior, however, has historically been more difficult for scientists and laypeople to imagine. Behavior, after all, leaves no fossils.

Darwin clearly envisioned his theory of natural selection as being just as applicable to behavior, including social behavior, as to physical structures. Several lines of evidence support this view. First, all behavior requires underlying physical structures. Bipedal locomotion is a behavior, for example, and requires the physical structures of two legs and a multitude of muscles to support those legs while the body is in an upright position. Second, species can be bred for certain behavioral characteristics using the principle of selection. Dogs, for example, can be bred (artificial selection) for aggressiveness or passivity. These lines of evidence all point to the conclusion that behavior is not exempt from the sculpting hand of evolution. The first major discipline to form around the study of behavior from an evolutionary perspective was the field of ethology, and one of the first phenomena the ethologists documented was imprinting.

Ducklings *imprint* on the first moving object they observe in life—forming an association during a critical period of development. Usually this object is the duck's mother. After imprinting, the baby ducks follow the object of their imprinting wherever it goes. Imprinting is clearly a form of learning—an association is formed between the duckling and the mother that was not there before the exposure to her motion. This form of learning, however, is "preprogrammed" and clearly part of the evolved structures of the duckling's biology. Although many have seen pictures of a line of baby ducks following their mother, if the first object a duck sees is a human leg, it will follow that person instead. Konrad Lorenz was the first to demonstrate this imprinting phenomenon by showing that baby birds would follow him for days rather than their own mother if exposed to his leg during the critical period shortly after birth. Lorenz (*Evolution*) started a new branch of evolutionary biology called *ethology*, and imprinting in birds was a vivid phenomenon used to launch this new field. Ethology is defined as "the study of the proximate mechanisms and adaptive value of animal behavior" (Alcock 548).

The ethology movement was in part a reaction to the extreme environmentalism in U.S. psychology. Ethologists were interested in four key issues, which have become known as the four "whys" of behavior advanced by one of the founders of ethology, Niko Tinbergen: (1) the *immediate influences* on behavior (e.g., the movement of the mother); (2) the *developmental influences* on behavior (e.g., the events during the duck's lifetime that cause changes); (3) the *function* of behavior, or the "adaptive purpose" it seems to fulfill (e.g., keeping the baby duck close to the mother, which helps it to survive); and (4) the *evolutionary* or *phylogenetic origins* of behavior (e.g., what sequence of evolutionary events led to the origins of an imprinting mechanism in the duck).

Ethologists developed an array of concepts to describe what they believed to be the innate properties of animals. *Fixed action patterns*, for example, are the stereotypic behavioral sequences an animal follows after being triggered by a well-defined stimulus (Tinbergen). Once a fixed action pattern is triggered, the animal performs it to completion. Showing certain male ducks a wooden facsimile of a female duck, for example, will trigger a rigid sequence of courting behavior. Concepts such as fixed action patterns were useful in allowing ethologists to partition the ongoing stream of behavior into discrete units for analysis.

The ethology movement went a long way toward orienting biologists to focus on the importance of adaptation. Indeed, the glimmerings of evolutionary psychology itself may be seen in the early writings of Lorenz, who wrote, "our cognitive and perceptual categories, given to us prior to individual experience, are adapted to the environment for the same reasons that the horse's hoof is suited for the plains before the horse is born, and the fin of a fish is adapted for water before the fish hatches from its egg" (Lorenz, "Vergleichende Begwegungsstudien," translated by and cited in Eibl-Eibesfeldt, *Human Ethology* 8).

Ethology ran into three problems, however. First, many descriptions acted more as "labels" for behavior patterns and did not really go very far in explaining them. Second, ethologists tended to focus on observable behavior—much like their behaviorist counterparts—and so did not look "inside the heads" of animals to the underlying mechanisms responsible for generating that behavior. And third, although ethology was concerned with adaptation (one of the four critical issues listed by Tinbergen), it did not develop rigorous criteria for discovering adaptations. Ethologists did, however, turn up many useful findings—for example, documenting the imprinting that occurs in a variety of bird species and stereotypic fixed action patterns that are released by particular stimuli. Ethology also forced psychologists to reconsider the role of biology in the study of human behavior. This set the stage for an important scientific revolution, brought about by a fundamental reformulation of Darwin's theory of natural selection.

The Inclusive Fitness Revolution

In the early 1960s a young graduate student named William D. Hamilton was working on his doctoral dissertation at University College, London. Hamilton proposed a radical new revision of evolutionary theory, which he termed "inclusive fitness theory." Legend has it that his professors failed to understand the dissertation or its significance (perhaps because it was highly mathematical), and so his work was initially rejected. When it was finally accepted and published in 1964 in the *Journal of Theoretical Biology*, however, Hamilton's theory sparked a revolution that transformed the entire field of biology.

Hamilton reasoned that *classical fitness*—the measure of an individual's direct reproductive success in passing on genes through the production of offspring—was too narrow to describe the process of evolution by selection. He theorized that natural selection favors characteristics that cause an organism's genes to be passed on, regardless of whether the organism produces offspring directly. Parental care—investing in your own children—was reinterpreted as merely a special case of caring for kin who carry copies of your genes in their bodies. An organism can also increase the reproduction of its genes by helping brothers, sisters, nieces, or nephews to survive and reproduce. All these relatives have some probability of carrying copies of the organism's genes. Hamilton's genius was in the recognition that the definition of classical fitness was too narrow and should be broadened to *inclusive fitness*.

Technically, inclusive fitness is not a property of an individual or an organism but rather a property of its *actions* or *effects*. Thus, inclusive fitness can be viewed as the sum of an individual's own reproductive success (classical fitness) *plus the effects* the individual's actions have on the reproductive success of his or her genetic relatives. For this second component the effects on relatives must be weighted by the appropriate degree of genetic relatedness to the target organism—for example, 0.50 for brothers and sisters (because they are genetically related by 50 percent with the target organism), 0.25 for grandparents and grandchildren (25 percent genetic relatedness), 0.125 for first cousins (12.5 percent genetic relatedness), and so on.

The inclusive fitness revolution marshaled a new era that may be called "gene's-eye thinking." If you were a gene, what would facilitate your replication? First, you might try to ensure the well-being of the "vehicle" or body in which you reside (survival). Second, you might try to induce the vehicle to reproduce. Third, you might want to help the survival and reproduction of vehicles that contain copies of you. Genes, of course, do not have thoughts, and none of this occurs with consciousness or intentionality. The key point is that the gene is the fundamental unit of inheritance, the unit that is passed on intact in the process of reproduction. Genes producing effects that increase their replicative success will replace other genes, producing evolution over time. Adaptations are selected and evolve because they promote inclusive fitness.

Thinking about selection from the perspective of the gene offered a wealth of insights to evolutionary biologists. The theory of inclusive fitness has profound consequences for how we think about the psychology of the family, altruism, helping, the formation of groups, and even aggression.

Clarifying Adaptation and Natural Selection

The rapid inclusive fitness revolution in evolutionary biology owes part of its debt to George C. Williams, who in 1966 published a now-classic work,

Adaptation and Natural Selection. This seminal book contributed to at least three key shifts in thinking in the field.

First, Williams challenged the prevailing endorsement of *group selection*, the notion that adaptations evolved for the benefit of the group through the differential survival and reproduction of groups (Wynne-Edwards), as opposed to benefit for the gene arising through the differential reproduction of genes. According to the theory of group selection, for example, an animal might limit its personal reproduction to keep the population low, thus avoiding the destruction of the food base on which the population relied. According to group selection theory, only species that possessed characteristics beneficial to their group survived. Those that acted more selfishly perished because of the overexploitation of the critical food resources on which the species relied. Williams argued persuasively that group selection, although theoretically possible, was likely to be an extraordinarily weak force in evolution, for the following reason. Imagine a bird species with two types of individuals—one that sacrifices itself by committing suicide so as not to deplete its food resources and the other that selfishly continues to eat the food, even when supplies are low. In the next generation, which type is likely to have descendants? The answer is that the suicidal birds will have died out and failed to reproduce, whereas those who refused to sacrifice themselves for the group will have survived and left descendants. Selection operating on individual differences *within* a species, in other words, undermines the power of selection operating at the level of the group. Within five years of the book's publication most biologists had relinquished their subscription to group selection, although recently there has been a resurgence of interest in the potential potency of group selection (Sober and Wilson; Wilson and Sober).

Williams's second contribution was in translating Hamilton's highly quantitative theory of inclusive fitness into clear prose that could be comprehended by everyone. Once biologists understood inclusive fitness, they began vigorously researching its implications. To mention one prominent example, inclusive fitness theory partially solved the "problem of altruism": How could altruism evolve—incurring reproductive costs to oneself to benefit the reproduction of others—if evolution favors genes that have the effect of self-replication? Inclusive fitness theory solved this problem (at least in part) because altruism could evolve if the recipients of help were one's genetic kin. Parents, for example, might sacrifice their own lives to save the lives of their children, who carry copies of the parents' genes within them. The same logic applies to making sacrifices for other genetic relatives, such as sisters or cousins. The benefit to one's relatives in fitness currencies must be greater than the costs to the self. If this condition is satisfied, then kin altruism can evolve.

The third contribution of *Adaptation and Natural Selection* was Williams's careful analysis of adaptation, which he referred to as "an onerous concept." *Adaptations* may be defined as evolved solutions to specific problems that contribute either directly or indirectly to successful reproduction. Sweat glands, for example, may be adaptations that help solve the survival problem of thermal regulation. Taste preferences may be adaptations that guide the successful consumption of nutritious food. Mate preferences may be adaptations that guide the successful selection of mates. The problem is how to determine which attributes of organisms are adaptations. Williams established several standards for invoking adaptation and believed that it should be invoked only when necessary to explain the phenomenon at hand. When a flying fish leaps out of a wave and falls back into the water, for example, we do not have to invoke an adaptation for "getting back to water." This behavior is explained more simply by the physical law of gravity, which explains why what goes up must come down.

In addition to providing conditions for which we should not invoke the concept of adaptation, Williams provided criteria for determining when we should invoke the concept: *reliability, efficiency,* and *economy.* Does the mechanism regularly develop in most or all members of the species across all "normal" environments and perform dependably in the contexts in which it is designed to function (reliability)? Does the mechanism solve a particular adaptive problem well (efficiency)? Does the mechanism solve the adaptive problem without extorting huge costs from the organism (economy)? In other words, adaptation is invoked not merely to explain the usefulness of a biological mechanism, but to explain *improbable usefulness* (i.e., too precisely functional to have arisen by chance alone) (Pinker, *How*). Hypotheses about adaptations are, in essence, probability statements about why a reliable, efficient, and economic set of design features could not have arisen by chance alone (Tooby and Cosmides, "Cognitive Adaptations," "Conceptual Foundations"; G. C. Williams).

Williams's book brought the scientific community one step closer to the Darwinian revolution by creating the downfall of group selection as a preferred and dominant explanation, by illuminating Hamilton's theory of inclusive fitness, and by putting the concept of adaptation on a more rigorous and scientific footing. Williams was extremely influential in showing that understanding adaptations requires being "gene-centered." As put eloquently by Helena Cronin in a recent volume dedicated to George Williams, "The purpose of adaptations is to further the replication of genes. . . . Genes have been designed by natural selection to exploit properties of the world that promote their self-replication; genes are ultimately machines for turning out more genes" ("Adaptation" 19–20).

Trivers's Seminal Theories

In the late 1960s and early 1970s a graduate student at Harvard University, Robert Trivers, studied Williams's 1966 book on adaptation. He was struck by the revolutionary consequences that gene-level thinking had for conceptualizing entire domains. A sentence or brief paragraph in Williams's book or Hamilton's articles might contain the seed of an idea that could blossom into a full theory if nurtured properly.

Trivers contributed three seminal papers, all published in the early 1970s. The first was the theory of reciprocal altruism among nonkin—the conditions under which mutually beneficial exchange relationships or transactions could evolve ("Evolution"). The second was parental investment theory, which provided a powerful statement of the conditions under which sexual selection would occur for each sex ("Parental Investment and Sexual Selection"). The third was the theory of parent–offspring conflict—the notion that even parents and their progeny will get into predictable sorts of conflicts because they share only 50 percent of their genes ("Parent–Offspring"). Parents may try to wean children before the children want to be weaned, for example, in order to free up resources to invest in other children. More generally, what might be optimal for a child (e.g., securing a larger share of parental resources) might not be optimal for the parents (e.g., distributing resources more equally across children). These theories have influenced literally thousands of empirical research projects, including many on humans.

The Sociobiology Controversy

Eleven years after Hamilton's pivotal paper on inclusive fitness was published, a Harvard biologist named Edward O. Wilson caused a scientific and public uproar that rivaled the outrage caused by Charles Darwin in 1859. Wilson's 1975 book, *Sociobiology: The New Synthesis*, was monumental in both size and scope, at nearly 700 double-column pages. It offered a synthesis of cellular biology, integrative neurophysiology, ethology, comparative psychology, population biology, and behavioral ecology. Further, it examined species from ants to humans, proclaiming that the same fundamental explanatory principles could be applied to all.

Sociobiology is not generally regarded as containing fundamentally new theoretical contributions to evolutionary theory. The bulk of its theoretical tools—such as inclusive fitness theory, parental investment theory, parent–offspring conflict theory, and reciprocal altruism theory—had already been developed by others (Hamilton, "Genetical Evolution"; Trivers, "Parental

Investment," "Parent–Offspring"). What it did do is synthesize under one umbrella a tremendous diversity of scientific endeavors and give the emerging field a visible name.

The chapter on humans, the last in the book and running a mere twenty-nine pages, created the most controversy. At public talks audience members shouted him down, and once a pitcher of water was dumped on his head. His work sparked attacks from Marxists, radicals, creationists, other scientists, and even members of his own department at Harvard. Part of the controversy stemmed from the nature of Wilson's claims. He asserted that sociobiology would "cannibalize psychology," which of course was not greeted with warmth by most psychologists. Further, he speculated that many cherished human phenomena, such as culture, religion, ethics, and even aesthetics, would ultimately be explained by the new synthesis. These assertions strongly contradicted the dominant theories in the social sciences. Culture, learning, socialization, rationality, and consciousness, not evolutionary biology, were presumed by most social scientists to explain the uniqueness of humans.

Despite Wilson's grand claims for a new synthesis that would explain human nature, he had little empirical evidence on humans to support his views. The bulk of the scientific evidence came from nonhuman animals, many far removed phylogenetically from humans. Most social scientists could not see what ants and fruit flies had to do with people. Although scientific revolutions always meet resistance, often from within the ranks of established scientists (Sulloway), Wilson's lack of relevant scientific data on humans did not help.

Part I

Evolution and Human Nature

THE THEORY OF EVOLUTION

2

On the Origin of Species

RECAPITULATION AND CONCLUSION

CHARLES DARWIN

Recapitulation of the difficulties on the theory of Natural Selection—Recapitulation of the general and special circumstances in its favour—Causes of the general belief in the immutability of species—How far the theory of natural selection may be extended—Effects of its adoption on the study of Natural history—Concluding remarks.

As this whole volume is one long argument, it may be convenient to the reader to have the leading facts and inferences briefly recapitulated. That many and grave objections may be advanced against the theory of descent with modification through natural selection, I do not deny. I have endeavoured to give to them their full force. Nothing at first can appear more difficult to believe than that the more complex organs and instincts should have been perfected, not by means superior to, though analogous with, human reason, but by the accumulation of innumerable slight variations, each good for the individual possessor. Nevertheless, this difficulty, though appearing to our imagination insuperably great, cannot be considered real if we admit the following propositions, namely,—that gradations in the perfection of any organ or instinct, which we may consider, either do now exist or could have existed, each good of its kind,—that all organs and instincts are, in ever so slight a degree, variable,—and, lastly, that there is a struggle for existence leading to the preservation of each profitable deviation of structure or instinct. The truth of these propositions cannot, I think, be disputed.

It is, no doubt, extremely difficult even to conjecture by what gradations many structures have been perfected, more especially amongst broken and failing groups of organic beings; but we see so many strange gradations in nature, as is proclaimed by the canon, "Natura non facit saltum," that we ought to be extremely cautious in saying that any organ or instinct, or any whole being, could not have arrived at its present state by many graduated steps.

On the view that species are only strongly marked and permanent varieties, and that each species first existed as a variety, we can see why it is that no line of demarcation can be drawn between species, commonly supposed to have been produced by special acts of creation, and varieties which are acknowledged to have been produced by secondary laws. On this same view we can understand how it is that in each region where many species of a genus have been produced, and where they now flourish, these same species should present many varieties; for where the manufactory of species has been active, we might expect, as a general rule, to find it still in action; and this is the case if varieties be incipient species. Moreover, the species of the large genera, which afford the greater number of varieties or incipient species, retain to a certain degree the character of varieties; for they differ from each other by a less amount of difference than do the species of smaller genera. The closely allied species also of the larger genera apparently have restricted ranges, and they are clustered in little groups round other species—in which respects they resemble varieties. These are strange relations on the view of each species having been independently created, but are intelligible if all species first existed as varieties.

As each species tends by its geometrical ratio of reproduction to increase inordinately in number; and as the modified descendants of each species will be enabled to increase by so much the more as they become more diversified in habits and structure, so as to be enabled to seize on many and widely different places in the economy of nature, there will be a constant tendency in natural selection to preserve the most divergent offspring of any one species. Hence during a long-continued course of modification, the slight differences, characteristic of varieties of the same species, tend to be augmented into the greater differences characteristic of species of the same genus. New and improved varieties will inevitably supplant and exterminate the older, less improved and intermediate varieties; and thus species are rendered to a large extent defined and distinct objects. Dominant species belonging to the larger groups tend to give birth to new and dominant forms; so that each large group tends to become still larger, and at the same time more divergent in character. But as all groups cannot thus succeed in increasing in size, for the world would not hold them, the more dominant groups beat the less dominant. This tendency in the large groups to go on increasing in size and diverging in character, together with the almost inevitable contingency of much extinction, explains the arrangement of all the forms of life, in groups subordinate to groups, all within a few great classes, which we now see everywhere around us, and which has prevailed throughout all time. This grand fact of the grouping of all organic beings seems to me utterly inexplicable on the theory of creation.

As natural selection acts solely by accumulating slight, successive, favourable variations, it can produce no great or sudden modification; it can act only by very short and slow steps. Hence the canon of "Natura non facit saltum," which every fresh addition to our knowledge tends to make more strictly correct, is on this theory simply intelligible. We can plainly see why nature is prodigal in variety, though niggard in innovation. But why this should be a law of nature if each species has been independently created, no man can explain.

Many other facts are, as it seems to me, explicable on this theory. How strange it is that a bird, under the form of woodpecker, should have been created to prey on insects on the ground; that upland geese, which never or rarely swim, should have been created with webbed feet; that a thrush should have been created to dive and feed on sub-aquatic insects; and that a petrel should have been created with habits and structure fitting it for the life of an auk or grebe! and so on in endless other cases. But on the view of each species constantly trying to increase in number, with natural selection always ready to adapt the slowly varying descendants of each to any unoccupied or ill-occupied place in nature, these facts cease to be strange, or perhaps might even have been anticipated.

As natural selection acts by competition, it adapts the inhabitants of each country only in relation to the degree of perfection of their associates; so that we need feel no surprise at the inhabitants of any one country, although on the ordinary view supposed to have been specially created and adapted for that country, being beaten and supplanted by the naturalised productions from another land. Nor ought we to marvel if all the contrivances in nature be not, as far as we can judge, absolutely perfect; and if some of them be abhorrent to our ideas of fitness. We need not marvel at the sting of the bee causing the bee's own death; at drones being produced in such vast numbers for one single act, and being then slaughtered by their sterile sisters; at the astonishing waste of pollen by our fir-trees; at the instinctive hatred of the queen bee for her own fertile daughters; at ichneumonidae feeding within the live bodies of caterpillars; and at other such cases. The wonder indeed is, on the theory of natural selection, that more cases of the want of absolute perfection have not been observed.

The complex and little known laws governing variation are the same, as far as we can see, with the laws which have governed the production of so-called specific forms. In both cases physical conditions seem to have produced but little direct effect; yet when varieties enter any zone, they occasionally assume some of the characters of the species proper to that zone. In both varieties and species, use and disuse seem to have produced some effect; for it is difficult to resist this conclusion when we look, for instance, at the logger-headed duck, which has wings incapable of flight, in nearly the same condition as in the domestic duck; or when we look at the burrowing tucutucu, which is occasionally blind, and then at certain moles, which are habitually blind and have their eyes covered with skin; or when we look at the blind animals inhabiting the dark caves of America and Europe. In both varieties and species correlation of growth seems to have played a most important part, so that when one part has been modified other parts are necessarily modified. In both varieties and species reversions to long-lost characters occur. How inexplicable on the theory of creation is the occasional appearance of stripes on the shoulder and legs of the several species of the horse-genus and in their hybrids! How simply is this fact explained if we believe that these species have descended from a striped progenitor, in the same manner as the several domestic breeds of pigeon have descended from the blue and barred rock-pigeon!

On the ordinary view of each species having been independently created, why should the specific characters, or those by which the species of the same genus differ from each other, be more variable than the generic characters in which they all agree? Why, for instance, should the colour of a flower be more likely to vary in any one species of a genus, if the other species, supposed to have been created independently, have differently coloured flowers, than if all the species of the genus have the same coloured flowers? If species are only well-marked varieties, of which the characters have become in a high degree permanent, we can understand this fact; for they have already varied since they branched off from a common progenitor in certain characters, by which they have come to be specifically distinct from each other; and therefore these same characters would be more likely still to be variable than the generic characters which have been inherited without change for an enormous period. It is inexplicable on the theory of creation why a part developed in a very unusual manner in any one species of a genus, and therefore, as we may naturally infer, of great importance to the species, should be eminently liable to variation; but, on my view, this part has undergone, since the several species branched off from a common progenitor, an unusual amount of variability and modification, and therefore we might expect this part generally to be still variable. But a part may be developed in the most unusual manner, like the wing of a bat, and yet not be more variable than any other structure, if the part be common to many subordinate forms, that is, if it has been inherited for a very long period; for in this case it will have been rendered constant by long-continued natural selection.

Glancing at instincts, marvellous as some are, they offer no greater difficulty than does corporeal structure on the theory of the natural selection of successive, slight, but profitable modifications. We can thus understand why nature moves by graduated steps in endowing different animals of the same class with their several instincts. I have attempted to show how much light the principle of gradation throws on the admirable architectural powers of the hive-bee. Habit no doubt sometimes comes into play in modifying instincts; but it certainly is not indispensable, as we see, in the case of neuter insects, which leave no progeny to inherit the effects of long-continued habit. On the view of all the species of the same genus having descended from a common parent, and having inherited much in common, we can understand how it is that allied species, when placed under considerably different conditions of life, yet should follow nearly the same instincts; why the thrush of South America, for instance, lines her nest with mud like our British species. On the view of instincts having been slowly acquired through natural selection we need not marvel at some instincts being apparently not perfect and liable to mistakes, and at many instincts causing other animals to suffer.

If species be only well-marked and permanent varieties, we can at once see why their crossed offspring should follow the same complex laws in their degrees and kinds of resemblance to their parents,—in being absorbed into each other by successive crosses, and in other such points,—as do the crossed offspring of acknowledged varieties. On the other hand, these would be strange facts if species have been independently created, and varieties have been produced by secondary laws.

If we admit that the geological record is imperfect in an extreme degree, then such facts as the record gives, support the theory of descent with modification. New species have come on the stage slowly and at successive intervals; and the amount of change, after equal intervals of time, is widely different in different groups. The extinction of species and of whole groups of species, which has played so conspicuous a part in the history of the organic world, almost inevitably follows on the principle of natural selection; for old forms will be supplanted by new and improved forms. Neither single species nor groups of species reappear when the chain of ordinary generation has once been broken. The gradual diffusion of dominant forms, with the slow modification of their descendants, causes the forms of life, after long intervals of time, to appear as if they had changed simultaneously throughout the world. The fact of the fossil remains of each formation being in some degree intermediate in character between the fossils in the formations above and below, is simply explained by their intermediate position in the chain of descent. The grand fact that all extinct organic beings belong to the same system with recent beings, falling either into the same or into intermediate groups, follows from the living and the extinct being the offspring of common parents. As the groups which have descended from an ancient progenitor have generally diverged in character, the progenitor with its early descendants will often be intermediate in character in comparison with its later descendants; and thus we can see why the more ancient a fossil is, the oftener it stands in some degree intermediate between existing and allied groups. Recent forms are generally looked at as being, in some vague sense, higher than ancient and extinct forms; and they are in so far higher as the later and more improved forms have conquered the older and less improved organic beings in the struggle for life. Lastly, the law of the long endurance of allied forms on the same continent,—of marsupials in Australia, of edentata in America, and other such cases,—is intelligible, for within a confined country, the recent and the extinct will naturally be allied by descent.

Looking to geographical distribution, if we admit that there has been during the long course of ages much migration from one part of the world to another, owing to former climatal and geographical changes and to the many occasional and unknown means of dispersal, then we can understand, on the theory of descent with modification, most of the great leading facts in Distribution. We can see why there should be so striking a parallelism in the distribution of organic beings throughout space, and in their geological succession throughout time; for in both cases the beings have been connected by the bond of ordinary generation, and the means of modification have been the same. We see the full meaning of the wonderful fact, which must have struck every traveller, namely, that on the same continent, under the most diverse conditions, under heat and cold, on mountain and lowland, on deserts and marshes, most of the inhabitants within each great class are plainly related; for they will generally be descendants of the same progenitors and early colonists. On this same principle of former migration, combined in most cases with modification, we can understand, by the aid of the Glacial period, the identity of some few plants, and the close alliance of many others, on the most distant mountains, under the most different climates; and likewise the close alliance of some of the inhabitants of the sea in the northern and southern temperate zones, though separated by the whole intertropical ocean. Although two areas may present the same physical conditions of life, we need feel no surprise at their inhabitants being widely different, if they have been for a long period completely separated from each other; for as the relation of organism to organism is the most important of all relations, and as the two areas will have received colonists from some third source or from each other, at various periods and in different proportions, the course of modification in the two areas will inevitably be different.

On this view of migration, with subsequent modification, we can see why oceanic islands should be inhabited by few species, but of these, that many should be peculiar. We can clearly see why those animals which cannot cross wide spaces of ocean, as frogs and terrestrial mammals, should not inhabit oceanic islands; and why, on the other hand, new and peculiar species of bats, which can traverse the ocean, should so often be found on islands far distant from any continent. Such facts as the presence of peculiar species of bats, and the absence of all other mammals, on oceanic islands, are utterly inexplicable on the theory of independent acts of creation.

The existence of closely allied or representative species in any two areas, implies, on the theory of descent with modification, that the same parents formerly inhabited both areas; and we almost invariably find that wherever many closely allied species inhabit two areas, some identical species common to both still exist. Wherever many closely allied yet distinct species occur, many doubtful forms and varieties of the same species likewise occur. It is a rule of high generality that the inhabitants of each area are related to the inhabitants of the nearest source whence immigrants might have been derived. We see this in nearly all the plants and animals of the Galapagos archipelago, of Juan Fernandez, and of the other American islands being related in the most striking manner to the plants and animals of the neighbouring American mainland; and those of the Cape de Verde archipelago and other African islands to the African mainland. It must be admitted that these facts receive no explanation on the theory of creation.

The fact, as we have seen, that all past and present organic beings constitute one grand natural system, with group subordinate to group, and with extinct groups often falling in between recent groups, is intelligible on the theory of natural selection with its contingencies of extinction and divergence of character. On these same principles we see how it is, that the mutual affinities of the species and genera within each class are so complex and circuitous. We see why certain characters are far more serviceable than others for classification;—why adaptive characters, though of paramount importance to the being, are of hardly any importance in classification; why characters derived from rudimentary parts, though of no service to the being, are often of high classificatory value; and why embryological characters are the most valuable of all. The real affinities of all organic beings are due to inheritance or community of descent. The natural system is a genealogical arrangement, in which we have to discover the lines of descent by the most permanent characters, however slight their vital importance may be.

The framework of bones being the same in the hand of a man, wing of a bat, fin of the porpoise, and leg of the horse,—the same number of vertebrae forming the neck of the giraffe and of the elephant,—and innumerable other such facts, at once explain themselves on the theory of descent with slow and slight successive modifications. The similarity of pattern in the wing and leg of a bat, though used for such different purpose,—in the jaws and legs of a crab,—in the petals, stamens, and pistils of a flower, is likewise intelligible on the view of the gradual modification of parts or organs, which were alike in the early progenitor of each class. On the principle of successive variations not always supervening at an early age, and being inherited at a corresponding not early period of life, we can clearly see why the embryos of mammals, birds, reptiles, and fishes should be so closely alike, and should be so unlike the adult forms. We may cease marvelling at the embryo of an air-breathing mammal or bird having branchial slits and arteries running in loops, like those in a fish which has to breathe the air dissolved in water, by the aid of well-developed branchiae.

Disuse, aided sometimes by natural selection, will often tend to reduce an organ, when it has become useless by changed habits or under changed conditions of life; and we can clearly understand on this view the meaning of rudimentary organs. But disuse and selection will generally act on each creature, when it has come to maturity and has to play its full part in the struggle for existence, and will thus have little power of acting on an organ during early life; hence the organ will not be much reduced or rendered rudimentary at this early age. The calf, for instance, has inherited teeth, which never cut through the gums of the upper jaw, from an early progenitor having well-developed teeth; and we may believe, that the teeth in the mature animal were reduced, during successive generations, by disuse or by the tongue and palate having been fitted by natural selection to browse without their aid; whereas in the calf, the teeth have been left untouched by selection or disuse, and on the principle of inheritance at corresponding ages have been inherited from a remote period to the present day. On the view of each organic being and each separate organ having been specially created, how utterly inexplicable it is that parts, like the teeth in the embryonic calf or like the shrivelled wings under the soldered wing-covers of some beetles, should thus so frequently bear the plain stamp of inutility! Nature may be said to have taken pains to reveal, by rudimentary organs and by homologous structures, her scheme of modification, which it seems that we wilfully will not understand.

I have now recapitulated the chief facts and considerations which have thoroughly convinced me that species have changed, and are still slowly changing by the preservation and accumulation of successive slight favourable variations. Why, it may be asked, have all the most eminent living naturalists and geologists rejected this view of the mutability of species? It cannot be asserted that organic beings in a state of nature are subject to no variation; it cannot be proved that the amount of variation in the course

of long ages is a limited quantity; no clear distinction has been, or can be, drawn between species and well-marked varieties. It cannot be maintained that species when intercrossed are invariably sterile, and varieties invariably fertile; or that sterility is a special endowment and sign of creation. The belief that species were immutable productions was almost unavoidable as long as the history of the world was thought to be of short duration; and now that we have acquired some idea of the lapse of time, we are too apt to assume, without proof, that the geological record is so perfect that it would have afforded us plain evidence of the mutation of species, if they had undergone mutation.

But the chief cause of our natural unwillingness to admit that one species has given birth to other and distinct species, is that we are always slow in admitting any great change of which we do not see the intermediate steps. The difficulty is the same as that felt by so many geologists, when Lyell first insisted that long lines of inland cliffs had been formed, and great valleys excavated, by the slow action of the coast-waves. The mind cannot possibly grasp the full meaning of the term of a hundred million years; it cannot add up and perceive the full effects of many slight variations, accumulated during an almost infinite number of generations.

Although I am fully convinced of the truth of the views given in this volume under the form of an abstract, I by no means expect to convince experienced naturalists whose minds are stocked with a multitude of facts all viewed, during a long course of years, from a point of view directly opposite to mine. It is so easy to hide our ignorance under such expressions as the "plan of creation," "unity of design," &c., and to think that we give an explanation when we only restate a fact. Any one whose disposition leads him to attach more weight to unexplained difficulties than to the explanation of a certain number of facts will certainly reject my theory. A few naturalists, endowed with much flexibility of mind, and who have already begun to doubt on the immutability of species, may be influenced by this volume; but I look with confidence to the future, to young and rising naturalists, who will be able to view both sides of the question with impartiality. Whoever is led to believe that species are mutable will do good service by conscientiously expressing his conviction; for only thus can the load of prejudice by which this subject is overwhelmed be removed.

Several eminent naturalists have of late published their belief that a multitude of reputed species in each genus are not real species; but that other species are real, that is, have been independently created. This seems to me a strange conclusion to arrive at. They admit that a multitude of forms, which till lately they themselves thought were special creations, and which are still thus looked at by the majority of naturalists, and which consequently have every external characteristic feature of true species,—they admit that these

have been produced by variation, but they refuse to extend the same view to other and very slightly different forms. Nevertheless they do not pretend that they can define, or even conjecture, which are the created forms of life, and which are those produced by secondary laws. They admit variation as a *vera causa* in one case, they arbitrarily reject it in another, without assigning any distinction in the two cases. The day will come when this will be given as a curious illustration of the blindness of preconceived opinion. These authors seem no more startled at a miraculous act of creation than at an ordinary birth. But do they really believe that at innumerable periods in the earth's history certain elemental atoms have been commanded suddenly to flash into living tissues? Do they believe that at each supposed act of creation one individual or many were produced? Were all the infinitely numerous kinds of animals and plants created as eggs or seed, or as full grown? and in the case of mammals, were they created bearing the false marks of nourishment from the mother's womb? Although naturalists very properly demand a full explanation of every difficulty from those who believe in the mutability of species, on their own side they ignore the whole subject of the first appearance of species in what they consider reverent silence.

It may be asked how far I extend the doctrine of the modification of species. The question is difficult to answer, because the more distinct the forms are which we may consider, by so much the arguments fall away in force. But some arguments of the greatest weight extend very far. All the members of whole classes can be connected together by chains of affinities, and all can be classified on the same principle, in groups subordinate to groups. Fossil remains sometimes tend to fill up very wide intervals between existing orders. Organs in a rudimentary condition plainly show that an early progenitor had the organ in a fully developed state; and this in some instances necessarily implies an enormous amount of modification in the descendants. Throughout whole classes various structures are formed on the same pattern, and at an embryonic age the species closely resemble each other. Therefore I cannot doubt that the theory of descent with modification embraces all the members of the same class. I believe that animals have descended from at most only four or five progenitors, and plants from an equal or lesser number.

Analogy would lead me one step further, namely, to the belief that all animals and plants have descended from some one prototype. But analogy may be a deceitful guide. Nevertheless all living things have much in common, in their chemical composition, their germinal vesicles, their cellular structure, and their laws of growth and reproduction. We see this even in so trifling a circumstance as that the same poison often similarly affects plants and animals; or that the poison secreted by the gall-fly produces monstrous growths

on the wild rose or oak-tree. Therefore I should infer from analogy that probably all the organic beings which have ever lived on this earth have descended from some one primordial form, into which life was first breathed.

When the views entertained in this volume on the origin of species, or when analogous views are generally admitted, we can dimly foresee that there will be a considerable revolution in natural history. Systematists will be able to pursue their labours as at present; but they will not be incessantly haunted by the shadowy doubt whether this or that form be in essence a species. This I feel sure, and I speak after experience, will be no slight relief. The endless disputes whether or not some fifty species of British brambles are true species will cease. Systematists will have only to decide (not that this will be easy) whether any form be sufficiently constant and distinct from other forms, to be capable of definition; and if definable, whether the differences be sufficiently important to deserve a specific name. This latter point will become a far more essential consideration than it is at present; for differences, however slight, between any two forms, if not blended by intermediate gradations, are looked at by most naturalists as sufficient to raise both forms to the rank of species. Hereafter we shall be compelled to acknowledge that the only distinction between species and well-marked varieties is, that the latter are known, or believed, to be connected at the present day by intermediate gradations, whereas species were formerly thus connected. Hence, without quite rejecting the consideration of the present existence of intermediate gradations between any two forms, we shall be led to weigh more carefully and to value higher the actual amount of difference between them. It is quite possible that forms now generally acknowledged to be merely varieties may hereafter be thought worthy of specific names, as with the primrose and cowslip; and in this case scientific and common language will come into accordance. In short, we shall have to treat species in the same manner as those naturalists treat genera, who admit that genera are merely artificial combinations made for convenience. This may not be a cheering prospect; but we shall at least be freed from the vain search for the undiscovered and undiscoverable essence of the term species.

The other and more general departments of natural history will rise greatly in interest. The terms used by naturalists of affinity, relationship, community of type, paternity, morphology, adaptive characters, rudimentary and aborted organs, etc., will cease to be metaphorical, and will have a plain signification. When we no longer look at an organic being as a savage looks

at a ship, as at something wholly beyond his comprehension; when we regard every production of nature as one which has had a history; when we contemplate every complex structure and instinct as the summing up of many contrivances, each useful to the possessor, nearly in the same way as when we look at any great mechanical invention as the summing up of the labour, the experience, the reason, and even the blunders of numerous workmen; when we thus view each organic being, how far more interesting, I speak from experience, will the study of natural history become!

A grand and almost untrodden field of inquiry will be opened, on the causes and laws of variation, on correlation of growth, on the effects of use and disuse, on the direct action of external conditions, and so forth. The study of domestic productions will rise immensely in value. A new variety raised by man will be a far more important and interesting subject for study than one more species added to the infinitude of already recorded species. Our classifications will come to be, as far as they can be so made, genealogies; and will then truly give what may be called the plan of creation. The rules for classifying will no doubt become simpler when we have a definite object in view. We possess no pedigrees or armorial bearings; and we have to discover and trace the many diverging lines of descent in our natural genealogies, by characters of any kind which have long been inherited. Rudimentary organs will speak infallibly with respect to the nature of long-lost structures. Species and groups of species, which are called aberrant, and which may fancifully be called living fossils, will aid us in forming a picture of the ancient forms of life. Embryology will reveal to us the structure, in some degree obscured, of the prototypes of each great class.

When we can feel assured that all the individuals of the same species, and all the closely allied species of most genera, have within a not very remote period descended from one parent, and have migrated from some one birthplace; and when we better know the many means of migration, then, by the light which geology now throws, and will continue to throw, on former changes of climate and of the level of the land, we shall surely be enabled to trace in an admirable manner the former migrations of the inhabitants of the whole world. Even at present, by comparing the differences of the inhabitants of the sea on the opposite sides of a continent, and the nature of the various inhabitants of that continent in relation to their apparent means of immigration, some light can be thrown on ancient geography.

The noble science of Geology loses glory from the extreme imperfection of the record. The crust of the earth with its embedded remains must not be looked at as a well-filled museum, but as a poor collection made at hazard and at rare intervals. The accumulation of each great fossiliferous formation will be recognised as having depended on an unusual concurrence of

circumstances, and the blank intervals between the successive stages as having been of vast duration. But we shall be able to gauge with some security the duration of these intervals by a comparison of the preceding and succeeding organic forms. We must be cautious in attempting to correlate as strictly contemporaneous two formations, which include few identical species, by the general succession of their forms of life. As species are produced and exterminated by slowly acting and still existing causes, and not by miraculous acts of creation and by catastrophes; and as the most important of all causes of organic change is one which is almost independent of altered and perhaps suddenly altered physical conditions, namely, the mutual relation of organism to organism,—the improvement of one being entailing the improvement or the extermination of others; it follows, that the amount of organic change in the fossils of consecutive formations probably serves as a fair measure of the lapse of actual time. A number of species, however, keeping in a body might remain for a long period unchanged, whilst within this same period, several of these species, by migrating into new countries and coming into competition with foreign associates, might become modified; so that we must not overrate the accuracy of organic change as a measure of time. During early periods of the earth's history, when the forms of life were probably fewer and simpler, the rate of change was probably slower; and at the first dawn of life, when very few forms of the simplest structure existed, the rate of change may have been slow in an extreme degree. The whole history of the world, as at present known, although of a length quite incomprehensible by us, will hereafter be recognised as a mere fragment of time, compared with the ages which have elapsed since the first creature, the progenitor of innumerable extinct and living descendants, was created.

In the distant future I see open fields for far more important researches. Psychology will be based on a new foundation, that of the necessary acquirement of each mental power and capacity by gradation. Light will be thrown on the origin of man and his history.

Authors of the highest eminence seem to be fully satisfied with the view that each species has been independently created. To my mind it accords better with what we know of the laws impressed on matter by the Creator, that the production and extinction of the past and present inhabitants of the world should have been due to secondary causes, like those determining the birth and death of the individual. When I view all beings not as special creations, but as the lineal descendants of some few beings which lived long before the first bed of the Silurian system was deposited, they seem to me to become ennobled. Judging from the past, we may safely infer that not one living species will transmit its unaltered likeness to a distant futurity. And of the species now living very few will transmit progeny of any kind to a far distant futurity; for the manner in which all organic beings are grouped, shows that the greater number of species of each genus, and all the species of many genera, have left no descendants, but have become utterly extinct. We can so far take a prophetic glance into futurity as to foretell that it will be the common and widely-spread species, belonging to the larger and dominant groups, which will ultimately prevail and procreate new and dominant species. As all the living forms of life are the lineal descendants of those which lived long before the Silurian epoch, we may feel certain that the ordinary succession by generation has never once been broken, and that no cataclysm has desolated the whole world. Hence we may look with some confidence to a secure future of equally inappreciable length. And as natural selection works solely by and for the good of each being, all corporeal and mental endowments will tend to progress towards perfection.

It is interesting to contemplate an entangled bank, clothed with many plants of many kinds, with birds singing on the bushes, with various insects flitting about, and with worms crawling through the damp earth, and to reflect that these elaborately constructed forms, so different from each other, and dependent on each other in so complex a manner, have all been produced by laws acting around us. These laws, taken in the largest sense, being Growth with Reproduction; Inheritance which is almost implied by reproduction; Variability from the indirect and direct action of the external conditions of life, and from use and disuse; a Ratio of Increase so high as to lead to a Struggle for Life, and as a consequence to Natural Selection, entailing Divergence of Character and the Extinction of less-improved forms. Thus, from the war of nature, from famine and death, the most exalted object which we are capable of conceiving, namely, the production of the higher animals, directly follows. There is grandeur in this view of life, with its several powers, having been originally breathed into a few forms or into one; and that, whilst this planet has gone cycling on according to the fixed law of gravity, from so simple a beginning endless forms most beautiful and most wonderful have been, and are being, evolved.

3

The Digital River

RICHARD DAWKINS

All peoples have epic legends about their tribal ancestors, and these legends often formalize themselves into religious cults. People revere and even worship their ancestors—as well they might, for it is real ancestors, not supernatural gods, that hold the key to understanding life. Of all organisms born, the majority die before they come of age. Of the minority that survive and breed, an even smaller minority will have a descendant alive a thousand generations hence. This tiny minority of a minority, this progenitorial élite, is all that future generations will be able to call ancestral. Ancestors are rare, descendants are common.

All organisms that have ever lived—every animal and plant, all bacteria and all fungi, every creeping thing, and all readers of this book—can look back at their ancestors and make the following proud claim: Not a single one of our ancestors died in infancy. They all reached adulthood, and every single one was capable of finding at least one heterosexual partner and of successfully copulating. (Strictly speaking, there are exceptions. Some animals, like aphids, reproduce without sex. Techniques such as artificial

---

From *River Out of Eden: A Darwinian View of Life* by Richard Dawkins.

fertilization make it possible for modern humans to have a child without copulating, and even—since eggs for *in vitro* fertilization could be taken from a female fetus—without reaching adulthood. But for most purposes the force of my point is undiminished.) Not a single one of our ancestors was felled by an enemy, or by a virus, or by a misjudged footstep on a cliff edge, before bringing at least one child into the world. Thousands of our ancestors' contemporaries failed in all these respects, but not a single solitary one of our ancestors failed in any of them. These statements are blindingly obvious, yet from them much follows: much that is curious and unexpected, much that explains and much that astonishes.

Since all organisms inherit all their genes from their ancestors, rather than from their ancestors' unsuccessful contemporaries, all organisms tend to possess successful genes. They have what it takes to become ancestors— and that means to survive and reproduce. This is why organisms tend to inherit genes with a propensity to build a well-designed machine—a body that actively works as if it is striving to become an ancestor. That is why birds are so good at flying, fish so good at swimming, monkeys so good at climbing, viruses so good at spreading. That is why we love life and love sex and love children. It is because we all, without a single exception, inherit all our genes from an unbroken line of successful ancestors. The world becomes full of organisms that have what it takes to become ancestors. That, in a sentence, is Darwinism. Of course, Darwin said much more than that, and nowadays there is much more we can say.

There is a natural, and deeply pernicious, way to misunderstand the previous paragraph. It is tempting to think that when ancestors did successful things, the genes they passed on to their children were, as a result, upgraded relative to the genes they had received from their parents. Something about their success rubbed off on their genes, and that is why their descendants are so good at flying, swimming, courting. Wrong, utterly wrong! Genes do not improve in the using, they are just passed on, unchanged except for very rare random errors. It is not success that makes good genes. It is good genes that make success, and nothing an individual does during its lifetime has any effect whatever upon its genes. Those individuals born with good genes are the most likely to grow up to become successful ancestors; therefore good genes are more likely than bad to get passed on to the future. Each generation is a filter, a sieve: good genes tend to fall through the sieve into the next generation; bad genes tend to end up in bodies that die young or without reproducing. Bad genes may pass through the sieve for a generation or two, perhaps because they have the luck to share a body with good genes. But you need more than luck to navigate successfully through a thousand sieves in succession, one sieve under the other. After a thousand successive generations, the genes that have made it through are likely to be the good ones.

I said that the genes that survive down the generations will be the ones that have succeeded in making ancestors. This is true, but there is one apparent exception I must deal with before the thought of it causes confusion. Some individuals are irrevocably sterile, yet they are seemingly designed to assist the passage of their genes into future generations. Worker ants, bees, wasps and termites are sterile. They labor not to become ancestors but so that their fertile relatives, usually sisters and brothers, will become ancestors. There are two points to understand here. First, in any kind of animal, sisters and brothers have a high probability of sharing copies of the same genes. Second, it is the environment, not the genes, that determines whether an individual termite, say, becomes a reproducer or a sterile worker. All termites contain genes capable of turning them into sterile workers under some environmental conditions, reproducers under other conditions. The reproducers pass on copies of the very same genes that make the sterile workers help them to do so. The sterile workers toil under the influence of genes, copies of which are sitting in the bodies of reproducers. The worker copies of those genes are striving to assist their own reproductive copies through the transgenerational sieve. Termite workers can be male or female; but in ants, bees and wasps the workers are all female; otherwise the principle is the same. In a watered-down form, it also applies to several species of birds, mammals and other animals that exhibit a certain amount of caring for young by elder brothers or sisters. To summarize, genes can buy their way through the sieve, not only by assisting their own body to become an ancestor but by assisting the body of a relation to become an ancestor.

The river of my title is a river of DNA, and it flows through time, not space. It is a river of information, not a river of bones and tissues: a river of abstract instructions for building bodies, not a river of solid bodies themselves. The information passes through bodies and affects them, but it is not affected by them on its way through. The river is not only uninfluenced by the experiences and achievements of the successive bodies through which it flows. It is also uninfluenced by a potential source of contamination that, on the face of it, is much more powerful: sex.

In every one of your cells, half your mother's genes rub shoulders with half your father's genes. Your maternal genes and your paternal genes conspire with one another most intimately to make you the subtle and indivisible amalgam you are. But the genes themselves do not blend. Only their effects do. The genes themselves have a flintlike integrity. When the time comes to move on to the next generation, a gene either goes into the body of a given child or it does not. Paternal genes and maternal genes do not blend; they recombine independently. A given gene in you came either from your mother or your father. It also came from one, and only one, of your four grandparents; from one, and only one, of your eight great-grandparents; and so on back.

I have spoken of a river of genes, but we could equally well speak of a band of good companions marching through geological time. All the genes of one breeding population are, in the long run, companions of each other. In the short run, they sit in individual bodies and are temporarily more intimate companions of the other genes sharing that body. Genes survive down the ages only if they are good at building bodies that are good at living and reproducing in the particular way of life chosen by the species. But there is more to it than this. To be good at surviving, a gene must be good at working together with the other genes in the same species—the same river. To survive in the long run, a gene must be a good companion. It must do well in the company of, or against the background of, the other genes in the same river. Genes of another species are in a different river. They do not have to get on well together—not in the same sense, anyway—for they do not have to share the same bodies.

The feature that defines a species is that all members of any one species have the same river of genes flowing through them, and all the genes in a species have to be prepared to be good companions of one another. A new species comes into existence when an existing species divides into two. The river of genes forks in time. From a gene's point of view, *speciation*, the origin of new species, is "the long goodbye." After a brief period of partial separation, the two rivers go their separate ways forever, or until one or the other dries extinct into the sand. Secure within the banks of either river, the water is mixed and remixed by sexual recombination. But water never leaps its banks to contaminate the other river. After a species has divided, the two sets of genes are no longer companions. They no longer meet in the same bodies and they are no longer required to get on well together. There is no longer any intercourse between them—and intercourse here means, literally, sexual intercourse between their temporary vehicles, their bodies.

Why should two species divide? What initiates the long goodbye of their genes? What provokes a river to split and the two branches to drift apart, never to meet again? The details are controversial, but nobody doubts that the most important ingredient is accidental geographical separation. The river of genes flows in time, but the physical repartnering of genes takes place in solid bodies, and bodies occupy a location in space. A gray squirrel in North America would be capable of breeding with a gray squirrel in England, if they ever met. But they are unlikely to meet. The river of gray-squirrel genes in North America is effectively separated, by three thousand miles of ocean, from the river of gray-squirrel genes in England. The two bands of genes are no longer companions in fact, although they are still presumably capable of acting as good companions should the opportunity arise. They have said farewell, though it is not an irrevocable goodbye—yet. But given another few thousand years of separation, it is probable that the two rivers will have drifted so far apart that if individual squirrels meet, they will no longer be able to exchange genes. "Drift apart" here means apart not in space but in compatibility.

Something like this almost certainly lies behind the older separation between gray squirrels and red squirrels. They cannot interbreed. They overlap geographically in parts of Europe and, although they meet and probably confront one another over disputed nuts from time to time, they cannot mate to produce fertile offspring. Their genetic rivers have drifted too far apart, which is to say that their genes are no longer well suited to cooperate with one another in bodies. Many generations ago, ancestors of gray squirrels and ancestors of red squirrels were one and the same individuals. But they became geographically separated—perhaps by a mountain range, perhaps by water, eventually by the Atlantic Ocean. And their genetic ensembles grew apart. Geographical separation bred a lack of compatibility. Good companions became poor companions (or they would turn out to be poor companions if put to the test in a mating encounter). Poor companions became poorer still, until now they are not companions at all. Their goodbye is final. The two rivers are separate and destined to become more and more separate. The same story underlies the much earlier separation between, say, our ancestors and the ancestors of elephants. Or between ostrich ancestors (which were also our ancestors) and the ancestors of scorpions.

There are now perhaps thirty million branches to the river of DNA, for that is an estimate of the number of species on earth. It has also been estimated that the surviving species constitute about 1 percent of the species that have ever lived. It would follow that there have been some three billion branches to the river of DNA altogether. Today's thirty million branch rivers are irrevocably separate. Many of them are destined to wither into nothing, for most species go extinct. If you follow the thirty million rivers (for brevity, I'll refer to the branch rivers as rivers) back into the past, you will find that, one by one, they join up with other rivers. The river of human genes joins with the river of chimpanzee genes at about the same time as the river of gorilla genes does, some seven million years ago. A few million years farther back, our shared African ape river is joined by the stream of orangutan genes. Farther back still, we are joined by a river of gibbon genes—a river that splits downstream into a number of separate species of gibbon and siamang. As we push on backward in time, our genetic river unites with rivers destined, if followed forward again, to branch into the Old World monkeys, the New World monkeys, and the lemurs of Madagascar. Even farther back, our river unites with those leading to other major groups of mammals: rodents, cats, bats, elephants. After that, we meet the streams leading to various kinds of reptiles, birds, amphibians, fish, invertebrates.

Now here is an important respect in which we have to be cautious about the river metaphor. When we think of the divide leading to all the mammals—as opposed to, say, the stream leading to the gray squirrel—it is tempting to imagine something on a grand, Mississippi/Missouri scale. The mammal branch is, after all, destined to branch and branch and branch again, until it produces all the mammals—from pigmy shrew to elephant, from moles underground to monkeys atop the canopy. The mammal branch of the river is destined to feed so many thousands of important trunk waterways, how could it be other than a massive, rolling torrent? But this image is deeply wrong. When the ancestors of all the modern mammals broke away from those that are not mammals, the event was no more momentous than any other speciation. It would have gone unremarked by any naturalist who happened to be around at the time. The new branch of the river of genes would have been a trickle, inhabiting a species of little nocturnal creature no more different from its nonmammalian cousins than a red squirrel is from a gray. It is only with hindsight that we see the ancestral mammal as a mammal at all. In those days, it would have been just another species of mammal-like reptile, not markedly different from perhaps a dozen other small, snouty, insectivorous morsels of dinosaur food.

The same lack of drama would have attended the earlier splits between the ancestors of all the great groups of animals: the vertebrates, the mollusks, the crustaceans, the insects, the segmented worms, the flatworms, the jellyfish and so on. When the river that was to lead to the mollusks (and others) parted from the river that was to lead to the vertebrates (and others), the two populations of (probably wormlike) creatures would have been so alike that they could have mated with one another. The only reason they didn't is that they had become accidentally separated by some geographical barrier, perhaps dry land separating previously united waters. Nobody could have guessed that one population was destined to spawn the mollusks and the other the vertebrates. The two rivers of DNA were streamlets barely parted, and the two groups of animals were all but indistinguishable.

Zoologists know all this, but they forget it sometimes when contemplating the really big animal groups, like mollusks and vertebrates. They are tempted to think of the divide between major groups as a momentous event. The reason zoologists may be so misled is that they have been brought up in the almost reverential belief that each of the great divisions of the animal kingdom is furnished with something deeply unique, often called by the German word *Bauplan*. Although this word just means "blueprint," it has become a recognized technical term, and I shall inflect it as an English word, even though (as I am slightly shocked to discover) it is not yet in the current edition of the Oxford English Dictionary. (Since I enjoy the word less than some of my colleagues do, I admit to a tiny *frisson* of *Schadenfreude* at its absence; those two foreign words *are* in the Dictionary, so there is no systematic prejudice against importation.) In its technical sense, bauplan is often translated as "fundamental body plan." The use of the word "fundamental" (or, equivalently, the self-conscious dropping into German to indicate profundity) is what causes the damage. It can lead zoologists to make serious errors.

One zoologist, for instance, has suggested that evolution in the Cambrian period (between about six hundred million and about five hundred million years ago) must have been a completely different kind of process from evolution in later times. His reasoning was that nowadays it is new species that are coming into existence, whereas in the Cambrian period major groups were appearing, such as the mollusks and the crustaceans. The fallacy is glaring! Even creatures as radically different from one another as mollusks and crustaceans were originally just geographically separated populations of the same species. For a while, they could have interbred if they had met, but they did not. After millions of years of separate evolution, they acquired the characteristics which we, with the hindsight of modern zoologists, now recognize as those of mollusks and crustaceans respectively. These characteristics are dignified with the grandiose title of "fundamental body plan" or "bauplan." But the major bauplans of the animal kingdom diverged from common origins by gradual degrees.

Admittedly, there is a minor, if much publicized, disagreement over quite *how* gradual or "jumpy" evolution is. But nobody, and I mean nobody, thinks that evolution has ever been jumpy enough to invent a whole new bauplan in one step. The author I quoted was writing in 1958. Few zoologists would explicitly take his position today, but they sometimes do so implicitly, speaking as though the major groups of animals arose spontaneously and perfectly formed like Athena from the head of Zeus, rather than by divergence of an ancestral population while in accidental geographical isolation. (Readers might like to keep these points in mind when consulting *Wonderful Life*, Stephen J. Gould's beautifully written account of the Burgess Shale Cambrian fauna.)

The study of molecular biology has, in any case, shown the great animal groups to be much closer to one another than we used to think. You can treat the genetic code as a dictionary in which sixty-four words in one language (the sixty-four possible triplets of a four-letter alphabet) are mapped onto twenty-one words in another language (twenty amino acids plus a punctuation mark). The odds of arriving at the same 64:21 mapping twice by chance are less than one in a million million million million million. Yet the genetic code is in fact literally identical in all animals, plants and bacteria that have ever been looked at. All earthly living things are certainly descended from a single ancestor. Nobody would dispute that, but some startlingly close resemblances between, for instance, insects and vertebrates are now showing up when people examine not just the code itself but detailed sequences of genetic information. There is a quite complicated genetic mechanism responsible for the segmented body plan of insects. An uncannily similar piece of genetic machinery has also been found in mammals. From a molecular point of view, all animals are pretty close relatives of one another and even of plants. You have to go to bacteria to find our distant cousins, and even then the genetic code itself is identical to ours. The reason it is possible to do such precise calculations on the genetic code but not on the anatomy of bauplans is that the genetic code is strictly digital, and digits are things you can count precisely. The river of genes is a digital river, and I must now explain what this engineering term means.

Engineers make an important distinction between digital and analog codes. Phonographs and tape recorders—and until recently most telephones—use analog codes. Compact disks, computers, and most modern telephone systems use digital codes. In an analog telephone system, continuously fluctuating waves of pressure in the air (sounds) are transduced into correspondingly fluctuating waves of voltage in a wire. A phonograph record works in a similar way: the wavy grooves cause a stylus to vibrate, and the movements of the stylus are transduced into corresponding fluctuations in voltage. At the other end of the line these voltage waves are reconverted, by a vibrating membrane in the telephone's earpiece or the phonograph's loudspeaker, back into the corresponding air-pressure waves, so that we can hear them. The code is a simple and direct one: electrical fluctuations in wire are proportional to pressure fluctuations in air. All possible voltages, within certain limits, may pass down the wire, and the differences between them matter.

In a digital telephone, only two possible voltages—or some other discrete number of possible voltages, such as 8 or 256—pass down the wire. The information lies not in the voltages themselves but in the patterning of the discrete levels. This is called Pulse Code Modulation. The actual voltage at any one time will seldom be exactly equal to any of the eight, say, nominal values, but the receiving apparatus will round it off to the nearest of the designated voltages, so that what emerges at the other end of the line is well-nigh perfect even if the transmission along the line is poor. All you have to do is set the discrete levels far enough apart so that random fluctuations can never be misinterpreted by the receiving instrument as the wrong level. This is the great virtue of digital codes, and it is why audio and video systems—and information technology generally—are increasingly going digital. Computers, of course, use digital codes for everything they do. For reasons of convenience, it is a binary code—that is, it has only two levels of voltage instead of 8 or 256.

Even in a digital telephone, the sounds entering the mouthpiece and leaving the earpiece are still analog fluctuations in air pressure. It is the information traveling exchange to exchange that is digital. Some kind of code has to be set up to translate analog values, microsecond by microsecond, into sequences of discrete pulses—digitally coded numbers. When you plead with your lover over the telephone, every nuance, every catch in the voice, every passionate sigh and yearning timbre is carried along the wire solely in the form of numbers. You can be moved to tears by numbers—provided they are encoded and decoded fast enough. Modern electronic switching gear is so fast that the line's time can be divided into slices, rather as a chess master may divide his time among twenty games in rotation. By this means, thousands of conversations can be slotted into the same telephone line, apparently simultaneously yet electronically segregated without interference. A trunk data line—many of them nowadays are not wires at all but radio beams, either transmitted directly from hilltop to hilltop or bounced off satellites—is a massive river of digits. But because of this ingenious electronic segregation, it is thousands of digital rivers, which share the same banks only in a superficial sense—like red squirrels and gray, who share the same trees but never intermingle their genes.

Back in the world of engineers, the deficiencies of analog signals don't matter too much, as long as they aren't copied repeatedly. A tape recording may have so little hiss on it that you hardly notice it—unless you amplify the sound, in which case you amplify the hiss and introduce some new noise too. But if you make a tape of the tape, then a tape of the tape of the tape, and so on and on, after a hundred "generations" a horrible hiss will be all that remains. Something like this was a problem in the days when telephones were all analog. Any telephone signal fades over a long wire and has to be boosted—reamplified—every hundred miles or so. In analog days this was a bugbear, because each amplification stage increased the proportion of background hiss. Digital signals, too, need boosting. But, for the reason we've seen, the boosting does not introduce any error: things can be set up so that the information gets through perfectly, no matter how many boosting stations intervene. Hiss does not increase even over hundreds and hundreds of miles.

When I was a small child, my mother explained to me that our nerve cells are the telephone wires of the body. But are they analog or digital? The answer is that they are an interesting mixture of both. A nerve cell is not like an electric wire. It is a long thin tube along which waves of chemical change pass, like a trail of gunpowder fizzing along the ground—except that, unlike a trail of gunpowder, the nerve soon recovers and can fizz again after a short rest period. The absolute magnitude of the wave—the temperature of the gunpowder—may fluctuate as it races along the nerve, but this is irrelevant. The code ignores it. Either the chemical pulse is there or it is not, like two discrete voltage levels in a digital telephone. To this extent, the nervous system is digital. But nerve impulses are not dragooned into bytes: they don't assemble into discrete code numbers. Instead, the strength of the message (the loudness of the sound, the brightness of the light, maybe even the agony of the emotion) is encoded as the rate of impulses. Engineers know this as Pulse Frequency Modulation, and it was popular with them before Pulse Code Modulation was adopted.

A pulse rate is an analog quantity, but the pulses themselves are digital: they are either there or they are not, with no half measures. And the nervous system reaps the same benefit from this as any digital system does. Because of the way nerve cells work, there is the equivalent of an amplifying booster, not every hundred miles but every millimeter—eight hundred boosting stations between the spinal cord and your fingertip. If the absolute height of the nerve impulse—the gunpowder wave—mattered, the message would be distorted beyond recognition over the length of a human arm, let alone a giraffe's neck. Each stage in the amplification would introduce more random error, like what happens when a tape recording is made of a tape recording eight hundred times over. Or when you Xerox a Xerox of a Xerox. After eight hundred photocopying "generations," all that's left is a gray blur. Digital coding offers the only solution to the nerve cell's problem, and natural selection has duly adopted it. The same is true of genes.

Francis Crick and James Watson, the unravelers of the molecular structure of the gene, should, I believe, be honored for as many centuries as Aristotle and Plato. Their Nobel Prizes were awarded "in physiology or medicine," and this is right but almost trivial. To talk of continuous revolution is almost a contradiction in terms, yet not only medicine but our whole understanding of life will go on being revolutionized again and again as a direct result of the change in thinking that those two young men initiated in 1953. Genes themselves, and genetic disease, are only the tip of the iceberg. What is truly revolutionary about molecular biology in the post–Watson-Crick era is that it has become digital.

After Watson and Crick, we know that genes themselves, within their minute internal structure, are long strings of pure digital information. What is more, they are truly digital, in the full and strong sense of computers and compact disks, not in the weak sense of the nervous system. The genetic code is not a binary code as in computers, nor an eight-level code as in some telephone systems, but a quaternary code, with four symbols. The machine code of the genes is uncannily computerlike. Apart from differences in jargon, the pages of a molecular-biology journal might be interchanged with those of a computer-engineering journal. Among many other consequences, this digital revolution at the very core of life has dealt the final, killing bow to vitalism—the belief that living material is deeply distinct from nonliving material. Up until 1953 it was still possible to believe that there was something fundamentally and irreducibly mysterious in living protoplasm. No longer. Even those philosophers who had been predisposed to a mechanistic view of life would not have dared hope for such total fulfillment of their wildest dreams.

The following science-fiction plot is feasible, given a technology that differs from today's only in being a little speeded up. Professor Jim Crickson has been kidnapped by an evil foreign power and forced to work in its biological-warfare labs. To save civilization it is vitally important that he should communicate some top-secret information to the outside world, but all normal channels of communication are denied him. Except one. The DNA code consists of sixty-four triplet "codons," enough for a complete upper- and lower-case English alphabet plus ten numerals, a space character and a full stop. Professor Crickson takes a virulent influenza virus off the laboratory shelf and engineers into its genome the complete text of his message to the outside world, in perfectly formed English sentences. He repeats his message over and over again in the engineered genome, adding an easily recognizable "flag" sequence—say, the first ten prime numbers. He then infects himself with the virus and sneezes in a room full of people. A wave of flu sweeps the world, and medical labs in distant lands set to work to sequence its genome in an attempt to design a vaccine. It soon becomes apparent that there is a strange repeated pattern in the genome. Alerted by the prime numbers—which cannot have arisen spontaneously—somebody tumbles to the idea of deploying code-breaking techniques. From there it would be short work to read the full English text of Professor Crickson's message, sneezed around the world.

Our genetic system, which is the universal system of all life on the planet, is digital to the core. With word-for-word accuracy, you could encode the whole of the New Testament in those parts of the human genome that are at present filled with "junk" DNA—that is, DNA not used, at least in the ordinary way, by the body. Every cell in your body contains the equivalent of forty-six immense data tapes, reeling off digital characters via numerous reading heads working simultaneously. In every cell, these tapes—the chromosomes—contain the same information, but the reading heads in different kinds of cells seek out different parts of the database for their own specialist purposes. That is why muscle cells are different from liver cells. There is no spirit-driven life force, no throbbing, heaving, pullulating, protoplasmic, mystic jelly. Life is just bytes and bytes and bytes of digital information.

Genes are pure information—information that can be encoded, recoded and decoded, without any degradation or change of meaning. Pure information can be copied and, since it is digital information, the fidelity of the copying can be immense. DNA characters are copied with an accuracy that rivals anything modern engineers can do. They are copied down the generations, with just enough occasional errors to introduce variety. Among this variety, those coded combinations that become more numerous in the world will obviously and automatically be the ones that, when decoded and obeyed inside bodies, make those bodies take active steps to preserve and propagate those same DNA messages. We—and that means all living things—are survival machines programmed to propagate the digital database that did the programming. Darwinism is now seen to be the survival of the survivors at the level of pure, digital code.

With hindsight, it could not have been otherwise. An analog genetic system could be imagined. But we have already seen what happens to analog information when it is recopied over successive generations. It is Chinese Whispers. Boosted telephone systems, recopied tapes, photocopies of photocopies—analog signals are so vulnerable to cumulative degradation that copying cannot be sustained beyond a limited number of generations. Genes, on the other hand, can self-copy for ten million generations and scarcely degrade at all. Darwinism works only because—apart from discrete mutations, which natural selection either weeds out or preserves—the copying process is perfect. Only a digital genetic system is capable of sustaining Darwinism over eons of geological time. Nineteen fifty-three, the year of the double helix, will come to be seen not only as the end of mystical and obscurantist views of life; Darwinians will see it as the year their subject went finally digital.

The river of pure digital information, majestically flowing through geological time and splitting into three billion branches, is a powerful image. But where does it leave the familiar features of life? Where does it leave bodies, hands and feet, eyes and brains and whiskers, leaves and trunks and roots? Where does it leave us and our parts? We—we animals, plants, protozoa, fungi and bacteria—are we just the banks through which rivulets of digital data flow? In one sense, yes. But there is, as I have implied, more to it than that. Genes don't only make copies of themselves, which flow on down the generations. They actually spend their time in bodies, and they influence the shape and behavior of the successive bodies in which they find themselves. Bodies are important too.

The body of, say, a polar bear is not just a pair of riverbanks for a digital streamlet. It is also a machine of bear-sized complexity. All the genes of the whole population of polar bears are a collective—good companions, jostling with one another through time. But they do not spend all the time in the company of all the other members of the collective: they change partners within the set that is the collective. The collective is defined as the set of genes that can potentially meet any other genes in the collective (but no member of any of the thirty million other collectives in the world). The actual meetings always take place inside a cell in a polar bear's body. And that body is not a passive receptacle for DNA.

For a start, the sheer number of cells, in every one of which is a complete set of genes, staggers the imagination: about nine hundred million million for a large male bear. If you lined up all the cells of a single polar bear in a row, the array would comfortably make the round trip from here to the moon and back. These cells are of a couple of hundred distinct types, essentially the same couple of hundred for all mammals: muscle cells, nerve cells, bone cells, skin cells and so on. Cells of any one of these distinct types are massed together to form tissues: muscle tissue, bone tissue and so on. All the different types of cells contain the genetic instructions needed to make any of the types. Only the genes appropriate to the tissue concerned are switched on. This is why cells of the different tissues are of different shapes and sizes. More interestingly, the genes switched on in the cells of a particular type cause those cells to grow their tissues into particular shapes. Bones are not shapeless masses of hard, rigid tissue. Bones have particular shapes, with hollow shafts, balls and sockets, spines and spurs. Cells are programmed, by the genes switched on inside them, to behave as if they know where they are in relation to their neighbouring cells, which is how they build their tissues up into the shape of ear lobes and heart valves, eye lenses and sphincter muscles.

The complexity of an organism such as a polar bear is many-layered. The body is a complex collection of precisely shaped organs, like livers and kidneys and bones. Each organ is a complex edifice fashioned from particular tissues whose building bricks are cells, often in layers or sheets but often in solid masses too. On a much smaller scale, each cell has a highly complex interior structure of folded membranes. These membranes, and the water between them, are the scene of intricate chemical reactions of very numerous distinct types. A chemical factory belonging to ICI or Union Carbide may have several hundred distinct chemical reactions going on inside it. These chemical reactions will be kept separate from one another by the walls of the flasks, tubes and so on. A living cell might have a similar number of chemical reactions going on inside it simultaneously. To some extent the membranes in a cell are like the glassware in a laboratory, but the analogy is not a good one for two reasons. First, although many of the chemical reactions go on between the membranes, a good many go on *within* the substance of the membranes themselves. Second, there is a more important way in which the different reactions are kept separate. Each reaction is catalyzed by its own special enzyme.

An enzyme is a very large molecule whose three-dimensional shape speeds up one particular kind of chemical reaction by providing a surface that encourages that reaction. Since what matters about biological molecules is their three-dimensional shape, we could regard an enzyme as a large machine tool, carefully jigged to turn out a production line of molecules of a particular shape. Any one cell, therefore, may have hundreds of separate chemical reactions going on inside it simultaneously and separately, on the surfaces of different enzyme molecules. Which particular chemical reactions go on in a given cell is determined by which particular kinds of enzyme molecules are present in large numbers. Each enzyme molecule, including its all-important shape, is assembled under the deterministic influence of a particular gene. To be specific, the precise sequence of several hundred code letters in the gene determines, by a set of rules that are totally known (the genetic code), the sequence of amino acids in the enzyme molecule. Every enzyme molecule is a linear chain of amino acids, and every linear chain of amino acids spontaneously coils up into a unique and particular three-dimensional structure, like a knot, in which parts of the chain form crosslinks with other parts of the chain. The exact three-dimensional structure of the knot is determined by the one-dimensional sequence of amino acids, and therefore by the one-dimensional sequence of code letters in the gene. And thus the chemical reactions that take place in a cell are determined by which genes are switched on.

What, then, determines which genes are switched on in a particular cell? The answer is the chemicals that are already present in the cell. There is an element of chicken-and-egg paradox here, but it is not insuperable. The solution to the paradox is actually very simple in principle, although complicated in detail. It is the solution that computer engineers know as bootstrapping. When I first started using computers, in the 1960s, all programs had to be loaded via paper tape. (American computers of the period often used punched cards, but the principle was the same.) Before you could load in the large tape of a serious program, you had to load in a smaller program called the bootstrap loader. The bootstrap loader was a program to do one thing: to tell the computer how to load paper tapes. But—here is the chicken-and-egg paradox—how was the bootstrap-loader tape itself loaded? In modern computers, the equivalent of the bootstrap loader is hardwired into the machine, but in those early days you had to begin by toggling switches in a ritually patterned sequence. This sequence told the computer how to begin to read the first part of the bootstrap-loader tape. The first part of the bootstrap-loader tape then told it a bit more about how to read the next part of the bootstrap-loader tape and so on. By the time the whole bootstrap loader had been sucked in, the computer knew how to read any paper tape, and it had become a useful computer.

When an embryo begins, a single cell, the fertilized egg, divides into two; the two divide to make four; the four divide to make eight, and so on. It takes only a few dozen generations to work the cell numbers up into the trillions, such is the power of exponential division. But, if this were all there was to it, the trillions of cells would all be the same. How, instead, do they differentiate (to use the technical term) into liver cells, kidney cells, muscle cells and so on, each with different genes turned on and different enzymes active? By bootstrapping, and it works like this. Although the egg looks like a sphere, it actually has polarity in its internal chemistry. It has a top and a bottom and, in many cases, a front and a rear (and therefore also a left and a right side) as well. These polarities show themselves in the form of gradients of chemicals. Concentrations of some chemicals steadily increase as you move from front to rear, others as you move from top to bottom. These early gradients are pretty simple, but they are enough to form the first stage in a bootstrapping operation.

When the egg has divided into, say, thirty-two cells—that is, after five divisions—some of those thirty-two cells will have more than their fair share of topside chemicals, others more than their fair share of bottomside chemicals. The cells may also be unbalanced with respect to the chemicals of the fore-and-aft gradient. These differences are enough to cause different combinations of genes to be turned on in different cells. Therefore different combinations of enzymes will be present in the cells of different parts of the early embryo. This will see to it that different combinations of further genes are turned on in different cells. Lineages of cells diverge, therefore, instead of remaining identical to their clone-ancestors within the embryo.

These divergences are very different from the divergences of species we talked about earlier. These cell divergences are programmed and predictable in detail, whereas those species divergences were the fortuitous results of geographical accidents and were unpredictable. Moreover, when species diverge, the genes themselves diverge, in what I fancifully called the long goodbye. When cell lineages within an embryo diverge, both divisions receive the same genes—all of them. But different cells receive different combinations of chemicals, which switch on different combinations of genes, and some genes work to switch other genes on or off. And so the bootstrapping continues, until we have the full repertoire of different kinds of cells.

The developing embryo doesn't just differentiate into a couple of hundred different types of cells. It also undergoes elegant dynamic changes in external and internal shape. Perhaps the most dramatic of these is one of the earliest: the process known as gastrulation. The distinguished embryologist Lewis Wolpert has gone so far as to say, "It is not birth, marriage, or death, but gastrulation which is truly the most important time in your life." What happens at gastrulation is that a hollow ball of cells buckles to form a cup with an inner lining. Essentially all embryologies throughout the animal kingdom undergo this same process of gastrulation. It is the uniform foundation on which the diversity of embryologies rests. Here I mention gastrulation as just one example—a particularly dramatic one—of the kind of restless, origami-like movement of whole sheets of cells that is often seen in embryonic development.

At the end of a virtuoso origami performance; after numerous foldings-in, pushings-out, bulgings and stretchings of layers of cells; after much dynamically orchestrated differential growth of parts of the embryo at the expense of other parts; after differentiation into hundreds of chemically and physically specialized kinds of cells; when the total number of cells has reached into the trillions, the final product is a baby. No, even the baby is not final, because the whole growth of the individual—again, with some parts growing faster than others—past adulthood into old age should be seen as an extension of the same process of embryology: total embryology.

Individuals vary because of differences in quantitative details in their total embryology. A layer of cells grows a little farther before folding in on itself, and the result is—What?—an aquiline rather than a retroussé nose; flat feet, which may save your life because they debar you from the Army; a particular conformation of the shoulder blade that predisposes you to be good at throwing spears (or hand grenades, or cricket balls, depending on your circumstances). Sometimes individual changes in the origami of cell layers can have tragic consequences, as when a baby is born with stumps for arms and no hands. Individual differences that do not manifest themselves in cell-layer origami but purely chemically may be no less important in their consequences: an inability to digest milk, a predisposition to homosexuality, or to peanut allergy, or to think that mangos taste offensively of turpentine.

Embryonic development is a very complicated physical and chemical performance. Change of detail at any point in its course can have remarkable consequences farther down the line. This is not so surprising, when you recall how heavily bootstrapped the process is. Many of the differences in the way individuals develop are due to differences in environment—oxygen starvation or exposure to thalidomide, for instance. Many other differences are due to differences in genes—not just genes considered in isolation but genes in interaction with other genes, and in interaction with environmental differences. Such a complicated, kaleidoscopic, intricately and reciprocally bootstrapped process as embryonic development is both robust and sensitive. It is robust in that it fights off many potential changes, to produce a living baby against odds that sometimes seem almost overwhelming; at the same time it is sensitive to changes in that no two individuals, not even identical twins, are literally identical in all their features.

And now for the point that this has all been leading up to. To the extent that differences between individuals are due to genes (which may be a large extent or a small one), natural selection can favor some quirks of embryological origami or embryological chemistry and disfavor others. To the extent that your throwing arm is influenced by genes, natural selection can favor it or disfavor it. If being able to throw well has an effect, however slight, on an individual's likelihood of surviving long enough to have children, to the extent that throwing ability is influenced by genes, those genes will have a correspondingly greater chance of winning through to the next generation. Any one individual may die for reasons having nothing to do with his ability to throw. But a gene that tends to make individuals better at throwing when it is present than when it is absent will inhabit lots of bodies, both good and bad, over many generations. From the point of view of the particular gene, the other causes of death will average out. From the gene's perspective, there is only the long-term outlook of the river of DNA flowing down through the generations, only temporarily housed in particular bodies, only temporarily sharing a body with companion genes that may be successful or unsuccessful.

In the long term, the river becomes full of genes that are good at surviving for their several reasons: slightly improving the ability to throw a spear, slightly improving the ability to taste poison, or whatever it may be. Genes that, on average, are less good at surviving—because they tend to cause astigmatic vision in their successive bodies, who are therefore less successful spear throwers; or because they make their successive bodies less attractive and therefore less likely to mate—such genes will tend to disappear from the river of genes. In all this, remember the point we made earlier: the genes that survive in the river will be the ones that are good at surviving in the average environment of the species, and perhaps the most important aspect of this average environment is the other genes of the species; the other genes with which a gene is likely to have to share a body; the other genes that swim through geological time in the same river.

Part I

Evolution and Human Nature

4

The Descent of Man

GENERAL SUMMARY AND CONCLUSION

CHARLES DARWIN

The main conclusion here arrived at, and now held by many naturalists who are well competent to form a sound judgment, is that man is descended from some less highly organised form. The grounds upon which this conclusion rests will never be shaken, for the close similarity between man and the lower animals in embryonic development, as well as in innumerable points of structure and constitution, both of high and of the most trifling importance,—the rudiments which he retains, and the abnormal reversions to which he is occasionally liable,—are facts which cannot be disputed. They have long been known, but until recently they told us nothing with respect to the origin of man. Now when viewed by the light of our knowledge of the whole organic world, their meaning is unmistakable. The great principle of evolution stands up clear and firm, when these groups or facts are considered in connection with others, such as the mutual affinities of the members of the same group, their geographical distribution in past and present times, and their geological succession. It is incredible that all these facts should speak falsely. He who is not content to look, like a savage, at the phenomena of nature as disconnected, cannot any longer believe that man is the work of a separate act of creation. He will be forced to admit that the close resemblance of the embryo of man to that, for instance, of a dog—the construction of his skull, limbs and whole frame on the same plan with that of other mammals, independently of the uses to which the parts may be put—the occasional re-appearance of various structures, for instance of several muscles, which man does not normally possess, but which are common to the Quadrumana—and a crowd of analogous facts—all point in the plainest manner to the conclusion that man is the co-descendant with other mammals of a common progenitor.

By considering the embryological structure of man,—the homologies which he presents with the lower animals,—the rudiments which he retains,—and the reversions to which he is liable, we can partly recall in imagination the former condition of our early progenitors; and can approximately place them in their proper place in the zoological series. We thus learn that man is descended from a hairy, tailed quadruped, probably arboreal in its habits, and an inhabitant of the Old World. This creature, if its whole structure had been examined by a naturalist, would have been classed amongst the Quadrumana, as surely as the still more ancient progenitor of the Old and New World monkeys. The Quadrumana and all the higher mammals are probably derived from an ancient marsupial animal, and this through a long line of diversified forms, from some amphibian-like creature, and this again from some fish-like animal. In the dim obscurity of the past we can see that the early progenitor of all the Vertebrata must have been an aquatic animal, provided with branchiae, with the two sexes united in the same individual, and with the most important organs of the body (such as the brain and heart) imperfectly or not at all developed. This animal seems to have been more like the larvae of the existing marine Ascidians than any other known form.

The high standard of our intellectual powers and moral disposition is the greatest difficulty which presents itself, after we have been driven to this conclusion on the origin of man. But every one who admits the principle of evolution, must see that the mental powers of the higher animals, which are the same in kind with those of man, though so different in degree, are capable of advancement. Thus the interval between the mental powers of one of the higher apes and of a fish, or between those of an ant and scale-insect, is immense; yet their development does not offer any special difficulty; for with our domesticated animals, the mental faculties are certainly variable, and the variations are inherited. No one doubts that they are of the utmost importance to animals in a state of nature. Therefore the conditions are favourable for their development through natural selection. The same conclusion may be extended to man; the intellect must have been all-important to him, even at a very remote period, as enabling him to invent and use language, to make weapons, tools, traps, etc., whereby with the aid of his social habits, he long ago became the most dominant of all living creatures.

A great stride in the development of the intellect will have followed, as soon as the half-art and half-instinct of language came into use; for the continued use of language will have reacted on the brain and produced an inherited effect; and this again will have reacted on the improvement of language. As Mr. Chauncey Wright has well remarked, the largeness of the brain in man relatively to his body, compared with the lower animals, may be attributed in chief part to the early use of some simple form of language,—that wonderful engine which affixes signs to all sorts of objects and qualities, and excites trains of thought which would never arise from the mere impression of the senses, or if they did arise could not be followed out (295). The higher intellectual powers of man, such as those of ratiocination, abstraction, self-consciousness, etc., probably follow from the continued improvement and exercise of the other mental faculties.

The development of the moral qualities is a more interesting problem. The foundation lies in the social instincts, including under this term the family ties. These instincts are highly complex, and in the case of the lower animals give special tendencies towards certain definite actions; but the more important elements are love, and the distinct emotion of sympathy. Animals endowed with the social instincts take pleasure in one another's company, warn one another of danger, defend and aid one another in many ways. These instincts do not extend to all the individuals of the species, but only to those of the same community. As they are highly beneficial to the species, they have in all probability been acquired through natural selection.

The main conclusion arrived at in this work, namely, that man is descended from some lowly organised form, will, I regret to think, be highly distasteful to many. But there can hardly be a doubt that we are descended from barbarians. The astonishment which I felt on first seeing a party of Fuegians on a wild and broken shore will never be forgotten by me, for the reflection at once rushed into my mind—such were our ancestors. These men were absolutely naked and bedaubed with paint, their long hair was tangled, their mouths frothed with excitement, and their expression was wild, startled, and distrustful. They possessed hardly any arts, and like wild animals lived on what they could catch; they had no government, and were merciless to every one not of their own small tribe. He who has seen a savage in his native land will not feel much shame, if forced to acknowledge that the blood of some more humble creature flows in his veins. For my own part I would as soon be descended from that heroic little monkey, who braved his dreaded enemy in order to save the life of his keeper, or from that old baboon, who descending from the mountains, carried away in triumph his young comrade from a crowd of astonished dogs—as from a savage who delights to torture his enemies, offers up bloody sacrifices, practices infanticide without remorse, treats his wives like slaves, knows no decency, and is haunted by the grossest superstitions.

Man may be excused for feeling some pride at having risen, though not through his own exertions, to the very summit of the organic scale; and the fact of his having thus risen, instead of having been aboriginally placed there, may give him hope for a still higher destiny in the distant future. But we are not here concerned with hopes or fears, only with the truth as far as our reason permits us to discover it; and I have given the evidence to the best of my ability. We must, however, acknowledge, as it seems to me, that man with all his noble qualities, with sympathy which feels for the most debased, with benevolence which extends not only to other men but to the humblest living creature, with his god-like intellect which has penetrated into the movements and constitution of the solar system—with all these exalted powers—Man still bears in his bodily frame the indelible stamp of his lowly origin.

5

Man

FROM SOCIOBIOLOGY TO SOCIOLOGY

Edward O. Wilson

Let us now consider man in the free spirit of natural history, as though we were zoologists from another planet completing a catalog of social species on Earth. In this macroscopic view the humanities and social sciences shrink to specialized branches of biology; history, biography, and fiction are the research protocols of human ethology; and anthropology and sociology together constitute the sociobiology of a single primate species.

Homo sapiens is ecologically a very peculiar species. It occupies the widest geographical range and maintains the highest local densities of any of the primates. An astute ecologist from another planet would not be surprised to find that only one species of *Homo* exists. Modern man has preempted all the conceivable hominid niches. Two or more species of hominids did coexist in the past, when the *Australopithecus* man-apes and possibly an early *Homo* lived in Africa. But only one evolving line survived into late Pleistocene times to participate in the emergence of the most advanced human social traits.

Modern man is anatomically unique. His erect posture and wholly bipedal locomotion are not even approached in other primates that occasionally walk on their hind legs, including the gorilla and chimpanzee. The skeleton has been profoundly modified to accommodate the change: the spine is curved to distribute the weight of the trunk more evenly down its length; the chest is flattened to move the center of gravity back toward the spine; the pelvis is broadened to serve as an attachment for the powerful striding muscles of the upper legs and reshaped into a basin to hold the viscera; the tail is eliminated, its vertebrae (now called the coccyx) curved inward to form part of the floor of the pelvic basin; the occipital condyles have rotated far beneath the skull so that the weight of the head is balanced on them; the face is shortened to assist this shift in gravity; the thumb is enlarged to give power to the hand; the leg is lengthened; and the foot is drastically narrowed and lengthened to facilitate striding. Other changes have taken place. Hair has been lost from most of the body. It is still not known why modern man is a "naked ape." One plausible explanation is that nakedness served as a device to cool the body during the strenuous pursuit of prey in the heat of the African plains. It is associated with man's exceptional reliance on sweating to reduce body heat; the human body contains from two to five million sweat glands, far more than in any other primate species.

The reproductive physiology and behavior of *Homo sapiens* have also undergone extraordinary evolution. In particular, the estrous cycle of the female has changed in two ways that affect sexual and social behavior. Menstruation has been intensified. The females of some other primate species experience slight bleeding, but only in women is there a heavy sloughing of the wall of the "disappointed womb" with consequent heavy bleeding. The estrus, or period of female "heat," has been replaced by virtually continuous sexual activity. Copulation is initiated not by response to the conventional primate signals of estrus, such as changes in color of the skin around the female sexual organs and the release of pheromones, but by extended foreplay entailing mutual stimulation by the partners. The traits of physical attraction are, moreover, fixed in nature. They include the pubic hair of both sexes and the protuberant breasts and buttocks of women. The flattened sexual cycle and continuous female attractiveness cement the close marriage bonds that are basic to human social life.

At a distance a perceptive Martian zoologist would regard the globular head as a most significant clue to human biology. The cerebrum of *Homo* was expanded enormously during a relatively short span of evolutionary time (figure 5.1). Three million years ago *Australopithecus* had an adult cranial capacity of 400–500 cubic centimeters, comparable to that of the chimpanzee and gorilla. Two million years later its presumptive descendant *Homo erectus* had a capacity of about 1000 cubic centimeters. The next million

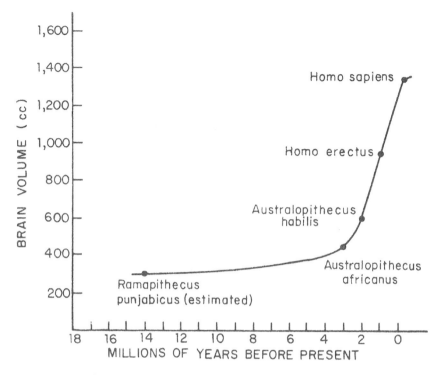

FIGURE 5.1 The increase in brain size during human evolution. (Redrawn from Pilbeam, *Ascent*)

years saw an increase to 1400–1700 cubic centimeters in Neanderthal man and 900–2000 cubic centimeters in modern *Homo sapiens*. The growth in intelligence that accompanied this enlargement was so great that it cannot yet be measured in any meaningful way. Human beings can be compared among themselves in terms of a few of the basic components of intelligence and creativity. But no scale has been invented that can objectively compare man with chimpanzees and other living primates.

We have leaped forward in mental evolution in a way that continues to defy self-analysis. The mental hypertrophy has distorted even the most basic primate social qualities into nearly unrecognizable forms. Individual species of Old World monkeys and apes have notably plastic social organizations; man has extended the trend into a protean ethnicity. Monkeys and apes utilize behavioral scaling to adjust aggressive and sexual interactions; in man the scales have become multidimensional, culturally adjustable, and almost endlessly subtle. Bonding and the practices of reciprocal altruism are rudimentary in other primates; man has expanded them into great networks where individuals consciously alter roles from hour to hour as if changing masks.

It is the task of comparative sociobiology to trace these and other human qualities as closely as possible back through time. Besides adding perspective and perhaps offering some sense of philosophical ease, the exercise will help to identify the behaviors and rules by which individual human beings increase their Darwinian fitness through the manipulation of society. In a phrase, we are searching for the human biogram (Count; Tiger and Fox). One of the key questions, never far from the thinking of anthropologists and biologists who pursue real theory, is to what extent the biogram represents an adaptation to modern cultural life and to what extent it is a phylogenetic vestige. Our civilizations were jerrybuilt around the biogram. How have they been influenced by it? Conversely, how much flexibility is there in the biogram, and in which parameters particularly? Experience with other animals indicates that when organs are hypertrophied, phylogeny is hard to reconstruct. This is the crux of the problem of the evolutionary analysis of human behavior.

6

The Universal People

DONALD E. BROWN

What do all people, all societies, all cultures, and all languages have in common? In the following pages I attempt to provide answers, in the form of a description of what I will call the Universal People (UP). Theirs is a description of every people or of people in general. Bear in mind the tentative nature of this chapter: as surely as it leaves out some universals it includes some that will prove in the long run not to be universal, and even more surely it divides up traits and complexes in ways that in time will give way to more accurate or meaningful divisions. At the end of the chapter I will discuss how it was put together and the ways in which it will change in the future.

Although humans are not unique in their possession of culture—patterns of doing and thinking that are passed on within and between generations by learning—they certainly are unique in the extent to which their thought and action are shaped by such patterns. The UP are aware of this uniqueness and posit a difference between their way—culture—and the way of nature.

A very significant portion of UP culture is embodied in their language, a system of communication without which their culture would necessarily

From *Human Universals* by Donald E. Brown. Philadelphia: Temple University Press, 1991. Copyright Donald E. Brown.

be very much simpler. With language the UP think about and discuss both their internal states and the world external to each individual (this is not to deny that they also think without language—surely they do). With language, the UP organize, respond to, and manipulate the behavior of their fellows. UP language is of strategic importance for those who wish to study the UP. This is so because their language is, if not precisely a mirror of, then at least a window into, their culture and into their minds and actions. Their language is not a perfect mirror or window, for there are often discrepancies between what the UP say, think, and do. But we would be very hard pressed to understand many aspects of the UP without access to their thinking through their language. Because their language is not a simple reflex of the way the world is, we need to distinguish their (emic) conceptualization of it from objective (etic) conceptualizations of the world.

The UP's language allows them to think and speak in abstractions, and about things or processes not physically present. If one of them is proficient in the use of language—particularly if it is a male—it gains him prestige, in part because good speech allows him to more effectively manipulate, for better or worse, the behavior of his fellows. An important means of verbal manipulation among the UP is gossip.

In their conversations the UP manage in many ways to express more than their mere words indicate. For example, shifts in tone, timing, and other features of speech indicate that one person is or is not ready for another to take a turn at speaking. UP speech is used to misinform as well as inform. Even if an individual among the UP does not tell lies, he understands the concept and watches for it in others. For some UP do lie, and they dissimulate and mislead in other ways too. UP use of language includes ways to be funny and ways to insult.

UP speech is highly symbolic. Let me explain how this is different from animal communication. Many bird species vocalize a danger warning. The vocalization is substantially the same for the species from one location to another. Indeed, it is somewhat similar from one species to another. Humans have cries of fright and warning that are in some ways analogous to these bird calls, but between many, many members of our species our routine vocalizations are meaningless. This is so because speech sounds and the things they signify have very little intrinsic connection. Sound and sense, as a rule, are only arbitrarily associated. Equally arbitrary is the way units of speech that are equivalent to our words get strung together to make sentences. But in spite of this arbitrariness there are features of language at all basic levels— phonemic, grammatical, and semantic—that are found in all languages.

Thus UP phonemes—their basic speech sounds—include a contrast between vocalics (sounds produced in or channeled through the oral cavity) and nonvocalics (e.g., nasals). UP language has contrasts between vowels and contrasts between stops and nonstops (a stop, e.g., English *p* or *b*, stops the flow of air during speech). The phonemes of UP speech form a system of contrasts, and the number of their phonemes goes neither above 70 nor below 10.

In time, their language undergoes change. So it follows that the UP do not speak the language of their more remote ancestors, though it may be quite similar.

However much grammar varies from language to language, some things are always present. For example, UP language includes a series of contrasting terms that theoretically could be phrased in three different ways, but that are only phrased two ways. To illustrate, they could talk about the "good" and the "bad" (two contrasting terms, neither with a marker added to express negation); or they could talk about the "good" and the "not good" (i.e., not having the word "bad" at all but expressing its meaning with a marked version of its opposite, the marking in this case to negate), or they could talk about the "bad" and the "not bad" (i.e., not having the word "good," etc.). Logically, these alternatives are identical: each arrangement conveys the same information. Similar possibilities exist for "deep" and "shallow," "wide" and "narrow," etc. But in each case the third possibility never occurs as the obligatory or common way of talking. So the UP are never forced to express, for lack of an alternative, the ideas of "good," "wide," "deep," and so on as negated versions of their opposites.

By virtue of its grammar UP language conveys some information redundantly. In English, for example, both subject and verb indicate number, while in Spanish both noun and adjective indicate gender.

Two final points about UP grammar are that it contains nouns and verbs, and the possessive. The latter is used both for what have been called the "intimate" or "inalienable" possessions, i.e., to talk about their fingers, your hands, and her thoughts, and for "loose" or "alienable" possessions too, e.g., my axe.

The UP have special forms of speech for special occasions. Thus they have poetic or rhetorical standards deemed appropriate to speech in particular settings. They use narrative to explain how things came to be and to tell stories. Their language includes figurative speech: metaphor is particularly prominent, and metonymy (the use of a word for that with which it is associated, e.g., crown for king) is always included too. The UP can speak onomatopoeically (using words that imitate sound, like "bowwow"), and from time to time they do. They have poetry in which lines, demarcated by pauses, are about 3 seconds in duration. The poetic lines are characterized by the repetition of some structural, semantic, or auditory elements but by free variation too.

Most of the specific elementary units of meaning in UP language—units that are sometimes but not always equivalent to words—are not found in all the rest of the languages of the world. This does not prevent us from translating much of the UP speech into our own or any other particular language: centimeters and inches are not the same entities, but we can translate one to another quite precisely; people who lack a word for "chin" and thus call it the "end of the jaw" still make sense.

A few words or meanings cut across all cultural boundaries and hence form a part of UP language. I am not saying, of course, that the UP make the same speech sounds as we English speakers do for these words, but rather that the meanings for these terms are expressed by the UP in their terms. For example, the UP have terms for black and white (equivalent to dark and light when no other basic colors are encoded) and for face, hand, and so on.

Certain semantic components are found in UP language, even if the terms in which they are employed are not. For example, UP kin terminology includes terms that distinguish male from female (and thus indicate the semantic component of sex) and some generations from others. If not explicit, durational time is semantically implicit in their language, and they have units of time—such as days, months, seasons, and years. In various ways there is a temporal cyclicity or rhythmicity to UP lives. The UP can distinguish past, present, and future.

UP language also classifies parts of the body, inner states (such as emotions, sensations, or thoughts), behavioral propensities, flora, fauna, weather conditions, tools, space (by which they give directions), and many other definite topics, though each of them does not necessarily constitute an emically distinct lexical domain. The UP language refers to such semantic categories as motion, speed, location, dimension, and other physical properties; to giving (including analogous actions, such as lending); and to affecting things or people.

As is implied in their use of metaphor and metonymy, UP words (or word equivalents) are sometimes polysemous, having more than one meaning. Their antonyms and synonyms are numerous. The words or word equivalents that the UP use more frequently are generally shorter, while those they use less frequently are longer.

UP language contains both proper names and pronouns. The latter include at least three persons and two categories of number. Their language contains numerals, though they may be as few as "one, two, and many."

The UP have separate terms for kin categories that include mother and father. That is, whereas some peoples include father and father's brothers in a single kin category, and lump mother with her sisters—so that it is obligatory or normal to refer to each of one's parents with terms that lump them with others—it is not obligatory among the UP to refer to their actual parents in ways that lump mother with father.

UP kinship terms are partially or wholly translatable by reference to the relationships inherent in procreation: mother, father, son, daughter. The UP have an age terminology that includes age grades in a linear sequence similar to the sequence child, adolescent, adult, etc. Our first reflex is to think that it could not be otherwise, but it could: an elderly person can be "like a child"; an age classification that had a term indicating "dependent age" could break from the normal pattern of linearity.

The UP have a sex terminology that is fundamentally dualistic, even when it comprises three or four categories. When there are three, one is a combination of the two basic sexes (e.g., a hermaphrodite), or one is a crossover sex (e.g., a man acting as a woman). When there are four there are then two normal sexes and two crossover sexes.

Naming and taxonomy are fundamental to UP cognition. Prominent elements in UP taxonomy and other aspects of their speech and thought are binary discriminations, forming contrasting terms or semantic components (a number of which have already been mentioned—black and white, nature and culture, male and female, good and bad, etc.). But the UP also can order continua, so they can indicate not only contrasts but polar extremes with gradations between them. Thus there are middles between their opposites, or ranked orders in their classifications. The UP are able to express the measure of things and distances, though not necessarily with uniform units.

The UP employ such elementary logical notions as "not," "and," "same," "equivalent," and "opposite." They distinguish the general from the particular and parts from wholes. Unfortunately, the UP overestimate the objectivity of their mode of thought (it is particularly unobjective when they compare their in-group with out-groups).

The UP use what has been called "conjectural" reasoning to, for example, deduce from minute clues the identification, presence, and behavior of animals, or from miscellaneous symptoms the presence of a particular disease that cannot in itself be observed and is a wholly abstract conception.

Language is not the only means of symbolic communication employed by the UP. They employ gestures too, especially with their hands and arms. Some of their nonverbal communication is somewhat one-sided, in that the message is received consciously but may be sent more or less spontaneously. For example, the squeals of children, cries of fright, and the like all send messages that UP watch closely or listen to carefully, even though the sender did not consciously intend them to communicate. The UP do not merely listen and watch what is on the surface, they interpret external behavior to grasp interior intention.

Communication with their faces is particularly complex among the UP, and some of their facial expressions are recognized everywhere. Thus UP faces show happiness, sadness, anger, fear, surprise, disgust, and contempt, in a manner entirely familiar from one society to another. When they smile while greeting persons it signifies friendly intentions. UP cry when they feel unhappiness or pain. A young woman acting coy or flirting with her eyes does it in a way you would recognize quite clearly. Although some facial communication is spontaneous, as noted earlier, the UP can mask, modify, and mimic otherwise spontaneous expressions. Whether by face, words, gesture, or otherwise, the UP can show affection as well as feel it.

The UP have a concept of the person in the psychological sense. They distinguish self from others, and they can see the self both as subject and object. They do not see the person as a wholly passive recipient of external action, nor do they see the self as wholly autonomous. To some degree, they see the person as responsible for his or her actions. They distinguish actions that are under control from those that are not. They understand the concept of intention. They know that people have a private inner life, have memories, make plans, choose between alternatives, and otherwise make decisions (not without ambivalent feeling sometimes). They know that people can feel pain and other emotions. They distinguish normal from abnormal mental states. The UP personality theory allows them to think of individuals departing from the pattern of behavior associated with whatever status(es) they occupy, and they can explain these departures in terms of the individual's character. The UP are spontaneously and intuitively able to, so to say, get in the minds of others to imagine how they are thinking and feeling.

In addition to the emotions that have already been mentioned, the UP are moved by sexual attraction; sometimes they are deeply disturbed by sexual jealousy. They also have childhood fears, including fear of loud noises and—particularly toward the end of the first year of life—of strangers (this is the apparent counterpart of a strong attachment to their caretaker at this time). The UP react emotionally—generally with fear—to snakes. With effort, the UP can overcome some of their fears. Because there is normally a man present to make a claim on a boy's mother, the Oedipus complex—in the sense of a little boy's possessiveness toward his mother and coolness toward her consort—is a part of male UP psychology.

The UP recognize individuals by their faces, and in this sense they most certainly have an implicit concept of the individual (however little they may explicitly conceptualize the individual apart from social statuses). They recognize individuals in other ways too.

The UP are quintessential tool makers: not simply because they make tools—some other animals do too—but because they make so many, so many different kinds of them, and are so dependent upon them. Unlike the other animals, the UP use tools to make tools. They make cutters that improve upon what they can do with their teeth or by tearing with their hands. They make pounders that improve upon what they can do with their teeth, fists, feet, knees, shoulders, elbows, and head. They make containers that allow them to hold more things at one time, to hold them more comfortably or continuously, and to hold them when they otherwise couldn't, as over a fire. Whether it be string, cord, sinew, vine, wire, or whatever, the UP have something to use to tie things together and make interlaced materials. They know and use the lever. Some of their tools are weapons, including the spear. The UP make many of their tools with such permanence that they can use them over and over again. They also make some of their tools in uniform patterns that are more or less arbitrary—thus we can often tell one people's tools from another's. Such patterns persist beyond any one person's lifetime. Since tools are so closely related to human hands, we might note in passing that most people among the UP are right-handed.

The UP may not know how to make fire, but they know how to use it. They use fire to cook food but for other purposes too. Tools and fire do much to make them more comfortable and secure. The UP have other ways to make themselves feel better (or different). These include substances they can take to alter their moods or feelings: stimulants, narcotics, or intoxicants. These are in addition to what they take for mere sustenance.

The UP always have some form of shelter from the elements. Further ways in which they attend to their material needs will be discussed later.

The UP have distinct patterns of preparation for birth, for giving birth, and for postnatal care. They also have a more or less standard pattern and time for weaning infants.

The UP are not solitary dwellers. They live part of their lives, if not the whole of them, in groups. One of their most important groups is the family, but it is not the only group among them. One or more of the UP groups maintains a unity even though the members are dispersed.

The UP have groups defined by locality or claiming a certain territory, even if they happen to live almost their entire lives as wanderers upon the sea. They are materially, cognitively, and emotionally adjusted to the environment in which they normally live (particularly with respect to some of its flora and fauna). A sense of being a distinct people characterizes the UP, and they judge other people in their own terms.

The core of a normal UP family is composed of a mother and children. The biological mother is usually expected to be the social mother and usually is. On a more or less permanent basis there is usually a man (or men) involved, too, and he (or they) serve minimally to give the children a status in the community and/or to be a consort to the mother. Marriage, in the sense of a "person" having a publicly recognized right of sexual access to a woman deemed eligible for childbearing, is institutionalized among the UP. While the person is almost always a male, it need not necessarily be a single individual, nor even a male. (Among some peoples, for example, a woman A may assume the status of a man, take a woman B as wife, and then arrange for the wife B to bear children to which A will be the social father.)

The UP have a pattern of socialization: children aren't just left to grow up on their own. Senior kin are expected to contribute substantially to socialization. One of the ways children learn among the UP is by watching elders and copying them. The socialization of UP children includes toilet training. Through practice, children and adults perfect what they learn. The UP learn some things by trial and error.

One's own children and other close kin are distinguished from more distant relatives or nonrelatives among the UP, and the UP favor their close kin in various contexts.

UP families and the relationships of their family members to each other and to outsiders are affected by their sexual regulations, which sharply delimit, if not eliminate, mating between the genetically close kin. Mating between mother and son, in particular, is unthinkable or taboo. Sex is a topic of great interest to the UP, though there may be contexts in which they will not discuss it.

Some groups among the UP achieve some of their order by division into socially significant categories or subgroups on the basis of kinship, sex, and age. Since the UP have kinship, sex, and age statuses, it follows, of course, that they have statuses and roles and hence a social structure. But they have statuses beyond those of sex, age, and kinship categories. And while these are largely ascribed statuses, they have achieved statuses too. There are rules of succession to some of their statuses.

Although it may be only another way of saying that they have statuses and roles, the UP recognize social personhood: social identities, including collective identities, that are distinguishable from the individuals who bear them. The distinction between persons and individuals involves the entification of the former; i.e., the UP speak of statuses as though they were entities that can act and be acted upon, such as we do when we say, for example, that "the legislature" (a social entity) "punished the university" (another social entity).

Prestige is differentially distributed among the UP, and the members of UP society are not all economically equal. They acknowledge inequalities of various sorts, but we cannot specify whether they approve or disapprove.

The UP have a division of labor, minimally based on the sex and age statuses already mentioned. For example, their women have more direct child-care duties than do their men. Children are not expected to, and typically do not, engage in the same activities in the same way that adults do. Related to this division of labor, men and women and adults and children are seen by the UP as having different natures. Their men are in fact on the average more physically aggressive than women and are more likely to commit lethal violence than women are.

In the public political sphere men form the dominant element among the UP. Women and children are correspondingly submissive or acquiescent, particularly, again, in the public political sphere.

In addition to their division of labor, whereby different kinds of people do different things, the UP have customs of cooperative labor, in which people jointly undertake essentially similar tasks. They use reciprocal exchanges, whether of labor, or goods, or services, in a variety of settings. Reciprocity—including its negative or retaliatory forms—is an important element in the conduct of their lives. The UP also engage in trade, that is, in nonreciprocal exchanges of goods and services (i.e., one kind of good or service for another). Whether reciprocally or not, they give gifts to one another too. In certain contexts they share food.

Whether in the conduct of family life, of subsistence activities, or other matters, the UP attempt to predict and plan for the future. Some of their plans involve the maintenance or manipulation of social relations. In this context it is important to note that the UP possess "triangular awareness," the ability to think not only of their own relationships to others but of the relationships between others in relation to themselves. Without such an ability they would be unable to form their ubiquitous coalitions.

The UP have government, in the sense that they have public affairs and these affairs are regulated, and in the sense that decisions binding on a collectivity are made. Some of the regulation takes place in a framework of corporate statuses (statuses with orderly procedures for perpetuating membership in them).

The UP have leaders, though they may be ephemeral or situational. The UP admire, or profess to admire, generosity, and this is particularly desired in a leader. No leader of the UP ever has complete power lodged in himself alone. UP leaders go beyond the limits of UP reason and morality. Since the UP never have complete democracy, and never have complete autocracy, they always have a de facto oligarchy.

The UP have law, at least in the sense of rules of membership in perpetual social units and in the sense of rights and obligations attached to persons or other statuses. Among the UP's laws are those that in certain situations proscribe violence and rape. Their laws also proscribe murder—unjustified taking of human life (though they may justify taking lives in some contexts).

They have sanctions for infractions, and these sanctions include removal of offenders from the social unit—whether by expulsion, incarceration, ostracism, or execution. They punish (or otherwise censure or condemn) certain acts that threaten the group or are alleged to do so.

Conflict is more familiar to the UP than they wish it were, and they have customary, though far from perfect, ways of dealing with it (their proscription of rape and other forms of violence, for example, does not eliminate them). They understand that wronged parties may seek redress. They employ consultation and mediation in some conflict cases.

Important conflicts are structured around in-group–out-group antagonisms that characterize the UP. These antagonisms both divide the UP as an ethnic group as well as set them off from other ethnic groups. An ethical dualism distinguishes the in-group from the out-group, so that, for example, cooperation is more expectable in the former than with the latter.

The UP distinguish right from wrong, and at least implicitly, as noted earlier, recognize responsibility and intentionality. They recognize and employ promises. Reciprocity, also mentioned earlier, is a key element in their morality. So, too, is their ability to empathize. Envy is ubiquitous among the UP, and they have symbolic means for coping with its unfortunate consequences.

Etiquette and hospitality are among UP ideals. They have customary greetings and customs of visiting kin or others who dwell elsewhere. They have standardized, preferred, or typical times of day to eat, and they have occasions on which to feast. In other ways, too, they have normal daily routines of activities and are fundamentally diurnal.

They have standards of sexual modesty—even though they might customarily go about naked. People, adults in particular, do not normally copulate in public, nor do they relieve themselves without some attempt to do it modestly. Among their other taboos are taboos on certain utterances and certain kinds of food. On the other hand, there are some kinds of food—sweets in particular—that they relish.

The UP have religious or supernatural beliefs in that they believe in something beyond the visible and palpable. They anthropomorphize and (some if not all of them) believe things that are demonstrably false. They also practice magic, and their magic is designed to do such things as to sustain and increase life and to win the attention of the opposite sex. They have theories of fortune and misfortune. They have ideas about how to explain disease and death. They see a connection between sickness and death. They try to heal the sick and have medicines for this purpose. The UP practice divination. And they try to control the weather.

The UP have rituals, and these include rites of passage that demarcate the transfer of an individual from one status to another. They mourn their dead.

Their ideas include a worldview—an understanding or conception of the world about them and their place in it. In some ways their worldview is structured by features of their minds. For example, from early infancy they have the ability to identify items that they know by one sense with the same items perceived in another sense, and so they see the world as a unity, not as different worlds imposed by our different sense modalities. Their worldview is a part of their supernatural and mythological beliefs. They have folklore too. The UP dream and attempt to interpret their dreams.

However spiritual they may be, the UP are materialists also. As indicated by their language having the possessive for use on "loose property," the UP have concepts of property, distinguishing what belongs—minimal though it may be—to the individual, or group, from what belongs to others. They also have rules for the inheritance of property.

In addition to their use of speech in poetic or polished ways, the UP have further aesthetic standards. However little clothing they wear, they nonetheless adorn their bodies in one way or another, including a distinctive way of maintaining or shaping their hair. They have standards of sexual attractiveness (including, for example, signs of good health and a clear male preference for the signs of early nubility rather than those of the postmenopausal state). Their decorative art is not confined to the body alone, for the UP apply it to their artifacts too. In addition to their patterns of grooming for essentially aesthetic reasons, they also have patterns of hygienic care.

The UP know how to dance and have music. At least some of their dance (and at least some of their religious activity) is accompanied by music. They include melody, rhythm, repetition, redundancy, and variation in their music, which is always seen as an art, a creation. Their music includes vocals, and the vocals include words—i.e., a conjunction of music and poetry. The UP have children's music.

The UP, particularly their youngsters, play and playfight. Their play, besides being fun, provides training in skills that will be useful in adulthood.

The materials presented in this chapter—essentially a list of absolute universals—draws heavily from Murdock, Tiger and Fox, and Hockett and also from many other sources. In some cases I have added items to the list because my own experience or that of a colleague or student has convinced me that the items ought to be there even though appropriate references could not be found. In a few cases I have counted something as a universal even though that required setting aside ethnographic testimony. There are, for example, some reports of societies in which getting into other people's minds (empathizing, divining intent or inner feeling, and the like) is not done or even conceived as possible. My assumption is that these reports may be emically correct but not etically. For example, Selby reports that the Zapotec, at least in some situations, do not think they can get into other people's minds (106–7, 109), but he gives a clear case of this happening (56). Similarly, to the Kaluli belief that "one cannot know what another thinks or feels," Ochs and Schieffelin comment that the Kaluli "obviously" do "interpret and assess one another's . . . internal states" (290).

In an equally few cases I omitted items from this chapter that nevertheless do appear in the bibliography—because I was not sufficiently convinced by the references. For example, after surveying ethnographic literature on abortion, Devereux felt the evidence was so strong for universality that he dismissed some reports of its absence (98). He may be correct, but his argument did not quite convince me and I decided to err on the side of caution and to count abortion as a near-universal. Similarly, Otterbein states in various places that the absolute universality of capital punishment is one of the major finds of his survey. But in other places in the same work he speaks more cautiously of the possibility that it is only a near-universal. I decided to accept the cautious judgment.

More important than uncertainties about the boundaries between universals and near-universals is the issue of adequate conceptualization or definition of particular universals. For example, the conceptualizations of marriage and the family that I presented are those that currently seem the most convincing to me; they have been differently conceived or defined in the past and may undergo further revision in the future.

There are also some general problems of conceptualization which, were they properly addressed, would have led to a different presentation than the one above. Some scholars distinguish between the surface (or substantive) universals and those that lie at some deeper level. This description has been more concerned with the former. A more serious pursuit of universals at the deeper level of process or innate mechanism may presumably unearth universals that are at present wholly unknown and almost certainly would produce hierarchical orders among some sets of universals, orders that distinguish the fundamental processes from their more superficial consequences.

Setting aside the issue of hierarchy, there are other problems with how the list of universals is ordered: which to start with, which to put in a set with which. Murdock took the easy way out and ordered his list alphabetically. While it seemed appropriate to me to begin with culture itself, and then to explore language, the order in which the remaining sets or clusters of items is presented is arbitrary. There is arbitrariness in each cluster, too, partly because I wanted to minimize repetitions. Repetitions do occur, and a fuller and truer account would include more repetitions or perhaps would show the interconnections between items by means of a diagram. For example, empathy (phrased in different ways but with the meaning of understanding another person's inner states) occurs in the description of the UP in the context of communication, morality, and psychological personhood—and is implicit elsewhere.

In sum, a fuller and truer account of the UP would in various ways show the relationships between the universals. But then a fuller and truer account of the UP would list their conditional universals (and *their* interrelationships and hierarchies) and would also offer explanations of the universals and their interrelationships. Anthropology has scarcely begun to illuminate the architecture of human universals. It is time to get on with the task.

7

Sociobiology at Century's End

Edward O. Wilson

Sociobiology was brought together as a coherent discipline in *Sociobiology: The New Synthesis* (1975), the book now reprinted, but it was originally conceived in my earlier work *The Insect Societies* (1971) as a union between entomology and population biology. This first step was entirely logical, and in retrospect, inevitable. In the 1950s and 1960s studies of the social insects had multiplied and attained a new but still unorganized level. My colleagues and I had worked out many of the principles of chemical communication, the evolution and physiological determinants of caste, and the dozen or so independent phylogenetic pathways along which the ants, termites, bees, and wasps had probably attained advanced sociality. The idea of kin selection, introduced by William D. Hamilton in 1963, was newly available as a key organizing concept. A rich database awaited integration. Also, more than 12,000 species of social insects were known and available for comparative studies to test the adaptiveness of colonial life, a great advantage over the

Reprinted by permission of the publisher from *Sociobiology: The New Synthesis* by Edward O. Wilson, pp. vi–viii, pp. 547–548, Cambridge, Mass.: The Belknap Press of Harvard University Press, Copyright © 1975, 2000 by the President and Fellows of Harvard College.

relatively species-poor vertebrates, of which only a few hundred are known to exhibit advanced social organization. And finally, because the social insects obey rigid instincts, there was little of the interplay of heredity and environment that confounds the study of vertebrates.

During roughly the same period, up to 1971, researchers achieved comparable advances in population biology. They devised richer models of the genetics and growth dynamics of populations, and linked demography more exactly to competition and symbiosis. In the 1967 synthesis *The Theory of Island Biogeography*, Robert H. MacArthur and I (if you will permit the continued autobiographical slant of this account) meshed principles of population biology with patterns of species biodiversity and distribution.

It was a natural step then to write *The Insect Societies* at the close of the 1960s as an attempt to reorganize the highly eclectic knowledge of the social insects on a base of population biology. Each insect colony is an assemblage of related organisms that grows, competes, and eventually dies in patterns that are consequences of the birth and death schedules of its members.

And what of the vertebrate societies? In the last chapter of *The Insect Societies*, entitled "The Prospect for a Unified Sociobiology," I made an optimistic projection to combine the two great phylads:

> In spite of the phylogenetic remoteness of vertebrates and insects and the basic distinction between their respective personal and impersonal systems of communication, these two groups of animals have evolved social behaviors that are similar in degree and complexity and convergent in many important details. This fact conveys a special promise that sociobiology can eventually be derived from the first principles of population and behavioral biology and developed into a single, mature science. The discipline can then be expected to increase our understanding of the unique qualities of social behavior in animals as opposed to those of man.

The sequel in this reasoning is contained in the book before you. Presented in this new release by Harvard University Press, it remains unchanged from the original. It provides verbatim the first effort to systematize the consilient links between termites and chimpanzees, the goal suggested in *The Insect Societies*, but it goes further, and extends the effort to human beings.

The response to *Sociobiology: The New Synthesis* in 1975 and the years immediately following was dramatically mixed. I think it fair to say that the zoology in the book, making up all but the first and last of its 27 chapters, was favorably received. The influence of this portion grew steadily, so much so that in a 1989 poll the officers and fellows of the international Animal Behavior Society rated *Sociobiology* the most important book on animal behavior of all time, edging out even Darwin's 1872 classic, *The Expression of the Emotions in Man and Animals.* By integrating the discoveries of many investigators into a single framework of cause-and-effect theory, it helped to change the study of animal behavior into a discipline connected broadly to mainstream evolutionary biology.

The brief segment of *Sociobiology* that addresses human behavior, comprising 30 out of the 575 total pages, was less well received. It ignited the most tumultuous academic controversy of the 1970s, one that spilled out of biology into the social sciences and humanities. The story has been told many times and many ways, including the account in my memoir, *Naturalist*, where I tried hard to maintain a decent sense of balance; and it will bear only a brief commentary here.

Although the large amount of commotion may suggest otherwise, adverse critics made up only a small minority of those who published reviews of *Sociobiology.* But they were very vocal and effective at the time. They were scandalized by what they saw as two grievous flaws. The first is inappropriate reductionism, in this case the proposal that human social behavior is ultimately reducible to biology. The second perceived flaw is genetic determinism, the belief that human nature is rooted in our genes.

It made little difference to those who chose to read the book this way that reductionism is the primary cutting tool of science, or that *Sociobiology* stresses not only reductionism but also synthesis and holism. It also mattered not at all that sociobiological explanations were never strictly reductionist, but interactionist. No serious scholar would think that human behavior is controlled the way animal instinct is, without the intervention of culture. In the interactionist view held by virtually all who study the subject, genomics biases mental development but cannot abolish culture. To suggest that I held such views, and it was suggested frequently, was to erect a straw man—to fabricate false testimony for rhetorical purposes.

Who were the critics, and why were they so offended? Their rank included the last of the Marxist intellectuals, most prominently represented by Stephen Jay Gould and Richard C. Lewontin. They disliked the idea, to put it mildly, that human nature could have any genetic basis at all. They championed the opposing view that the developing human brain is a tabula rasa. The only human nature, they said, is an indefinitely flexible mind. Theirs was the standard political position taken by Marxists from the late 1920s forward: the ideal political economy is socialism, and the tabula rasa mind of people can be fitted to it. A mind arising from a genetic human nature might not prove conformable. Since socialism is the supreme good to be sought, a tabula rasa it must be. As Lewontin, Steven Rose, and Leon J. Kamin frankly expressed the matter in *Not in Our Genes*: "We share a commitment to the prospect of the creation of a more socially just—a socialist—society. And we recognize that a critical science is an integral part of the struggle to create that society, just as we also believe that the social function of much of today's science is to hinder the creation of that society by acting to preserve the interests of the dominant class, gender, and race" (ix–x).

That was in 1984—an apposite Orwellian date. The argument for a political test of scientific knowledge lost its strength with the collapse of world social-ism and the end of the Cold War. To my knowledge it has not been heard since.

In the 1970s, when the human sociobiology controversy still waxed hot, however, the Old Marxists were joined and greatly strengthened by members of the New Left in a second objection, this time centered on social justice. If genes prescribe human nature, they said, then it follows that ineradicable differences in personality and ability also might exist. Such a possibility can-not be tolerated. At least, its discussion cannot be tolerated, said the critics, because it tilts thinking onto a slippery slope down which humankind eas-ily descends to racism, sexism, class oppression, colonialism, and—perhaps worst of all—capitalism! As the century closes, this dispute has been settled. Genetically based variation in individual personality and intelligence has been conclusively demonstrated, although statistical racial differences, if any, remain unproven. At the same time, all of the projected evils except capital-ism have begun to diminish worldwide. None of the change can be ascribed to human behavioral genetics or sociobiology. Capitalism may yet fall—who can predict history?—but, given the overwhelming evidence at hand, the hereditary framework of human nature seems permanently secure.

Among many social scientists and humanities scholars a deeper and less ideological source of skepticism was expressed, and remains. It is based on the belief that culture is the sole artisan of the human mind. This perception is also a tabula rasa hypothesis that denies biology, or at least simply ignores biology. It too is being replaced by acceptance of the interaction of biology and culture as the determinant of mental development.

Overall, there is a tendency as the century closes to accept that *Homo sapiens* is an ascendant primate, and that biology matters.

The path is not smooth, however. The slowness with which human socio-biology (nowadays also called evolutionary psychology) has spread is due not merely to ideology and inertia, but also and more fundamentally to the traditional divide between the great branches of learning. Since the early nineteenth century it has been generally assumed that the natural sciences, the social sciences, and the humanities are epistemologically disjunct from one another, requiring different vocabularies, modes of analysis, and rules of validation. The perceived dividing line is essentially the same as that between the scientific and literary cultures defined by C. P. Snow in 1959. It still frag-ments the intellectual landscape.

The solution to the problem now evident is the recognition that the line between the great branches of learning is not a line at all, but instead a broad, mostly unexplored domain awaiting cooperative exploration from both sides. Four borderland disciplines are expanding into this domain from the natural sciences side:

Cognitive neuroscience, also known as the brain sciences, maps brain activity with increasingly fine resolution in space and time. Neural pathways, some correlated with complex and sophisticated patterns of thought, can now be traced. Mental disorders are routinely diagnosed by this means, and the effects of drugs and hormone surges can be assessed almost directly. Neuroscientists are able to construct replicas of mental activity that, while still grossly incomplete, go far beyond the philosophical speculations of the past. They can then coordinate these with experiments and models from cognitive psychology, thus drawing down on independent reservoirs from yet another discipline bridging the natural and social sciences. As a result, one of the major gaps of the intellectual terrain, that between body and mind, may soon be closed.

In human genetics, with base pair sequences and genetic maps far advanced and near completion, a direct approach to the heredity of human behavior has opened up. A total genomics, which includes the molecular steps of epigenesis and the norms of reaction in gene–environment interaction, is still far off. But the technical means to attain it are being developed. A large portion of research in molecular and cellular biology is devoted to that very end. The implications for consilience are profound: each advance in neuropsychological genomics narrows the mind–body gap still further.

Where cognitive neuroscience aims to explain *how* the brains of animals and humans work, and genetics how heredity works, evolutionary biology aims to explain *why* brains work, or more precisely, in light of natural selection theory, what adaptations if any led to the assembly of their respective parts and processes. During the past 25 years an impressive body of ethnographic data has been marshaled to test adaptation hypotheses, especially those emanating from kin-selection and ecological optimization models. Much of the research, conducted by both biologists and social scientists, has been reported in the journals *Behavioral Ecology and Sociobiology, Evolution and Human Behavior* (formerly *Ethology and Sociobiology*), *Human Nature*, the *Journal of Social and Biological Structures*, and others, as well as in excellent summary collections such as *The Adapted Mind: Evolutionary Psychology and the Generation of Culture* (Barkow, Cosmides, and Tooby) and *Human Nature: A Critical Reader* (Betzig).

As a result we now possess a much clearer understanding of ethnicity, kin classification, bridewealth, marriage customs, incest taboos, and other staples of the human sciences. New models of conflict and cooperation, extending from Robert L. Trivers's original parent–offspring conflict theory of the 1970s and from ingenious applications of game theory, have been applied fruitfully to developmental psychology and an astonishing diversity of other fields—embryology, for example, pediatrics, and the study of genomic imprinting. Comparisons with the social behavior of the nonhuman primates, now a major concern of biological anthropology, have proven valuable in the analysis of human behavioral phenomena that are cryptic or complex.

Sociobiology is a flourishing discipline in zoology, but its ultimately greatest importance will surely be the furtherance of consilience among the great branches of learning. Why is this conjunction important? Because it offers the prospect of characterizing human nature with greater objectivity and precision, an exactitude that is the key to self-understanding. The intuitive grasp of human nature has been the substance of the creative arts. It is the ultimate underpinning of the social sciences and a beckoning mystery to the natural sciences. To grasp human nature objectively, to explore it to the depths scientifically, and to comprehend its ramifications by cause-and-effect explanations leading from biology into culture, would be to approach if not attain the grail of scholarship, and to fulfill the dreams of the Enlightenment.

The objective meaning of human nature is attainable in the borderland disciplines. We have come to understand that human nature is not the genes that prescribe it. Nor is it the cultural universals, such as the incest taboos and rites of passage, which are its products. Rather, human nature is the epigenetic rules, the inherited regularities of mental development. These rules are the genetic biases in the way our senses perceive the world, the symbolic coding by which our brains represent the world, the options we open to ourselves, and the responses we find easiest and most rewarding to make. In ways that are being clarified at the physiological and even in a few cases the genetic level, the epigenetic rules alter the way we see and intrinsically classify color. They cause us to evaluate the aesthetics of artistic design according to elementary abstract shapes and the degree of complexity. They lead us differentially to acquire fears and phobias concerning dangers in the ancient environment of humanity (such as snakes and heights), to communicate with certain facial expressions and forms of body language, to bond with infants, to bond conjugally, and so on across a wide spread of categories in behavior and thought. Most of these rules are evidently very ancient, dating back millions of years in mammalian ancestry. Others, like the ontogenetic steps of linguistic development in children, are uniquely human and probably only hundreds of thousands of years old.

The epigenetic rules have been the subject of many studies during the past quarter century in biology and the social sciences, reviewed for example in my extended essays *On Human Nature* (1978) and *Consilience: The Unity of Knowledge* (1998), as well as in *The Adapted Mind*, edited by Barkow et al. (1992). This body of work makes it evident that in the creation of human nature, genetic evolution and cultural evolution have together produced a closely interwoven product. We are only beginning to obtain a glimmer of how the process works. We know that cultural evolution is biased substantially by biology, and that biological evolution of the brain, especially the neocortex, has occurred in a social context. But the principles and the details are the great challenge in the emerging borderland disciplines just described. The exact process of gene–culture coevolution is the central problem of the social sciences and much of the humanities, and it is one of the great remaining problems of the natural sciences. Solving it is the obvious means by which the great branches of learning can be foundationally united.

Finally, during the past quarter century another discipline to which I have devoted a good part of my life, conservation biology, has been tied more closely to human sociobiology. Human nature—the epigenetic rules—did not originate in cities and croplands, which are too recent in human history to have driven significant amounts of genetic evolution. They arose in natural environments, especially the savannas and transitional woodlands of Africa, where *Homo sapiens* and its antecedents evolved over hundreds of thousands of years. What we call the natural environment or wilderness today was home then—the environment that cradled humanity. Before agriculture the lives of people depended on their intimate familiarity with wild biodiversity, both the surrounding ecosystems and the plants and animals composing them.

The link was, on a scale of evolutionary time, abruptly weakened by the invention and spread of agriculture and then nearly erased by the implosion of a large part of the agricultural population into the cities during the industrial and postindustrial revolutions. As global culture advanced into the new, technoscientific age, human nature stayed back in the Paleolithic era.

Hence the ambivalent stance taken by modern *Homo sapiens* to the natural environment. Natural environments are cherished at the same time they are subdued and converted. The ideal planet for the human psyche seems to be one that offers an endless expanse of fertile, unoccupied wilderness to be churned up for the production of more people. But Earth is finite, and its still exponentially growing human population is rapidly running out of productive land for conversion. Clearly humanity must find a way simultaneously to stabilize its population and to attain a universal decent standard of living while preserving as much of Earth's natural environment and biodiversity as possible.

Conservation, I have long believed, is ultimately an ethical issue. Moral precepts in turn must be based on a sound, objective knowledge of human nature. In 1984 I combined my two intellectual passions, sociobiology and the study of biodiversity, in the book *Biophilia*. Its central argument was that the epigenetic rules of mental development are likely to include deep adaptive responses to the natural environment. This theme was largely speculation. There was no organized discipline of ecological psychology that addressed such a hypothesis. Still, plenty of evidence pointed to its validity. In *Biophilia* I reviewed information then newly provided by Gordon Orians that points to innately preferred habitation (on a prominence overlooking a savanna and body of water), the remarkable influence of snakes and serpent images on culture, and other mental predispositions likely to have been adaptive during the evolution of the human brain.

Since 1984 the evidence favoring biophilia has grown stronger, but the subject is still in its infancy and few principles have been definitively established (Kellert and Wilson). I am persuaded that as the need to stabilize and protect the environment grows more urgent in the coming decades, the linking of the two natures—human nature and wild Nature—will become a central intellectual concern.

8

Evolution and Explanation

STEVEN PINKER

For many years after I decided to become a psychologist I was frustrated by my chosen field, and fantasized about a day when it would satisfy the curiosity that first led me to devote my professional life to studying the mind. As with many psychology students, the frustration began with my very first class, in which the instructor performed the ritual that begins every introduction to psychology course: disabusing students of the expectation that they would learn about any of the topics that attracted them to the subject. Forget about love and hate, family dynamics, and jokes and their relation to the unconscious, they said. Psychology was a rigorous science which investigated quantifiable laboratory phenomena; it had nothing to do with self-absorption on an analyst's couch or the prurient topics of daytime talk shows. And in fact the course confined itself to "perception," which meant psychophysics, and "learning," which meant rats, and "the brain," which meant neurons, and "memory," which meant nonsense syllables, and "intelligence," which meant IQ tests, and "personality," which meant personality tests.

Foreword to *The Handbook of Evolutionary Psychology* by David M. Buss. Hoboken, N.J.: John Wiley & Sons, xi–xvi. Reprinted with permission of John Wiley & Sons, Inc.

When I proceeded to more advanced courses, they only deepened the disappointment by revealing that the psychology canon was a laundry list of unrelated phenomena. The course on perception began with Weber's Law and Fechner's Law and proceeded to an assortment of illusions and aftereffects familiar to readers of cereal boxes. There was no there there—no conception of what perception *is* or of what it is for. Cognitive psychology, too, consisted of laboratory curiosities analyzed in terms of dichotomies such as serial/parallel, discrete/analog, and top-down/bottom-up (inspiring Alan Newell's famous jeremiad, "You can't play twenty questions with nature and win"). To this day, social psychology is driven not by systematic questions about the nature of sociality in the human animal but by a collection of situations in which people behave in strange ways.

But the biggest frustration was that psychology seemed to lack any sense of *explanation*. Like the talk show guest on *Monty Python's Flying Circus* whose theory of the brontosaurus was that "the brontosaurus is skinny at one end; much, much thicker in the middle; and skinny at the other end," psychologists were content to "explain" a phenomenon by redescribing it. A student rarely enjoyed the flash of insight which tapped deeper principles to show why something *had* to be the way it is, as opposed to some other way it could have been.

My gold standard for a scientific explanation was set when I was a graduate student—not by anything I learned in graduate *school*, mind you, but by a plumber who came to fix the pipes in my dilapidated apartment and elucidated why they had sprung a leak. Water, he explained, obeys Newton's second law. Water is dense. Water is incompressible. When you shut off a tap, a large incompressible mass moving at high speed has to decelerate quickly. This imparts a big force to the pipes, like a car slamming into a wall, which eventually damages the threads and causes a leak. To deal with this problem, plumbers used to install a closed vertical section of pipe, a pipe riser, near each faucet. When the faucet is shut, the decelerating water compresses the column of air in the riser, which acts like a shock absorber, protecting the pipe joints. Unfortunately, this is a perfect opportunity for Henry's Law to apply, namely, that gas under pressure is absorbed by a liquid. Over time, the air in the column dissolves into the water, filling the pipe riser and rendering it useless. So every once in a while, a plumber should bleed the system and let air back into the risers, a bit of preventive maintenance the landlord had neglected. I only wished that psychology could meet that standard of explanatory elegance and show how a seemingly capricious occurrence falls out of laws of greater generality.

It's not that psychologists never tried to rationalize their findings. But when they did, they tended to recycle a handful of factors like similarity, frequency, difficulty, salience, and regularity. Each of these so-called explanations is, in the words of the philosopher Nelson Goodman, "a pretender, an impostor, a quack." Similarity (and frequency and difficulty and the rest) are in the eye of the beholder, and it is the eye of the beholder that psychologists should be trying to explain.

This dissatisfaction pushed me to the broader interdisciplinary field called cognitive science, where I found that other disciplines were stepping into the breach. From linguistics I came across Noam Chomsky's criteria for an adequate theory of language. At the lowest level was observational adequacy, the mere ability to account for linguistic behavior; this was the level at which most of psychology was stuck. Then there was descriptive adequacy, the ability to account for behavior in terms of the underlying mental representations that organize it. At the highest level was explanatory adequacy, the ability of a theory to show why *those* mental representations, and not some other ones, took root in the mind. In the case of linguistics, Chomsky continued, explanatory adequacy was rooted in the ability of a theory to solve the problem of language acquisition, explaining how children can learn an infinite language from a finite sample of sentences uttered by their parents. An explanatory theory must characterize Universal Grammar, a part of the innate structure of the mind. This faculty forces the child to analyze speech in particular ways, those consistent with the way human languages work, rather than in any of the countless logically possible ways that are consistent with the input but dead ends in terms of becoming an expressive language user (e.g., memorizing every sentence or combining nouns and verbs promiscuously). As a result, a person's knowledge of language is not just any old set of rules, but ones that conform to an algorithm powerful enough to have acquired an infinite language from a finite slice of the environment. For example, locality conditions on movement rules in syntax—the fact that you can say, "What do you believe he saw?" but not, "What do you believe the claim that he saw?"—allow children to acquire a language from the kinds of simple sentences that are available in parental speech. In this way, a psychological phenomenon (the distribution of well-formed and malformed questions) could be explained in terms of what was necessary to solve the key problem faced by a human child in this domain.

Artificial intelligence, too, set a high standard of explanation via the work of the vision scientist David Marr. A theory of vision, he suggested, ought to characterize visual processing at three levels: the neurophysiological mechanism, the algorithm implemented by this mechanism, and, crucially, a "theory of the computation" for that domain. A theory of the computation is a formal demonstration that an algorithm can, in principle, compute the desired result, given certain assumptions about the way the world works.

And the desired result, in turn, should be characterized in terms of the overall "goal" of the visual system, namely to compute a useful description of the world from the two-dimensional array of intensity and wavelength values falling on the retina. For example, the subsystem that computes the perception of shape from shading (as when we perceive the contours of a cheek or the roundness of a Ping-Pong ball) relies on a fact of physics that governs how the intensity of light reflecting off a surface depends on the relative angles of the illuminant, the surface, and the observer, and on the physical properties of the surface. A perceptual algorithm can exploit this bit of physics to "work backward" from the array of light intensities, together with certain assumptions about typical illuminants and surfaces in a terrestrial environment, and compute the tangent angle of each point on a surface, yielding a representation of its shape. Many perceptual phenomena, from the way makeup changes the appearance of a face to the fact that turning a picture of craters upside down makes it look like a picture of bumps, can be explained as byproducts of this shape-from-shading mechanism. Most perception scientists quickly realized that conceiving the faculty of vision as a system of well-designed neural computers that supply the rest of the brain with an accurate description of the visible environment was a big advance over the traditional treatment of perception as a ragbag of illusions, aftereffects, and psychophysical laws.

Language and perception, alas, are just two of our many talents and faculties, and it was unsatisfying to think of the eyes and ears as pouring information into some void that constituted the rest of the brain. Might there be some comparable framework for the rest of psychology, I wondered, that addressed the engaging phenomena of mental and social life, covered its subject matter systematically rather than collecting oddities like butterflies, and explained its phenomena in terms of deeper principles? The explanations in language and vision appealed to the *function* of those faculties: in linguistics, acquiring the language of one's community; in vision, constructing an accurate description of the visible world. Both are extraordinarily difficult computational problems (as yet unsolvable by any artificial intelligence system) but ones that any child can perform with ease. And both are not esoteric hobbies but essential talents for members of our species, affording obvious advantages to their well-being. Couldn't other areas of psychology, I wondered, benefit from an understanding of the problems our mental faculties solve—in a word, what they are *for*?

When I discovered evolutionary psychology in the 1980s through the work of Donald Symons, Leda Cosmides, and John Tooby, I realized my wait was over. Evolutionary psychology was the organizing framework—the source of "explanatory adequacy" or a "theory of the computation"—that the science of psychology had been missing. Like vision and language, our emotions and cognitive faculties are complex, useful, and nonrandomly organized, which means that they must be a product of the only physical process capable of generating complex, useful, non-random organization, namely, natural selection. An appeal to evolution was already implicit in the metatheoretical directives of Marr and Chomsky, with their appeal to the function of a mental faculty, and evolutionary psychology simply shows how to apply that logic to the rest of the mind.

Just as important, the appeal to function in evolutionary psychology is itself constrained by an external body of principles—those of the modern, replicator-centered theory of selection from evolutionary biology—rather than being made up on the spot. Not just any old goal can count as the function of a system shaped by natural selection, that is, an adaptation. Evolutionary biology rules out, for example, adaptations that work toward the good of the species, the harmony of the ecosystem, beauty for its own sake, benefits to entities other than the replicators that create the adaptations (e.g., horses that evolve saddles), functional complexity without reproductive benefit (e.g., an adaptation to compute the digits of pi), and anachronistic adaptations that benefit the organism in a kind of environment other than the one in which it evolved (e.g., an innate ability to read or an innate concept of *carburetor* or *trombone*). Natural selection also has a positive function in psychological discovery, impelling psychologists to test new hypotheses about the possible functionality of aspects of the mind that previously seemed functionless. For example, the social and moral emotions (sympathy, trust, guilt, anger, gratitude) appear to be adaptations for policing reciprocity in nonzero sum games; an eye for beauty appears to be an adaptation for detecting health and fertility in potential mates. None of this research would be possible if psychologists had satisfied themselves with a naive notion of function instead of the one licensed by modern biology.

Evolutionary psychology also provides a motivated research agenda for psychology, freeing it from its chase of laboratory curiosities. An explanatory hypothesis for some emotion or cognitive faculty must begin with a theory of how that faculty would, on average, have enhanced the reproductive chances of the bearer of that faculty in an ancestral environment. Crucially, the advantage must be demonstrable by some independently motivated causal consequence of the putative adaptation. That is, laws of physics or chemistry or engineering or physiology, or some other set of laws independent of the part of our psychology being explained must suffice to establish that the trait is useful in attaining some reproduction-related goal. For example, using projective geometry one can show that an algorithm can compare images from two adjacent cameras and calculate the depth of a distant object using the disparity of the two images. If you write out the specs for computing depth in this way—what engineers would specify if they were building a robot that had to see in depth—you can then examine human stereoscopic depth perception and ascertain whether humans (and other primates) obey those specs. The closer the empirical facts about our psychology are to the engineering specs for a well-designed system, the greater our confidence that we have explained the psychological faculty in functional terms. A similar example comes from the wariness of snakes found in humans and many other primates. We know from herpetology that snakes were prevalent in Africa during the time of our evolution and that getting bitten by a snake is harmful because of the chemistry of snake venom. Crucially, these are not facts of psychology. But they help to establish that something that *is* a fact of psychology, namely the fear of snakes, is a plausible adaptation. In a similar manner, robotics can help explain motor control, game theory can explain aggression and appeasement, economics can explain punishment of free riders, and mammalian physiology (in combination with the evolutionary biology of parental investment) makes predictions about sex differences in sexuality. In each case, a "theory of the computation" is provided by an optimality analysis using a set of laws outside the part of the mind we are trying to explain. This is what entitles us to feel that we have explained the operation of that part of the mind in a noncircular way.

In contrast, it's not clear what the adaptive function of music is, or of religion. The popular hypothesis that the function of music is to keep the community together may be true, but it is not an *explanation* of why we like music, because it just begs the question of why sequences of tones in rhythmic and harmonic relations should keep the group together. Generating and sensing sequences of sounds is not an independently motivated solution to the problem of maintaining group solidarity, in the way that, say, the emotion of empathy, or a motive to punish free riders, is part of such a solution. A similar problem infects the "explanation" that people are prone to believe in incredible religious doctrines because those doctrines are comforting—in other words, that the doctrines of a benevolent shepherd, a universal plan, an afterlife, and divine retribution ease the pain of being a human. There's an element of truth to each of these suggestions, but they are not legitimate adaptationist explanations, because they beg the question of *why* the mind should find comfort in beliefs that it is capable of perceiving as false. In these and other cases, a failure to find an adaptationist explanation does not mean that no explanation is forthcoming at all. Recent books by Pascal Boyer (*Naturalness*) and Scott Atran have insightfully explained the phenomenon of religious belief as a byproduct of adaptations (such as a Theory of Mind module and free-rider detection mechanisms) that are demonstrably useful for solving *other* adaptive problems.

Evolutionary psychology is the cure for one last problem ailing traditional psychology: its student-disillusioning avoidance of the most fascinating aspects of human mental and social life. Even if evolutionary psychology had not provided psychology with standards of explanatory adequacy, it has proved its worth by opening up research in areas of human experience that have always been fascinating to reflective people but that had been absent from the psychology curriculum for decades. It is no exaggeration to say that contemporary research on topics like sex, attraction, jealousy, love, food, disgust, status, dominance, friendship, religion, art, fiction, morality, motherhood, fatherhood, sibling rivalry, and cooperation has been opened up and guided by ideas from evolutionary psychology. Even in more traditional topics in psychology, evolutionary psychology is changing the face of theories, making them into better depictions of the real people we encounter in our lives, and making the science more consonant with common sense and the wisdom of the ages. Before the advent of evolutionary thinking in psychology, theories of memory and reasoning typically didn't distinguish thoughts about people from thoughts about rocks or houses. Theories of emotion didn't distinguish fear from anger, jealousy, or love. And theories of social relations didn't distinguish among the way people treat family, friends, lovers, enemies, and strangers.

For many reasons, then, this *Handbook* represents a remarkable milestone in the science of psychology. The theoretical rigor and empirical richness showcased in the *Handbook* have more than fulfilled evolutionary psychology's initial promise, and they demolish lazy accusations that the field is mired in speculative story-telling or rationalizations of reactionary politics. The chapters don't, of course, summarize a firm consensus or present the final word in any of the areas they cover. But in topics from parenting to fiction, from predation to religion, they deliver subtle and deep analyses, genuinely new ideas, and eye-opening discoveries. *The Handbook of Evolutionary Psychology* is far more than a summary of the state of the art of evolutionary psychology. It is the realization of the hope that psychology can be a systematic and explanatory science of the human condition.

9

Evolutionary Social Constructivism

David Sloan Wilson

Evolutionary theory has been controversial throughout its history for reasons that go beyond religious matters. Even among nonbelievers, something momentous and contentious appears to be at stake. The controversy also transcends knowledge of the subject. It has not quieted over the decades, despite tremendous advances in knowledge, and it currently divides the foremost authorities on evolution, as the pages of professional journals and popular intellectual forums such as the *New York Review of Books* attest.

Among the sophisticates, the controversy does not center on the basic fact of evolution but on certain consequences, such as the importance of natural selection and especially the relevance of evolution to human affairs. The intellectual positions most fiercely opposed to "sociobiology" and "evolutionary psychology" include social constructivism, postmodernism, and deconstructionism. These positions are different from each other but united in their commitment to the idea that individuals and societies have

From "Evolution and Social Constructivism" from *The Literary Animal: Evolution and the Nature of Narrative*. Edited by Jonathan Gottschall and David Sloan Wilson, pp. 20–29, 35.

enormous flexibility in what they can become, in contrast to the inflexibility and determinism attributed to evolutionary approaches to human behavior. I will refer to this core idea as social constructivism, with apologies for obscuring the differences between the positions referred to above that are important in other contexts.

These debates usually become so polarized that they reveal the worst aspects of tribalism in our species. Each side regards the other as the enemy whose position has no substance or rational basis, other than being ideologically driven. The middle ground becomes a no-man's-land into which no one dares to venture. Given this kind of intellectual trench warfare, it is no wonder that ideas can stagnate for years, decades, and even centuries.

This essay attempts a more productive exploration of the middle ground. I will try to show that the heart of social constructivism can be given an evolutionary formulation. Social constructivists have more to gain from adopting an evolutionary perspective than by avoiding it, and sociobiologists need to incorporate large elements of social constructivism into their own framework. This is not an exercise in empty diplomacy in which everyone continues to think as they did before. Instead, it is an attempt to genuinely occupy the middle ground that requires fundamental movement on both sides.

Why should a chapter on evolutionary social constructivism appear in a book on evolution and literature? One reason is that literary studies have historically been dominated by social constructivist perspectives. Skepticism toward a genre of evolutionary literary studies is fueled by the larger issues, with literature the battleground rather than the battle. Another reason is that evolutionary social constructivism relies fundamentally upon narrative. The reason that individuals and societies have a capacity for change is largely because of the importance of stories in psychological and cultural evolutionary processes.

Three Evolutionary and Two Social Constructivist Positions

I will begin by outlining three evolutionary (E) and two social constructivist (S) positions. The evolutionary positions are as follows:

E1. The minds of all organisms are genetically adapted to their ancestral environments. Because there are many adaptive problems to solve, all minds consist of a collection of specialized adaptations rather than a single all-purpose adaptation. Understanding the human mind is complicated by the fact that genetic evolution has not kept pace with the social and environmental changes brought about by the advent of agriculture. We therefore often behave maladaptively in our current environments, much as a rain forest lizard would behave maladaptively when transported into the desert. To understand the human mind and its products, we need to examine their adaptedness in ancestral environments, not in modern environments. This is the position most often associated with the term "evolutionary psychology" (Barkow, Cosmides, and Tooby). When related to literature, a typical hypothesis emanating from this position might be "sex differences and sexual relationships in literature should reflect adaptive male and female reproductive strategies that evolved in ancestral environments and are part of our human nature."

E2. There is more to evolution than genetic evolution. Physiological, psychological, and cultural processes can also be evolutionary in the sense that alternatives are created and selected on the basis of given criteria. The immune system is a well-known example of a physiological evolutionary process. Antibodies are created at random, and those that successfully bind to antigens replicate faster than those that don't. The late social psychologist Donald Campbell never tired of using the phrase "blind variation and selective retention" to describe the essence of evolution and its relevance to psychological and cultural processes, including the process of scientific inquiry, in addition to genetic evolution ("Blind Variation," "Evolutionary Epistemology"). When related to literature, a hypothesis emanating from this position might be "narratives have a powerful effect on human behavior and adaptation to current environments proceeds in part through the creation and selection of alternative narratives."

E3. There is more to evolution than adaptation. Evolving systems are often poorly adapted to their environments for a host of reasons, including genetic drift; phylogenetic, developmental, and genetic constraints; and more, all of which have counterparts in nongenetic evolutionary processes (Gould and Lewontin). When related to literature, a hypothesis emanating from this position might be "literature is a form of play in humans, and adult play exists not as an adaptation in its own right but as part of selection for juvenile characters in general (neoteny). Neotonous behaviors such as play got dragged along with neotonous morphological characters that enabled us to have big heads and stand upright."

Here are the social constructivist positions.

S1. Individuals and societies have enormous flexibility in what they can become, which is largely unconstrained by human biology. This flexibility is reflected in the diversity of behaviors that we observe within and among societies around the world and throughout history. People have almost no instincts and obtain their behaviors through learning and cultural transmission. Current inequities that are often justified as part of human nature, therefore inevitable, are nothing of the sort and usually reflect the efforts of powerful elements of society to dominate less powerful elements. When related to literature, a typical hypothesis emanating from this position is "the association of witchcraft with females, including the representation of witches in literature, is an effort to limit the social role of females to that of the good wife" (Brauner).

S2. Individuals and societies have such enormous flexibility that anything—absolutely anything—goes. For example, all possible combinations of sex roles can exist and have been observed in societies around the world. Words for categories that seem weird to us, such as "red hats worn on the same day that broccoli is eaten" or "second cousins who commit unspeakable acts with barnyard animals," can be found in other cultures, just as our words and categories appear weird to them. When related to literature, it is difficult to provide a hypothesis emanating from this position, precisely because anything goes.

Positions S1 and S2 agree about flexibility but disagree about whether it leads to sensible versus nonsensical outcomes. Critics of social constructivism often portray the S2 version but I think that a more sympathetic reading is closer to S1. Social constructivists are first and foremost trying to imagine and implement a better world. What they imagine may strike some as naively optimistic or wrongheaded, but it is perfectly sensible, even in biological terms—equality, respect, basic necessities for all, the end of repression, and so on. When social constructivists say that anything goes, it is usually in the context of saying that their desired outcome is possible. This is the form of social constructivism that I will defend, and indeed the only form that I think is worth defending. Who would want to defend a view in which absolutely nothing winnows the functional from the dysfunctional?

Our question therefore becomes, what is the potential for incorporating the social constructivist position S1 into the three evolutionary positions E1, E2, and E3? Ever since the publication of *Sociobiology* in 1975, critics have taken refuge primarily in E3 (Levins and Lewontin). The reason that first sociobiology and then evolutionary psychology are fatally flawed, say the critics, is because they rely excessively on adaptationism. I regard this as an unfortunate wrong turn on the part of the critics. As an evolutionary biologist, I am perfectly comfortable with the fact that there is more to evolution than adaptation. It is definitely the middle ground that needs exploring as far as the general subject of evolution is concerned. However, the nonadaptive side of evolution provides little comfort for S1, the social constructivist position that we are trying to place on an evolutionary foundation. Since S1 involves the achievement of goals that are desirable, therefore adaptive at least in the everyday sense of the word, biologically nonadaptive processes can only accomplish these goals as a happy coincidence, by accident so to speak. Before we rely excessively on a happy accident argument, let us see if a stronger foundation can be found in E1 and E2.

It might surprise some readers to learn that E1, the position most closely associated with sociobiology and evolutionary psychology, provides substantial support for S1, even before we proceed to E2. The key concept that provides a link between E1 and S1 is *behavioral flexibility*, also called *phenotypic plasticity*. No organism is so simple that it is instructed by its genes to "do *x*." Even bacteria and protozoa are genetically endowed with a set of if–then rules of the form "do *x* in situation 1," "do *y* in situation 2," and so on. These rules enable organisms to do the right thing at the right time, not only behaviorally but physiologically and morphologically. The literature on nonhumans is full of wonderful examples of caterpillars that look like twigs in spring and leaves in summer, fish that grow streamlined bodies in the absence of predators but flattened bodies in their presence to exceed the gape of their jaws, frog eggs designed to hatch prematurely at the approach of a snake, salamanders that morph into big-jawed cannibals when food becomes short, and on and on (West-Eberhard). In all of these cases, information from the environment is combined with a set of predetermined if–then rules to determine the structure and behavior of the organism, much as your tax-preparation software branches off in different directions depending upon the information that it prompts you for.

This kind of adaptive behavioral flexibility provides an intriguing twist to the concept of genetic determinism. Let's assume for the moment that we are driven by our genes to obey the following set of if–then rules.

IN THIS SITUATION . . . BEHAVE THIS WAY

A	A'
B	B'
C	C'
Etc.	Etc.

Each behavior is adaptive for its respective situation and maladaptive for the other situations. For these if–then rules to evolve, all of the situations must exist in the overall ancestral environment. For example, birds that evolve in environments where predators may or may not be present have evolved to modify their behavior accordingly. In contrast, birds that evolve in environments where predators are always absent do not behave appropriately when the first ones appear. The first sailors to set foot on the Galápagos Islands were surprised when the birds acted as if they were trees rather than predators (Weiner). To pick an example more relevant to humans, we might be psychologically adapted to live in groups that vary in size from ten to a thousand but genetically unprepared for the megagroups of modern life (Dunbar, *Grooming*).

As surprising as it might seem, the genetic determinism of if–then rules provides at least a partial foundation for the social constructivist position S1. Suppose that we regard behavior C′ in the above list as socially desirable. We will never achieve behavior C′ in situations A and B, but behavior C′ will be inevitable if we can implement situation C. The key to achieving the desired social outcome is therefore to change the *situation*, an environmental intervention more reminiscent of social constructivism than genetic determinism as it is usually imagined.

My favorite example of this important concept is a study by evolutionary psychologists Margo Wilson and Martin Daly on risk-taking in men and age of first reproduction in women in the city of Chicago ("Life Expectancy"). Unlike most cities, whose neighborhoods are subject to a rapid turnover of residents, Chicago neighborhoods tend to be demographically stable. They also vary greatly in their quality of life, which is reflected in life expectancies that range from the mid-fifties for the worst neighborhoods to the mid-seventies for the best neighborhoods. Wilson and Daly showed that violent risk-taking in men and age of first reproduction in women correlate very strongly with life expectancy. Of course, both of these are perceived as social problems. Politicians talk endlessly about reducing violence and teenage pregnancies, especially in our inner cities. However, when women from the worst neighborhoods were asked why they had children so young, they gave an answer that can only invoke sympathy: they said that they wanted their mothers to see their grandchildren and in turn wanted to see their own grandchildren. They used the term "weathering" to refer to the aging process that they observed in themselves and their loved ones all around them. If everyone around you was weathering and dying at an average age of fifty-five, wouldn't *you* want to start having children early (as a female) or take great risks to obtain the status and resources required to reproduce (as a male)?

We can portray this example in terms of hypothetical genetically determined if–then rules as follows:

Low life expectancy	Reproduce early (women)
	Take high risks (men)
	Heavily discount the future (both)
High life expectancy	Reproduce later (women)
	Take fewer risks (men)
	Long-range planning (both)

The female and male strategies both fall under the more general categories of discounting the future when life expectancies are low (because there might not be a future) as opposed to forgoing short-term benefits in favor of future benefits when life expectancies are high. If we provisionally accept these if–then rules as the dictates of our genes, then we can derive a straightforward prediction and plan of action: to solve the problems of early pregnancy and violence in our worst neighborhoods, increase life expectancy and otherwise provide a stable social environment with a future to plan for. Of course, this is the kind of solution that a self-described social constructivist and critic of biological determinism might advise.

Genetic determinism contributes positively to social constructivism in this hypothetical scenario, at least in some respects. It leads to a clear plan of action, in contrast to the "anything goes" version of social constructivism. To achieve any given behavior in the right column, simply create the corresponding situation in the left column. In addition, the "anything goes" version of social constructivism can lead to scary outcomes, such as brainwashing people in our worst neighborhoods to be docile and childless. As many critics of social constructivism have observed, it is naive and illogical to think that "anything goes" leads consistently to "socially desirable." The idea of an evolved human nature that fights tenaciously for adaptive outcomes provides a firmer foundation for the optimistic brand of social constructivism (S1) than the "anything goes" portrayal of human nature.

I am not the first person to point out that adaptive behavioral flexibility turns the implications of genetic determinism topsy-turvy. Numerous self-described sociobiologists and evolutionary psychologists have made the same points and justly feel misunderstood by their social constructivist critics who continue to associate evolution with inflexibility (Gaulin and McBurney). Here, then, is an important meeting ground in which social constructivism can be placed on an evolutionary foundation. However, I will argue that it does not go far enough. The evolutionary position that I have designated E2 provides even more scope for social constructivism.

Innate Psychology and Nongenetic Evolutionary Processes

"Learning" and "culture" have always been the alternatives to "evolution" for those who reject evolutionary approaches to human behavior. However, learning and cultural change are themselves evolutionary in the sense that alternative behaviors are created and selected according to certain criteria. They are "blind variation and selective retention" processes, as Campbell put it ("Blind Variation"). What separates learning and culture from genetic evolution is not their evolutionary character but their speed. Learning and cultural evolution adapt organisms to their environment quickly, while genetic evolution is so slow that its products are essentially fixed over the time scales that matter most in contemporary human affairs. Another potential difference involves the criteria for selection. Perhaps nongenetic evolutionary processes favor the same behaviors that would evolve by genetic evolution, given enough time, but perhaps they favor a different set of behaviors (Plotkin).

For much of the twentieth century, learning and cultural evolution were invoked so heavily to explain human behavior that genetic evolution seemed irrelevant. If people can be made to do anything with the appropriate reinforcement and enculturation, who cares what happened during the Stone Age? Long before sociobiology and evolutionary psychology made the scene, cognitive psychologists were dismantling the notion of the blank slate by revealing the enormously complicated circuitry that was required to perform such "simple" acts as seeing, hearing, and remembering. It went without saying that this circuitry was largely innate and a product of genetic evolution. However, no cognitive psychologist to my knowledge has interpreted this kind of innateness as denying the existence of nongenetic evolutionary processes.

The cognitive revolution in psychology tended to focus on basic faculties such as vision, hearing, memory, language, and so on. These traits (with the exception of language) are obviously required for survival and reproduction, but what about other traits such as mating, foraging, cooperation, aggression, and migration? According to evolutionary psychologists such as Leda Cosmides and John Tooby, these traits are like vision in their requirement for an elaborate innate circuitry. Just as different circuits are required for vision and hearing (although they must also be integrated with each other), different circuits are required for the evaluation of long-term mates, the evaluation of short-term mates, response to infidelity, the detection of cheaters in social exchange, and so on. The list of specialized cognitive adaptations is not endless but runs into the hundreds and thousands, covering all of the important behaviors that helped us to survive and reproduce in ancestral environments (Tooby and Cosmides, "Psychological Foundations").

This is a startlingly different conception of the mind that will be important even if only partially correct. In its extreme form, however, it has led to the denial of learning and culture as open-ended evolutionary processes in their own right. If this were true, then modern evolutionary theory would provide justification for E1 but not E2, and the only evolutionary foundation for social constructivism would be the innate if–then rules described in the previous section.

The argument upon which the denial is based goes like this: All cognitive adaptations must be specialized to be smart. The first artificial intelligence researchers naively thought that they could build smart general-purpose learning machines, but they soon discovered that the only way to make a machine smart is to make it specialized for a particular task. Chess-playing computers are smart at playing chess but can't do anything else. Similarly, your tax-preparation software can calculate your taxes only if you give it exactly the right information, which it is designed to process in exactly the right way. The world is so full of potential information, which can be processed in so many ways, that all cognitive adaptations must be like your tax-preparation software in its specialized perception and processing of information.

To see why this argument fails, consider the mammalian immune system. Just like the mind, it can be regarded as a collection of genetically evolved mechanisms for helping us to survive and reproduce in our ancestral environment. The number and sophistication of mechanisms that comprise the immune system are mind-boggling when understood in detail (Sompayrac). Nevertheless, the centerpiece of the immune system is an open-ended process of blind variation and selective retention. Antibodies are produced at random, and those that successfully fight invading disease organisms are selected. Diseases are so numerous and evolve so fast with their short generation times that the only way to fight them is with another evolutionary process.

The immune system shows that genetic evolution and elaborate innateness do not invariably lead to the kind of modularity that excludes open-ended processes. Indeed, when the pace of environmental change becomes too fast and the number of challenges too great, genetically fixed if–then rules break down and must be supplemented by rapid nongenetic evolutionary processes that generate and select new solutions to current problems. As for the immune system, so also for psychological and cultural processes.

These observations are elementary but profound in their implications for placing social constructivism on an evolutionary foundation. They mean that *whatever the virtues of the evolutionary position outlined in E1, they do not exclude the evolutionary position outlined in E2.* Put another way, all metaphors make a connection between two things that are valid in some respects but not others. My love is a rose even though she is not red and thorny. The blank slate metaphor might be a total failure as a mechanistic conception of the mind but still be perfectly valid with respect to the open-ended nature of individual and societal change.

It would be a mistake to take this reasoning too far. Our eating behaviors provide fine examples of evolved predispositions that were adaptive in ancestral environments, have become maladaptive in modern environments, and are difficult to change. Religions encourage and often achieve altruism at a scale that never existed in ancestral environments, but they don't say that it's easy. There is a difference between *potential* for individual and societal change and *equi-potential*. If by "blank slate" we mean "anything can be written with equal ease," then that part of the metaphor is false.

My argument for placing social constructivism on an evolutionary foundation can be summarized as follows: Those who feel strongly about the potential for individual and societal change need not feel threatened by evolutionary theory. Even the elaborate innateness of the immune system does not exclude and indeed makes possible the potential for open-ended change, leading to new solutions to current problems. However, fulfilling the valid aspects of the blank slate metaphor requires abandoning the invalid aspects. "Potential" does not mean "equi-potential." Realizing potential can be facilitated by a detailed understanding of the mechanisms of genetic evolution and nongenetic evolutionary processes both built by and partially constrained by genetic evolution. In short, the way forward for social constructivism is to become sophisticated about evolution, not to deny its relevance to human affairs.

Evolutionary biologists interested in human behavior, in turn, must realize that there is more to human evolution than genetic adaptation to ancestral environments. The position that I have outlined as E1 does not exclude the position I have outlined as E2, however valid in other respects. Part of our genetic endowment is the capacity for rapid individual and societal adaptation to current environments, which is the heart of social constructivism.

Evolutionary Social Constructivism and Literature

Years ago I asked Napoleon Chagnon, one of the first anthropologists to call himself a sociobiologist, what he found so insightful about evolution. "Because it tells anthropologists to study reproduction instead of pottery!" he snapped back. The simplicity of Chagnon's answer took me aback. He was saying that the study of humans should be centered upon survival and reproduction—and indeed survival only to the extent that it leads to reproduction—just like any other species. We might be playing the reproduction game differently from other species in some respects, but we are playing the same game.

As for anthropology, so also for psychology. David Buss's textbook *Evolutionary Psychology: The New Science of the Mind* has section headings unlike any other psychology textbook: "Problems of Survival," "Challenges of Sex and Mating," "Challenges of Parenting and Kinship," and "Problems of Group Living." This organization reflects the fact that we evolved to do certain things well and that the study of psychology should be organized around those things. As the evolutionary biologist George Williams is often quoted as saying: "Is it not reasonable that our understanding of the human mind would be aided greatly by knowing the purpose for which it was designed?" (16).

In many respects, the study of literature from an evolutionary perspective needs to begin with the same refocusing of attention that is already taking place in anthropology and psychology. As Daniel Nettle puts it, if we ask what themes would most interest a nonhuman primate, those are the themes that are most prominently featured in Shakespeare and indeed all literature ("What Happens"). I once tested this proposition for myself during a trip to Japan by asking my hosts to provide me with a list of classic Japanese novels, short stories, and plays, which I purchased in their English translations and read during the course of my trip. Even though Japanese culture is often said to be different from Western culture, especially during the times when some of the older works on my list were written, the evolutionary themes leapt off the pages and would have interested our nonhuman primate as much as Shakespeare.

If this volume succeeds in refocusing attention for the study of literature on a par with anthropology and psychology, then it will have accomplished an important task. However, it will not have gone far enough. Stories do more than reflect the ancient concerns of our species, which we hold largely in common with other species. Narratives play an integral role in the nongenetic evolutionary processes outlined in E2 and perhaps even the innate flexibility outlined in E1. Unless we appreciate the importance of narratives in adapting us to our current environments, we will not have a fully developed genre of evolutionary literary studies.

Welcome to the Middle Ground

I began this essay by saying that it is not an exercise in idle diplomacy but a serious attempt to find the common ground between evolutionary theory and social constructivism. The heart of social constructivism is an optimistic belief that people and societies can become better in the future than in the present or past. This belief is not threatened by evolutionary theory. Indeed, evolution is all about change, and only by the strangest of secondary assumptions can it be interpreted as implying an incapacity for change. In particular, if we restrict evolution to genetic evolution and ignore the concept of adaptive behavioral flexibility, then evolution indeed implies an incapacity for change over the time scales most relevant to contemporary human affairs. However, no sophisticated evolutionary biologist would accept both of these restrictive assumptions. The adaptive behavioral flexibility that already occupies center stage in sociobiology and evolutionary psychology provides some scope for the optimistic spirit of social constructivism. Nongenetic evolutionary processes provide even more scope.

Not only should social constructivists feel unthreatened by evolutionary theory, but they should actively learn to use it to achieve their objectives. The blank slate metaphor and the concept of "learning" and "culture" as generic alternatives to "evolution" may crudely capture the spirit of social constructivism, but they fail in every other respect. Evolution is a complicated process, and the factors that constrain adaptation (E3) lurk around every corner. Understanding E2 in conjunction with E1 is even more complicated. After we decide that evolutionary theory is a vehicle that can take us where we want to go, we need to learn how to drive it. The only way forward for social constructivism in a practical sense is to master and advance our knowledge of evolution, the only known process that can create islands of function out of the sea of entropy.

As for evolutionists, it takes an insider to appreciate the diversity of opinion and lack of integration gathered under that term, ranging from nearly exclusive focus on one of the three positions that I have outlined to those who attempt to occupy the middle ground. Perhaps we can understand and sympathize with an excessive focus on E1 as a reaction to its denial in psychology and the social sciences during most of the twentieth century. However, those of us who broadly use the term "evolutionary psychology" think of it not as a counterweight but as a framework for explaining all aspects of psychology from an evolutionary perspective. Evolutionary psychology in its current form, therefore, must take back some of what has been rejected as part of the "standard social science model," in particular open-ended, nongenetic evolutionary processes that adapt individuals and groups to their current environments.

Einstein's quote "It is the theory that decides what we can observe" might seem to imply that every theoretical perspective is like a mask with narrow slits for eyes, providing only a partial view of the world. Perhaps this is true in some sense, but I don't think it explains the kind of narrowness that has existed in the past and need not exist in the future for this subject. Clearly, the middle ground that we have been discussing has remained unoccupied because of perceived implications, not just because of intellectual difficulty. Intellectually it is fully possible to achieve a theory of evolution that acknowledges the importance of all three positions and their relationships with each other and that serves as a resource for individual and societal change. Perhaps before long we will be able to say that the evolution wars are over and the task of reconstruction has begun.

Part II

The Riddle of Art

10

Art and Adaptation

STEVEN PINKER

Man does not live by bread alone, nor by know-how, safety, children, or sex. People everywhere spend as much time as they can afford on activities that, in the struggle to survive and reproduce, seem pointless. In all cultures, people tell stories and recite poetry. They joke, laugh, and tease. They sing and dance. They decorate surfaces. They perform rituals. They wonder about the causes of fortune and misfortune, and hold beliefs about the supernatural that contradict everything else they know about the world. They concoct theories of the universe and their place within it (Brown, *Human Universals*; Eibl-Eibesfeldt, *Human Ethology*).

As if that weren't enough of a puzzle, the more biologically frivolous and vain the activity, the more people exalt it. Art, literature, music, wit, religion, and philosophy are thought to be not just pleasurable but noble. They are the mind's best work, what makes life worth living. Why do we pursue the trivial and futile and experience them as sublime? To many educated people the question seems horribly philistine, even immoral. But it is unavoidable

---

for anyone interested in the biological makeup of *Homo sapiens*. Members of our species do mad deeds like taking vows of celibacy, living for their music, selling their blood to buy movie tickets, and going to graduate school (Tooby and Cosmides, "Past"). Why? How might we understand the psychology of the arts, humor, religion, and philosophy within the theme of this book, that the mind is a naturally selected neural computer?

Every college has a faculty of arts, which usually dominates the institution in numbers and in the public eye. But the tens of thousands of scholars and millions of pages of scholarship have shed almost no light on the question of why people pursue the arts at all. The function of the arts is almost defiantly obscure, and I think there are several reasons why.

One is that the arts engage not only the psychology of aesthetics but the psychology of status (Bell; Wolfe). The very uselessness of art that makes it so incomprehensible to evolutionary biology makes it all too comprehensible to economics and social psychology. What better proof that you have money to spare than your being able to spend it on doodads and stunts that don't fill the belly or keep the rain out but that require precious materials, years of practice, a command of obscure texts, or intimacy with the elite? Thorstein Veblen's and Quentin Bell's analyses of taste and fashion, in which an elite's conspicuous displays of consumption, leisure, and outrage are emulated by the rabble, sending the elite off in search of new inimitable displays, nicely explains the otherwise inexplicable oddities of the arts. The grand styles of one century become tacky in the next, as we see in words that are both period labels and terms of abuse (*gothic, mannerist, baroque, rococo*). The steadfast patrons of the arts are the aristocracy and those who want to join them. Most people would lose their taste for a musical recording if they learned it was being sold at supermarket checkout counters or on late-night television, and even the work of relatively prestigious artists, such as Pierre-Auguste Renoir, draws derisive reviews when it is shown in a popular "blockbuster" museum show. The value of art is largely unrelated to aesthetics: a priceless masterpiece becomes worthless if it is found to be a forgery; soup cans and comic strips become high art when the art world says they are, and then command conspicuously wasteful prices. Modern and postmodern works are intended not to give pleasure but to confirm or confound the theories of a guild of critics and analysts, to *épater la bourgeoisie*, or to baffle the rubes in Peoria.

The banality that the psychology of the arts is partly the psychology of status has been repeatedly pointed out, not just by cynics and barbarians but by erudite social commentators such as Quentin Bell and Tom Wolfe (Bell; Brockman). But in the modern university, it is unmentioned, indeed, unmentionable. Academics and intellectuals are culture vultures.

In a gathering of today's elite, it is perfectly acceptable to laugh that you barely passed "Physics for Poets" and "Rocks for Jocks" and have remained ignorant of science ever since, despite the obvious importance of scientific literacy to informed choices about personal health and public policy. But saying that you have never heard of James Joyce or that you tried listening to Mozart once but prefer Andrew Lloyd Webber is as shocking as blowing your nose on your sleeve or announcing that you employ children in your sweatshop, despite the obvious *un*importance of your tastes in leisure-time activity to just about anything. The blending in people's minds of art, status, and virtue is an extension of Bell's principle of sartorial morality: people find dignity in the signs of an honorably futile existence removed from all menial necessities.

I mention these facts not to denigrate the arts but to clarify my topic. I want you to look at the psychology of the arts with the disinterested eye of an alien biologist trying to make sense of the human species rather than as a member of the species with a stake in how the arts are portrayed. *Of course* we find pleasure and enlightenment in contemplating the products of the arts, and not all of it is a pride in sharing the tastes of the beautiful people. But to understand the psychology of the arts that remains when we subtract out the psychology of status, we must leave at the door our terror of being mistaken for the kind of person who prefers Andrew Lloyd Webber to Mozart. We need to begin with folk songs, pulp fiction, and paintings on black velvet, not Mahler, Eliot, and Kandinsky. And that does *not* mean compensating for our slumming by dressing up the lowly subject matter in highfalutin "theory" (a semiotic analysis of *Peanuts*, a psychoanalytic exegesis of Archie Bunker, a deconstruction of *Vogue*). It means asking a simple question: What is it about the mind that lets people take pleasure in shapes and colors and sounds and jokes and stories and myths?

That question might be answerable, whereas questions about art in general are not. Theories of art carry the seeds of their own destruction. In an age when any Joe can buy CDs, paintings, and novels, artists make their careers by finding ways to avoid the hackneyed, to challenge jaded tastes, to differentiate the cognoscenti from the dilettantes, and to flout the current wisdom about what art is (hence the fruitless attempts over the decades to define art). Any discussion that fails to recognize that dynamic is doomed to sterility. It can never explain why music pleases the ear, because "music" will be defined to encompass atonal jazz, chromatic compositions, and other intellectual exercises. It will never understand the bawdy laughs and convivial banter that are so important in people's lives because it will define humor as the arch wit of an Oscar Wilde. Excellence and the avant-garde are designed for the sophisticated palate, a product of years of immersion in a genre and a familiarity with its conventions and clichés. They rely on one-upmanship and arcane allusions and displays of virtuosity. However fascinating and worthy of our support they are, they tend to obscure the psychology of aesthetics, not to illuminate it.

⎯⎯ ⧂⧂⧂ ⎯⎯

Another reason the psychology of the arts is obscure is that they are not adaptive in the biologist's sense of the word. This book has been about the adaptive design of the major components of the mind, but that does not mean that I believe that everything the mind does is biologically adaptive. The mind is a neural computer, fitted by natural selection with combinatorial algorithms for causal and probabilistic reasoning about plants, animals, objects, and people. It is driven by goal states that served biological fitness in ancestral environments, such as food, sex, safety, parenthood, friendship, status, and knowledge. That toolbox, however, can be used to assemble Sunday afternoon projects of dubious adaptive value.

Some parts of the mind register the attainment of increments of fitness by giving us a sensation of pleasure. Other parts use a knowledge of cause and effect to bring about goals. Put them together and you get a mind that rises to a biologically pointless challenge: figuring out how to get at the pleasure circuits of the brain and deliver little jolts of enjoyment without the inconvenience of wringing bona fide fitness increments from the harsh world. When a rat has access to a lever that sends electrical impulses to an electrode implanted in its medial forebrain bundle, it presses the lever furiously until it drops of exhaustion, forgoing opportunities to eat, drink, and have sex. People don't yet undergo elective neurosurgery to have electrodes implanted in their pleasure centers, but they have found ways to stimulate them by other means. An obvious example is recreational drugs, which seep into the chemical junctions of the pleasure circuits.

Another route to the pleasure circuits is via the senses, which stimulate the circuits when they are in environments that would have led to fitness in past generations. Of course a fitness-promoting environment cannot announce itself directly. It gives off patterns of sounds, sights, smells, tastes, and feels that the senses are designed to register. Now, if the intellectual faculties could identify the pleasure-giving patterns, purify them, and concentrate them, the brain could stimulate itself without the messiness of electrodes or drugs. It could give itself intense artificial doses of the sights and sounds and smells that ordinarily are given off by healthful environments. We enjoy strawberry cheesecake, but not because we evolved a taste for it.

We evolved circuits that gave us trickles of enjoyment from the sweet taste of ripe fruit, the creamy mouth feel of fats and oils from nuts and meat, and the coolness of fresh water. Cheesecake packs a sensual wallop unlike anything in the natural world because it is a brew of megadoses of agreeable stimuli which we concocted for the express purpose of pressing our pleasure buttons. Pornography is another pleasure technology. In this chapter I will suggest that the arts are a third.

There is another way that the design of the mind can throw off fascinating but biologically functionless activities. The intellect evolved to crack the defenses of things in the natural and social world. It is made up of modules for reasoning about how objects, artifacts, living things, animals, and other human minds work. There are problems in the universe other than those: where the universe came from, how physical flesh can give rise to sentient minds, why bad things happen to good people, what happens to our thoughts and feelings when we die. The mind can pose such questions but may not be equipped to answer them, even if the questions have answers. Given that the mind is a product of natural selection, it should not have a miraculous ability to commune with all truths; it should have a mere ability to solve problems that are sufficiently similar to the mundane survival challenges of our ancestors. According to a saying, if you give a boy a hammer, the whole world becomes a nail. If you give a species an elementary grasp of mechanics, biology, and psychology, the whole world becomes a machine, a jungle, and a society. I will suggest that religion and philosophy are in part the application of mental tools to problems they were not designed to solve.

Some readers may be surprised to learn that after seven chapters of reverse-engineering the major parts of the mind, I will conclude by arguing that some of the activities we consider most profound are nonadaptive by-products. But both kinds of argument come from a single standard, the criteria for biological adaptation. For the same reason that it is wrong to write off language, stereo vision, and the emotions as evolutionary accidents—namely, their universal, complex, reliably developing, well-engineered, reproduction-promoting design—it is wrong to invent functions for activities that lack that design merely because we want to ennoble them with the imprimatur of biological adaptiveness. Many writers have said that the "function" of the arts is to bring the community together, to help us see the world in new ways, to give us a sense of harmony with the cosmos, to allow us to experience the sublime, and so on. All these claims are true, but none is about adaptation in the technical sense that has organized this book: a mechanism that brings about effects that would have increased the number of copies of the genes building that mechanism in the environment in which we evolved. Some aspects of the arts, I think, do have functions in this sense, but most do not.

⸻ ∞ ⸻

"The fact is I am quite happy in a movie, even a bad movie. Other people, so I have read, treasure memorable moments in their lives." At least the narrator of Walker Percy's novel *The Moviegoer* acknowledges the difference. Television stations get mail from soap-opera viewers with death threats for the evil characters, advice to the lovelorn ones, and booties for the babies. Mexican moviegoers have been known to riddle the screen with bullets. Actors complain that fans confuse them with their roles; Leonard Nimoy wrote a memoir called *I Am Not Spock*, then gave up and wrote another one called *I Am Spock*. These anecdotes appear regularly in the newspapers, usually to insinuate that people today are boobs who cannot distinguish fantasy from reality. I suspect that the people are not literally deluded but are going to extremes to enhance the pleasure we all get from losing ourselves in fiction. Where does this motive, found in all peoples, come from?

Horace wrote that the purpose of literature is "to delight and instruct" ("Ars Poetica" l. 343), a function echoed centuries later by John Dryden when he defined a play as "a just and lively image of human nature, representing its passions and humours, and the changes of fortune to which it is subject; for the delight and instruction of mankind" (quoted in Carroll, *Evolution* 170). It's helpful to distinguish the delight, perhaps the product of a useless technology for pressing our pleasure buttons, from the instruction, perhaps a product of a cognitive adaptation.

The technology of fiction delivers a simulation of life that an audience can enter in the comfort of their cave, couch, or theater seat. Words can evoke mental images, which can activate the parts of the brain that register the world when we actually perceive it. Other technologies violate the assumptions of our perceptual apparatus and trick us with illusions that partly duplicate the experience of seeing and hearing real events (Hobbs, *Literature*; Tan). They include costumes, makeup, sets, sound effects, cinematography, and animation. Perhaps in the near future we can add virtual reality to the list, and in the more distant future the feelies of *Brave New World*.

When the illusions work, there is no mystery to the question "Why do people enjoy fiction?" It is identical to the question "Why do people enjoy life?" When we are absorbed in a book or a movie, we get to see breathtaking landscapes, hobnob with important people, fall in love with ravishing men and women, protect loved ones, attain impossible goals, and defeat wicked enemies. Not a bad deal for seven dollars and fifty cents!

Of course, not all stories have happy endings. Why would we pay seven dollars and fifty cents for a simulation of life that makes us miserable? Sometimes, as with art films, it is to gain status through cultural machismo. We endure a pummeling of the emotions to differentiate ourselves from the crass philistines who actually go to the movies to enjoy themselves. Sometimes it is the price we pay to satisfy two incompatible desires: stories with happy endings and stories with unpredictable endings, which preserve the illusion of a real world. There have to be some stories in which the murderer does catch up with the heroine in the basement, or we would never feel suspense and relief in the stories in which she escapes. The economist Steven Landsburg observes that happy endings predominate when no director is willing to sacrifice the popularity of his or her film for the greater good of more suspense in the movies in general.

But then how can we explain the tearjerker, aimed at a market of moviegoers who *enjoy* being defrauded into grief? The psychologist Paul Rozin lumps tearjerkers with other examples of benign masochism like smoking, riding on roller coasters, eating hot chili peppers, and sitting in saunas. Benign masochism, recall, is like the drive of Tom Wolfe's test pilots to push the outside of the envelope (*Right Stuff*). It expands the range of options in life by testing, in small increments, how closely one can approach a brink of disaster without falling over it. Of course the theory would be vacuous if it offered a glib explanation for every inexplicable act, and it would be false if it predicted that people would pay to have needles stuck under their fingernails. But the idea is more subtle. Benign masochists must be confident that no serious harm will befall them. They must bring on the pain or fear in measured increments. And they must have an opportunity to control and mitigate the damage. The technology of tearjerkers seems to fit. Moviegoers know the whole time that when they leave the theater they will find their loved ones unharmed. The heroine is done in by a progressive disease, not a heart attack or a piece of hot dog stuck in the throat, so we can prepare our emotions for the tragedy. We only have to accept the abstract premise that the heroine will die; we are excused from witnessing the disagreeable details. (Greta Garbo, Ali MacGraw, and Debra Winger all looked quite lovely as they wasted away from consumption and cancer.) And the viewer must identify with the next of kin, empathize with their struggle to cope, and feel confident that life will go on. Tearjerkers simulate a triumph over tragedy.

Even following the foibles of ordinary virtual people as they live their lives can press a pleasure button, the one labeled "gossip." Gossip is a favorite pastime in all human societies because knowledge is power. Knowing who needs a favor and who is in a position to offer one, who is trustworthy and who is a liar, who is available (or soon to become available) and who is under the protection of a jealous spouse or family—all give obvious strategic advantages in the games of life. That is especially true when the information is not yet widely known and one can be the first to exploit an opportunity, the social equivalent of insider trading. In the small bands in which our minds evolved, everyone knew everyone else, so all gossip was useful (Barkow). Today, when we peer into the private lives of fictitious characters, we are giving ourselves the same buzz.

Literature, though, not only delights but instructs. The computer scientist Jerry Hobbs has tried to reverse-engineer the fictional narrative in an essay he was tempted to call "Will Robots Ever Have Literature?" Novels, he concluded, work like experiments. The author places a fictitious character in a hypothetical situation in an otherwise real world where ordinary facts and laws hold, and allows the reader to explore the consequences. We can imagine that there was a person in Dublin named Leopold Bloom with the personality, family, and occupation that James Joyce attributed to him, but we would object if we were suddenly to learn that the British sovereign at the time was not Queen Victoria but King Victor. Even in science fiction, we are asked to suspend belief in a few laws of physics, say to get the heroes to the next galaxy, but the events should otherwise unfold according to lawful causes and effects. A surreal story like Kafka's *Metamorphosis* begins with one counterfactual premise—a man can turn into an insect—and plays out the consequences in a world where everything else is the same. The hero retains his human consciousness, and we follow him as he makes his way and people react to him as real people would react to a giant insect. Only in fiction that is *about* logic and reality, such as *Alice's Adventures in Wonderland*, can any strange thing happen (Hobbs, *Literature*; Turner, *Reading Minds*).

Once the fictitious world is set up, the protagonist is given a goal and we watch as he or she pursues it in the face of obstacles. It is no coincidence that this standard definition of plot is identical to the definition of intelligence I suggested in a previous chapter. Characters in a fictitious world do exactly what our intelligence allows us to do in the real world. We watch what happens to them and mentally take notes on the outcomes of the strategies and tactics they use in pursuing their goals (Carroll, *Evolution*; Hobbs, *Literature*).

What are those goals? A Darwinian would say that ultimately organisms have only two: to survive and to reproduce. And those are precisely the goals that drive the human organisms in fiction. Most of the thirty-six plots in Georges Polti's catalogue are defined by love or sex or a threat to the safety of the protagonist or his kin (for example, "Mistaken jealousy," "Vengeance taken for kindred upon kindred," and "Discovery of the dishonor of a loved one"). The difference between fiction for children and fiction for adults is commonly summed up in two words: sex and violence. Woody Allen's homage to Russian literature was entitled *Love and Death*. Pauline Kael got the title for one of her books of movie criticism from an Italian movie poster that she said contained "the briefest statement imaginable of the basic appeal of the movies": *Kiss Kiss Bang Bang*.

Sex and violence are not just the obsessions of pulp fiction and trash TV. The language maven Richard Lederer and the computer programmer Michael Gilleland present the following tabloid headlines:

CHICAGO CHAUFFEUR SMOTHERS BOSS'S DAUGHTER, THEN CUTS HER UP AND STUFFS HER IN FURNACE

DOCTOR'S WIFE AND LOCAL MINISTER EXPOSED FOR CONCEIVING ILLEGITIMATE DAUGHTER

TEENAGERS COMMIT DOUBLE SUICIDE; FAMILIES VOW TO END VENDETTA

STUDENT CONFESSES TO AXE MURDER OF LOCAL PAWNBROKER AND ASSISTANT

GARAGE OWNER STALKS AFFLUENT BUSINESSMAN, THEN SHOTGUNS HIM IN HIS SWIMMING POOL

MADWOMAN LONG IMPRISONED IN ATTIC SETS HOUSE ON FIRE, THEN LEAPS TO DEATH

FORMER SCHOOLTEACHER, FOUND TO HAVE BEEN PROSTITUTE, COMMITTED TO INSANE ASYLUM

PRINCE ACQUITTED OF KILLING MOTHER IN REVENGE FOR MURDER OF HIS FATHER

Sound familiar? The plots are from *Native Son* by Richard Wright; *The Scarlett Letter* by Nathaniel Hawthorne; *Romeo and Juliet* by William Shakespeare; *Crime and Punishment* by Fyodor Dostoevsky; *The Great Gatsby* by F. Scott Fitzgerald; *Jane Eyre* by Charlotte Brontë; *A Streetcar Named Desire* by Tennessee Williams; and *Eumenides* by Aeschylus.

Fiction is especially compelling when the obstacles to the protagonist's goals are other people in pursuit of incompatible goals. Life is like chess, and plots are like those books of famous chess games that serious players study so they will be prepared if they ever find themselves in similar straits. The books are handy because chess is combinatorial; at any stage there are too many possible sequences of moves and countermoves for them all to be played out in one's mind. General strategies like "Get your Queen out early" are too vague to be of much use, given the trillions of situations the rules permit. A good training regime is to build up a mental catalogue of tens of thousands of game challenges and the moves that allowed good players to do well in them. In artificial intelligence, it is called case-based reasoning (Schank).

Life has even more moves than chess. People are always, to some extent, in conflict, and their moves and countermoves multiply out to an unimaginably vast set of interactions. Partners, like the prisoners in the hypothetical dilemma, can either cooperate or defect, on this move and on subsequent moves. Parents, offspring, and siblings, because of their partial genetic overlap, have both common and competing interests, and any deed that one party directs toward another may be selfless, selfish, or a mixture of the two. When boy meets girl, either or both may see the other as a spouse, as a one-night stand, or neither. Spouses may be faithful or adulterous. Friends may be false friends. Allies may assume less than their fair share of the risk, or may defect as the finger of fate turns toward them. Strangers may be competitors or outright enemies. These games are taken into higher dimensions by the possibility of deception, which allows words and deeds to be either true or false, and self-deception, which allows *sincere* words and deeds to be either true or false. They are expanded into still higher dimensions by rounds of paradoxical tactics and countertactics, in which a person's usual goals—control, reason, and knowledge—are voluntarily surrendered to make the person unthreatenable, trustworthy, or too dangerous to challenge.

The intrigues of people in conflict can multiply out in so many ways that no one could possibly play out the consequences of all courses of action in the mind's eye. Fictional narratives supply us with a mental catalogue of the fatal conundrums we might face someday and the outcomes of strategies we could deploy in them. What are the options if I were to suspect that my uncle killed my father, took his position, and married my mother? If my hapless older brother got no respect in the family, are there circumstances that might lead him to betray me? What's the worst that could happen if I were seduced by a client while my wife and daughter were away for the weekend? What's the worst that could happen if I had an affair to spice up my boring life as the wife of a country doctor? How can I avoid a suicidal confrontation with raiders who want my land today without looking like a coward and thereby ceding it to them tomorrow? The answers are to be found in any bookstore or video shop. The cliché that life imitates art is true because the function of some kinds of art is for life to imitate it (*Hamlet, The Godfather, Fatal Attraction, Madame Bovary, Shane*).

11

The Arts and Their Interpretation

EDWARD O. WILSON

In many respects, the most interesting challenge to consilient explanation is the transit from science to the arts. By the latter I mean the creative arts, the personal productions of literature, visual arts, drama, music, and dance marked by those qualities which for lack of better words (and better words may never be coined) we call the true and beautiful.

The arts are sometimes taken to mean all the humanities, which include not only the creative arts but also, following the recommendations of the 1979–80 Commission on the Humanities, the core subjects of history, philosophy, languages, and comparative literature, plus jurisprudence, the comparative study of religions, and "those aspects of the social sciences which have humanistic content and employ humanistic methods" (Commission on the Humanities). Nevertheless, the arts in the primary and intuitively creative sense, *ars gratia artis*, remain the definition most widely and usefully employed.

Reflection leads us to two questions about the arts: where they come from, in both history and personal experience, and how their essential qualities of truth and beauty are to be described through ordinary language. These matters are the central concern of interpretation, the scholarly analysis and criticism of the arts. Interpretation is itself partly an art, since it expresses not just the factual expertise of the critic but also his character and aesthetic judgment. When of high quality, criticism can be as inspired and idiosyncratic as the work it addresses. Further, as I now hope to show, it can also be part of science, and science part of it. Interpretation will be the more powerful when braided together from history, biography, personal confession—and science.

The profane word now having been spoken on hallowed ground, a quick disclaimer is in order. While it is true that science advances by reducing phenomena to their working elements—by dissecting brains into neurons, for example, and neurons into molecules—it does not aim to diminish the integrity of the whole. On the contrary, synthesis of the elements to re-create their original assembly is the other half of scientific procedure. In fact, it is the ultimate goal of science.

Nor is there any reason to suppose that the arts will decline as science flourishes. They are not, as suggested recently by the distinguished literary critic George Steiner, in a twilight, past high noon in Western civilization, thus unlikely to witness the reappearance of a Dante, a Michelangelo, or a Mozart. I can conceive of no intrinsic limit to future originality and brilliance in the arts as the consequence of the reductionist understanding of the creative process in the arts and science. On the contrary, an alliance is overdue, and can be achieved through the medium of interpretation. Neither science nor the arts can be complete without combining their separate strengths. Science needs the intuition and metaphorical power of the arts, and the arts need the fresh blood of science.

Scholars in the humanities should lift the anathema placed on reductionism. Scientists are not conquistadors out to melt the Inca gold. Science is free and the arts are free, and the two domains, despite the similarities in their creative spirit, have radically different goals and methods. The key to the exchange between them is not hybridization, not some unpleasantly self-conscious form of scientific art or artistic science, but reinvigoration of interpretation with the knowledge of science and its proprietary sense of the future. Interpretation is the logical channel of consilient explanation between science and the arts.

⸙

For a promising example out of many that might be chosen, consider the episode in *Paradise Lost*–Book IV, when, in a riveting narrative, Milton sends

Satan to Eden. Upon arrival the arch-felon and grand thief leaps a barrier of impenetrable bramble and a high wall and settles "like a cormorant" in the branches of the Tree of Life. He waits for the fall of night, when he can enter the dreams of innocent Eve. Milton now unleashes his imaginative powers to tell us what humanity i s about to lose. All around the roosting schemer is the environment designed by God to aesthetic perfection: "Crispèd brooks, rolling on orient pearl and sands of gold" descend to "a lake, that to the fringèd bank with myrtle crowned her crystal mirror holds." All through the blessed oasis grow "flowers of all hue and without thorn the rose."

Milton, though now blind, has retained a fine sense of biophilia, the innate pleasure from living abundance and diversity, particularly as manifested by the human impulse to imitate Nature with gardens. But he is far from satisfied with the mere dream of natural harmony. In eight lines of astonishing symphonic power he tries to capture the mythic core of paradise:

> Not that fair field
> Of Enna, where Proserpin gathering flowers,
> Herself a fairer flower, by gloomy Dis
> Was gathered, which cost Ceres all that pain
> To seek her through the world, nor that sweet grove
> Of Daphne, by Orontes and the inspired
> Castalian spring, might with this Paradise
> Of Eden strive.

How can anyone hope to express Creation's heart at the dawn of time? Milton tries. He summons archetypes that have descended undiminished from ancient Greece and Rome to his own time, and thereafter to ours. They are of a kind that are also innate to the human mental process. He shadows beauty with a hint of tragedy, giving us the untrammeled and fertile world awaiting corruption. He transforms the beauty of the garden into that of a young woman, Proserpine, about to be seized and taken away to the underworld by the god Dis. She, as Nature's beauty, will be concealed in darkness because of conflict between gods. Ceres, Proserpine's mother and goddess of agriculture, turns in grief from her duties and the world plunges into famine. The passion of Apollo for beautiful Daphne is unrequited; in order to escape she turns into a tree, a laurel, in a garden of her own.

Milton means to play on the emotions of readers of his own time, the seventeenth century, when Hellenic mythology was second nature to the educated mind. He counterposes emotions to magnify their force. Beauty clashes with darkness, freedom with fate, passion with denial. Building tension, he leads us through lesser paradises to arrive, suddenly, at the mystical prototype of Eden. In yet another well-grounded artifice, reliance on authority, Milton chooses allusions not to his own time, not for example to Cromwell and Charles II and the Restoration, from which he himself has narrowly escaped death (he had championed revolution and the Commonwealth), but to ancient texts of another civilization, ancient Greece and Rome, robust enough to have survived in remembrance across centuries. He conveys by their use that what we are not told, we must know nevertheless to be true.

The defining quality of the arts is the expression of the human condition by mood and feeling, calling into play all the senses, evoking both order and disorder. From where then does the ability to create art arise? Not cold logic based on fact. Not God's guidance of Milton's thoughts, as the poet himself believed. Nor is there any evidence of a unique spark that ignites such genius as is evident in *Paradise Lost*. Experiments using brain imaging, for example, have failed to disclose singular neurobiological traits in musically gifted people. Instead, they show engagement of a broader area of the same parts of the brain used by those less able (Schlaug et al., "Increased," "In Vivo Evidence"). History supports this incremental hypothesis. Behind Shakespeare, Leonardo, Mozart, and others in the foremost rank are a vast legion whose realized powers form a descending continuum to those who are merely competent. What the masters of the Western canon, and those of other high cultures, possessed in common was a combination of exceptional knowledge, technical skill, originality, sensitivity to detail, ambition, boldness, and drive.

They were obsessed; they burned within. But they also had an intuitive grasp of inborn human nature accurate enough to select commanding images from the mostly inferior thoughts that stream through the minds of all of us. The talent they wielded may have been only incrementally greater, but their creations appeared to others to be qualitatively new. They acquired enough influence and longevity to translate into lasting fame, not by magic, not by divine benefaction, but by a quantitative edge in powers shared in smaller degree with those less gifted. They gathered enough lifting speed to soar above the rest.

Artistic inspiration common to everyone in varying degree rises from the artesian wells of human nature. Its creations are meant to be delivered directly to the sensibilities of the beholder without analytic explanation. Creativity is therefore humanistic in the fullest sense. Works of enduring value are those truest to these origins. It follows that even the greatest works of art might be understood fundamentally with knowledge of the biologically evolved epigenetic rules that guided them.

This is not the prevailing view of the arts. Academic theorists have paid little attention to biology; consilience is not in their vocabulary. To varying degrees they have been more influenced by postmodernism, the competing hypothesis that denies the existence of a universal human nature. Applied to literary criticism, the extreme manifestation of postmodernism is the deconstructive philosophy formulated most provocatively by Jacques Derrida and Paul de Man. In this view, truth is relative and personal. Each person creates his own inner world by acceptance or rejection of endlessly shifting linguistic signs. There is no privileged point, no lodestar, to guide literary intelligence. And given that science is just another way of looking at the world, there is no scientifically constructible map of human nature from which the deep meaning of texts can be drawn. There is only unlimited opportunity for the reader to invent interpretations and commentaries out of the world he himself constructs. "The author is dead" is a favorite maxim of the deconstructionists.

Deconstructionist scholars search instead for contradictions and ambiguities. They conceive and analyze what is left out by the author. The missing elements allow for personalized commentary in the postmodernist style. Postmodernists who add political ideology to the mix also regard the traditional literary canon as little more than a collection confirming the world view of ruling groups, and in particular that of Western white males.

The postmodernist hypothesis does not conform well to the evidence. It is blissfully free of existing information on how the mind works. Yet there is surely *some* reason for the popularity of postmodernism other than a love of chaos. If the competing biological approach is correct, its widespread appeal must be rooted in human nature. Postmodernism in the arts is more than a School of Resentment—Harold Bloom's indictment in *The Western Canon*—and more than the eunuch's spite, to borrow a phrase from Alexander Pope, and it is sustained by more than the pathetic reverence commonly given Gallic obscurantism by American academics. There is also a surge of revolutionary spirit in postmodernism, generated by the real—not deconstructed—fact that large segments of the population, most notably women, have unique talents and emotional lives that have been relatively neglected for centuries, and are only now beginning to find full expression within the mainstream culture.

If we are to believe evidence from the biological and behavioral sciences gathered especially during the past quarter century, women differ genetically from men in ways other than reproductive anatomy. In aggregate, on average, with wide statistical overlap, and in many venues of social experience, they speak with a different voice. Today it is being heard loud and clear. But I do not read the welcome triumph of feminism, social, economic, and creative, as a brief for postmodernism. The advance, while opening new avenues of expression and liberating deep pools of talent, has not exploded human nature into little pieces. Instead, it has set the stage for a fuller exploration of the universal traits that unite humanity.

⣿⣿

Can the opposed Apollonian and Dionysian impulses, cool reason against passionate abandonment, which drive the mood swings of the arts and criticism, be reconciled? This is, I believe, an empirical question. Its answer depends on the existence or nonexistence of an inborn human nature. The evidence accumulated to date leaves little room for doubt. Human nature exists, and it is both deep and highly structured.

If that much is granted, the relation of science to interpretation of the arts can be made clearer, as follows. Interpretation has multiple dimensions, namely history, biography, linguistics, and aesthetic judgment. At the foundation of them all lie the material processes of the human mind. Theoretically inclined critics of the past have tried many avenues into that subterranean realm, including most prominently psychoanalysis and postmodernist solipsism. These approaches, which are guided largely by unaided intuition about the way the brain works, have fared badly. In the absence of a compass based on sound material knowledge, they make too many wrong turns into blind ends. If the brain is ever to be charted, and an enduring theory of the arts created as part of the enterprise, it will be by stepwise and consilient contributions from the brain sciences, psychology, and evolutionary biology. And if during this process the creative mind is to be understood, it will need collaboration between scientists and humanities scholars.

The collaboration, now in its early stages, is likely to conclude that innovation is a concrete biological process founded upon an intricacy of nerve circuitry and neurotransmitter release. It is not the outpouring of symbols by an all-purpose generator or any conjuration therein by ethereal agents. To fathom the origin of innovation in the arts will make a great deal of difference in the way we interpret its creations. The natural sciences have begun to form a picture of the mind, including some of the elements of the creative process itself. Although they are still considerably far from the ultimate goal, they cannot help in the end but strengthen interpretation of the arts.

The growing evidence of an overall structured and powerful human nature, channeling development of the mind, favors a more traditionalist view of the arts. The arts are not solely shaped by errant genius out of historical circumstances and idiosyncratic personal experience. The roots of their inspiration date back in deep history to the genetic origins of the human brain, and are permanent.

While biology has an important part to play in scholarly interpretation, the creative arts themselves can never be locked in by this or any other discipline of science. The reason is that the exclusive role of the arts is the transmission of the intricate details of human experience by artifice to intensify aesthetic and emotional response. Works of art communicate feeling directly from mind to mind, with no intent to explain why the impact occurs. In this defining quality, the arts are the antithesis of science.

When addressing human behavior, science is coarse-grained and encompassing, as opposed to the arts, which are fine-grained and interstitial. That is, science aims to create principles and use them in human biology to define the diagnostic qualities of the species; the arts use fine details to flesh out and make strikingly clear by implication those same qualities. Works of art that prove enduring are intensely humanistic. Born in the imagination of individuals, they nevertheless touch upon what was universally endowed by human evolution. Even when, as part of fantasy, they imagine worlds that cannot possibly exist, they stay anchored to their human origins. As Kurt Vonnegut, Jr., master fantasist, once pointed out, the arts place humanity in the center of the universe, whether we belong there or not.

If the arts are steered by inborn rules of mental development, they are end products not just of conventional history but also of genetic evolution. The question remains: Were the genetic guides mere byproducts—epiphenomena—of that evolution, or were they adaptations that directly improved survival and reproduction? And if adaptations, what exactly were the advantages conferred? The answers, some scholars believe, can be found in artifacts preserved from the dawn of art. They can be tested further with knowledge of the artifacts and customs of present-day hunter-gatherers.

This is the picture of the origin of the arts that appears to be emerging. The most distinctive qualities of the human species are extremely high intelligence, language, culture, and reliance on long-term social contracts. In combination they gave early *Homo sapiens* a decisive edge over all competing animal species, but they also exacted a price we continue to pay, composed of the shocking recognition of the self, of the finiteness of personal existence, and of the chaos of the environment.

These revelations, not disobedience to the gods, are what drove humankind from paradise. *Homo sapiens* is the only species to suffer psychological exile. All animals, while capable of some degree of specialized learning, are instinct-driven, guided by simple cues from the environment that trigger complex behavior patterns. The great apes have the power of self-recognition, but there is no evidence that they can reflect on their own birth and eventual death. Or on the meaning of existence—the complexity of the universe means nothing to them. They and other animals are exquisitely adapted to just those parts of the environment on which their lives depend, and they pay little or no attention to the rest.

The dominating influence that spawned the arts was the need to impose order on the confusion caused by intelligence. In the era prior to mental expansion, the ancestral prehuman populations evolved like any other animal species. They lived by instinctive responses that sustained survival and reproductive success. When *Homo*-level intelligence was attained, it widened that advantage by processing information well beyond the releaser cues. It permitted flexibility of response and the creation of mental scenarios that reached to distant places and far into the future. The evolving brain, nevertheless, could not convert to general intelligence alone; it could not turn into an all-purpose computer. So in the course of evolution the animal instincts of survival and reproduction were transformed into the epigenetic algorithms of human nature. It was necessary to keep in place these inborn programs for the rapid acquisition of language, sexual conduct, and other processes of mental development. Had the algorithms been erased, the species would have faced extinction. The reason is that the lifetime of an individual human being is not long enough to sort out experiences by means of generalized, unchanneled learning. Yet the algorithms were jerry-built: They worked adequately but not superbly well. Because of the slowness of natural selection, which requires tens or hundreds of generations to substitute new genes for old, there was not enough time for human heredity to cope with the vastness of new contingent possibilities revealed by high intelligence. Algorithms could be built, but they weren't numerous and precise enough to respond automatically and optimally to every possible event.

The arts filled the gap. Early humans invented them in an attempt to express and control through magic the abundance of the environment, the power of solidarity, and other forces in their lives that mattered most to survival and reproduction. The arts were the means by which these forces could be ritualized and expressed in a new, simulated reality. They drew consistency from their faithfulness to human nature, to the emotion-guided epigenetic rules—the algorithms—of mental development. They achieved that fidelity by selecting the most evocative words, images, and rhythms, conforming to the emotional guides of the epigenetic rules, making the right moves. The arts still perform this primal function, and in much the same ancient way. Their quality is measured by their humanness, by the precision of their adherence to human nature. To an overwhelming degree that is what we mean when we speak of the true and beautiful in the arts.

12

Art and Intimacy:

HOW THE ARTS BEGAN

ELLEN DISSANAYAKE

Art and Infancy

The biological phenomenon of love is originally manifested—expressed and exchanged—by means of emotionally meaningful "rhythms and modes" that are jointly created and sustained by mothers and their infants in ritualized, evolved interactions. From these rudimentary and unlikely beginnings grow adult expressions of love, both sexual and generally affiliative, *and the arts.* That is to say, in their origins in ourselves and in our species, love and art are, I suggest, inherently related.

A number of recent psychological studies of early interactions between mothers and infants modify the sweeping assumption that pervasive selfishness lies at the heart of human nature. Evolutionary science has yet to recognize and take account of their implications. The studies establish unequivocally the extent to which mutuality between mother and infant, and its influence

---

upon other consequent or related psychological needs, developed over hundreds of thousands of years of human evolution as a crucial motivating force that enabled our ancestors to survive in the earliest human "life-style"—that of small bands of foragers and hunters on the African savannah.

The foraging-hunting way of life of our hominid ancestors required not only resourceful, competitive individuals but also strongly bonded social groups that could work together with confidence and loyalty, convinced of the efficacy of their joint actions. The usual view of humans as selfish, cooperating only so they could advance their own interests, cannot account for the resilience and responsivity of the skeins of mutuality.

Early hominids obviously possessed the motivations and social reinforcements of their primate cousins, which enabled them to care for their offspring, too. And so do most of us today, who admit to undeniable satisfaction while caring for a small helpless thing—especially when it is our own. Neurotransmitters such as oxytocin are released in mothers before childbirth and with suckling, so that "maternal affect," if nothing interferes with it psychologically or physically, is a demonstrable biological reality. And apart from that, our brain circuitry has evolved to respond with tenderness and positive emotion to such signs of lovability as small size, a round head that is large in proportion to the body, big eyes, plump cheeks, downiness, softness, and other indicators of infantility—in baby animals and in pets, stuffed animals, cartoon characters, and advertising images as well as in our own kind.

Although most people take human mother love for granted, it was an important evolutionary adaptation. Until the 1960s, psychologists generally thought of it as fairly straightforward: human mothers, like other animals, had "maternal emotion," and babies—through conditioning, like pets—gradually came to love the person who fed and looked after them.

In 1969 in England, John Bowlby, a child psychiatrist with an interest in ethology, challenged this rather simplistic idea in the first volume of a path-breaking treatise called *Attachment and Loss*. He was acquainted with the reactions of young children who for various reasons—illness, death, wartime dispersals, abandonment—had been separated from their mothers, and he was led to propose that there is a positive need for infants to form what he called *attachment* with caretakers. By the age of about eight months, especially in circumstances of uncertainty, children in all cultures do similar things to attract and sustain their mothers' attention: they cry when separated, lift their arms to be picked up, cling to her body, stay near her, and even when playing happily look at her frequently. They do this whether or not the mother has shown them affection. In orphanages, young children often choose one staff person as a favorite, even if other individuals feed and tend them. Contrary to previous assumptions, the tendency to attach was observably separate from simple conditioning to a positive stimulus such as food or care.

Bowlby suggested that the evolutionary value of attachment was that the helpless hunter-gatherer's baby would not wander off, and when frightened or alone it would cry, reach out, move toward, or otherwise try to resume contact with a specific protective figure rather than remain vulnerable to predators or accidents. Many helpless young birds and mammals have comparable behaviors.

In the years since Bowlby's formulation, research with much younger infants has enriched his pioneering work, showing quite remarkable and unexpected earlier abilities and proclivities for interaction and intimacy. These suggest that attachment—which in Bowlby's scheme appears at about the time the baby is first able to move about on its own and is concerned primarily with the infant's physical safety through "proximity-seeking"— should be viewed as a late-appearing consequence of a prior, equally innate and adaptive predisposition to engage in relationship and emotional communion, over and above any need for protection.

University of Edinburgh psychologist Colwyn Trevarthen has called this predisposition *innate intersubjectivity* ("Communication," "Concept"). He sees it as a fundamental inborn readiness of the baby to seek, respond to, and affect the mother's provision of not only physical protection and care but also emotional regulation and support—that is, her provision of companionship. Trevarthen's studies, like many others, show clearly how the mother–infant pair together engages in a mutually improvised interaction based on innate competencies and sensitivities—an interaction, sometimes called "baby talk," whose importance was for years overlooked if not altogether dismissed. (I use the term "baby talk" specifically to refer to the interactive behavior between adults and nonverbal infants under approximately six months of age. As I use it, the term does not include the imperfect speech directed at older infants and toddlers [for example, "get your blankie and let's go night-night"].) Long before the attachment described by Bowlby takes place, this common pastime, which falsely seems to be both trivial and inane, provides enjoyment and intimacy for both participants and significant developmental benefits for the infant.

The Complexities of Baby Talk

From the first weeks, in all cultures, human mothers (and even other adults) behave differently with infants than with adults or even older children. In most cases a mother's vocalizations to the baby and her facial expressions, gestures, and head and body movements are exaggerated—made clear and rhythmic. Babies in turn respond with corresponding sounds, expressions, and movements of their own, and over the first months a mutual multimedia ritual performance emerges and develops. Exquisitely satisfying to both participants, it inundates both mother and baby with a special pleasure that is all the more powerful because it is not just felt alone (like the interest, excitement, or joy felt while privately thinking about or watching one's baby) but is mirrored or shared.

This mother–infant interaction has been well studied in a number of different cultures over the past quarter century (Brazelton; Callaghan; Fajado and Freedman; Feiring and Lewis; Field and Widmayer; Lewis and Ban; Nakano). Its beginnings are now seen to be at birth, although as the weeks and months pass it becomes more and more a consciously improvised and improved-upon duet. All over the world people (especially mothers) talk to babies (and generalizing, often to any smaller creature) in a special vocal register: a higher, softer, breathier, singsong tone of voice. The contours—the ups and downs—of these utterances are much more labile and exaggerated than the contours of ordinary speech to other adults (Fernald).

Mothers in many cultures speak to small babies as if they expect a reply— "*Too* much milk? You've had *too* much milk? Ohhhhhh!"—even when they realize that infants cannot understand words. In some societies there is no tradition of *talking* to babies, but other rhythmically regular noises such as tongue-clicking, hissing, grunting, or lip-smacking may be used and supplemented by physical movements and exaggerated facial expressions.

While talking or making sounds, mothers rock or pat the baby as well as look at or gaze into its face. They usually smile. The things they say are structured in time, like poetry or song: if transcribed they reveal formal segments like stanzas, often based on one theme, with variations, that have to do with the looks or actions of the baby (frequently its digestion: burps, hiccups, and poops) or something about its lovability—for example, "Mommy loves you. Yes. Yes. Did you know Mommy loves you? Yes she does. She does. She loves you."

As in this segment, the words are organized into phrases, each (whether having one or seven syllables) about three and a half to five seconds in length (Lynch et al.; Turner, "Neural Lyre"). The utterances are rhythmic and highly repetitive. At the end of one stanza, after a pause, another subject or theme may suggest itself: "You sleepy? You sleepy now? Come on, don't sleep yet. Noooo. Don't sleep yet. Come on. Come on. All right, then. All right."

Up to about eight weeks, the human mother's baby talk is primarily soothing and fondly affectionate: it provides a tender singsong jokiness: "*Talk* to me, *talk* to me, won't you *talk* to me?" Gradually she will more insistently engage the infant's attention, attracting and sustaining it by stops and starts, moving her head closer and farther away or raising her chin and letting it drop suddenly while making an exaggerated, wide-eyed expression with pursed lips and perhaps a "tch" or tongue click. (One can observe strangers doing these things to babies of the appropriate age in supermarket checkout lines and other public places.) Such antics generally induce the baby to respond with wriggles, kicks, and smiles. Without an expected infant response (and one's own wish to provoke it) there would be no reason to behave in this otherwise inexplicable way.

A temporal analysis of the sounds, movements, and facial expressions of these interactive sequences reveals an amazing attunement, a synchronization of interaction. Both mother and baby adjust their responses to each other within seconds or fractions of seconds, according to discernible "rules" of mutual regulation that are made up as they go along (Beebe). (There may be, of course, "misattunements," as when one or the other member of a pair is unresponsive or when a mother is "intrusive," unable to read the baby's signals of overstimulation and its wish to "tune out" or disengage.) What is particularly interesting is that small infants are not only supremely sensitive and responsive to rhythmic properties of their mothers' sounds, facial expressions, and body movements but are also able to perceive these "cross-modally." For example, three-week-olds can perceive the similarity between bright colors and loud sounds (Lewkowicz and Turkewitz). At six months babies recognize that a pulsing tone (heard) and a dotted line (seen) are alike, as are a continuous tone and an unbroken line (Wagner et al.). Thus in early interactions, behaviors are not only directly mirrored or imitated but also may be *matched* by either partner in supramodal qualities such as intensity, contour, duration, or rhythm—qualities that apply to any sense modality. That is to say, the loudness of a sound may be matched by a strong arm or leg movement (or vice versa); the downward contour of a head movement may be matched by a downward fall of the voice (or vice versa) (Beebe and Gerstman; Eimas; Marks, Hammeal, and Bornstein; Stern, *Interpersonal World*).

Interestingly, the three-and-a-half- to five-second segmental length of a typical utterance in baby talk corresponds to the temporal length of a poetic line, a musical phrase, and a phrase of speech in adults (Lynch et al.; Turner, "Neural Lyre"). That infants are supremely sensitive to this universal measure and its rhythmic subdivisions in syllables and syllable clusters, and that mothers spontaneously produce it, argues that both the creation and the experience of the temporal arts of poetry, music, and dance (the movements of which accord with music) inhere in our fundamental psychobiology—our inner brain sense of rhythm and melody.

Neurologists (e.g., Schore 80) describe how external indicators of internal states, such as vocal, facial, and gestural expressions of adoration and pleasure, trigger physiological arousal in both the "receiver" and the "sender," whether infant and mother or beloved and lover. Thus, producing the signal additionally generates or reinforces the emotion (Ekman, "Facial Expressions"; Ekman, Davidson, and Friesen; Ekman, Levenson, and Friesen). (Such feedback effects have received popular attention from studies of hospital patients who, by consciously trying to smile and laugh, recover from illnesses sooner than people who are bad-tempered or gloomy [Cousins, *Anatomy*, *Healing Heart*].) Simply making the facial or vocal correlate of an emotion tends to release the brain chemicals that cause people to *feel* the emotion, so we can assume that early hominid mothers who performed and even exaggerated affiliative signals to their infants would have produced in themselves more affiliative and loving feelings than if they had remained poker-faced, stiff-bodied, and silent.

Baby-talk interactions also provide intellectual, linguistic, and cultural practice. By anticipating what comes next in a familiar sequence, the baby "hypothesizes" or predicts when a climax will occur and experiences its fulfillment. Being able to recognize pattern in the behavior of others—what psychologists call "sequencing"—is essential to eventual social and intellectual competence, making it possible to comprehend and predict others' behavior (Greenspan 6, 67).

The existence of evolved mechanisms in mothers for spontaneously producing these rhythmically coordinated and patterned signals, and the evolved sensitivities of newborns to recognize and reciprocate them (at a stage of development when they have few other psychological capabilities), argues for their primal importance in infancy. Even older children and young adults with profound mental handicap (for example, restricted mobility, limited ability to use language) can participate in and enjoy such interactions with sensitive caregivers, strongly indicating "a biologically robust system of basic emotional communication" (Burford 189).

Baby talk, as just described, has nothing to do with the exchange of verbal information about the world and everything to do with participating in an impromptu expression of accord and a narrative of feelings, ideas, and impulses to act. It is this wish to share emotional experience that motivates early vocalization (or "talk before speech") and sets a child on "the path to spoken language," as the neurobiologist of language John L. Locke nicely described it—not the instrumental need to request or name things, which comes later. In this view, language emerges from and first expresses emotional needs of mutuality and belonging, although it will eventually become also an instrument of symbolic reasoning and intellectual analysis.

Although humans are predisposed for rhythmic-modal expressions of mutuality with caretakers and other close familiars, and although such interchanges are even necessary to infant survival, they eventually take their place as means to other ends. Expressions of mother–infant mutuality occur in brief bouts that may last only a few seconds, and in hunter-gatherer and other similar societies, they can take place with several or even many other people. They are not exclusively with one person and soon are replaced by similarly structured and longer-lasting group events that unite numbers of people.

In small-scale (and probably ancestral) societies these expressions are transformed into feelings of belonging to a group and sharing its ideals. In heterogeneous modern societies, where we have relationships of various kinds (not necessarily even face-to-face) with a variety of people who know us only in certain roles or at certain times of our lives and who often do not know one another, mutuality and belonging are less inclusive and well-defined. Perhaps this disruption to our evolved nature accounts at least in part for the obsession with sexual or romantic love in modern society's popular songs and pastimes.

Rhythmic and modal elements such as synchronizing, turn-taking, imitating or matching, and sequentially patterning movements and vocalizations are also the stuff of ceremonial ritual—a universal and age-old practice. Just as it was essential during human evolutionary history for mothers and infants to establish mutuality, so was it essential for members of groups to work together in confidence and harmony rather than to act individually, selfishly, or haphazardly, without regard to tradition and communal purpose.

Art and Meaning

Over the millennia of human evolution, the mind increasingly became a "making-sense organ": interrelated powers of memory, foresight, and imagination gradually developed and allowed humans to stabilize and confine the stream of life by making connections between past, present, and future, or among experiences or observations. Rather than taking the world on its own terms of significance and value (the basic survival needs, sought and recognized by instinct), people came more and more to systematize or order it and act upon it. Eventually, this powerful and deep-rooted desire to make sense of the world became part of what it meant to be human—to *impose* sense or order and thereby give the world additional (what we now call "cultural") meaning.

What cultures systematize and value derives from the basic biological requirements for survival and well-being—such essentials as finding, preparing, and assuring the continuance of food; rearing children; and maintaining social relationships, social practices, health, safety, prosperity, and competence. Biological meaning—significance or value—implies that we have emotional investment in these fundamental things: that is, we have evolved to care about them. Cultures have in turn evolved to assure that we care by appropriately emphasizing what we need to care about most. Cultural knowledge and practices direct our attention to particular biologically significant things and help us know what to think and do about them.

Nonhuman animals are born generally knowing what to do: they build nests, seek food, find mates, and bear and nourish young, largely without being taught. Humans, by contrast, have to learn the rules and schemes that make their world orderly—comprehensible and manageable. We are not born knowing such survival skills as how to make a shelter, find and prepare food, or care for infants. We must learn social skills, how to act and what to think, so that we will be regarded positively by those with whom we live, those we love. Compared with humans, other animals have a much more straightforward path to maturity. Yet the rhythms and modes of human infancy and childhood predispose us to acquire systematic and storied accounts of the world into which we are born.

Although ways and worldviews differ from one group to another, all cultures, like the Yekuana (a tribe of the Amazon Basin), have, or have had, frameworks or rules for living, along with stories about themselves and why things are as they are. The mythical, social, and practical orders are inextricably interwoven. Adherents of an individual culture's systems and stories consider them to be foreordained, unequivocally right and meaningful, logical and aesthetic. The Yekuana call themselves So'to, meaning "human" or "person," as distinguished from any other species, human or nonhuman. This word refers to their unique heritage of common culture and language, their "identity." David Guss tells us that the Yekuana, like other small-scale cultural groups, have utter confidence in the propriety of their way of life as opposed to any other.

No matter how isolated or technologically impoverished, every human culture has, like the Yekuana, an account of the cosmos, its creation and maintenance, and the origin of themselves and other beings. The cosmic order includes the natural order—life and nonlife, male and female, human and animal—along with the divine and the mundane. There are notions of an "other world," different in some way from this one, concepts of souls and spirits, things unseen as well as seen, and an eschatology of what happens at and after death.

From a modern standpoint, the cosmologies and supernatural beliefs of traditional peoples are "myths," meaning fabrications. Yet for their adherents, they satisfy the need to explain, within an accepted order and system, why and how things are as they are. Our more abstract metaphysical systems—whether theological, philosophical, or scientific—are similarly ways of ordering and explaining the world, of understanding ultimate reality.

Along with metaphysical systems come rules that regulate social, familial, and sexual relations. Every society recognizes duties and benefits that accompany different ages and stages of life, with acknowledged procedures—rites of passage—for entering or leaving these stages or roles: ceremonies for weaning, first menstruation, circumcision, marriage, graduation, or inauguration, to name a few.

As humans are natural system-builders, they are also natural storytellers. To some degree we tell stories and hear those of others as part of everyday interaction—"The Visit with Susie's Teacher," "Car Trouble on the Way to Work," "The Dreaded Job Interview," "The Trip to Hawaii." Such stories may well be informative, but they are more than that. Using devices such as vivid detail or suspense that will engage the hearer, they call for and receive empathy and interaction.

We have evolved to respond to archetypal images and narrative structures that touch upon fundamental existential themes and interests: being and nonbeing, birth and death, the life course from infancy to old age, males and females, mothers and fathers, sons and daughters, wise elders, helpless children, heroes and villains, wild animals, and bodies and bodily functions such as eating, drinking, urinating, defecating, and copulating.

From fundamental human struggles and outcomes—failing or succeeding (escaping harm, overcoming, losing or gaining), giving, loving, being demeaned, admired, or rejected (Lazarus 307)—arise such common emotions as joy, grief, desire, anger, frustration, and anxiety; experiences of plenty and want, love and loss; and expressions of violence, duty, obligation, shame, and redemption. While such features are the stuff of popular entertainment and everyday conversation, they are also the substrate of age-old myths, epics, and scriptures.

Stories move people in ways unlike other uses of language. They provide a certain kind of knowledge: a person has done this or that, and this is what came of it (Burkert 56). Children at two and a half years of age recognize that to tell a story is different from other uses of language (Olson).

Children notoriously like to hear the same story over and over again. So do adults, when the story is a "good" one. We wait expectantly for the punch line or the denouement, even when we already know it—and then laugh or feel the appropriate emotions of sadness, disbelief, outrage, amazement.

A good tale is easy to remember, even if heard only once. We do not remember or reproduce it as a sequence of sounds or words (like a melody or an address) but rather as a sequence of events and actions that, even with different words, *makes sense*. A story is, as Walter Burkert described it, a structure of sense (58).

Literature, whether from the ancient oral traditions of people without writing or from the latest bestseller, has always dealt with the vital interests and concerns of humans—such prosaic but elemental and inexhaustibly myriad matters as staying alive and well, being accepted and thought well of by associates, developing normally both physically and socially, learning the ways of one's fellows, engaging in activity that is materially, emotionally, and socially rewarding, finding a mate and mating, successfully producing and caring for offspring, helping one's offspring thrive, overcoming threats, and otherwise affecting the outcomes of things one cares about. Whether they appear in Homer or Lady Murasaki, Dostoyevski or Danielle Steele, grand opera or soap opera, news features or television commercials, we are attracted by these and other humanly relevant themes that derive from evolved needs and interests. They invariably capture our fascinated attention.

If literacy and the alphabet have changed the way humans think, events in western Europe and the Americas following the Renaissance and Industrial Revolution have changed the way we and the rest of the people in the world live. Minds and lives, both, have tasted of the fruits of the tree of possibility as well as of knowledge, trading meaning for meanings. The logico-aesthetic integration of our ancestors' lives and their certainty of knowing what is true and good have gradually dissolved, along with culturewide acceptance of divine providence or a destiny that shapes our ends.

Ironically, despite all the knowledge and sophisticated technological mastery that have arisen from contemporary science, its most advanced ideas reveal fundamental principles of uncertainty, relativity, and the infinite—whether of outer space or inner matter—that inescapably confront our wish to understand with what is humanly inconceivable. Although natural science separated itself from theology centuries ago, the entities with which it now concerns itself—superstrings, quantum gravity, and black holes, for example—seem as mystically alluring and beyond human understanding as intimations of the divine. Uncertainty and relativity have become articles of faith even in the humanities, as professors teach that we are incapable of knowing what is true or real, and doctors of philosophy maintain that all knowledge and reality, being mere interpretations of cultures and individuals, are unavoidably partial.

Although the larger implications of science may give rise to wonder and even mystical speculation, as a tool it addresses quite specific questions about nature, such as how things work and what they are made of. Although its findings have given us order and powerful ways to affect the world, it remains the case that human minds evolved to require from life more than discovery and problem solving. Analytic thinking requires an unnaturally high level of training and vigilance and can easily be overridden by wish, desire, anxiety, or strong authority. It does not address or answer questions of ultimate meaning, purpose, intent, and justification for the ways things are; instead, it narrows its focus to description, subsumption, and probability.

Yet judging from ancestral and traditional examples, the systems and stories that keep society orderly and individuals secure work best when they are vividly presented as part of a compelling belief system and irradiated with convictions of transcendent truth. What we call "religion" and "art" were for countless centuries intrinsic to the order of ordinary life and to the motivations of human minds. They were not optional practices to be indulged one morning each week or when there was nothing better to do, nor were they superfluous pastimes that could be rejected altogether. Pueblo Indian cosmology may seem quaint and dubious compared with Los Alamos physics, but it satisfies human needs and addresses nonscientific questions to which people want answers.

Where do we come from? What are we? Where are we going? Why do the innocent suffer? What should we do to make things better? These are questions answered succinctly and without hesitation by religious and political ideologues, but not by the American Association for the Advancement of Science. The former are more popularly convincing about the truth of their convictions than scientists are about the probability of their hypotheses or the applicability of their conclusions.

For important reasons, scientific and other academic "-ologies" present their findings dispassionately, in abstract terms. Even when discussing the birth of stars, the beginning of life and the finale of death, or the rise and fall of empires, their accounts describe impersonal antecedents and consequents, unpicturable principles and processes. Although such analyses allow a useful, rationally expressed general understanding, they deliberately distance themselves from a humanly relevant world with its mysteries, wonders, and dramatic personal conflicts or satisfactions. Faced with a painful disease, people may be instructed about viruses or defective genes. But these "explanations" still don't answer the more urgent question, "Why me?" Sorcery or witchcraft provides a more tangible explanation, with a built-in course of action.

Human minds evolved, like those of other animals, to pursue and find meaning in their own interests (which include their own kin and close associates). Appeals to concern for the common good are typically less effective than demonstrations of how policies will affect your paycheck or the health of your family. It remains all too true that for most human minds it seems insufficient simply to find and make order, nor are our minds easily convinced of something unless it is presented with emotionally appealing personalized relevance.

Ten thousand years after the invention of writing, the kind of disembedded, critical thinking that writing makes possible remains difficult and even irrelevant for many, as is evident to anyone who observes general public discourse and its assessments of complex social issues. It is not difficult for "evidence" from one's feelings and senses to outweigh arguments addressed to one's reason. Juries find it difficult to separate the personal and dramatic from the evidential and legally admissible and may base their verdicts more on the former than on the latter.

The mythopoetic expressions that we evolved to respond to and require are concrete and vivid, embedded in the world they explain, and based on analogies with human actions and emotions: they catch us up in dramatic, descriptive narrative. There are heroes, villains, good and evil, transgression and retribution. No wonder stories about amazing adventures and celebrities' lives appeal far more to human minds than descriptions of the complex history of a political crisis or careful analyses of global problems and economic trends.

We view with alarm many traditional beliefs and their attendant practices—such as subjugation, scarification, or mutilation—and consider them to be ignorant or enslaving. Kenneth Maddock, in a sympathetic book about Australian Aborigines, points out that what we regard as illusions and even oppression often are inherent in beliefs and practices whose *absence* removes meaning from life and renders it mediocre or even unsupportable (183). Such meaning-rich practices frequently involve rhythmic-modal elaborations.

In the laudable post-Enlightenment endeavor of dispelling illusions and casting off oppression, it is easy to leave a desolate void that is all too ready to be filled by other emotion-laden explanatory schemes that satisfy the needs for belonging and meaning—say, conspiracy theories, obedience to mind-controlling cults, or fanatical adherence to fundamentalist doctrines both secular and divine.

13

Arts of Seduction

Geoffrey Miller

Art has always been a puzzle for evolutionists. Michelangelo's *David* seems singularly resistant to the universal acid of Darwinism, which is otherwise so efficient at dissolving the cultural into the biological. Like any nouveau-riche connoisseur, we are both proud of our art and ashamed of our ancestry, and the two seem impossible to reconcile.

The evolution of art is hard to explain through survival selection, but is a pretty easy target for sexual selection. The production of useless ornamentation that looks mysteriously aesthetic is just what sexual selection is good at. Artistic ornamentation beyond the body is a natural extension of the penises, beards, breasts, and buttocks that adorn the body itself. We shall begin our tour of the human mind with a look at our artistic instincts for producing and appreciating aesthetic ornamentation that is made by the hands rather than grown on the body.

Art as an Adaptation

In her books *What Is Art For?* and *Homo Aestheticus*, anthropologist Ellen Dissanayake made one of the first serious attempts to analyze art as a human adaptation that must have evolved for an evolutionary purpose. She argued that human art shows three important features as a biological adaptation. First, it is ubiquitous across all human groups. Every culture creates and responds to clothing, carving, decorating, and image-making. Second, the arts are sources of pleasure for both the artist and the viewer, and evolution tends to make pleasurable those behaviors that are adaptive. Finally, artistic production entails effort, and effort is rarely expended without some adaptive rationale. Art is ubiquitous, and costly, so is unlikely to be a biological accident.

Art fits most of the other criteria that evolutionary psychology has developed for distinguishing genuine human adaptations from non-adaptations. It is relatively fun and easy to learn. Given access to materials, children's painting and drawing abilities unfold spontaneously along a standard series of developmental stages. Humans are much better at producing and judging art than is any artificial intelligence program or any other primate. Of course, just as our universal human capacity for language allows us to learn distinct languages in different cultures, our universal capacity for art allows us to learn different techniques and styles of aesthetic display in different cultures. Like most human mental adaptations, the ability to produce and appreciate art is not present at birth. Very little of our psychology is "innate" in this sense, because human babies do not have to do very much. Our genetically evolved adaptations emerge when they are needed to deal with particular stages of survival and reproduction. They do not appear at birth just so psychologists can conveniently distinguish the evolved from the cultural. Beards have evolved, but they grow only after puberty, so are they "innate"? Is menopause "innate"? "Innateness" is a relatively useless concept that has little relevance in modern evolutionary theory or behavior genetics.

Some archeologists have argued that art only emerged 35,000 years ago in the Upper Paleolithic period, when the first cave paintings and Venus figurines were made in Europe. They follow archeologist John Pfeiffer's suggestion that this period marks a "creative explosion" when human art, language, burial ceremonies, religion, and creativity first emerged. This is a remarkably Eurocentric view. The Aborigines colonized Australia at least 50,000 years ago, and have apparently been making paintings on rock ever since. If art were an invention of the Upper Paleolithic 35,000 years ago in Europe, how could art be a human universal (Sandars; Taylor)? There is evidence from Africa of red ocher being used for body ornamentation over 100,000 years

ago (Knight, Power, and Watts; Low, "Sexual Selection"; Power). This is about the latest possible time that art could have evolved, since it is around the time that modern *Homo sapiens* spread out from Africa. Had it evolved later, it is unclear how it could have become universal across human groups.*

The Functions of Art

The aesthetic has often been defined in opposition to the pragmatic. If we view art as something that transcends our immediate material needs, it looks hard to explain in an evolutionary way. Selection is usually assumed to favor behaviors that promote survival, but almost no art theorist has ever proposed that art directly promotes survival. It costs too much time and energy and does too little. This problem was recognized very early in evolutionary theorizing about art. In his 1897 book *The Beginnings of Art*, Ernst Grosse commented on art's wastefulness, claiming that natural selection would "long ago have rejected the peoples which wasted their force in so purposeless a way, in favor of other peoples of practical talents; and art could not possibly have been developed so highly and richly as it has been" (312). He struggled, like many after him, to find a hidden survival function for art.

To Darwin, high cost, apparent uselessness, and manifest beauty usually indicated that a behavior had a hidden courtship function. But to most art theorists, art's high cost and apparent uselessness has usually implied that a Darwinian approach is inappropriate, that art is uniquely exempt from selection's cost-cutting frugality. This has led to a large number of rather weak theories of art's biological functions. I shall briefly consider their difficulties before attempting to bring art back into the evolutionary framework.

ART FOR ART'S SAKE

Ever since the German Romanticism of Schiller and Goethe in the early nineteenth century, many have viewed art as a utopian escape from reality, a zone of selfless self-expression, a higher plane of being where genius sprouts lotus-like above the petty concerns of the world. This Romantic view opposes art to nature, but also opposes art to popular culture, art to market commodity, art to social convention, art to decoration, and art to

---

*For more recent research locating the dispersal from Africa at about 55,000 years before the present, see Wade, *Before*. Eds.

practical design. It has often presented the artist as a male genius shunning the female temptresses that would sap the vital fluids that sustain his creativity (Dijkstra). Thus, artistic success has also been seen as opposed to sexual reproduction.

Perhaps it is not surprising that many modern artists have adopted the ideology of these German philosophers. Romanticism makes excellent status-boosting rhetoric for artists. It presents them as simultaneously overcoming their instincts, avoiding banality, striving against capitalism, rebelling against society, and transcending the ornamental. The genius's need to shun sexual temptation also provides a ready excuse for avoiding sleeping with one's less attractive admirers. But this Romantic view makes no attempt to offer a scientific analysis of art—indeed, it actively rejects the possibility.

The kernel of truth in the Romantic view is that art is pleasurable to make and to look at, and this pleasure can seem a sufficient reason for art's existence. Its pleasure-giving power can seem to justify art despite its apparent uselessness. But from a Darwinian perspective, pleasure is usually an indication of biological significance. Subjectively, everything an animal does may appear to be done simply to experience pleasure or avoid pain. If we did not understand that animals need energy, we might say that they eat for the pleasure of eating. But we do understand that they need energy, so we say instead that they have evolved a mechanism called hunger that makes it feel pleasurable to eat (Tiger; Tooby and Cosmides, "Past"). The Romantic view of art fails to take this step, to ask why we evolved a motivational system that makes it pleasurable to make and see good art. Pleasure explains nothing; it is what needs explaining.

SOCIAL SOLIDARITY, CULTURAL IDENTITY, AND RELIGIOUS POWER

Many anthropologists view art, like ritual, religion, music, and dance, as a social glue that holds groups together. This hypothesis dates back to the early twentieth century and the "functionalist" views of Emile Durkheim, Bronislaw Malinowski, A. R. Radcliffe-Brown, and Talcott Parsons (Haviland; Knight). For them, a behavior's function meant its function in sustaining social order and cultural stability, rather than its function in propagating an individual's genes. The social functions postulated for art were usually along the lines of "expressing cultural identity," "reflecting cultural values," "merging the individual into the collective," "sustaining social cohesion," "creating a collective consciousness," and "socializing the young." It is not easy to be sure what any of these phrases really means, and in any case

these putative social functions are not easy to relate to legitimate biological functions in evolution.

Primate groups work perfectly well without any of these mechanisms. Chimpanzees don't need to express their cultural identities or create a collective consciousness in order to live in groups. They need only a few social instincts to form dominance hierarchies, make peace after quarrels, and remember their relationships. Humans do not seem any worse at these things than chimpanzees, so there seems no reason why we should need art or ritual to help us "bond" into groups. Human groups may be larger than chimpanzee groups, but Robin Dunbar has argued convincingly that language is the principal way in which humans manage the more complex social relationships within our larger groups (*Grooming*).

The view that art conveys cultural values and socializes the young seems plausible at first glance. It could be called the propaganda theory of art. The trouble with propaganda is that it is usually produced only by large institutions that can pay propagandists. In small prehistoric bands, who would have any incentive to spend the time and energy producing group propaganda? It would be an altruistic act in the technical biological sense: a behavior with high costs to the individual and diffuse benefits to the group. Such altruism is not usually favored by evolution. Evolution can sometimes favor group-benefiting behaviors, if individuals can attain higher social and sexual status for producing them. But such opportunities are relatively rare, and one would have to show that art is well designed as a propaganda tool to create norms and ideals that benefit the group. Language is surely a much more efficient tool for telling people what to do and what not to do. The best commands are imperative sentences, not works of art.

A popular variant of the cultural-value idea is the hypothesis that most art during human evolution served a "religious function." Museum collections of art from primitive societies routinely label almost every item a fertility god, an ancestral figure, a fetish, or an altarpiece. Until recently, archeologists routinely described every Late Paleolithic statue of a naked woman as either a "goddess" or a "fertility symbol." Usually, there is no evidence supporting such an interpretation. It would be equally plausible to call them "Paleolithic pornography" (Hersey; Taylor). The importance of church-commissioned art in European art history may have led archeologists to attribute religious content to most prehistoric art.

In any case, religious functions for art don't make much Darwinian sense. Some anthropologists have suggested that the principal function of art during human evolution was to appease gods and dead ancestors, and to put people in touch with animal spirits. In his textbook *The Anthropology of Art*, Robert Layton claimed that the function of Kalabari sculpture in Africa is "a

pragmatic one of manipulating spiritual forces" (7). This overlooks the possibility that gods, ancestral ghosts, and animal spirits may not really exist. If they do not exist, there is no survival or reproductive advantage to be gained from appeasing or contacting them. Some artists may believe that making a certain kind of statue will give them "spiritual powers." Scientifically, we have to take the view that they might be deluded. Their delusion, on its own, is not evolutionarily stable, because it costs them time and energy and the "spiritual powers" probably cannot deliver what is hoped for. However, if an individual's production or possession of a putatively religious object brings them higher social or sexual status, then it can be favored by evolution. A person can spend hours hacking at a piece of wood, making a fetish, and telling people about their extraordinary spiritual powers. If others grant the religiously imaginative individual higher status or reproductive opportunities, such behavior can be sustained by sexual selection.

The same argument applies to art that has the alleged function of curing disease, such as some Navajo sand-paintings. Navajo artists could speculate that the human capacity for making sand-paintings must have evolved through survival selection for curing diseases. If sand-paintings were proven medically effective in double-blind randomized clinical trials, they would have a good argument. But the sand-paintings probably have nothing more than a placebo effect. Like "appeasing the gods," "curing disease" works as an evolutionary explanation only if the trait in question actually does what is claimed.

Evolution is not a cultural relativist that shows equal respect for every ideological system. If an artistic image intended to control spirits or cure disease does not actually improve survival prospects, evolution has no way to favor its production except through sexual selection. Evolutionary psychologists should accept ideologies like religion and traditional medicine as human behavioral phenomena that need explaining somehow. This does not mean that we have to give them any credence as world-views. For scientists, science has epistemological priority.

There are important differences between the social functions of art (which may support religious, political, or military organizations), the conscious individual motivations for producing art (which may include making money, achieving social status, or going to heaven), and the unconscious biological functions of producing art (which must concern survival or reproduction). Darwinian theories of the origins of our capacity for art cannot hope to account for all of the social functions and various forms of art that happen to have emerged in diverse human cultures throughout history. Evolutionary psychology tries to answer only a tiny number of questions about human art, such as "What psychological adaptations have evolved for producing and appreciating art?" and "What selection pressures shaped those adaptations?" These are important questions, but they are by no means the only interesting ones. All the other questions about art will remain in the domain of art history and aesthetics, where a Darwinian perspective may offer some illumination, but never a complete explanation. We shall still need cultural, historical, and social explanations to account for the influences of Greek and Indian traditions on Gandhara sculpture, or the way in which Albert Hoffman's serendipitous discovery of LSD in 1943 led to the "happenings" organized by the Fluxus group in the 1960s. As we shall see, the human capacity for art is a particularly flexible and creative endowment, and identifying its evolutionary origins by no means undermines the delights of art history, or limits the range or richness of artistic expression.

A Bottom-Up View of Art

None of the standardly proposed "functions" of art are legitimate evolutionary functions that could actually shape a genetically inherited adaptation. As Steven Pinker has observed,

> Many writers have said that the "function" of the arts is to bring the community together, to help us see the world in new ways, to give us a sense of harmony with the cosmos, to allow us to appreciate the sublime, and so on. All these claims are true, but none is about adaptation in the technical sense. (*How* 535–36)

If this is right, then what are we to do? The human capacity for art shows evidence of adaptive design, but its function remains obscure. Perhaps we need a broader view of art, inspired by more biologically relevant examples.

There are two strategies science can take in trying to understand the evolutionary origins of art: top-down or bottom-up. The top-down strategy focuses on the fine arts and their elite world of museums, galleries, auction houses, art history textbooks, and aesthetic theory. The bottom-up strategy surveys the visual ornamentation of other species, of diverse human societies, and of various subcultures within our society. In this broader view, the fine arts are a relatively unpopular and recent manifestation of a universal human instinct for making visual ornamentation. Most scientists, being anxious to display their cultural credentials as members of the educated middle class, feel obligated to take a top-down approach. There is a temptation to display one's familiarity with the canon of Great Art, to counter the stereotype that scientists are so obsessed with truth that they have forgotten beauty. One may even feel obliged to start with a hackneyed example of Italian Renaissance sculpture, as I have done in this chapter.

But what if we step back from the fine arts and ask ourselves what engagement ordinary humans have with visual ornamentation, once they step outside the dim museums of Florence and return to their real lives. Our opportunities to appreciate the fine arts typically arise during vacations and weekend trips to local museums. But visual ornamentation surrounds us every day. We wear clothing and jewelry. We buy the biggest, most beautiful houses we can afford. We decorate our homes with furniture, rugs, prints, and gardens. We drive finely designed, brightly colored automobiles, which we choose for their aesthetic appeal as much as their fuel efficiency. We may even paint the odd watercolor. This sort of everyday aesthetic behavior comes quite naturally, in every human culture and at every moment in history.

There is no clear line between fashion and art, between ornamenting our bodies and beautifying our lives. Body-painting, jewelry, and clothing were probably the first art forms, since they are the most common across cultures. Nor is there a clear line between art and craft—as William Morris argued when founding the Arts and Crafts movement in Victorian England. Fine art may be strictly useless in pragmatic terms, while good design merely makes beautiful that which is already useful. When we address the evolution of human art, we need to explain both the aesthetic made useless and the useful made aesthetic. Even apparently pragmatic tools like *Homo erectus* handaxes may have evolved in part through sexual selection as displays of manual skill.

In this chapter I take a bottom-up approach to analyzing the evolutionary origins of art, ornamentation, and aesthetics. This makes it easier to trace the adaptive function of these seemingly useless biological luxuries. Most of the visual ornamentation in nature is a product of sexual selection. The peacock's tail is a natural work of art evolved through the aesthetic preferences of peahens. We have also seen that some of our bodily organs, including hair, faces, breasts, buttocks, penises, and muscles, evolved partly as visual ornaments. It seems reasonable to ask how far we can get with the simplest possible hypothesis for art: that it evolved, at least originally, to attract sexual partners by playing upon their senses and displaying one's fitness. To see how this idea could work, let's consider an example of sexual selection for art in another animal species.

BOWERBIRDS

Human ornamentation is distinctive because most of it is made consciously with our hands rather than grown unconsciously on our bodies. However, this does not mean that its original adaptive function was different. The only other animals that spend significant time and energy constructing purely aesthetic displays beyond their own bodies are the male bowerbirds of Australia and New Guinea. Their displays are obvious products of female sexual choice (Andersson 172–74; Borgia, "Complex Male Display," "Sexual Selection"; Diamond).

Each of the 18 existing species constructs a different style of nest. They are constructed only by males, and only for courtship. Each male constructs his nest by himself, then tries to attract females to copulate with him inside it. Males that build superior bowers can mate up to ten times a day with different females. Once inseminated, the females go off, build their own small cup-shaped nests, lay their eggs, and raise their offspring by themselves with no male support, rather like Picasso's mistresses. By contrast, the male nests are enormous, sometimes large enough for David Attenborough to crawl inside. The Golden Bowerbird of northern Australia, though only nine inches long, builds a sort of roofed gazebo up to nine feet high. A hut built by a human male to similar proportions would top 70 feet and weigh several tons.

Males of most species decorate their bowers with mosses, ferns, orchids, snail shells, berries, and bark. They fly around searching for the most brilliantly colored natural objects, bring them back to their bowers, and arrange them carefully in clusters of uniform color. When the orchids and berries lose their color, the males replace them with fresh material. Males often try to steal ornaments, especially blue feathers, from the bowers of other males. They also try to destroy the bowers of rivals. The strength to defend their delicate work is a precondition of their artistry. Females appear to favor bowers that are sturdy, symmetrical, and well-ornamented with color.

Regent and Satin Bowerbirds go an astonishing step further in their decorative efforts. They construct avenue-shaped bowers consisting of a walkway flanked by two long walls. Then they use bluish regurgitated fruit residues to paint the inner walls of their bowers, sometimes using a wad of leaves or bark held in the beak. This bower-painting is one of the few examples of tool use by birds under natural conditions. Presumably the females have favored the best male painters for many generations.

Sexual selection for ornamental bower-building has not replaced sexual selection for the more usual kinds of display. Males of many bowerbird species are much more brightly colored than females, and they dance in front of the bowers when females arrive. They also sing, producing guttural wheezes and cries, and good imitations of the songs of other bird species. However, male bowerbirds are not nearly as spectacular as their relatives, the birds-of-paradise, the most gorgeous animals in the world. Somehow, having evolved from a drab crow-like form, the female ancestors of the bowerbirds and birds-of-paradise developed an incredible aesthetic sense. In the birds-of-paradise, their sexual choices resulted in an efflorescence of plumage in

40 species. In the bowerbirds, they resulted in a proliferation of ornamental nests in 18 species.

The bowerbirds create the closest thing to human art in a non-human species. Their art is a product of sexual selection through female choice. The males contribute nothing but their genes when breeding, and their art serves no survival or parental function outside courtship. The bowers' large size, symmetric form, and bright colors may reflect female sensory biases. However, the bowers also have high costs that make them good fitness indicators. It takes time, energy, and skill to construct the enormous bower, to gather the ornaments, to replace them when they fade, to defend them against theft and vandalism by rivals, and to attract female attention to them by singing and dancing. During the breeding season, males spend virtually all day, every day, building and maintaining their bowers.

If you could interview a male Satin Bowerbird for *Artforum* magazine, he might say something like "I find this implacable urge for self-expression, for playing with color and form for their own sake, quite inexplicable. I cannot remember when I first developed this raging thirst to present richly saturated color-fields within a monumental yet minimalist stage-set, but I feel connected to something beyond myself when I indulge these passions. When I see a beautiful orchid high in a tree, I simply must have it for my own. When I see a single shell out of place in my creation, I must put it right. Birds-of-paradise may grow lovely feathers, but there is no aesthetic mind at work there, only a body's brute instinct. It is a happy coincidence that females sometimes come to my gallery openings and appreciate my work, but it would be an insult to suggest that I create in order to procreate. We live in a post-Freudian, post-modernist era in which crude sexual meta-narratives are no longer credible as explanations of our artistic impulses."

Fortunately, bowerbirds cannot talk, so we are free to use sexual selection to explain their work, without them begging to differ. With human artists things are rather different. They usually view their drive to artistic self-expression not as something that demands an evolutionary explanation, but as an alternative to any such explanation. They resist a "biologically reductionist" view of art. Or they buy into a simplistic Freudian view of art as sublimated sexuality, as when Picasso repeated Renoir's quip that he painted with his penis. My sexual choice theory, however, is neither biologically nor psychologically reductionist. It views our aesthetic preferences and artistic abilities as complex psychological adaptations in their own right, not as side-effects of a sex drive. Bowerbirds have evolved instincts to construct bowers that are distinct from the instinct to copulate once a female approves of the bower. We humans have evolved instincts to create ornaments and works of art that are distinct from the sexual instincts behind copulatory courtship. Yet both types of instinct may have evolved through sexual selection.

The bowerbirds show the evolutionary continuity between body ornamentation and art. They happen to construct their courtship displays out of twigs and orchids instead of growing them from feathers like their cousins, the birds-of-paradise. We happen to apply colored patterns to rock or canvas. Biologists no longer draw a boundary around the body and assume that anything beyond the body is beyond the reach of evolution. In *The Extended Phenotype*, Richard Dawkins argued that genes are often selected for effects that spread outside the body into the environment. It is meaningful to talk about genes for a spider's web, a termite's mound, and a beaver's dam. Some genes even reach into the brains of other individuals to influence their behavior for the genes' own benefit. All sexual ornaments do that, by reaching into the mate choice systems of other individuals. At the biochemical level, genes only make proteins, but at the level of evolutionary functions they can construct eyes, organize brains, activate behaviors, build bowers, and create status hierarchies. Whereas an organism's "phenotype" is just its body, its "extended phenotype" is the total reach of its genes into the environment.

In this extended-phenotype view, bipedalism freed our hands for making not just tools, but sexual ornaments and works of art. Some of our ornaments are worn on the body, while others may be quite distant, connected to us only by memory and reputation. We ornament the skin directly with ocher, pigments, tattoos, or scars. We apply makeup to the face. We braid, dye, or cut our hair. We drape the body with jewelry and clothing. We even borrow the sexual ornamentation of other species, killing birds for their feathers, mammals for their hides, and plants for their flowers. At a greater distance, we ornament our residences, be they caves, huts, or palaces. We make our useful objects with as much style and ornament as we can afford, and make useless objects with purely aesthetic appeal.

The Rise and Fall of Sexual Art

The idea that art emerged through sexual selection was fairly common a century ago, and seems to have fallen out of favor through neglect rather than disproof. Darwin viewed human ornamentation and clothing as natural outcomes of sexual selection. In *The Descent of Man* he cited the popularity across tribal peoples of nail colors, eyelid colors, hair dyes, hair cutting and braiding, head shaving, teeth staining, tooth removal, tattooing, scarification, skull deformations, and piercings of the nose, ears, and lips. Darwin observed that "self-adornment, vanity, and the admiration of others, seem to be the commonest motives" for self-ornamentation (2:343). He also noted that in most cultures men ornament themselves more than women, as sexual selection theory would predict. Anticipating the handicap principle, Darwin also stressed the pain costs of aesthetic mutilations such as scarification, and the time costs of acquiring rare pigments for body decoration (Ludvico and Kurland; Singh and Bronstad). Finally, he argued against a cultural explanation of ornamentation, observing that "it is extremely improbable that these practices which are followed by so many distinct nations are due to tradition from any common source" (2:343). Darwin believed the instinct for self-ornamentation to have evolved through sexual selection as a universal part of human nature, more often expressed by males than by females.

Throughout the late 1800s, Herbert Spencer argued that Darwin's sexual selection process accounts for most of what humans consider beautiful, including bird plumage and song, flowers, human bodies, and the aesthetic features of music, drama, fiction, and poetry. In his 1895 book *Paradoxes*, Max Nordau attributed sexual emotions and artistic productivity to a hypothetical part of the brain he called the generative center. Freud viewed art as sublimated sexuality ("Leonardo").

However, these speculations did not lead very far because sexual selection theory was not very well developed at the beginning of the twentieth century. By 1908, aesthetic theorist Felix Clay had grown weary of the facile equating of artistic production with reproduction. In *The Origin of the Sense of Beauty*, he complained:

> How the pleasure in some stately piece of beautifully proportioned architecture, the thrill produced by solemn music, or the calm sweetness of a summer landscape in the evening, is to be attributed to the feeling of sex only, it is hard to see; they have in common a pleasurable emotion, and that is all. That a very large part of art is directly inspired by erotic motives is perfectly true, and that various forms of art play an important part in love songs and courtship is obvious; but this is so because beauty produced by art has in itself the power of arousing emotion, and is therefore naturally made use of to heighten the total pleasure. That love has provided the opportunity and incentive to innumerable works of art, that it has added to the pleasure and enjoyment of countless beauties, need not be denied; but we cannot admit that it is due to the sex feeling that rhythm, symmetry, harmony, and beautiful colour are capable of giving us a pleasurable feeling. (162–63)

In reading some of these century-old works, it is impressive how sophisticated and earnest their use of sexual selection theory was, and how favorably they compare to some current theories of art's evolution. Nevertheless, they repeat Freud's cardinal error, as Clay does here, of confusing sexual functions with sexual motivations. Art does not have to be about sex to serve the purposes of attracting a mate—it can be about anything at all, or about nothing, as in the geometric art of Islam, or Donald Judd's stainless-steel minimalist sculpture. As we saw with the bowerbirds, a sexually selected instinct for making ornamentation need not have any motivational or emotional connection with a sexually selected desire to copulate. The displayer does not need to keep track of the fact that beautiful displays often lead to successful reproduction. Evolution keeps track for us.

To be reliable, fitness indicators must be difficult for low-fitness individuals to produce. Applied to human art, this suggests that beauty equals difficulty and high cost (Gombrich, *Sense*; Power; Zahavi). We find attractive those things that could have been produced only by people with attractive, high-fitness qualities such as health, energy, endurance, hand–eye coordination, fine motor control, intelligence, creativity, access to rare materials, the ability to learn difficult skills, and lots of free time. Also, like bowerbirds, Pleistocene artists must have been physically strong enough to defend their delicate creations against theft and vandalism by sexual rivals.

The beauty of a work of art reveals the artist's virtuosity. This is a very old-fashioned view of aesthetics, but that does not make it wrong. Throughout most of human history, the perceived beauty of an object has depended very much on its cost. That cost could be measured in time, energy, skill, or money. Objects that were cheap and easy to produce were almost never considered beautiful. As Veblen pointed out in *The Theory of the Leisure Class*, "The marks of expensiveness come to be accepted as beautiful features of the expensive articles." Our sense of beauty was shaped by evolution to embody an awareness of what is difficult as opposed to easy, rare as opposed to common, costly as opposed to cheap, skillful as opposed to talentless, and fit as opposed to unfit.

In her books on the evolution of art, Ellen Dissanayake pointed out that the human arts depend on "making things special" to set them apart from ordinary, utilitarian functions (*Homo*). Making things special can be done in many ways: using special materials, special forms, special decorations, special sizes, special colors, or special styles. Indicator theory suggests that making things special means making them hard to do, so that they reveal something special about the maker. This explains why almost any object can be made aesthetically: anything can be made with special care that would be difficult to imitate by one who was not so careful. From an evolution-

ary point of view, the fundamental challenge facing artists is to demonstrate their fitness by making something that lower-fitness competitors could not make, thus proving themselves more socially and sexually attractive. This challenge arises not only in the visual arts, but also in music, storytelling, humor, and many other behaviors. The principles of fitness-display are similar across different display domains, and this is why so many aesthetic principles are similar.

Anthropologist Franz Boas insisted that in most cultures he studied, the artist's virtuosity was fundamental to artistic beauty. In *Primitive Art*, he observed that "the enjoyment of form may have an elevating effect upon the mind, but this is not its primary effect. Its source is in part the pleasure of the virtuoso who overcomes technical difficulties that baffle his cleverness" (349). For Boas, works of art were principally indicators of skill, valued as such in almost every culture. He added, "Among primitive peoples . . . goodness and beauty are the same" (356). Whatever people make, they tend to ornament. He spent a good deal of *Primitive Art* trying to show that most of the aesthetic preferences of tribal peoples can be traced to the appreciation of patience, careful execution, and technical perfection. In his view, this thirst for virtuosity explains our preferences for regular form, symmetry, perfectly repeated decorative motifs, smooth surfaces, and uniform color fields. Art historian Ernst Gombrich made powerful arguments along similar lines in his book *The Sense of Order*, which viewed the decorative arts as displays of skill that play upon our perceptual biases.

Beauty conveys truth, but not the way we thought. Aesthetic significance does not deliver truth about the human condition in general: it delivers truth about the condition of a particular human, the artist. The aesthetic features of art make sense mainly as displays of the artist's skill and creativity, not as vehicles of transcendental enlightenment, religious inspiration, social commentary, psychoanalytic revelation, or political revolution. Plato and Hegel derogated art for failing to deliver the same sort of truth that they thought philosophy could produce. They misunderstood the point of art. It is unfair to expect a medium that evolved to display biological fitness to be well adapted for communicating abstract philosophical truths.

This fitness indicator theory helps us to understand why "art" is an honorific term that connotes superiority, exclusiveness, and high achievement. When mathematicians talk about the "art" of theorem-proving, they are recognizing that good theorems are often beautiful theorems, and beautiful theorems are often the products of minds with high fitness. It is a claim for the social and sexual status of their favorite display medium. Likewise for the "arts" of warfare, chess, football, cooking, gardening, teaching, and sex itself. In each case, art implies that application of skill beyond the pragmatically

necessary. Anyone who wishes to imply superiority in their particular line of work is apt to style themselves an artist. The imperatives of fitness display allow us to understand the passion with which people debate whether something is or is not an art. A claim that one's work is art is a claim for sexual and social status.

Cyrano's Panache

The verbal fireworks of male courtship are personified in the title character from Edmond Rostand's 1897 play *Cyrano de Bergerac*. Cyrano had a big nose, a big sword, and a big vocabulary. Much of the play concerns Cyrano's mission to convince his bookish, beautiful cousin Roxane to commit herself to the inarticulate but handsome baron Christian de Neuvillette. In preparing a translation of *Cyrano* for the New York stage in 1971, novelist Anthony Burgess noted of Roxane that "she loves Christian, and yet she rebuffs him because he cannot woo her in witty and poetic language. This must seem very improbable in an age that finds a virtue in sincere inarticulacy, and I was told to find an excuse for this near-pathological dismissal of a good wordless soldier whose beauty, on her own admission, fills Roxane's heart with ravishment" (Rostand ii). Our modern verbal displays remain a pale imitation of classic French wit—Cyrano's quatrains have given way to our anodyne psychobabble, self-help platitudes, and management buzzwords. We can be linguistically lazy now because we are surrounded by professional wordsmiths who entertain our sexual partners on our behalf: television, movie, comedy, and novel writers. We may never know whether our Pleistocene ancestors favored French-style wit, English-style irony, or German-style engineering. But they apparently favored some verbal fluency beyond the demands of flint-knapping and berry-picking.

The Cyrano story really illustrates verbal display by five males. First, the historical Cyrano de Bergerac: large-nosed seventeenth-century political satirist, wounded veteran, dramatist, free-thinking materialist, ridiculer of religious authority, and master of baroque prose and bold metaphors, whose *A Voyage to the Moon* of 1754 was arguably the first science-fiction novel. Second, the nineteenth-century playwright Edmond Rostand, whose dazzling versification throughout five acts of rhymed alexandrines secured his literary status. Third, Rostand's fictional character Cyrano, whose astonishing poetic fluency won Roxane's heart. Fourth, the play's translator, Anthony Burgess. Perhaps their lovers were equally fluent in private conversation, but we do not know, for they were not so motivated to broadcast their verbal genius to such wide audiences. The fifth male displayer is, of course, me, since I'm

writing about Cyrano here. These endless chains of male verbal display constitute most of human literature and science.

Facing death at the end of the play, Cyrano's final words emphasized the similarities between ornamental bird plumage in nature, the white feather in his hat, and the style of his language:

> There is one thing goes with me when tonight
> I enter my last lodging, sweeping the bright
> Stars from the blue threshold of my salute.
> A thing unstained, unsullied by the brute
> Broken nails of the world, by death, by doom
> Unfingered—See it there, a white plume
> Over the battle—A diamond in the ash
> Of the ultimate combustion—My panache.

(ROSTAND 174–75)

His reputation for wit and valor will outlast his death—as would his genes for those virtues, if Roxane had not secluded herself in that convent. His death-speech is a rather moving evolutionary metaphor, with the white plume of sexual selection flying high above the battleground of natural selection. This is not to suggest that Rostand of 1897 had read Darwin of 1871, only that both recognized that there is more to life than swords and noses, and more to female choice than lust for good wordless soldiers.

Poetic Handicaps

Cyrano's panache was manifest in his poetry. Literary souls sometimes praise poetry as a zone of linguistic freedom where words can swirl in dazzling flocks above the gray cityscape of pragmatic communication. A sexual selection viewpoint suggests a different interpretation. Poetry, in my view, is a system of handicaps.

Meter, rhythm, and rhyme make communication harder, not easier. They impose additional constraints on speakers. One must not only find the words to express meaning, but, to appropriate Coleridge, the right words with the right sounds in the right order and the right rhythm. These constraints make poetry more impressive than prose as a display of verbal intelligence and creativity. For example, literary scholar John Constable has noted that poetic meter is a kind of handicap in Zahavi's sense. A metric line must have a regular number of syllables. Across different poetic styles, languages, and cultures, this number is usually between six and twelve syllables. Constable

showed that even successful writers such as George Eliot have trouble composing metric poetry. His evidence shows that on average they use shorter words when writing metric poetry than when writing prose, because shorter words are easier to fit together into regular line lengths. Meter imposes a measurable cost on the writer's verbal efforts, which makes it a good verbal handicap. Only those with verbal capacity to spare can write good metric lines.

Often, poetry demands a regular rhythm of stressed and unstressed syllables. This requires selecting words not only for their meaning and syllable number, but also for their stress pattern. Meter and rhythm are usually combined to form a double handicap. In iambic pentameter, for example, each line must be of exactly ten syllables, with alternating stresses on successive syllables. Moreover, poetry in many languages is expected to rhyme. Words must be selected so the last few phonemes (sound units) match across different lines. Rap musicians develop reputations largely for the ingenuity of their rhymes, especially the rhyming of rare, multi-syllabic words. Some poetic forms such as haiku, limericks, and sonnets also have constraints for the total number of lines (three, five, and fourteen, respectively). The most highly respected poetic forms such as the sonnet are the most difficult, because they combine all four rules, creating a quadruple handicap under which the poet must labor. Some poetic handicaps such as meter, rhythm, and rhyme are fairly universal across cultures, suggesting that our minds may have evolved some verbal adaptations for dealing with them. Specific forms of poetry are, of course, cultural inventions.

Good prose enhances the speaker's status. Good poetry is an even better indicator of verbal intelligence. This is why Cyrano was so impressive: we are clever enough to comprehend his wit, while acknowledging that we would have extraordinary difficulty matching it. If I had written this book in sonnets at Shakespeare's standard, you would not have understood human mental evolution any better, but you might have a higher opinion of my verbal ability.

In most cultures a substantial proportion of poetry is love poetry, closely associated with courtship effort. Poetry often overlaps with musical display, as in folk music with rhyming lyrics. Sung poetry demands the additional skill of holding a melody while maintaining meter, rhythm, rhyme, and line-number norms. In modern societies, poets who publish their work are little read, but poets who sing their work, backed up by guitars and sequencers, sell millions of albums and attract thousands of groupies. In considering whether ancestral poetry would have been considered sexually attractive, do not visualize Wallace Stevens, my favorite modernist poet, a drab New Haven insurance executive who wrote in the evenings after work. Instead,

visualize Frank Sinatra, Jim Morrison, Courtney Love, or whichever song-writer/vocalist happens to be fashionable when you are reading this.

Our capacity for poetic language probably evolved after our capacity for prose. If the ability to produce good love poetry had been strongly selected during courtship ever since modern *Homo sapiens* originated 100,000 years ago, we would be much better at it. We would speak effortlessly in rhyming couplets, and find that quatrains of trochaic septameter take only a little effort. But we have not yet evolved the ability to handle multiple poetic handicaps very easily. Indeed, some among us may still believe that Keats rhymes with Yeats. Of course, if we had all evolved to the standard of Cyrano, then sexual selection would raise its standard again, perhaps favoring only those whose trochaic-septameter quatrains were composed of alliterative word-triplets. The exact nature and number of poetic handicaps do not matter. What counts is that they function as proper biological handicaps, discriminating between those whose verbal displays can follow the rules, and those without sufficient verbal intelligence to play these bizarre word-games. At the moment, the meter, rhythm, and rhyme handicaps are sufficient hurdles that few of us can clear them.

Clearly, this analysis of poetry as a system of sexually selected handicaps aims to explain why poetry originated; it does not claim to account for poetry's content or contemporary human significance. Good poetry offers emotionally moving insights into the human condition, the natural world, and the transience of life. These psychologically appealing aspects may make it a more effective courtship display than if it droned on about nothing more than sex. (Indeed, because courtship is a way to arouse sexual interest in someone who is not already interested, courtship displays that make explicit reference to sex may be particularly unappealing.) Because humans are fascinated by many things, courtship displays can successfully appeal to human interests by talking about almost anything under the sun. This Darwinian account of poetry does not drain poetry of its meaning—on the contrary, it shows why its meaning is free to range over the entirety of human experience.

14

Does Beauty Build Adapted Minds?

TOWARD AN EVOLUTIONARY THEORY OF
AESTHETICS, FICTION, AND THE ARTS

JOHN TOOBY AND LEDA COSMIDES

Opening Mysteries: The Anomaly of the Arts in the Evolutionary Landscape

According to the logical framework derived from modern Darwinism, all features of a species' cognitive or neural architecture are either adaptations, byproducts, or genetic noise. Adaptations are present because they were selected to perform a function that ultimately contributed to genetic propagation. For example, the system that identifies snake-ness in the visual array, coupled to fear-releasing circuits, is an adaptation that lowered deaths due to venomous snake bites among our ancestors. Byproducts perform no function of their own, but are present because they are causally coupled to traits that were selected for: avoiding harmless snakes is a byproduct of adaptations for avoiding venomous ones. Noise was injected by the stochastic components of evolution—for example, genetic drift appears responsible for the capricious fact that a small percentage of humans sneeze as a reflex when exposed to sunlight.

John Tooby and Leda Cosmides. "Does Beauty Build Adapted Minds? Toward an Evolutionary Theory of Aesthetics, Fiction, and the Arts." Originally published in *SubStance* 30, nos. 1 and 2 (2001): 6–27. © by the Board of Regents of the University of Wisconsin System. Reproduced by the permission of the University of Wisconsin Press.

Like modern physics, Darwinian theory is subtle, strange, inhuman, and enormously successful as an explanatory system. It has successfully explained why a great many of the functional design features of species, including humans, have the forms that they do. As evolutionary psychologists have found, evolutionary theory also accounts precisely for a large number of features of the human mind, and their expressions in human behavior. Evolutionary theories of function elegantly predict and account for the existence and detailed structure of cooperation, aggression, sexual desire, love for one's children and family, the dimensions of conflict or tension inside the family, sexual jealousy, the avoidance of incest, the formation of in-groups, and hundreds of other major phenomena that organize human life in all cultures (see, e.g., Barkow, Cosmides and Tooby).

Nevertheless, unlike the case for nonhumans, there are large realms of human behavior and experience that have resisted any easy or straightforward explanation in Darwinian terms. Indeed, leaving aside history, philosophy, and linguistics (whose findings and objects of study are generally consistent with Darwinism), almost all of the phenomena that are central to the humanities are puzzling anomalies from an evolutionary perspective. Chief among these are the human attraction to fictional experience (in all media and genres) and other products of the imagination. (We will be using the word *fiction* in its broadest sense, to refer to any representation intended to be understood as nonveridical, whether story, drama, film, painting, sculpture, and so on.) If these phenomena did not exist, no evolutionary psychologist, at our present level of understanding, would have felt compelled to look for or predict them. If they were rare or culturally limited phenomena, their existence would not pose a theoretical problem. (No one feels the need to develop an evolutionary explanation for rocketry or Esperanto.) However, aesthetically driven activities are not marginal phenomena or elite behavior without significance in ordinary life. Humans in all cultures spend a significant amount of time engaged in activities such as listening to or telling fictional stories (Brown, *Human Universals*; Hernadi; Scalise Sugiyama, "Narrative Theory," "On the Origins"), participating in various forms of imaginative pretense (Leslie, "Pretense"), thinking about imaginary worlds, experiencing the imaginary creations of others, and creating public representations designed to communicate fictional experiences to others. First, *involvement in fictional, imagined worlds appears to be a cross-culturally universal, species-typical phenomenon.*

Second, *involvement in the imaginative arts appears to be an intrinsically rewarding activity, without apparent utilitarian payoff.* With rare exceptions (e.g., Hamlet entrapping his uncle, or Maoist revolutionary opera), this activity is not easily explained as instrumental in any ordinary sense. Third, although fiction seems to be processed as surrogate experience, some psychological subsystems reliably react to it as if it were real, while others reliably do not. In particular, *fictional worlds engage emotion systems while disengaging action systems* (just as dreams do). An absorbing series of fictional events will draw out of our mental mechanisms a rich array of emotional responses—the same responses that would be appropriate to those same events and persons if they were real. We care about the people involved, we identify our welfare with one or more of the characters, we may be afraid, or disgusted, or shattered, as if (in the emotional channel) those events were happening to us. We feel richly but act not at all, indeed losing awareness of our bodies and nonrelevant senses and activities in proportion to how absorbing the fictional input is. A real lion actually lunging at us would evoke terror *and* flight—the emotion program and behavior are linked (Cosmides and Tooby, "Evolutionary Psychology"). But while a cinematic version of the lion may evoke terror, the flight behavior that terror is ordinarily designed to produce is disengaged: We do not run from the theater. The experience may give us new weightings on the fearfulness of lions or the dark, but these weightings express themselves in real behavior elicited by real situations subsequently, not in behavior directed toward the fictional event. This selectivity in how our mental subsystems respond suggests functional design.

Fourth, it appears as if *humans have evolved specialized cognitive machinery that allows us to enter and participate in imagined worlds*, including pretense (Leslie, "Pretense") and fiction (Cosmides and Tooby, "Consider"). Pretend play is now recognized as so fundamental an expression of the human cognitive architecture that its absence in a toddler is seen as diagnostic of a neurological impairment (that is, autism [Frith, "Autism," *Autism*]). The machinery that permits pretense can be selectively impaired while other faculties are spared, suggesting that it is the product of a specialized subsystem, and not simply a byproduct of general intelligence. There are children with autism who cannot pretend despite having a normal IQ, and correspondingly children who can pretend despite being saddled with very severe cognitive dysfunctions (Baron-Cohen, *Mindblindness*). Behaviorally, pretend play appears in all normally developing children in all cultures around eighteen months of age, about the time that infants become maturationally equipped to engage in sophisticated social activity that acknowledges the existence of other minds. The cognitive machinery underlying pretend play includes specialized forms of representation (metarepresentations), which decouple the pretense from one's store of world knowledge (Leslie, "Pretense"). These decoupling mechanisms appear to be adaptations, whose function is to protect our knowledge stores from being corrupted by the flood of false information ("fictions") that the ability to engage in imaginative activities allows (Cosmides and Tooby, "Consider"; Leslie, "Pretense").

In short, pretend play is a reliably developing feature of the human species, which appears to be complexly designed to perform certain activities. Moreover, the existence of adaptations designed to prevent the data corruption problems that would otherwise be caused by fictional information implies that there was some benefit to being able to entertain fictions. But what is the adaptively functional end-product that this machinery evolved to serve? How did this contribute, under ancestral conditions, to gene propagation? This is a major mystery. Even more intriguingly, pretend play parallels adult involvement in fictional worlds in a number of key respects. It is intrinsically rewarding, non-instrumental, and it centrally involves the mental representation of states of affairs known to be false to the individual carrying out the mental activity. Moreover, it appears to involve the same cognitive design features that protect children in pretend play from confusing fiction and reality (Cosmides and Tooby, "Consider"). Indeed, from a cognitive or computational point of view, they seem to be fundamentally the same activity, and so discovering the function of pretend play may elucidate the function of participating in imagined worlds for adults.

The anomaly posed to evolutionary psychologists by the arts (and pretend play) can now be stated. Our species-typical neural architecture is equipped with motivational and cognitive programs that appear to be specially designed to input fictional experiences and engage in other artistic activities (Cosmides and Tooby, "Consider"). Yet the evolved function or selective benefits that would favor the evolution of such adaptations remains obscure. Natural selection is relentlessly utilitarian according to evolution's bizarre and narrow standards of utility, and does not construct complex neural machinery unless that machinery promoted, among our ancestors, the genetic propagation of the traits involved. So, why are these neurocognitive programs built in to human nature?

From an evolutionary perspective, acceptable answers are limited to three:

1. The human engagement in fictional experience, pretend play, and other aesthetic activities is the functional product of adaptations that are designed to produce this engagement. Therefore, engagement in fictional experience and other aesthetic activities must have contributed to the survival and reproduction of our hunter-gatherer ancestors, even though we do not presently know how.

2. The human engagement in fictional experience, pretend play, and other aesthetic activities is an accidental and functionless byproduct—a susceptibility—of adaptations that evolved to serve functions that have nothing to do with the arts *per se*. According to this hypothesis, engagement in

the arts is like catching a disease or becoming addicted to drugs. It is not something that humans were designed to do, but something they are vulnerable to. Or as W. H. Auden put it, "Poetry makes nothing happen."

3. The psychological basis of these activities is the result of genes that spread by chance during evolution. (We consider the cognitive and motivational features related to aesthetic experience and pretense to be too well-organized and reliably developing to be explicable as chance fixation of neutral alleles, and will not consider this hypothesis further.)

If the first hypothesis is true, it is important to identify the function or functions (i.e., the adaptive consequences) of these adaptations, which caused them to be selected for among our ancestors. Discovering the function of an adaptation is worthwhile because it usually opens the door to a large series of additional discoveries about the structure and organization of the phenomena in question. In this paper, we would like to outline some hypotheses about the possible functions and significance of immersing oneself in aesthetic activities, fictional worlds, and pretend play (Cosmides and Tooby, "Consider"; Steen and Owens).

For many years, we considered the second hypothesis—the byproduct hypothesis—to be the correct one. Until a decade ago we routinely used various artistic behaviors unproblematically as examples of evolutionary byproducts in our lectures. We still consider the byproduct hypothesis to be the default hypothesis, with a great body of logic and evidence in favor of it. Steven Pinker has recently argued this position with great cogency, suggesting that many of the arts are technologies that "pick the locks" that safeguard the brain's pleasure circuits (*How*). Pinker sketched out how many well-known features of the visual arts, music, and literature take advantage of design features of the mind that were targets of selection not because they caused enjoyment of the arts, but because they solved other adaptive problems such as interpreting visual arrays, understanding language, or negotiating the social world.

Although this argument is a powerful one, and we remain persuaded that it successfully explains many features of the arts, we nevertheless have gradually come to the conclusion that there is much that it leaves unexplained. We think that the human mind is permeated by an additional layer of adaptations that were selected to involve humans in aesthetic experiences and imagined worlds, even though these activities superficially appear to be nonfunctional and even extravagantly non-utilitarian (Cosmides and Tooby, "Consider"; Steen and Owens; see also Cosmides and Tooby, "Evolutionary Psychology"; Pinker, *How* 541–43). This seemingly unlikely possibility is the hypothesis we would like to explore in the remainder of this article.

To give one reason why we think the byproduct explanation is inadequate, consider one straightforward expectation about the human mind derived from evolutionary analyses. Successful action depends on access to accurate information about the world. Accordingly, organisms should have an appetite for obtaining accurate information, and the distinction between true information and false information should be important in determining whether the information is absorbed or disregarded. This "appetite for the true" model spectacularly fails to predict large components of the human appetite for information. When given a choice, most individuals prefer to read novels over textbooks, and prefer films depicting fictional events over documentaries. That is, they remain intensely interested in communications that are explicitly marked as false.

The familiarity of this phenomenon hides its fundamental strangeness. Novels and films are not an accidental side effect of attempts to manufacture accurate informational packages. They are not near misses. And the most basic design feature you would expect to be built into a reward system for inputting information—an appreciation for its truth—seems to be completely switched off in a wide variety of circumstances.

Decoupling as the Doorway into a Wider Aesthetic Universe

Humans are radically different from other species in the degree to which we use contingently true information—information that allows the regulation of improvised behavior that is successfully tailored to local conditions (Cosmides and Tooby, "Consider"). When hominids evolved or elaborated cognitive adaptations that could use information based on relationships that were only "true" temporarily, locally, or contingently rather than stably and across the species range, this opened up a new and vastly enlarged universe of potentially representable information for use. This new universe makes possible the identification of an immensely more varied set of advantageous behaviors than other species employ, giving human life its distinctive complexity, variety, and relative success.

This advance, however, was purchased at a high price. The exploitation of this exploding universe of contingent information created a vastly expanded risk of possible misapplications, in which information that may be usefully descriptive in a narrow arena of conditions is false, misleading, or harmful outside the scope of those conditions. Because information is being input into the human mind that is only applicable temporarily or locally, the success of this new hominid computational strategy depends on continually monitoring and re-establishing the boundaries within which each set of representations remains useful. In short, one price that humans paid for this new knowledge was surrendering the naïve realism that is the birthright of other species—species untroubled by the need to piece together belief systems and struggle with the question of how much to trust them and when to abandon them. (On this view, both *Alice's Adventures in Wonderland* and *Hamlet* are works of literature focused on an evolutionarily ancient but quintessentially human problem, the struggle for coherence and sanity amid radical uncertainty.)

As a result, humans live with and within large new libraries of representations that are not simply stored as true information. These are the new worlds of the might-be-true, the true-over-there, the once-was-true, the what-others-believe-is-true, the true-only-if-I-did-that, the not-true-here, the what-they-want-me-to-believe-is-true, the will-someday-be-true, the certainly-is-not-true, the what-he-told-me, the seems-true-on-the-basis-of-these claims, and on and on. Managing these new types of information adaptively required the evolution of a large set of specialized cognitive adaptations. For example, it involved the evolution of new information formats, based on what we call scope syntax, that tag and track the boundaries within which a given set of representations can safely be used for inference or action (for a fuller discussion of these issues, see Cosmides and Tooby, "Consider").

Elements of this syntax include operations that *decouple* or cognitively quarantine sets of representations from each other, so that they do not interact with each other promiscuously—that is, without respect to the scope boundaries within which they are applicable (Leslie, "Pretense"). This decoupling allows us to solve problems by supposing and by reasoning counterfactually: evolved inference engines can be vigorously applied to propositions, and possible outcomes evaluated, without the risk that either the counterfactual premise or any conditional downstream inferences will be stored in our encyclopedia of world knowledge as unqualifiedly true. It would be catastrophic if, for example, the human cognitive architecture did not distinguish what actually was true from what others believe to be true, or from what would be true under certain conditions. Viewed against this background, fiction as "false" information no longer seems quite so strange or different from the other types of information humans are designed to actively represent and maintain. Indeed, if other facts are scope-bound to particular situations, contingencies, times, places, plans, conditional futures, or minds, fiction is only a limiting case, in which the real-world scope wherein the bundled set of representations reigns as true has shrunk to nothing. You might think that a well-engineered mind should therefore discard or disregard such representations, but this would only be a good choice if inputting such representations had no positive organizational impact on the brain or mind. We think that fiction consists of representations in a special format, *the narrative*, that are attended to, valued, preserved, and transmitted because the mind detects that such bundles of representations have a powerfully organizing effect on our neurocognitive adaptations, even though the representations are not literally true (for a fuller discussion of the evolutionary function of fiction, and possible cognitive adaptations for fiction, see Cosmides and Tooby, "Consider").

The presence in the human mind of scope syntax and the machinery for decoupling representations allows humans to greatly widen the contexts within which adaptation-organizing experiences can occur. Through scope syntax and other design features, activities that organize an adaptation can be liberated from the constraints of having to encounter and practice the actual task, which may be very limited, dangerous, or may simply not contain the informative feedback or revelatory data necessary by the time the organism needs the adaptation to be functioning effectively. For example, as Steen and Owens point out, chase play may develop flight skills that could not be advantageously developed purely in the context of actual instances of predator escape.

Humans, being social and communicative organisms equipped with decoupling, are no longer limited by the slow and unreliable flow of actual experience. Instead, we can immerse ourselves in the comparatively rapid flow of vicarious, orchestrated, imagined, or fictional experience. A hunter-gatherer band might contain scores or even hundreds of lifetimes' worth of experience whose summary can be tapped into if it can be communicated. So, vicarious experience of especially interesting events, communicated from others, should be aesthetically rewarding. Moreover, these do not need to be accurate portrayals of real events to have valuable organizing effects.

For example, the precultural animals ancestral to humans evolved emotion programs that were activated by the detection of certain images or situation-cues. The organism only had to know how to respond when it *actually* faced the situation. What should be done when a snake is detected? What is it like when a sibling is attacked? Because humans are descended from animals whose psychological architectures were organized in this way, a great deal of information about value-weightings is locked up as emotional responses, waiting to be triggered by exposure to the correct constellation of situation-cues (Cosmides and Tooby, "Evolutionary Psychology"). Fictional information input as a form of simulated or imagined experience presents various constellations of situation-cues, unlocking these responses, and making this value information available to systems that produce foresight, planning, and empathy. With fiction unleashing our reactions to potential lives and realities, we feel more richly and adaptively about what we have not actually experienced. This allows us not only to understand others' choices and inner lives better, but to feel our way more foresightfully to adaptively better choices ourselves (Cosmides and Tooby, "Consider," "Evolutionary Psychology"). How would I feel if I acted in a cowardly fashion, and my community knew it (*Lord Jim*)? How would I feel if my sister died, and I were responsible?

Indeed, we evolved not so long ago from organisms whose sole source of (non-innate) information was the individual's own experience. Therefore, even now our richest systems for information extraction and learning are designed to operate on our own experience. It seems therefore inevitable, now that we can receive information through communication from others, that we should still process it more deeply when we receive it in a form that resembles individual experience, even though there is no extrinsic reason why communicated information needs to be formatted in such a way. That is, we extract more information from inputs structured in such a form. What form is this? People prefer to receive information in the form of stories. Textbooks, which are full of true information, but which typically lack a narrative structure, are almost never read for pleasure. We prefer accounts to have one or more persons from whose perspective we can vicariously experience the unfolding receipt of information, expressed in terms of temporally sequenced events (as experience actually comes to us), with an agent's actions causing and caused by events (as we experience ourselves), in pursuit of intelligible purposes. Scalise Sugiyama ("Narrative Theory") and Abbott have made the interesting proposal that the narrative form is itself a cognitive adaptation. Whether or not this is true, we think that stories are told in a way that mimics the format in which experienced events are mentally represented and stored in memory, in order to make them acceptable to the machinery the mind uses to extract meaning from experience. We are designed, for example, to extract new information from episodic memory, even though it lacks full sensory detail, and our preferences for narrative inputs may owe a great deal to our ability to process this schematically condensed simulacrum of experience.

According to this account, the kind of truth conveyed in art is not propositional or referential in the ordinary sense. It consists of the increased mental organization that our minds extract from experiencing art, which is why this form of truth has seemed so elusive, so difficult to articulate or explicitly define. This organization consists mostly of what might, for want of a better word, be called skills: skills of understanding and skills of valuing, skills of feeling and skills of perceiving, skills of knowing and skills of moving. Picasso's paradox—that "art is a lie which makes us see the truth"—turns out not to be so paradoxical after all. To call art "lies" simply acknowledges that a simulacrum of individual experience has been manufactured largely out of false propositions or orchestrated appearances. Such falsities can convey truth because they are not processed as propositions with truth values, but as an experience whose false particulars are (in effect) thrown away. The truth inheres in what the experience builds in us.

In sum, we think that art is universal because each human was designed by evolution to be an artist, driving her own mental development according to evolved aesthetic principles. From infancy, self-orchestrated experiences are the original artistic medium, and the self is the original and primary audience. Although others cannot experience the great majority of our self-generated aesthetic experiences, from running and jumping to imagined scenarios, there are some avenues of expression that can be experienced by the creator and others. Sounds generated for aesthetic purposes—whether in the form of music, mimicry, or words—are a major example. Once others can experience an individual's aesthetic productions as members of the audience beyond the self, art becomes social, and motives for its production may become mixed. Moreover, the ability to remember (or to record on some kind of medium) how to produce an experience allows the experience to be re-performed, and hence successively elaborated and perfected. Music and stories are two forms of art that require no technology, in which the audience can include more than the creator, and for which memory (and skill-acquisition) are sufficient to record enough of the experience to allow its repeated performance and hence improvement. This allows a history of elaboration to develop for a work of art both for a given performer, and over successive performers (Sperber). Art forms that meet these requirements become cultural as well as individual, creating the cultural category "art." The invention of additional recording media (paint, clay, film, etc.) has allowed a steady expansion of audience-accessible art forms over human history. We suggest, however, that the socially recognized arts are only a small part of the realm of human aesthetics, even though our ability to record performances permanently has caused the body of such audience-directed efforts to become massive, and its best exemplars overpowering. These works are compelling because their experimentally tested social elaboration has led them through long sequences of improvement. But still, they stand on a base of an evolved psychology that uses aesthetic experience throughout the lifecycle to guide our minds into becoming more fully realized.

15

The Uses of Fiction

Denis Dutton

The ability to imagine scenarios and states of affairs not present to direct consciousness must have had adaptive power in human prehistory, as it does in today's world. We imagine where we were last month, what we plan to do tomorrow, or next year. We imagine what it might be like to visit that restaurant, or what it would have been like had I taken that job or married that person. We try to imagine where we left that prescription or what the rough distance between Boston and New York must be. Imagination allows the weighing of indirect evidence, making chains of inference for what might have been or what might come to be. It allows for intellectual simulations and forecasting, the working out of solutions to problems without high-cost experimentation in actual practice. Pentagon strategists engage in elaborate war games, in which imagined plans of action are proposed and outcomes induced. In this respect, war-gamers are in a line that extends back to the Pleistocene hunter-gatherer band that imagined what food was likely available in the next valley and weighed heading in that direction against

From *The Art Instinct: Beauty, Pleasure, and Human Evolution* by Denis Dutton. Published by Bloomsbury Press, New York, 2009. Copyright © 2009 by Denis Dutton.

envisioned risks and opportunity costs of doing so. This capacity for strategic, prudential, conditional thinking gave to such bands a vast adaptive advantage over groups that could not plan with imaginative detail.

By allowing us to confront the world not just as naïve realists who respond directly to immediate threats or opportunities (the general condition of other animals) but as imaginative supposition-makers and thought-experimenters, human beings attained one of their greatest evolved cognitive assets. For Tooby and Cosmides, "it appears as if *humans have evolved specialized cognitive machinery that allows us to enter and participate in imagined worlds*" ("Does" 9; see also Cosmides and Tooby, "Evolutionary Psychology"). They call this capacity *decoupled* cognition. Like Aristotle, who in his *Poetics* both refers to the universality of children's imitative play and to the universality of the pleasure felt in seeing imitations, Tooby and Cosmides lay stress on the genesis of this mental process in children's pretend play. Pretend play, exhibited as a part of normal child development, requires breathtakingly subtle mechanisms to decouple the play world from the real world, and from other play worlds. It is not that the true/false distinction is abandoned in play, but that the play world is bracketed off, and truth becomes truth-for-the-play-world.

The intricate imaginative capacities of small children should impress us not simply in terms of creative range—the ability to have pretend games around any domain of the child's real experience. At least as remarkable is the way children can invoke consistent rules and limitations within freely invented yet coherent fantasy worlds. What's more, children are also able with remarkable accuracy to keep fantasy worlds separate from one another, and to quarantine multiple imaginary worlds from the actual life of the real world. They can "tag" the play-world facts for separate play-worlds, and distinguish them from real-world activities. In this way, the natural appearance of pretend play is accompanied with an equally natural ability to distinguish fantasy from reality. If human beings did not possess this capacity, which develops spontaneously in very young children, the mind's ability to process information about reality would be systematically undercut and confused by the workings of imaginative fantasy.

In this respect, children's pretend play is of a piece with another realm of spontaneous pleasure, the adult experience of fiction. In Tooby and Cosmides's account, fictions are made up out of "sets of propositions" that are all tied together but are walled off from what the child knows to be true of the real world and prevented from migrating into factual knowledge and corrupting it. Thus, a child knows what is true or false of the real world of her bedroom or her school and knows what is true or false of the fictional world of *Goldilocks and the Three Bears*. The child who has independently learned about bears is unlikely to generalize from *Goldilocks* to, for instance, the notion that real bears have furniture in their caves and cook porridge.

The same capacity for pretend play we instinctively developed as children is exercised every time we pick up a novel or sit down to watch a television drama: we then enter into an imaginative set of coherent, internally related ideas, including ontologies and rules of inference, that make their own world. The imagined world may be as gritty as that portrayed by a Scorsese movie, as remote as the wine-dark seas of Odysseus, or as comically surreal as a Road Runner cartoon. But overt resemblance to the world of mundane experience is not the issue: many compelling examples of that most adult of art forms, grand opera, present scenes and events that are no more "real" than a tea party with teddy bears.

Pretend play predictably occurs among children of all cultures at around eighteen months to two years—about the time that they begin to talk and engage socially. This fact, along with the high sophistication of decoupling mechanisms that isolate real from pretend worlds, stands as evidence that pretend play and fiction-making are isolable, evolved adaptations, forms of specialized intellectual machinery.

The evolutionary basis for these mechanisms is also supported by another fact: the universal human fascination with imagined fictions is not what we should predict as a properly evolved function for the human mind. Suppose for a moment that human beings only and exclusively took pleasure in what they took to be true narratives, factual reports that describe the real world. Evolutionary theory would then have no difficulty in attributing adaptive utility to the pleasure: the human love of the true, we would say, had survival value in the Pleistocene. We might further argue that just as early *Homo sapiens* needed to hew sharp adzes and know the ways of game animals, so they needed to employ language accurately to describe themselves and their environment and to communicate facts to each other. If humans loved *only* true stories, there would be no philosophical "problem of fiction," because there would be no intentionally constructed fiction in human life: the only alternatives to universally desired truth would be unintentional mistakes or intentional lies. Such Pleistocene Gradgrinds, as evolution might have developed them, would have been about as eager to waste their narrative efforts creating fables and fictions as they would have been to waste their manual labor in producing dull adzes. If this strict devotion to the truth, and with it a revulsion for fantasy and fiction, had developed and had shown clear adaptive value in the Pleistocene, making the Gradgrinds our direct ancestors, then we would not be the people we are today. We could be expected to react to known-to-be-untrue stories and made-up fantasy much as we react to uselessly dull knives or, worse, the smell of rotting meat.

Clearly, such Gradgrinds could not possibly have been our most genetically contributing ancestors, given the present make-up of the human personality. Human beings across the globe expend staggering amounts of time and resources on creating and experiencing fantasies and fictions. The human fascination with fiction is so intense that it can amount to a virtual addiction. A government study in Britain—likely typical of the industrialized world—showed not long ago that if you add together annual attendances in plays and cinema with hours watching television drama, the average Briton spends roughly 6 percent of all waking life attending to fictional dramatic performances (Nettle, "What Happens" 56). And that figure does not even include books and magazines: further vast numbers of hours spent reading short stories, bodice-rippers, mysteries, and thrillers, as well as so-called serious fiction, old and new. Stories told, read, and dramatically or poetically performed are independently invented in all known cultures, literate or not, having advanced technologies or not. Where writing arrives, it is used to record fictions. When printing shows up, it is used to reproduce fictions more efficiently. Wherever television appears in the world, soap operas soon show up on the daily schedules. The love of fiction—a fiction instinct—is as universal as hierarchies, marriage, jokes, religion, sweet, fat, and the incest taboo.

For evolutionary aesthetics, the problem is to reverse-engineer the improbable but manifestly human urge to both make fictions and enjoy them. Any solution to this problem must take into account Steven Pinker's caution: it is easy but empty to observe that fiction, for instance, helps us to cope with the world, increases our capacity for cooperation, or comforts the sick, unless we can plausibly hypothesize that such coping or cooperation or comfort evolved as adaptive functions, conferring survival advantages in the ancestral environment (*How*). A thorough-going Darwinism makes a specific demand: nothing can be proposed as an adaptive function of fiction unless it explains how the human appetite for fictional narratives acted to increase, however marginally, the chances of our Pleistocene forebears surviving and procreating.

Keeping this requirement in mind, here are three interconnected kinds of adaptive advantage that might explain the pervasiveness of fictions in life, now and as far back as history can record. "Fictions" includes oral story traditions and pre-literate mythologies in the ancestral environment and, by implication after the invention of writing, such enhancements, extensions and intensifications as novels, plays, operas, movies, and video games:

1. Stories provide low-cost, low-risk surrogate experience. They satisfy a need to experiment with answers to "what if?" questions that focus on the problems, threats, and opportunities life might have thrown before our

ancestors, or might throw before us, both as individuals and as collectives. Fictions are preparations for life and its surprises.

2. Stories—whether overtly fictional, mythological, or representing real events—can be richly instructive sources of factual (or putatively factual) information. The didactic purpose of storytelling is diminished in literate cultures, but by providing a vivid and memorable way of communicating information, it likely had actual survival benefits in the Pleistocene.

3. Stories encourage us to explore the points-of-view, beliefs, motivations, and values of other human minds, inculcating potentially adaptive interpersonal and social capacities. They extend mind-reading capabilities that begin in infancy and come into full flower in adult sociality. Stories provide regulation for social behavior.

Although he generally regards the arts as by-products of adaptations, Pinker nevertheless grants that in all likelihood fictional storytelling directly evolved as an adaptation. The mind uses fiction to explore and solve life problems in the imagination: "The cliché that life imitates art is true because the function of some kinds of art is for life to imitate it." Pinker's most striking metaphor for this function of storytelling is drawn from chess (*How* 542–43). We might expect that the countless potential situations and possibilities of chess would require a reliance by chess grandmasters on strategic rules—maxims and algorithms—for play. This, however, is not the case. "Keep your king safe" or "Get your queen out early" are not much help beyond beginner's chess, and experienced players do not have available to them higher-level algorithms to rely on: chess games become too complicated too fast for mere maxims to be of much use. So instead, expert players study the published moves in countless actual games, building up a mental catalogue of the ways good players have reacted to challenges and situations. These games, studied and memorized in part or in whole in their thousands by chess grandmasters, provide the templates and heuristics that make chess victory possible in uniquely new games. "Chess experience" for a mature player does not consist only in what the player knows from games personally played, but from all the games the player has played or has studied or observed.

Although 32 pieces on 64 squares provides an enormous number of combinatorial possibilities—untold trillions of moves—they would be dwarfed in range by the contingencies served up by daily experience in a human lifetime. Here too, rules and maxims such as "A penny saved is a penny earned" or "Look before you leap" have their elementary uses, but in actual life provide about as much utility as "try to control the center of the board" has to an experienced chess player. Imaginative stories, pleasurably experienced and remembered by both maturing and adult human beings, provide a far more complex and useful set of templates and examples to guide and inspire human action. To a large extent, they are built into life experience.

Pinker's chess analogy for the adaptive value of fiction and storytelling is, I think, richly suggestive. Chess, after all, is not a game with which the human mind engages at the level of considering all possible moves for any one situation. That in fact is exactly how computers have classically treated chess: they use brute computational force to out-calculate a human opponent, including eliminating vast numbers of moves so bad a human being would never consider them. Chess has an intrinsic teleology not found in a computer's combinatorial bad-move elimination; the game has as its object that each player purposively tries to outsmart an opponent—taking into account the opponent's assessment of each situation—in order to achieve checkmate. This in any event is how the human mind engages chess, and it is how the human mind engages the strategic teleology of life, except that instead of the single checkmating purpose, in life it must engage an indefinitely large number of purposes.

The adaptive value of storytelling in the ancestral environment would lie neither in deriving and applying relatively vacuous advisories such as "The early bird gets the worm," nor in specific instructions about avoiding road-rage incidents, lest I kill my father and end up inadvertently marrying my mother. Adaptiveness derives from the capacity of the human mind to build a store of experience in terms of individual, concrete cases—not just the actual lived and self-described life experiences of an individual, but the narratives accumulated in memory that make up storytelling traditions: vivid gossip, mythologies, technical know-how, and moral fables—in general, what would have been the lore of a hunter-gatherer society. Whatever the brain processes that enable it, the human mind acquires and organizes a vast knowledge-store in terms of dramatic episodes of clan history, war stories, hunting anecdotes, near misses, tales of forbidden love, foolhardy actions with tragic outcomes, and so forth. These cases are used analogically and in terms of shared and differentiated features to make sense of new situations and to interpret past experience. (In this connection, it is worth noting that the ability to recognize and use analogies effectively remains a staple basis of contemporary intelligence tests.)

Cross-culturally today and through all of known cultural history, stories are about problems and conflict: competing human interests for power or love are prime topics, as well as natural threats to life and limb (in which case a dangerous animal may well be an animal and not a surrogate person). In this way, the most abstract characterization that can be given of stories is that they involve (1) a human will and (2) some kind of resistance to it. That is why "Mary was hungry, Mary ate dinner," narrates a sequence of events that does not yet seem like a story, while "John was starving, but the pantry was empty . . . ," does sound like the beginning of one. Obstacles—to life, wealth, ambition, love, comfort, status, or power—are one central element in the fundamental idea of a story; the other element is how a human will triumphantly overcomes—or tragically fails to overcome—obstacles. Stories are intrinsically about how the minds of real or fictional characters attempt to surmount problems, which means stories not only take their audiences into fictional settings, but they take them into the inner lives of imaginary people.

Just as there would have been a major adaptive advantage in learning from stories to deal with the threats and opportunities of the external physical world, so for an intensely social species such as *Homo sapiens* there was an advantage in the ancestral environment in honing an ability to navigate in the endlessly complex mental worlds people shared with their hunter-gatherer compatriots. The features of a stable human nature revolve around human relationships of every variety: social coalitions of kinship or tribal affinity; issues of status; reciprocal exchange; the complexities of sex and child-rearing; struggles over resources; benevolence and hostility; friendship and nepotism; conformity and independence; moral obligations, altruism, and selfishness; and so on. These themes and issues constitute the major themes and subjects of literature and its oral antecedents. Stories are universally constituted in this way because of the role storytelling can play in helping individuals and groups develop and deepen their own grasp of human social and emotional experience. The teller of a story has in the nature of the storytelling art direct access to the inner mental experience of the story's characters. This access is impossible to develop in other arts—music, dance, painting, and sculpture—to anything like the extent that it is available to oral or literary narrative.

If the storytelling traditions of the world express pleasures and capacities that evolved long before the invention of writing—a storytelling instinct—can anything be further said about innate tendencies in the *structure of stories*? This is a vexed question throughout the history of criticism. Aristotle identified plot (Greek *mythos*) as the "structure of incidents" in a story; he claimed that of all the elements of drama, plot was the most important and the hardest to achieve effectively: it was the first principle and "soul" of tragedy. Plots organize fiction, and while they can be reasonably sorted into a manageable number of types, that in itself does not explain why plots exist. A plot is a structure of actions and events. The events may be fortuitous, but the actions must be motivated, produced by the causality of human intention. Everything that Oedipus or any other fictional character does is done for a reason (known, unclear, or unknown), whether or not the results of the actions are foreseen. It is therefore normal in criticism to disparage an author for portraying a character's action as "unmotivated." In fact, a story wholly without motivation would test the very idea of a story in the first place. A play in which a man brews a cup of tea, throws it down the drain without tasting it, makes another and throws it out too, and another repeatedly on to the end, might be a Dadaist experiment, or an illustration of an obsessive disorder, but it would be better described as an anti-story than a story. A character's motivation involves the expression of will, normally toward the fulfillment of a desire, and against resistance or obstruction of some kind.

That a plot was a causal structure, an arrangement of motivations—in essence, desire against obstruction—was grasped by Aristotle, who was not interested in plot types beyond comedy and tragedy. Write a story about a character, Aristotle argued, and you face only so many logical alternatives. In tragedy, for instance, either bad things will happen to a good person (unjust and repugnant) or bad things happen to a bad person (just, but boring). Or good things happen to a bad person (unjust again). The poignancy of tragedy requires that bad things happen to a good but flawed person: though he may not have deserved his awful fate, through his arrogance Oedipus was asking for it. In the same rational spirit, Aristotle works out dramatic relations. A conflict between strangers or natural enemies is of little concern to us. What arouses interest is a hate-filled struggle between people who ought to love each other—the mother who murders her children to punish her husband, or two brothers who fight to the death. Aristotle knew this for the drama of his age as much as soap-opera writers know it today.

Aristotle's thoughts on stories and drama endure because of his basic insight that the substance of drama is action and the emotions that animate it. The themes of Greek drama are built on such emotions as erotic and familial love, a sense of honor, civic loyalty, the inevitability of death, and so forth. The basic themes and situations of fiction are a product of fundamental, evolved interests human beings have in love, death, adventure, family, justice, and overcoming adversity. "Reproduction and survival" is the evolutionary slogan, which in fiction is translated straight into the eternal themes of love and death for tragedy, and love and marriage for comedy. Stories are populated with character-types relevant to these themes: beautiful young women, handsome strong men, courageous leaders, children needing protection, and wise old people. Add to this the threats and obstacles to the fulfillment of love and fortune, including bad luck, villains, and mere misunderstanding, and you have the makings of literature. Story plots are not, therefore, unconscious archetypes, but structures that inevitably follow, as Aristotle realized and Darwinian aesthetics can explain, from an instinctual desire to tell stories about the basic features of the human predicament.

⸙

Modern technologies for creating and enjoying fictions are a long way from sitting around a Pleistocene campfire listening to the storyteller's tale. However, these technological developments don't mean as much as we might imagine. The first ancient innovation in storytelling was probably the decision of a single creative storyteller to imitate various voices for different characters in a narrative. Later on, dramatic acting, with different people taking the parts of characters, would have been a natural development. Moving forward some thousands of years, the next great innovation for fictional entertainment would have been the invention of writing. To be able to write down a story frees it from the limits of a storyteller's memory; this vastly increased the variety and complexity of fictional narratives. It was another ten thousand years after the invention of writing that the growth of literature in our modern sense was accelerated by Gutenberg's printing press. Reading and acting were therefore the essential means of telling stories in literate cultures from before the *Epic of Gilgamesh* through the nineteenth century, and with movies on into the twenty-first. Non-literate cultures continued to rely—as they had for many thousands of years—on telling and acting as their means of enjoying fiction.

Because film and video loom so large in our experience today, it is easy to overestimate their importance. People who are enchanted by movies and electronic media but are strangers to traditional live theatre usually have little idea of the visual spectacle theatrical production could offer audiences as far back as the Renaissance, let alone after the introduction of electric lights in the nineteenth century. Visual extravaganzas did not begin with Hollywood, but very likely amazed audiences in Paleolithic caves, with firelight and cave echoes providing their special effects.

Fans today also overrate the contribution of video games to storytelling as an art form. Video games are complicated and visually arresting forms of make-believe that allow viewers to jump onto the stage and participate in the action. This is regarded by video-game enthusiasts as an earth-shaking advance. On the contrary, it is less an extension of storytelling art than a regression to its precursors. While the themes and content of video games may be complex and adult, the logic of viewer participation in the story reverts back to the child's tea party with teddy bears. Video games are first and foremost *games*, with rules for participants and the attention-maintaining quality of uncertain outcomes (Tavinor). They somewhat resemble fiction, but so do chess, poker, and Monopoly, by giving players a defined role in a rule-governed, imaginative world. Video games do not create a new kind of make-believe entertainment, or much improve on the older kinds of games and fictions, except by the addition of intense visual or virtual-reality effects. (Someday, a video-game version of *King Lear* may allow players to step into the action, even to save poor Cordelia if they play skillfully enough. Whatever the fun, this will not be an enhancement of Shakespeare.)

Some 2,500 years before the first stop-action King Kong made his way up a model of the Empire State Building, Aristotle argued that "spectacle"—the eye-popping effects of scenery, costumes, and stage machinery (the *deus ex machina*) that were already in use in the theatres of Greece—was despite its popularity the least important aspect of drama. The most important aspect was an arresting plot, which was the hardest thing for a dramatist to achieve—ahead even of interesting characters and the poetic use of language. In cinema today, it is still the story told that makes the greatest films. In this respect, little has changed since our ancestors sat around a fire listening to a storyteller. Hollywood is engaged in a special-effects arms race, where every movie has to have bigger explosions, uglier villains, more frenzied, realistic violence, ear-splitting noises, and ever-expanding battle scenes. No computer-generated crowd, according to current rules, is allowed to be smaller than the crowds in last month's blockbuster.

Such effects can give pleasure to many, but they do not replace the fundamental attraction of a rational, coherent story well told. The appeal of the story is an evolved, innate adaptation: universal, intensely pleasurable, emerging spontaneously in childhood in ways (like the deep structure of speech) that are already so logically complex even in children's play that they require an explanation in terms of innate capacities. That it remains possible today to be engaged, amused, or moved by a story told by a single speaker—next to a campfire, around a water cooler, or across a dinner table—shows us that we are with regard to fiction the same people as our prehistoric ancestors. Good stories, well told, compel our attention. So do good storytellers.

Part III

Literature, Film, and Evolution

THEORY

16

Getting It All Wrong

BIOCULTURE CRITIQUES CULTURAL CRITIQUE

BRIAN BOYD

We love stories, and we will continue to love them. But for more than 30 years, as Theory has established itself as "the new hegemony in literary studies" (to echo the title of Tony Hilfer's cogent critique), university literature departments in the English-speaking world have often done their best to stifle this thoroughly human emotion. Every year, heavy hitters in the academic literary world sum up the state of the discipline in the Modern Language Association of America's annual, *Profession*. In *Profession 2005*, Louis Menand, Harvard English professor, Pulitzer Prize–winning author, and *New Yorker* essayist, writes that university literature departments "could use some younger people who think that the grownups got it all wrong" ("Dangers" 12). He has no hunch about what they should say his generation got wrong, but he deplores the absence of a challenge to the reigning ideas in the discipline. He laments the "culture of conformity" in professors and graduate students alike. He notes with regret that the profession "is not reproducing itself so much as cloning itself" (13).

But then, curiously, he insists that what humanities departments should definitely not seek is "consilience, which is a bargain with the devil."

From the *American Scholar* 75, no. 4, Autumn 2006. Copyright © 2006 by Brian Boyd.

Consilience, in biologist E. O. Wilson's book of that name, is the idea that the sciences, the humanities, and the arts should be connected with each other, so that science (most immediately, the life sciences) can inform the humanities and the arts, and vice versa. Menand claims that he wants someone to say "You are wrong," but he rules out anyone challenging the position in which he and his generation have entrenched themselves. For they are certain there is at least one thing that just *cannot* be wrong: that the sciences, especially the life sciences, have no place in the study of the human world. Well, Professor Menand, you, and those you speak for, are wrong.

The position you represent has neither the intellectual nor the moral high ground you are so sure it occupies. Until literature departments take into account that humans are not just cultural or textual phenomena but something more complex, English and related disciplines will continue to be the laughingstock of the academic world that they have been for years because of their obscurantist dogmatism and their coddled and preening pseudo-radicalism. Until they listen to searching criticism of their doctrine, rather than dismissing it as the language of the devil, literature will continue to be betrayed in academe, and academic literary departments will continue to lose students and to isolate themselves from the intellectual advances of our time.

Not everything in human lives is culture. There is also biology. Human senses, emotions, and thought existed before language, and as a consequence of biological evolution. Though deeply inflected by language, they are not the product of language. Language, on the contrary, is a product of them: if creatures had not evolved to sense, feel, and think, none would ever have evolved to speak. In his presidential address to the 2004 MLA convention, the distinguished critic Robert Scholes offered an overview of the problems and prospects for literary studies. When another critic, Harold Fromm, challenged him in a letter in *PMLA* for ignoring biology, Scholes answered: "Yes, we were natural for eons before we were cultural . . . but so what? We are cultural now, and culture is the domain of the humanities" (Fromm and Scholes 297–98). We *were* natural? Have we ceased to be so? Why do Scholes, Menand, and the MLA see culture as ousting nature rather than as enriching it? Don't they know that over the last couple of decades biology has discovered culture—knowledge transmitted nongenetically and subject to innovation and fashion—in birds, whales, and dolphins, and among primates other than ourselves, at least in chimpanzees, orangutans, and gorillas? Do they not see that without our own species' special biology, culture could not be as important to us as it is?

Menand forcefully expresses his sense of the dramatic change in literary studies that began in 1966. A "greatest generation" of iconoclasts established two fundamental principles: first, *anti-foundationalism*, the idea that there is no secure basis for knowledge; and, second, *difference*, the idea that any universal claims or attempts to discuss universal features of human nature are instead merely the product of local standards, often serving the vested interests of the status quo, and should be critiqued, dismantled, overturned.

Menand and those he speaks for believe that the French poststructuralists, beginning with Jacques Derrida, offered an unprecedentedly profound challenge to the history of thought, a challenge since summed up as Theory. Despite admitting that the humanities are now sick, Menand nevertheless exhorts them not to retrench but "to colonize." As the critic Christopher Ricks notes, "Theory's empire [is] an empire zealously inquisitorial about every form of empire but its own" (181).

Like others of his era, Menand is sure that: (1) the "greatest generation" secured for its "disciples" (these are his terms) the intellectual and moral high ground; (2) the insights of anti-foundationalism would be accepted by all other disciplines, if only they would listen; and (3) the crusade made possible by an understanding of "difference" must continue. He ends by saying that when these positions are challenged, academics are being invited "to assist in the construction of the intellectual armature of the status quo. This is an invitation we should decline without regrets" ("Dangers" 17).

I, like others who think that humans need to be understood as more than cultural or textual entities, do not wish to entrench the status quo. But in the four decades since Menand's "greatest generation," science and technology have altered the status quo far more radically than anything literature professors have managed. By increasing the world's food output dramatically, scientists have saved hundreds of millions of people from hunger. Their labor-saving devices have freed scores of millions from domestic drudgery and allowed countless women into the paid work force. They have raised life expectancy around the world. And if knowledge is indeed power, as Michel Foucault says, then through the Internet, scientists have made possible the greatest democratization of power ever. True, there is much more to be accomplished, but the triumphalist defeatism that has been so dominant in the profession of literary studies (the *practice* has actually been less narrow) seems unlikely to help.

The "we" to and for whom Menand speaks imagine that they have the intellectual high ground because of their anti-foundationalism, the cornerstone of Theory. Anti-foundationalism is an idea uncongenial to common sense: that we have no secure foundation for knowledge. But the fact that it is uncongenial does not make it wrong. Indeed there are good reasons, however troubling, to think it right.

Yet the particular brand of anti-foundationalism Derrida offered in the late 1960s was not the challenge to the whole history of Western thought that he supposed or that literary scholars assumed it must be. Derrida insisted that an untenable "metaphysics of presence" pervaded Western thought—in less grandiose terms, a yearning for certainty. Unless meaning or knowledge could be founded in the intention or the referent of the speaker or in some other unshakable way—ultimately, perhaps, in the authority of God or gods—it would have to be endlessly referred or deferred to other terms or experiences, themselves part of an endless chain of referral or deferral.

If they had been less parochial, the literary scholars awed by Derrida's assault on the whole edifice of Western thought would have seen beyond the provincialism of this claim. They would have known that skepticism has a long heritage in Western thought and that science, the most successful branch of human knowledge, had for decades accepted anti-foundationalism, in the wake of Karl Popper's *Logik der Forschung* (*The Logic of Scientific Discovery*, 1935) and especially after Popper's 1945 move to England, where he was influential among leading scientists. They should have known that a century before Derrida, Darwin's theory of evolution by natural selection—hardly an obscure corner of Western thought—had made anti-foundationalism almost an inevitable consequence. I say "parochial" because Derrida and his disciples think only in terms of humans, of language, and of a small pantheon of French philosophers and their approved forebears, especially the linguist Ferdinand de Saussure. There was some excuse for Derrida in 1966, but there is none for the disciples in 2006, after decades of scientific work on infant and animal cognition.

Just where is the problem in the supposedly devastating insight that meaning or knowledge has to be referred or deferred to other terms or experiences, themselves part of an endless chain of referral or deferral? How could things be otherwise? This state is not only to be expected, but in an evolutionary perspective can be *explained* without apocalyptic paroxysms. In a biological view, our understanding of the world always depends on earlier and less-developed forms of understanding, on simpler modes of knowledge. Knowledge registers regularities in the environment (shape, orientation, light, color, and so on), qualities, therefore, not contained within the moment of perception but repeatedly similar enough to previous circumstances to produce similar effects. Knowledge is also registered by emerging regularities in the senses and the brains that process their input, through capacities, therefore, that have been developed by minute increments over thousands of generations.

Repetition applies not only to objects of knowledge and organs of knowledge but also to communication, through a process that biologists call ritualization. When it is typically to the advantage of one member of a species for another to understand its behavior in circumstances like courtship or threat, key patterns of action gradually become formalized, intensified, exaggerated, and contrasted sharply with other behaviors in order to maximize distinctions and minimize confusion. For the French heirs of Saussure, the principle of phonemic opposition in language (that *b* and *v*, say, generate significant distinctions in English, as in *bat* and *vat*, but not necessarily in another language, such as Spanish) is supposed to suggest an arbitrariness at the base of all thought (Tallis, *Not Saussure*). But this principle can easily be understood as merely another case of ritualized behavior that helps a given community keep its signals straight.

How could concepts or communication *not* be endlessly deferred or referred back, once we accept the fact of evolution, once we move beyond language to consider how human understanding slowly emerged? If we are evolved creatures, our brains are not guarantors of truth, citadels of reason, or shadows of the mind of God but simply organs of survival, built to cope with the immediate environment and perhaps to develop some capacity to recall and anticipate. Evolution has no foresight and no aims, least of all an aim like truth. It simply registers what suffices, what allows some organisms to last and reproduce better than others.

Because accurate information is costly, evolution must economize. A bacterium does not need to know its environment in detail, but only which nearby substances harbor opportunities and dangers. So too for humans. We do not need the long-distance visual acuity of hawks or the fine canine sense of smell or the high or low hearing ranges of bats or elephants. These extra capacities might be handy, but not at the expense of the range of senses that most often allow us to cope better.

Evolution has equipped us with fast and frugal heuristics, rough ways of knowing that suffice for our mode of life (Gigerenzer and Todd). We can expect imprecision and even systematic error in our "knowledge" if they help us to survive. We therefore have, for instance, a systematic bias toward over-interpreting objects as agents, in case that thing moving on the ground is a snake and not just the shadow of a branch, and we have a bias in memory toward recency, so that we recall more easily something encountered yesterday—and therefore likelier to recur today—than something from two decades ago.

Human minds are as they are because they evolved from earlier forms. Being ultimately biological, knowledge is likely to be imperfect, affording no firm foundation, no "originary" moment, in Derrida diction. Reality is enormously complex and vast. If we want to go beyond the familiar, beyond the immediate world of midsized objects that our senses were shaped to understand, beyond the inferences our minds naturally make, all we can do is guess, grope, or jump from whatever starting points we happen to have reached. Almost all our attempts at deeper explanations are likely to be flawed and skewed, as the hundred thousand religious explanations of the world suggest.

The best we can do is generate new hunches, test them, and reject those found wanting in the clearest, most decisive tests we can concoct. Of course we may not be predisposed to devise severe tests for ideas we have become attached to through the long cumulative processes of evolutionary, cultural, or individual trial and error. And it is not easy to discern *what* can be tested, let alone *how* it can be tested, especially in the case of "truths" we have long accepted. But in a milieu that rewards challenges to received notions, others will test our conclusions if we do not. If exacting tests contradict our predictions, we may be motivated to seek new explanations or to find flaws in the critics' tests (Popper, *Conjectures*, *Logik*, *Objective Knowledge*). The discovery of possible error can prompt us to look for less inadequate answers, even if there is no guarantee that the next round of hypotheses will fare better. Most, indeed, will again prove flawed—yet one or two may just inaugurate new routes to discovery.

Some people find that such a view of science amounts to extreme skepticism, and some scientists suppose much in science is conclusively confirmed. But Newton's laws seemed to have been confirmed endlessly, until Einstein showed that they were not universally valid, that Newtonian motion was only a special case of a much larger picture, a deeper truth. Or to take an even simpler case: the stability of species seems to be confirmed every time we see another sparrow, swan, or duck. Yet after Darwin, that turns out to be wrong. We just do not know where something that appears to be repeatedly confirmed may prove to be inadequate, even drastically so, in a larger perspective.

Every day seems to confirm the stability of the earth and the "fact" that the sun goes around the earth. Francis Bacon, the first great theorist of science, thought it unbelievably perverse to imagine that the earth revolved around the sun and rotated on its own axis: we would feel the motion, and since we don't, experience *proves* every moment that the earth is unmoving. In later decades, once the findings of Copernicus and Galileo were assimilated, it was assumed that if the universe was no longer geocentric, it was heliocentric. Then it was discovered that, no, the sun is just one star within the galaxy. Then it was discovered that the galaxy was just one among hundreds of galaxies, no, wait, millions, no, wait, billions. Now we are wondering why ninety percent of the matter we think is in the universe (and in our own galaxy) is invisible. Who knows how the quest for dark matter will turn out, and what new understanding of our planet, solar system, galaxy, and universe we will have?

A biological view of our knowledge shows both its insecurity and its dependence on older and poorer forms of knowing, while also explaining the possibility of the growth of knowledge. Derrida's challenge to the basis of knowledge seems bold, but it cannot explain advances in understanding, evident in the slow gradient from single cells to societies and the steep one from smoke signals to cell phones. Evolutionary biology offers a far deeper critique of *and* explanation of the origins and development of knowledge, as something, in Derrida's terms, endlessly deferred, yet also, as biology and history show, recurrently enlarged.

Recognizing our uncertainty helps us in our search to understand more. But those in the humanities who have become "disciples" of the "greatest generation" argue against the possibility of knowledge or truth, since meaning is forever deferred. *That* is the knowledge or truth, however self-contradictory and self-defeating, that they insist on imparting. Their commitment to undermining the possibility of knowledge, even while claiming this as bracing new knowledge, explains much of the stasis of the Theorized humanities that Menand deplores.

A biocultural perspective, by contrast, can explain how evolution has made knowledge possible, albeit imperfect, and how it has made the quest for better knowledge possible. The process offers no guarantee of truth, only the prospect of our collectively learning from one another through both cooperation (*sharing* ideas) and competition (*challenging* ideas). In that sense, an evolutionary epistemology is progressive, but far from naïvely optimistic, for every apparent advance in knowledge may turn out to be flawed in its turn, although even to discover this advances our knowledge. Derrida announced an anti-foundationalist epistemology in a spirit of revolutionary self-congratulation. He did not know, any more than his acolytes, that the sciences had already begun to accept a much less flawed anti-foundationalism, based not on parochialism and arrogance, contradiction and despair, but on humility and hope.

Just as Menand thinks the "greatest generation" secured the intellectual high ground for its "disciples" by virtue of establishing anti-foundationalism, so he supposes it secured the moral high ground by establishing what he usually calls *différance*, and occasionally situatedness or a critique of ethnocentrism. The "greatest generation," in this view, introduced the humbling recognition that any claims to universal truth, or to universal human nature, are merely local ideas, often in the service of those in power, even where they attempt to be universal and self-evident. All claims to objective truth, it pointed out, are situated in a particular social origin.

At this point the anti-foundationalism of Theory segues into Cultural Critique. The massive *Norton Anthology of Theory and Criticism* (Leitch et al.) offers, as Menand notes, a kind of "guided tour" for North American graduate programs in literature. With the characteristic provincialism and hubris of recent literary theory, it claims "Theory" as its empire, as if all theory were literary, as if the theories of gravity, evolution, and relativity were nugatory compared with the anti-foundationalist truths of the "greatest generation." And as the *Norton Anthology* itself suggests, the approach that unites them should now be called Cultural Critique, in consequence of the emphasis on difference established in Barthes, Derrida, Foucault, and after.

Menand writes: "Humanities departments have turned into the little boy who cries, 'Difference!' Humanities professors are right: there *is* difference, it always *is* more complicated, concepts *are* constructed" ("Dangers" 13). Although he complains that they often go no further, Menand nevertheless does not lament their insistence on difference, for he adds that "humanities departments do not need to retrench; they need, on the contrary, to colonize" (14). He claims that the insight of the "greatest generation" into difference has been resisted because it challenges ethnocentrism.

There are many problems in this account. First, the simple logical one. The idea that there is no universal truth runs into crippling difficulties straightaway, since it claims to be a universal truth. The idea that all is difference, merely local and situated, must apply, if true, to itself, and if this disqualifies its claim to truth, as the implication seems to be, then it contradicts itself. The only way out of the muddle of such paradoxes is by assuming that the propositions are false: then no self-contradictions arise.

A second problem arises from the attempt to define difference as uniquely human. "Culture . . . is constitutive of species identity" ("Dangers" 15), writes Menand, meaning *human* species identity. The implied corollary is that culture is always local, always marked by difference. Actually, culture by itself

is not uniquely constitutive of human identity, for many other species have culture. Every known group of wild chimpanzees has its own unique complex of cultural traditions and could not survive without them (de Waal, *Ape*). But apart from being unable to distinguish humans as he thinks it can, Menand's declaration also contradicts his claim of difference, since it presupposes a distinctive, species-typical trait, a common feature, as he thinks, uniting all humans and only humans. Yet this is exactly what the doctrine that all is difference purports to deny: that there are some features common to all human natures.

In fact, not everything in human lives is difference. Commonalities also exist, and without those commonalities between people, *culture* could not exist, since it could not pass from one person to another or one tradition to another. Cultural Critique wants to stress the "situatedness" of all that is human, but wants to define that situation only in terms of particular cultures. But why not also the unique situation of being human, with the special powers evolution has made possible in us?

The idea that all claims are situated, that they have particular origins, is surely not one many would argue with. But it does not follow that if a claim has a specific origin, this proves its error or incompleteness or nonobjectivity. If a Cretan, or Baron von Münchhausen, or Pinocchio, says "This is a cow," and it is, it is no less true than if George Washington says it. An idea may derive from observation, tradition, a dream, a guess, intoxication, or hallucination, or any combination of these. No origin guarantees the validity or invalidity of the idea.

Of course at a given time, in a given place, certain thoughts are less likely to arise than in other places. But significant new truth remains difficult to reach from any starting point. Nevertheless this does not mean it cannot be attained or approached wherever one begins. Take the example of evolution. Ideas of evolution preceded Darwin by centuries. They became more likely after the establishment of systematic taxonomy and the search for new species around the world, after Malthus's work on population, after Lyell's laying down the principles of modern geology and showing how imperceptibly small changes could accrete into major differences, and after the explanation for and systematic collection of fossils. Evolution would have been still more readily explicable had Mendel's hypothesis of particulate inheritance been known to Darwin and Wallace—yet even without that, both developed a powerful explanation for evolution that seemed to contradict the apparent observed stability of species.

Menand supposes that others resist the claim of difference because they resist the challenge to ethnocentrism. In fact challenges to ethnocentrism had been widespread long before Derrida's seminal paper of 1966, in the recoil from the horrors of Nazi racism, in the hope for a better world that led to the founding of the United Nations and to anti-colonial independence movements; in the American recognition of the role of African Americans in World War II, the Civil Rights movement, the increasing acceptance of and interest in African-American musical cultures in the 1950s and 1960s; and in anthropology, from early in the twentieth century. It is a strange fantasy to suppose that the humanities, inspired by the "greatest generation," have led the attack against ethnocentrism that was already well established in both intellectual and political culture.

What others resist in Cultural Critique is not critiques of ethnocentrism but the self-contradictory and defeatist claim that all knowledge, except the knowledge of the situatedness of all knowledge, is situated and therefore flawed. A corollary, making the idea even less inviting, has been that if all claims to universality and transparency of knowledge are false, then the appropriate response is to challenge the claims obscurely. Hence, in part, the vogue for bad writing, the self-confessedly exclusionary opacity of much writing inspired by Theory (Meltzer 470).

Of course there have been many claims of universal truths or of universal human nature that reflect the partial vision of particular cultures or interest groups within them. But rejecting false claims to the universality of a particular set of views does not entail a need to reject universality altogether. As Ghanian-American philosopher Kwame Anthony Appiah notes, what postcolonial opponents of universals actually object to in these cases "is the posture which conceals its privileging of one national (or racial) tradition against others in false talk of the Human Condition": "those who pose as anti-universalists . . . use the term 'universalism' as if it meant 'pseudouniversalism'; and the fact is that their complaint is not with universalism at all. What they truly object to—and who would not?—is Eurocentric hegemony *posing* as universalism" (*In* 91–92).

Indeed to reject claims of a common human nature, far from securing the moral high ground, is to undermine the grounds for treating other human beings as equals. One of the most extreme advocates of difference was Hitler, with his sense of the special destiny of the Aryan people and the German nation, and of the utter difference between Aryan and Jew. This is not to

equate Cultural Critique with Nazism—although those who have tried to critique Cultural Critique by considering human nature in terms of biology have themselves been accused of Nazism (Gottschall and Wilson xx)—but merely to stress that claims of utter human difference are not themselves ethically sufficient. We need to accept both the commonality of human nature and the differences between individuals and peoples. If we reject all claims to commonality, we risk denying a sufficient basis for concern for other humans.

For most of the twentieth century, anthropology has stressed the difference between peoples, since anthropologists earn attention by reporting on the exoticism of other ways of life. But that does not mean that human universals are not there, as the anthropologist Donald E. Brown and others have been able to document extensively (*Human Universals*). And indeed the universals of human nature, the factors that made it possible to understand another people in depth, had simply been taken for granted, even ignored, in the emphasis on difference.

In all species, from bacteria up, communication within the species is possible because of shared senses and interests. In the human case, we can understand one another, even across cultures, because of a range of intraspecies similarities. And we can understand one another especially well because humans are geared to learn from one another through joint attention, the expressiveness of the human facial musculature, the precision of human pointing (all of which develop before language, and make it possible), and language. Our capacity for social learning, for acquiring our own culture, also makes it possible to appreciate and enjoy the culture of others.

The idea that there is only cultural difference between peoples discourages cultural contact and cultural sharing, which has been of benefit to all, over the years, from stone tools to the Internet. The insistence on difference, on refusing to see similarities, inhibits dialogue and the chance to learn from, understand, and appreciate others. This is particularly disturbing in the case of art and literary studies. Menand writes that "a nineteenth-century novel is a report on the nineteenth century; it is not an advice manual for life out here on the twenty-first-century street" ("Dangers" 15). So all those who read *Pride and Prejudice* in the twentieth century and since and felt that it showed something about the dangers of first impressions and the error of equating social ease with merit and social stiffness with coldness or disdain have been wrong?

Appiah offers a much more attractive and defensible attitude toward the arts of other times and places. Because he sees us all as humans, and not as defined

primarily by group differences, he argues that cultural patrimonies should be seen as part of the whole human heritage, not the exclusive property of a single place or people. In his recent *Cosmopolitanism* (2005), he takes as his example the Nok sculptures of the sixth century B.C.E., from an area now part of Nigeria:

> If they are of cultural value—as the Nok sculptures undoubtedly are—it strikes me that it would be better for [the Nigerian government and citizens] to think of themselves as trustees for humanity. While the government of Nigeria reasonably exercises trusteeship, the Nok sculptures belong in the deepest sense to all of us. "Belong" here is a metaphor: I just mean that the Nok sculptures are of potential value to all human beings. . . . It is the value of the cultural property to people and not to peoples that matters. It isn't peoples who experience and value art; it's men and women. ("Whose Culture?" 39)

Further, Appiah advances the claim for an artistic connection "not through identity but *despite* difference. We can respond to art that is not ours; indeed, we can only fully respond to 'our' art if we move beyond thinking of it as ours and start to respond to it as art. . . . My people—human beings—made the Great Wall of China, the Sistine Chapel, the Chrysler Building: these things were made by creatures like me, through the exercise of skill and imagination" (41).

Yet all too often in the academy today, the literature of other times and places is taught only as a demonstration of difference, of the local, false, constructed, and oppressive or contested nature of the concepts of those times and places. But as a species uniquely capable of social learning, of learning from others—and remember, this is what makes our deep immersion in culture possible—we are capable also of learning from, responding to, feeling a kinship with, times and places other than our own.

A biocultural perspective on the human offers the strongest possible reasons to take into account artistic accomplishment in all areas and cultures, and the strongest reasons for considering local difference in terms of a genuinely broad understanding of species-wide commonalities and differences. It is the least likely to fix on an artistic canon within a particular language or region, a particular cultural level ("high" art versus "low," say), a particular state of civilization. In a biocultural view, the Paleolithic and the present, the hunter-gatherer and the cosmopolitan, orature and opera are all part of our human repertoire.

Not only does the stress on difference discourage the study of works of art outside the present, *except* as demonstrations of the truths of Theory, it dis-

courages attention to works of art *tout court*. Menand writes disapprovingly that "there is talk of a return to the literary and to sterile topics like beauty—the very things that the greatest generation rescued us from" ("Dangers" 12). Why should literary studies think they have been fortunate to be rescued from the literary? Would Menand—who, recall, advocates that the humanities should colonize other areas—deem it a success if medicine were rescued from the medical, perhaps by a Theory-inspired denial of the possibility of knowledge (think of all the money that could be saved on cancer research) and a Cultural Critique insistence on difference (genital mutilation? diagnosis by divination? cures by incantation?)?

It is deeply troubling that those teaching any branch of the arts should find beauty sterile, rather than something to enjoy, to explain, and to augment. Is Menand indifferent to his wife's looks? Does he dress shabbily or write sloppily? Does he refuse to decorate his home? Does he disdain the lively concern that people like the Wodaabe or the Nuba or the Maasai have for physical beauty? Does he scorn, indeed, the vast majority of people everywhere who take an interest in personal, scenic, or artistic beauty, and who try to decorate their homes or their lives accordingly? I recently saw a photograph of a Haitian woman who had just eaten a mud pie—yes, just mud—because she could afford no more, yet she clutched a transistor radio as she danced to its music: the beauty of sound, at least, she could have some share in. Does Menand not see that an interest in beauty is a real part of humanity (as of other species), and of the humanities, and that it needs to be explained rather than dismissed? Does he not realize that his dismissal of the sterility of beauty in theory, if not in his personal practice, is all too typical of the pharisaic hypocrisy of Theory?

Our shared sense of beauty is one of our surest avenues to cross-cultural understanding and enrichment. Does the real, deep beauty of the Nok figurines that Appiah discusses and illustrates not say more, and more swiftly, for the cultural creativity, and the justified pride in their achievement, of a people otherwise unknown to us? Dürer in the 1520s, encountering elaborate Mexican craftsmanship, commented that he had never in all his life "seen . . . anything that has moved my heart so much." Goethe, reading Chinese novels, observed that "these people think and feel much as we do." Japanese audiences respond to Shakespeare and Beethoven with rapture. And if audiences appreciate, artists appropriate. Over a millennium ago, the makers of the Book of Kells sublimely synthesized calligraphic and pictorial traditions from Europe and around the Mediterranean. In the nineteenth and twentieth centuries, Māori and New Guinea carvers picked up Western tools and techniques as keenly as Gauguin or Picasso borrowed from non-Western cultures.

Menand wrongly assumes that the recent insistence on cultural difference in university literature departments has helped to undermine whatever is most deplorable in the status quo. I suspect it has done little to undermine anything except student interest in academic literary study, while it has shored up the status of English professors who enlisted as disciples of the "greatest generation" and their conviction that they are in the intellectual and moral right.

Evolution has made knowledge possible. Not necessarily reliable knowledge, but knowledge good enough, on average, to confer a benefit. Evolution has developed sociality to the point where members of many species can transfer knowledge across time: culture, in other words. As comparative and developmental psychology have shown, evolution has developed the human brain's capacity to understand false belief—to understand that others, or we ourselves, might be mistaken about a situation—and hence has driven our quest for better knowledge (on false belief and Theory of Mind, see Perner; Wellman and Gelman). Both human culture and our human awareness of the possibility of being mistaken have eventually given rise to science, to the systematic challenging of our own ideas. The methods of science make relatively rapid change and improvement possible—as well, of course, as unforeseen new problems. They offer no guarantee of the validity of individual ideas we propose, but they do offer the prospect of our collectively learning from one another.

In the long perspective of evolution, testing proposals systematically, as science does, is a very new step for all humanity. Anyone, regardless of origins, can participate in the process, which harvests the natural strengths of our twin tendencies to compete and cooperate. But in order to work, science requires a commitment to the possibility that we can improve our thinking. Insisting that no ideas are valid except the idea that all ideas are invalid, or that all ideas are merely local, except this one idea, is the least likely route to genuine change.

17

Imagining Human Nature

JOSEPH CARROLL, JONATHAN GOTTSCHALL,
JOHN JOHNSON, AND DANIEL KRUGER

Human Nature and the Arts

Until fairly recently in literary history, most writers and literary theorists presupposed that human nature was their subject and their central point of reference. Dryden following Horace, who follows others, offers a representative formulation. In "Of Dramatic Poesy," Dryden's spokesman, Lisideius, defines a play as "a just and lively image of human nature, representing its passions and humours, and the changes of fortune to which it is subject; for the delight and instruction of mankind" (25) (for other such examples, see Carroll, *Evolution* 170; Pinker, *Blank Slate* 404–20). The understanding of human nature in literature is the most articulate form of what evolutionists call "folk psychology" (Boyer, "Specialised Inference"; Dunbar, "Why"; Geary, *Origin*; Mithen; Sterelny). When writers invoke human nature or ordinary people say, "Oh, that's just human nature," what do they have in mind? They almost always have in mind the basic animal and social motives: self-preservation, sexual desire, jealousy, maternal love, favoring kin, belonging to a social group, and desiring prestige. Usually, they also have in mind basic forms of social morality: resentment against wrongs, gratitude for kindness, honesty in fulfilling contracts, disgust at cheating, and the sense of justice in its simplest forms—reciprocation and revenge. All these substantive motives are complicated by the ideas that enter into the folk understanding of ego psychology: the primacy of self-interest and the prevalence of self-serving delusion, manipulative deceit, vanity, and hypocrisy. Such notions of ego psychology have a cynical tinge, but they all imply failures in more positive aspects of human nature: honesty, fairness, and impulses of self-sacrifice for kin, friends, or the common good.

Postmodernists have put all such ideas of human nature out of play. Evolutionary social scientists, fortunately, have taken a different path. While literary theorists were immersing themselves in speculative theoretical systems such as phenomenology, psychoanalysis, deconstruction, and Marxism, the evolutionists were gradually developing an empirically based model of human nature, including childhood development, family dynamics, sexual relations, social dynamics, and cognition.

In the early days of sociobiology, through the 1980s, evolutionary theorists of human nature concentrated on "inclusive fitness"—passing on genes through offspring or other kin (Dunbar and Barrett, "Evolutionary Psychology"; Laland and Brown). In the 1990s, "evolutionary psychologists" distinguished themselves from sociobiologists by emphasizing "proximate mechanisms" that mediate reproductive success, but they still did not produce a whole, usable model of human nature. Instead, they compiled open-ended and unorganized lists of "modules," dedicated bits of neural machinery that were supposed to have solved specific adaptive problems in ancestral environments. Modules were postulated for sense perceptions, various forms of subsistence activity, categorizing plants and animals, selecting mates, detecting cheaters, recognizing emotions, avoiding predators, "and so on" (Carroll, *Literary Darwinism* 106–7). As a complement to lists of modules in evolutionary psychology, Donald Brown offered a list of "human universals"—that is, practices found in all known cultures and thus presumably constrained by the evolved and genetically transmitted features of human nature. The ideas of human universals and domain specificity have remained important in human evolutionary theory, but over the past decade or so behavioral ecologists and developmental psychologists have finally supplied the crucial idea that had been missing from these lists—a total systemic organization in human nature. A scholar or scientist adopting a systemic perspective envisions all the parts of a system as functionally interactive. Variation in one component affects relations among all the components. As a concept of structure, this idea is essentially the same as that of "organic unity" espoused by Samuel Taylor Coleridge and other Romantics.

The most comprehensive concepts for the systemic organization of the parts of human nature derive from "human life history theory" (Hill; Kaplan and Gangestad; Low, *Why*). All species have a "life history," a species-typical pattern for birth, growth, reproduction, social relations (if the species is social), and death. For each species, the pattern of life history forms a reproductive cycle. "Human nature" is the set of species-typical characteristics regulated by the human reproductive cycle. This concept of human nature assimilates the sociobiological insight into the "ultimate" importance of inclusive fitness as a regulative principle, and it accords proximal mechanisms a functional place within the human life cycle. Early models of "the adapted mind," concentrating too exclusively on "modularity," had excluded the idea of flexible general intelligence (Mithen). Using human life history as a systemic concept enables evolutionists to integrate domain specificity with a flexible general intelligence (Geary, *Origin*; Kaplan and Gangestad 122; MacDonald).

Human beings have a life history that is similar in some ways to that of their nearest relatives, the chimpanzees, but humans also have unique species characteristics deriving from their larger brains and more highly developed forms of social organization. Unlike chimpanzees and most other mammals, humans display pair-bonded male–female parenting; and unlike all other animals, they combine pair bonding with complex social organizations involving cooperative groups of males (Flinn and Ward; Geary and Flinn). Humans take longer to grow up, allowing time for their brains to mature and their social skills to develop. And, finally, culture has an importance for humans that it does not have for other species. Culture consists of information transmitted in nongenetic ways: arts, technologies, literature, myths, religions, ideologies, philosophies, and science. From the evolutionary perspective, culture does not stand apart from the genetically transmitted dispositions of human nature. It is, rather, the medium through which humans organize those dispositions into systems that regulate public behavior and inform private thoughts. Culture translates human nature into social norms and shared imaginative structures. The genetically mediated dispositions of human nature—survival, mating, kinship, friendship, dominance, cooperative group endeavor, and intergroup competition—have evolved in a reciprocally causal relationship with the cognitive and behavioral dispositions for producing and consuming imaginative representations. That causal interdependence is part of the evolutionary process that evolutionists denote as "gene–culture co-evolution" (Barrett, Dunbar, and Lycett 351–83; Lumsden and Wilson; Richerson and Boyd).

We live in the imagination. No action or event is, for humans, ever just itself. It is always a component in mental representations of the natural and social order, extending over time. All our actions take place within imaginative structures that include our vision of the world and our place in the world—our internal conflicts and concerns, our relations to other people, and our connections to nature and to whatever spiritual forces we imagine might exist. We live in communities that contain not just the people with whom we come directly into contact, but also our memories of the dead, the traditions of our ancestors, our sense of connection with generations yet unborn, and every person, living or dead, who joins with us in imaginative structures—social, ideological, religious, or philosophical—that subordinate our individual selves to a collective body. Our sense of ourselves derives from our myths and artistic traditions, from the stories we tell, the songs we sing, and the visual images that surround us. We do not have the option of living outside our own imaginative constructs. "Meaning" for us is always part of some imaginative structure, and art works constantly at forming and re-forming those structures.

Human Nature as a Basis of Shared Understanding

Whether traditionally humanistic or poststructuralist in orientation, literary criticism over the past century has spread itself along a continuum between two poles. At one pole, eclectic general knowledge provides a framework for impressionistic and improvisatory commentary. At the other pole, an established school of thought, in a domain not specifically literary, supplies a more systematic vocabulary for the description and analysis of literary texts. The most influential schools have been those that use Marxist social theory, Freudian psychology, Jungian psychology, phenomenological metaphysics, deconstructive linguistic philosophy, and feminist gender theory. Poststructuralist literary criticism operates through a synthetic vocabulary that integrates deconstructive epistemology, postmodern Freudian analysis (especially that of Jacques Lacan), and postmodern Marxism (especially that of Louis Althusser, as mediated by Fredric Jameson). Outside literary study proper, the various source theories of poststructuralism converge most comprehensively in the cultural histories of Michel Foucault, and since the 1980s, Foucauldian cultural critique has been overwhelmingly the dominant conceptual matrix of literary study. Foucault is the patron saint of New Historicism. Postcolonialist criticism is a subset of historicist criticism that employs its synthetic vocabulary chiefly to contest Western hegemony. Queer theory is a subset of historicist criticism that employs the poststructuralist vocabulary chiefly to challenge the normative character of heterosexuality. Most contemporary feminist criticism is conducted within the matrix of Foucauldian cultural critique and dedicates itself to challenging patriarchy—the social and political predominance of males.

Each of the vocabulary sets that have come into prominence in literary criticism has been adopted because it gives access to a significant aspect of the human experience depicted in literature: class conflicts and the material base for imaginative superstructures; the psycho-symbolic dimensions of parent–child relations and the continuing active force of repressed impulses; universal "mythic" images derived from the ancestral experience of the human race; elemental forms in the organization of time, space, and consciousness; the irrepressible conflicts lying dormant within all partial resolutions; or social gender identity. All these larger frameworks have enabled some insights not readily available through other means. They have nonetheless been flawed or limited in one crucial respect. None of them has come to terms with the reality of an evolved and adapted human nature.

Humanist critics do not often overtly repudiate the idea of human nature, but they do not typically seek causal explanations in evolutionary theory, either. In the thematic reductions of humanist criticism, characters typically appear as allegorical embodiments of humanist norms—metaphysical, ethical, political, psychological, or aesthetic. In the thematic reductions of postmodern criticism, characters appear as allegorical embodiments of the terms within the source theories that produce the standard postmodern blend—most importantly, deconstruction, feminism, psychoanalysis, and Marxism. In their postmodern form, all these component theories emphasize the exclusively cultural character of symbolic constructs. "Nature" and "human nature," in this conception, are themselves cultural artifacts. Because they are contained in and produced by culture, they can exercise no constraining force on culture. Hence Fredric Jameson's dictum that "postmodernism is what you have when the modernization process is complete and nature is gone for good" (ix). From the postmodern perspective, any appeal to "human nature" would necessarily appear as a delusory reification of a specific cultural formation. By self-consciously distancing itself from the folk understanding of human nature, postmodern criticism loses touch with both biological reality and the imaginative structures that authors share with their projected audience. In both the biological and the folk understanding, there is a world outside the text. From an evolutionary perspective, the human senses and the human mind have access to reality because they have evolved in adaptive relation to a physical and social environment about which the organism urgently needs to acquire information. An evolutionary approach shares with the humanist a respect for the common understanding, and it shares with the postmodern a drive to explicit theoretical reduction. From an evolutionary perspective, folk perceptions offer insight into important features of human nature, and evolutionary theory makes it possible to situate those features within broader biological processes that encompass humans and all other living organisms.

Emotion and Genre: Getting the Reader into the Picture

The highest level in the formal organization of specifically literary categories is "genre." Elements of form and content can be combined in different ways to constitute diverse systems of genre (Fowler). No one system has yet succeeded in establishing itself as a "natural" classification, but most theories incorporate some version of basic emotions. The most influential taxonomy yet contrived is that of Northrop Frye in *Anatomy of Criticism*. The main elements in Frye's system are social relations and their corresponding emotions. By taking human life history as a theoretical framework, we can assimilate Frye's insights, place them on a stronger empirical foundation, and locate them in the causal, explanatory context of an evolved human nature.

Novelists and playwrights do not just invent meaningful order in human life. They isolate the basic motives that shape our lives and evoke the subjective feeling states that activate these motives. One central purpose of novelists is to illuminate the deep structures of experience and make them available to our imagination. In the organization of human experience, three basic genres seem to constitute something like "natural kinds": comedy, tragedy, and satire. Both comedy and tragedy engage affiliative dispositions, enabling readers either to empathize happily with the good fortunes of a protagonist—some character they like and admire—or to feel sorrow for the unhappiness of the protagonist. Satire, in contrast, is designed to ridicule and is thus hostile in intent. It activates contempt and anger, usually modulated by amusement.

These three basic emotional configurations can be integrated with plotlines derived from basic motives. The species-typical needs of an evolved and adapted human nature center on sexual and familial bonds within a community. Romantic comedy typically concludes with a marriage and thus affirms and celebrates the social organization of reproductive interests within a given culture (Frye). In tragedy, sexual and familial relations become pathological, and social bonds disintegrate. *A Midsummer Night's Dream* ends with reconciliations, multiple marriages, and festivity among the rulers of the land. *King Lear* concludes with an abdicated king dying in anguish amid the bodies of his children and friends. Satire engages a fundamental social disposition for detecting and exposing duplicity and delusion. This disposition evolved in tandem with human dispositions for cooperative behavior and manipulative deceit. Sustaining a cooperative social group depends on being able to expose and punish "free-riders" and cheats (Boehm; Richerson and Boyd; Wilson, *Evolution*).

Very early in any given narrative, authors typically send multiple signals that establish generic expectations. These signals serve as something like an implicit tonal contract with readers. Readers might well feel anxiety about the fortunes of a protagonist in one of Jane Austen's novels, but they would be truly shocked, even outraged, if one of Austen's novels took a turn, toward the end, to tragedy. The closest Austen comes to tragedy is the prospect, real enough, that Anne Elliot in *Persuasion* will be condemned to a lonely old age as a spinster. Since that prospect is clearly registered at the beginning of the novel, Anne's rejuvenation at the end has an imaginative effect something like that produced by the magical transformations at the end of some fairy tales—the beast becoming a handsome prince, the old crone turning into a beautiful maiden. It would not be shocking if Anne's story ended unhappily, but there is not the slightest chance that she would ever be raped or hanged, like Thomas Hardy's Tess. That would be outside the bounds of the tonal contract Austen establishes with her readers. When Tess's fate is finally accomplished, few readers are "happy" at the outcome, but most readers feel a specifically aesthetic satisfaction that derives from Hardy's faithful completion of a tonal contract. The terms of that contract are established very early in the novel, in scenes like that in which Tess looks up at the stars and informs her little brother that, by sheer mischance, the world in which they themselves live is a blighted one.

The tonal signals characterizing genres establish a range of emotional expectations for readers but do not impose cookie-cutter shapes on the thematic and tonal structures of particular works. Novels engage universal themes of human experience, but while depicting basic motives and evoking basic emotions, they organize those universal elements in ways that answer to the distinctive artistic visions of individual writers. As Henry James observes, "the deepest quality of a work of art will always be the quality of the mind of the producer" (64). All human beings have species-typical characteristics, but all also display unique individual differences shaped both by serendipitous recombinations of DNA and by circumstances that necessarily differ, in however slight a degree, for every individual. We can add one further level of distinction. Each writer has a unique identity, and each work by each writer has its own artistic character (Boyd, *On the Origin*). All great novels can be located in one or another genre, but none of them is "generic." They all display peculiarities of thematic and tonal organization that combine with elements of style to produce unique artistic structures.

The Circulation of an Ethos

Novelists and playwrights present characters as persons intent on achieving goals (Bower and Morrow; Scalise Sugiyama, "Reverse-Engineering Narrative"; Turner, *Literary Mind*). The success or failure of the character in achieving his or her goals is the main action in the story—broadly, the "plot." Goals are the end-objects of motives—for instance, the desire to survive, to get married, to make friends, to obtain education, or to assist one's friends. Readers recognize characters as agents with goals and have emotional responses to the characters. In an obvious sense, an author is the first causal force in this sequence. The author creates characters and designates their features and fortunes. For a main character, the novelist or playwright fabricates a situation, identifies the hopes and fears of the character, invents a sequence of actions organized around those hopes and fears, and determines the outcome for that sequence of actions. In all of this, outside of recognizing what the writer has stipulated, the auditor—reader or viewer—has no part. The auditor must take it as the author gives it. But in giving it, the author does not neglect to consult the auditor, at least prospectively. The author anticipates the effects that his or her designs will have on the minds and emotions of auditors (on cognitive adaptations for perceiving goal-directed behavior, see Premack and Premack; Rizzolatti and Fogassi; Sterelny; Tomasello et al.).

Despite the power exercised by authors, the causal force between an author and his or her auditors does not move in only one direction—from author to auditors. In anticipating the effects that their designations will have on auditors, novelists and playwrights are themselves the cunning servants of their auditors. They are themselves constrained in constructing meaning by their own sense of what auditors expect and demand. Dickens's revision of the end of *Great Expectations* offers a case in point. Having done a little judicious prepublication market testing by consulting a savvy friend, Dickens decided that the original, unhappy conclusion he had written for his novel would not sell nearly so well as a hopefully upbeat ending, and he changed the ending accordingly. The author's ability to manipulate the responses of his or her audience depends on keeping his or her depictions within the range of the audience's expectations or desires. Writers rule, but only because they provide their subjects with what the subjects want. Authors dominate the feelings and thoughts of their audience, but only because they allow the feelings and thoughts of the audience at least partially to determine the parameters within which they work.

Great literary authors do not just passively reflect the established and conventional values and beliefs of their culture. That conception of the inert passivity of the authorial mind is, in our view, an important limitation in Foucauldian cultural theory and the New Historicist literary criticism that flows from it. Great novelists and playwrights tap into the deepest levels of the human psyche, connect their contemporary cultural forms with basic human passions, and give their own idiosyncratic and distinctive stamp to the world they envision. Despite his willingness to play to his audience, Dickens is still "the inimitable Dickens." Great and original authors create new possibilities of understanding, but no matter how original and independent they might be, all writers feed off the meanings that are available within their culture: the literary forms and traditions with which they work and the forms of cultural imagination—ideological, religious, and philosophical—in which they participate. Authors, readers, and the larger culture are all locked into an interdependent relationship.

18

Two Worlds

THE GHOST AND THE MACHINE

Edward Slingerland

The university today is, as we know, divided into two broad magisteria, the humanities and the natural sciences, usually located on opposites sides of campus, served by separate funding agencies, and characterized by radically different methodologies and background theoretical assumptions. Although rarely explicitly acknowledged in our secular age, the primary rationale behind this division is a rather old-fashioned and decidedly metaphysical belief: that there are two utterly different types of substances in the world, mind and matter, which operate according to distinct principles. The humanities study the products of the free and unconstrained spirit or mind—literature, religion, art, history— while the natural sciences concern themselves with the deterministic laws governing the inert kingdom of dumb objects. This relationship of metaphysics to institutional structure is expressed most honestly in German, where the sciences of mechanistic nature (*Naturwissenschaften*) are distinguished from the sciences of the elusive human *Geist* (*Geisteswissenschaften*)—*Geist* being a cognate of the

English "ghost," and alternately translatable as "ghost," "mind," or "spirit." German also helpfully provides us with technical terms, always hovering somewhere in the background of contemporary humanistic debate, to distinguish clearly between the two types of knowing appropriate to each domain. The natural world is subject to *Erklären*, or "explanation," which is necessarily reductive, explaining complex physical phenomena in terms of simpler ones. Products of the human mind, however, can only be grasped by means of the mysterious communication that occurs when one *Geist* opens itself up to the presence of another *Geist*. This process is known as *Verstehen*, or "understanding," and it is seen as an *event*, requiring sensitivity, openness, and a kind of commitment on the part of one spirit to another. This is the fundamental intuition motivating the common conviction that only trained humanists can seriously engage in humanistic inquiry. It is also the framework behind the common charge that any attempt to explain a human-level phenomenon in terms of more basic principles is "reductionistic": the understood spirit must be able to see itself reflected, in terms that it recognizes, in the product of the understanding spirit.

Vertical Integration

In place of what has turned into a jealously guarded division of labor between the humanities and the natural sciences, I argue for an integrated, "embodied" approach to the study of human culture. While the humanities do concern themselves with human-level structures of meaning characterized by emergent structures irreducible (at least in practice) to the lower-level structures of meaning studied by the natural sciences, they are not completely *sui generis*. If we are to take the humanities beyond dualistic metaphysics, these human-level structures of meaning need to be seen as grounded in the lower levels of meaning studied by the natural sciences, rather than hovering magically above them. Understood in this way, human-level reality can be seen as eminently *explainable*. Practically speaking, this means that humanists need to start taking seriously discoveries about human cognition being provided by neuroscientists and psychologists, which have a constraining function to play in the formulation of humanistic theories—calling into question, for instance, such deeply entrenched dogmas as the "blank slate" theory of human nature, strong versions of social constructivism and linguistic determinism, and the ideal of disembodied reason. Bringing the humanities and the natural sciences together into a single, integrated chain seems to me the only way to clear up the current miasma of endlessly contingent discourses and representations of representations that currently hampers humanistic inquiry. By the same token, as natural scientists begin poking their noses into areas traditionally studied by the humanities—the nature of ethics, literature, consciousness, emotions, or aesthetics—they are sorely in need of humanistic expertise if they are to effectively decide what sorts of questions to ask, how to frame these questions, and what sorts of stories to tell in interpreting their data.

THE TROUBLE WITH EMBODIMENT

Much of the resistance to integrating the humanities and natural sciences arises out of concerns about crude reductionism, or worries about the politically and morally unsavory manner in which essentialist claims about human nature have been employed in the past. The primary justification and intellectual rallying point for this resistance, however, is theoretical, and emerges from a cluster of theories that can be referred to as "postmodernism." "Postmodernist" is, of course, a notoriously vague adjective, being applied nowadays to everything from poststructuralist French literary theory to trendy styles of living room furniture. I think, however, that it has not entirely lost its usefulness as a signifier. What I see as the core of "postmodern relativism" is an approach to the study of culture that assumes that humans are fundamentally linguistic/cultural beings, and that our experience of the world is therefore mediated by language and/or culture *all the way down*. That is, we have no direct cognitive access to reality, and things in the world are meaningful to us only through the filter of linguistically or culturally mediated preconceptions. Inevitable corollaries of this stance are a strong linguistic/cultural relativism, epistemological skepticism, and a "blank slate" view of human nature: we are nothing until inscribed by the discourse into which we are socialized, and therefore nothing significant about the way in which we think or act is a direct result of our biological endowment (Brown, *Human Universals* 9–38; Pinker, *Blank Slate* 5–29; Tooby and Cosmides, "Psychological Foundations" 24–32). As I argued above, this approach has served as the background theoretical stance in most fields of the humanities for the past several decades, and even a cursory perusal of the annual conference schedules of the American Academy of Religion, Modern Language Association, or the American Anthropological Association will show that it continues to serve as the default approach in these fields.

ACCOUNTING FOR TASTE:
THE EMBODIED APPROACH TO AESTHETICS

For a quick suggestion of what taking embodiment seriously might mean for the study of the arts, let us consider Pierre Bourdieu's account of taste.

For Bourdieu, taste is a socially constructed and socially defined signifier: "what is commonly called distinction, that is, a certain quality of bearing and manners, most often considered innate (one speaks of distinction naturelle, 'natural refinement'), is nothing other than *difference*, a gap, a distinctive feature, in short, a relational property existing only in and through its relation with other properties" (*Distinction* 6). Like any good postmodernist, he is celebrating radical difference and taking aim at the Enlightenment; Bourdieu's particular *bête noire* here is Kant, with his notion of transcendental aesthetic standards. Bourdieu does a brilliant job of showing how the ideal of "natural taste" has been used to legitimize and reinforce class distinctions, and how aesthetic judgments and lifestyle decisions in superficially unrelated "fields" in fact cohere in proclaiming to the world, and to oneself, a particular class identity: the preponderance of sleek new Volvo wagons in the Ikea parking lot is not purely fortuitous, nor is the fact that the owners of these cars also tend to buy their clothes at Banana Republic, read the *New Yorker*, and profess a passion for Pan-Asian fusion cuisine paired with a nice Gewürtztraminer. (I think a case could be made, however, that once a culture has acquired pan-Asian fusion cuisine and nice Gewürtztraminers, pairing the two is an inevitably emergent "Good Trick.")

We have here, though, a classic example of postmodern "slippage" in my sense of the term: the slide from a perfectly reasonable claim that questions overly naïve objectivist models into unjustified cultural relativism. Many things typically or formerly viewed as natural are, in fact, not; however, it does not necessarily follow from this fact that *nothing* is natural. Bourdieu's style of analysis works best at picking out fine and arbitrary distinctions, the sort that matter especially when one is considering, say, postmodern art. The idea of invoking transcendental Kantian judgments is patently absurd when we want to explain why a shark floating in formaldehyde is merely a specimen in a naturalist's lab, but is "art"—and quite expensive art at that—when Damien Hirst places it in an imposing temple of postmodern aesthetics.

However, there is considerable evidence that human taste is *not* a completely contingent social construct, but is grounded in a set of fairly robust innate dispositional propensities and perceptual capacities. There are important commonalities underlying human taste—both literal and metaphorical. We appear to have evolved preferences for particular types of sights, tastes, sounds, and sensations, and even the strangest manifestations of postmodern art work precisely because they plug into the power of these preferences—if only to negate them. An "evolutionary Kantian" approach to taste would want to go deeper than Bourdieu's analyses, asking what, if any, constraints on taste formation there might be, or why people might want to mark class distinctions in the first place. Bourdieu defines *habitus* at one point as "embodied history," an unconscious but nonetheless "active presence of the whole past of which it is the product" (*Logic* 56). Perhaps one way of understanding his limitations—and the limitations of the entire social constructivist paradigm within which he is working—is to see him as not going *far enough back* in his conception of history. The deposited layers of history that form our schemas of perception and motivation go much deeper than the reaction against L'École des Beaux-Arts or the rise of the salon: they go back into *evolutionary* time, into the history of interactions between creatures more and more like us trying to make their way through a complex world. So, in an important sense we might say that the problem with post-modernism is not that it is overly historicist, but that it has an overly super-ficial and myopic conception of what history *is*. Why should Baudelaire's proclamation of the absolute distinction between art and morality count as a historical incident capable of contributing to *habitus*, and not the mutation in some nameless mammalian ancestor's genes that caused it to value sym-metrical faces over others?

Obviously we still need the sort of high-level, subtle analysis Bourdieu is able to provide in order to fully understand the functioning of "fields" such as artistic production: no amount of theorizing by a team of cognitive scientists and evolutionary psychologists is *alone* sufficient to explain the appeal of, say, Manet's *The Absinthe Drinker*. But Bourdieu's account of the symbolic revolution brought about by the Impressionists' break with the academic system only makes sense when understood against the background of evolved human preferences and motivations. Similarly, his attack on the myth of the "fresh eye" in art appreciation is predicated on the claim that "there is no perception which does not involve an unconscious code" (*Field* 217), and this is no doubt true. But again he is too shallow in his history: it has to be realized that the sort of culturally constructed code that he focuses upon is merely a rather superficial modification of a deep and complex perceptual system that comes as standardized equipment in creatures like us. Against Bourdieu, then, it seems that in order to fully understand human culture and cognition we *do* need something like a naturalistic version of Kant's *a priori* categories of understanding and judgment. The tools of vertical integration tell a plausible story about how the sorts of novel, idiosyncratic, high-level cultural- or class-specific distinctions analyzed and catalogued with such nuance by Bourdieu and his students could be seen as grounded in basic and universal human capacities and dispositions.

19

Consilient Literary Interpretation

MARCUS NORDLUND

This is an exciting time in the history of human self-knowledge. Like the two souls in Plato's *Symposium*, the life sciences and the human sciences are slowly coming to terms with their painful divorce and are increasingly on speaking terms. Thanks to important developments across a broad range of academic disciplines from biology and neuroscience to anthropology and psychology, we are starting to glimpse the possibility of a new conceptual integration that may have important consequences for our understanding of human nature. After hundreds of years of virtuoso variations on the theme of philosophical and scientific dualism, we have come some way towards bridging the gap between mind and matter, biology and culture, science and the arts.

From the perspective of literary studies, two recent developments have been especially valuable. First, social scientists have increasingly recognized the explanatory force of evolutionary theory, and, as a consequence, the ultimate bankruptcy of the nature–nurture opposition that forces biology

From *Philosophy and Literature* 26, no. 2, October 2002, pp. 312–333, by Marcus Nordlund.

and culture into different compartments. Since the social sciences have always supplied literary studies with food for thought, it would be strange if such developments did not reflect on literature, too. Second, scholars in the humanities have become increasingly impatient with the radical skepticism, antiscientism, and dogmatic agnosticism that turn both science and literature into mere self-validating narratives, acknowledging no significant difference between descriptive and normative statements. Together, these theoretical shifts have slowly begun to pave the way for a new dialogue between those who study biology, psychology, and cultural artifacts like literature. Since the mid-1990s, a number of literary scholars have taken the first steps towards merging evolutionary and cultural explanation in the study of literature.

A few years ago, in his landmark work *Consilience*, Edward O. Wilson formulated an impressive argument in favour of integrating science and literary study. According to Wilson, there is "only one way to unite the great branches of learning and end the culture wars. It is to view the boundary between the scientific and literary cultures not as a territorial line but as a broad and mostly unexplored terrain awaiting cooperative entry from both sides" (138). While he concedes that the two cultures can and should be characterized by "radically different goals and methods," he also stresses that interpretation provides a "logical channel of consilient explanation between science and the arts" (234).

We have good reason to be optimistic about the potential of this joint endeavor, but it nonetheless presents massive challenges. The current track record in interdisciplinary literary studies is uneven, to say the least. Strange confusions about the nature and status of theory abound, giving rise to spontaneous raids into foreign conceptual territories that are bound to baffle the practitioners of the host discipline. As Richard Levin points out, there are numerous reasons why current interdisciplinary literary theories tend to become self-validating (*New Readings* 13–44). There are no negative tests; critics tend to choose theories on the basis of ideological preferences rather than the criterion of truth; and theorists commonly assume that theories can simply be transformed willfully whenever they do not meet the requirements of the interpreter.

Many of the problems Levin points to can, I think, be remedied by a change in attitude within the discipline of literature itself. For example, something has obviously gone seriously wrong when the automatic spouting of political *ad hominem* arguments and genetic fallacies has become second nature to so many critics. But there are other and more fundamental problems that seem intrinsic to the whole question of interdisciplinarity. One of these, according to Wilson, is that as we move forward in the hierarchy of sciences, from physics to biology to literature, we encounter a rapidly accelerating complexity: "It is far easier to go backward through the branching corridors than to go forward. . . . Biology is almost unimaginably more complex than physics, and the arts equivalently more complex than biology" (*Consilience* 72–73).

In order to explain how evolutionary thinking is relevant to literature, evolutionists have had to work out linkages between basic evolutionary principles and the inescapable complexities of literary meaning. While seeking to meet this challenge, daunting enough in itself, evolutionists have also been forced to defend scientific epistemology against various forms of radical constructivism. In theoretical and critical formulations designed to meet these challenges, scientific and literary epistemologies have often coexisted uneasily. Being "a realist about one class of entity does not commit one to being a realist about every class" (Trigg 21), and it is perfectly possible (at least in theory) to be a scientific realist as well as a literary relativist. While affirming a scientific epistemology, evolutionary literary critics have often neglected more general hermeneutical questions. They have reflected too little about the nature and scope of interpretation.

Many evolutionary critics and theorists have not figured out whether they want to engage in poetics, concerned with the evolutionary function of art and the cognitive predispositions it activates, or thematics, concerned with the meanings of literary works, such as how texts relate to human universals. While writers like Robert Storey and Joseph Carroll have enjoyed the pioneer's liberty of ranging across wide literary and cognitive expanses—with many fruitful results—future theorists will probably be wise to limit the scope of their analyses. Evolutionary and cognitive theory can offer deeply intriguing perspectives on literary universals (poetics) as well as human universals in literature (thematics). But we need to be careful to separate the two methodologically.

A distinction of this kind has already been called for by Tony Jackson. I have reservations about his assumption that we are now faced with yet another historic battle between essentialism and relativism, and I think he fails to appreciate the substantial concessions that most evolutionary critics make to the shaping force of human culture. Most literary evolutionists believe that the biological concept of organism and environment more or less forces us to take a *biocultural* approach that synthesizes biological and cultural explanation (Easterlin, "Voyages" 65). In spite of these oversights, Jackson makes a welcome (if implicit) call for the kind of distinction I am after. As he puts it, there are two major paths that cognitive and evolutionary critics can take; one that explores "*what* literature is by showing *how* it operates in relation to cognition and evolution" and another that concerns itself with "explaining the meanings of specific examples of literature, that is, with interpretation itself" (336).

Not everyone would agree that poetics and thematics are equally promising. Drawing on a previous discussion of literary universals by Patrick Colm Hogan ("Literary Universals"), cognitive critic Alan Richardson has suggested recently that the evolutionary approach to literature "will fare badly if it becomes a matter of simplistically mapping the human universals posited by evolutionary psychology or sociobiology onto the content of fictive texts" ("Rethinking" 569). Such an approach does not respect the necessarily skewed relation between text and world: "Like dreams, fictive works can bear a number of different relations to the rules and regularities of daily experience, often giving us the inverse of the lived world" (561). To avoid this problem, Richardson argues, critics should instead focus on "formal features and constraints, rhetorical and prosodic devices, and questions of genre and narrative, treating thematic elements or other aspects of content more sparingly and with special caution" (569–70). Literary thematics is to be subordinated to poetics and formal analysis.

All of us can, I think, agree that any form of "simplistic mapping" is inherently undesirable, and that any treatment of human universals in literary texts must be conducted with the utmost care. Even if many political objections tend to confuse the justified study of statistical and empirical universals with spurious claims for absolute or normative universality (Hogan, "Literary Universals"), there is little to be gained from riding roughshod over cultural and historical difference. But it is another thing altogether to turn this problem into a case for the fundamental primacy of poetics and formalist analysis over thematics, and to say that the study of "literary universals" is somehow more important or theoretically defensible than "human universals in literature." Surely, the humanities will always be interested in what texts have to tell us about human beings and their place in the world, areas where evolutionary theory is offering increasingly compelling perspectives. A cursory glance at any major literary journal will demonstrate that the vast majority of literary interpreters are still closet thematicians. Indeed, as soon as we even ask what a text is *about*, we are already doing thematics. The real problem, therefore, is not thematics itself, but how we can best apply human universals to literary texts without plowing historical, cultural, and literary specificity into the ground.

In what follows, I will argue—contra Richardson, and in answer to Jackson—that evolutionary theory can make substantial contributions to thematic literary interpretation. But my strategy will not be to give striking examples of how literary texts can be illuminated by evolutionary theory. Instead, I will assume that any such contribution must take a more fundamental question—the *"unimaginable complexity" of interpretation*—as its point of departure. We must recognize what we cannot know (at least for certain) *a priori*, and we must recognize the point at which an accelerating complexity renders theoretical prediction tenuous or even useless.

The more specific questions that any theory must answer are: What can we know about literary interpretation? What can we know *a priori*, and what can only be decided in concrete interpretive contexts? And finally, what are the consequences of such knowledge or its absence? Taking on questions this big requires a simplified model, one located at the most basic level of abstraction. So what are the absolute minimum requirements for any act of reading or interpretation? We need a *reader*, a *text*, and a *world* in which to do the reading. This interpretive triad can be visualized as an equilateral triangle whose corners each represent one of the three factors. The model is so simple that it should not require an actual diagram, but the reader is invited to draw the figure in the margin or on a paper napkin for easy reference.

The first thing that must be pointed out is that at least two of these categories are characterized by an indefinite complexity. That is, they involve a complexity that cannot be grasped by a finite, temporal mind (or even, I would suspect, the most advanced of computers). The Reader, first of all, is capable of projecting an indefinite number of assumptions, facts, and values onto the text as he or she reads it. Moreover, the historical and cultural locations of readers differ, and with every such difference the potential projections expand. The category of the World, secondly, denotes *everything that exists or has existed independently of either the individual text or the individual reader—everything that has become the object of the reader's cognition.* The reason why this category must be so inclusive is that there can be no *a priori*, qualitative distinction between what is relevant and what is not in relation to a given text. Anything that I know, or think that I know, about the world *could* be relevant and throw important new light on the text (including folk psychology or the science of hydraulics). It is only the third category, the Text, that can be regarded as a reasonably finite creation with a fixed content, but this finitude gives way to an open-ended potential as soon as it interacts with one of the other categories.

To many readers, the interpretive triad (reader/text/world) might seem to involve an obvious omission in a list of minimally necessary elements in a model of literature. To complete the schema, do we not also need an author? In my conception, the author must be regarded as a contextual entity—part of the "world." Let me explain. Our knowledge of authors is always contextual because we do not have direct, unmediated access to their intentions. In fact, we have good reason to believe that authors do not have such unmediated access to their own minds, and even if they did, they might lie to us. Since there can be no direct access to authorial intention, we can either attempt to deduce it intratextually, which immediately leads to the mental projection of a contextual author who is separate from his text, or we can piece together contextual information from the outside world. Such information might include anything from the author's general social environment to biographical idiosyncrasies. It is, moreover, perfectly possible—as formalists, hermeneuticians, and poststructuralists have all argued—to read texts without active recourse to the figure of the author as an adjudicating principle. But in spite of such caveats, it seems reasonable to regard the author as the most immediate, and therefore the most central, of all contextual explanations. For instance, if contextual research can demonstrate (1) that many Elizabethans believed in spontaneous combustion, or (2) that Shakespeare believed in spontaneous combustion, then the latter point will naturally be more compelling because it is more proximate to the text.

In this conception, then, the author is part of the reader's world. What else can we say *a priori* about this world? One of the most necessary, intriguing, and complex conceptions of reality that a reader projects onto literary narratives is that of human nature. After all, the improbable alternative would be to construct a new theory from scratch to account for every single character in every single narrative. Since this is clearly impracticable, all readers automatically employ generalized expectations based on their conceptions of how humans function. (Whether a literary character actually *meets* these expectations is, again, a different matter.) Significantly, theories that actively deny the existence of a human nature actually constitute theories of human nature, as Mary Midgley notes: "All moral doctrines, all practical suggestions about how we ought to live, depend on some belief about what human nature is like. . . . This includes doctrines that 'we have no nature,' since that means that we are—naturally—quite plastic. Very often these beliefs are wrong. When they are so, they are often evolutionarily implausible" (166).

A salient prejudice of literary critics is the belief that the evolutionary and biological challenge to the literary-philosophical "linguistic turn" denies the shaping force of either language or history in the construction of human identity. On the contrary, evolution and biology radicalize them both. Stressing the formative role of language as verbal communication or as linguistically mediated cognition offers important insights, but it represents merely the tip of an enormous iceberg. All human beings engage in an astoundingly intricate process of *semiosis* that begins at the moment of conception, picks up speed during the period of gestation, and then carries on throughout our lives. It ranges from the genetic coding of proteins into brain cells to the biological as well as cultural structuring of a brain that decodes the meaning of traffic lights and worries about the meaning of Wittgenstein. Likewise, to look only at the social structures and conceptual frameworks that have evolved in the mere thousands of years we call history is to disregard the much larger historical process that produced the very creature that now turns around self-consciously and explores its own pedigree. From the perspective of *historical* time, someone like Shakespeare is a distant stranger, enveloped in his dark backward and abysm of time. But from the perspective of *evolutionary* time, he is more like our next-door neighbor.

The evolutionary perspective serves as a theoretical foil that makes claims for cultural and historical specificity meaningful, sharpens their precision considerably, and opens them up to critical examination. To take an example from my own research, literary critics can no longer maintain that jealousy or romantic love in Shakespeare is simply a "social construction" if it turns out that variants of these phenomena are universal across all known human cultures and also have convincing evolutionary functions (*Shakespeare*). Once we have established this dialectic between evolutionary time and historical time, or what evolutionists call distal and proximal explanation, we can go on to replace the abortive and misguided battle between essentialism and constructivism with more productive questions. We can address how a particular historical situation or a particular kind of social structure may interact with an evolved feature of human nature—in the example above, the capacity for erotic attraction—and how that interaction is reflected, recast, expressed, challenged, or violated in the form of fiction.

The unimaginable complexity of interpretation does not land us in a sea of endlessly deferred signification and uncertainty. Once we adopt a biocultural perspective, things simultaneously become a little easier (as we gradually structure and clarify our methods) and a little harder (as we sentence ourselves to wide reading in previously unfamiliar fields). Literary critics must familiarize themselves with the debates over evolution and human behaviour in the biological and social sciences and then sort out the literary consequences for themselves. The concept of human nature is now—at last—a shared object of scientific and humanistic exploration.

20

Humanism and Human Nature in the Renaissance

ROBIN HEADLAM WELLS

A vision of a future social order [must] be based on a concept of human nature. If, in fact, man is an indefinitely malleable, completely plastic being, with no innate structures of mind and no intrinsic needs of a cultural or social character, then he is a fit subject for the "shaping of behavior" by the State authority, the corporate manager, the technocrat, or the central committee. Those with some confidence in the human species will hope that this is not so and will try to determine the intrinsic human characteristics that provide the framework for intellectual development, the growth of moral consciousness, cultural achievement, and participation in a free community. . . . We must break away, sharply and radically, from much of modern social and behavioral science if we are to move towards a deeper understanding of these matters.

NOAM CHOMSKY, "LANGUAGE AND FREEDOM"

Postmodernists do not share Noam Chomsky's views on human nature. Cultural Materialists and New Historicists believe that talk of innate structures of mind or intrinsic human needs is no more than ideological mystification; in reality there are as many forms of human nature as there are human societies. "Constructionism," writes one leading American Shakespeare scholar, "is one of the basic propositions by which new historicism as a way of reading has distinguished itself from humanism. Where humanism assumes a core essence that unites people otherwise separated in time and social circumstances new historicism insists on cultural differences" (Smith, *Shakespeare* 132–32).

The belief that our minds are shaped largely by sensory experience is not a new one. John Locke famously declared that at birth the mind was "a white sheet, void of all characters, without any ideas" (*Essay* 105). But in denying the existence of innate ideas Locke did not reject the principle of a universal human nature. He argued that, although we may not come into the world with ready-made notions of, let's say, truth or justice, we are nevertheless equipped with faculties that enable us to learn what we need to know as human beings, and it's those inborn faculties that define our humanity (91). What concerned Chomsky was not the notion of the mind as a *tabula rasa* passively absorbing experience—though psychologists now dispute that idea—but the claim that we inherit no species-specific mental characteristics of any description. It was in the early decades of the last century that it became fashionable to argue that human nature was inherently unstable. "On or about December 1910 human character changed," wrote Virginia Woolf in 1924 ("Character" 4). "There is nothing that can be changed more completely than human nature when the job is taken in hand early enough," declared Bernard Shaw ten years later (166). Woolf and Shaw were being deliberately provocative. But the new denial of human nature wasn't just a matter of novelists and playwrights rhetorically asserting a modernist sense of cultural crisis. Anthropologists from Margaret Mead to Clifford Geertz agreed that human nature was infinitely malleable; even the central nervous system was thought to be a cultural artefact. Sometimes referred to as the Standard Social Science Model, this constructionist view of humanity was the orthodox theory of mind in university social science departments for much of the twentieth century (Tooby and Cosmides, "Psychological Foundations").

For Chomsky there was something profoundly disturbing in the prospect of an Orwellian world in which human nature is fabricated by the state and truth merely an effect of power. It was also bad science. But since the 1970s there has been a revolution in the psychological and biological sciences. Where "humanity" was once seen as a purely cultural construct, a consensus is now emerging among psychologists and neuroscientists that our minds are the product of a complex interaction between genetically determined predispositions and an environment that has itself been shaped by generations of human culture. The zoologist and polymath Edward O. Wilson has a phrase that sums it up well: we are, he says, the products of "gene–culture coevolution" (*Consilience* 139).

But literary postmodernists are suspicious of the truth claims of science and remain ideologically committed to the principle that the mind, and even gender, is shaped exclusively by social forces and owes nothing to our biological nature. It's true that some of anti-humanism's most passionate former champions have now modified their constructionist theories. But in doing so they have effectively abandoned the core principle of postmodern literary theory. In her seminal *Critical Practice*, Catherine Belsey rejected any notion of "an essential human nature" based on "a quasi-biological theory of instincts" (131). Yet in a more recent essay she describes the interaction between biology and culture in a way that is virtually indistinguishable from that of an evolutionary biologist like Edward O. Wilson: "the biology that constitutes human beings always interacts with the relatively autonomous culture their evolved brains make possible, and culture too exercises determinations" ("Biology").

To acknowledge that human beings are the product of "gene–culture coevolution," to use Wilson's phrase, is to challenge one of the defining principles of postmodern theory. As the neo-Marxist critic Jean Howard explains, central to the New Historicist project is "the attack on the notion that man possesses a transhistorical core of being. Rather, everything from 'maternal instinct' to conceptions of the self are now seen to be the products of specific discourses and social processes" (20–21). Postmodernists insist that we bring into the world no inherited predispositions that are typical of our species. It's not just a question of the infant mind being a blank sheet devoid of innate mental content; for the postmodernist there are none of the built-in rules that Locke thought were essential for processing experience. If there is nothing in our mental constitution that can be said to be intrinsically human, any Lockean notions of universal human rights (*Two Treatises* 348) evaporate and we are left with a cipher waiting to be given shape and form by society. As Howard puts it, "*nothing exists* before the human subject is created by history" (21). Stephen Greenblatt spells out this key principle of New Historicist criticism in one of his most influential essays: "The very idea of a 'defining human essence' is precisely what new historicists find vacuous and untenable" ("Resonance" 165).

Anti-essentialism is as fundamental to Cultural Materialism as it is to New Historicism. Alan Sinfield speaks for a whole generation of post-structuralist Marxist critics when he writes: "as a cultural materialist I don't believe in common humanity" (10). Reviewing the critical developments of the past twenty years, Jonathan Dollimore has recently reminded us that Cultural Materialism has always been "resolute" in its rejection of "universal humanism" and "essentialist individualism" (xxv).

Postmodernists believe that the notion of a transhistorical essence of human nature is an invention of the modern world. Citing Foucault—"before the end of the eighteenth century, man did not exist" (*Order* 308)—Cultural Materialists and New Historicists argue that to attribute essentialist ideas of human nature to Shakespeare and his contemporaries is an historical anachronism. (Foucault himself meant something very different from what his followers took him to mean.) In one of the truly seminal critical books of the late twentieth century—*Radical Tragedy*—Jonathan Dollimore declared that it wasn't until the Enlightenment that "essentialist humanism" first made its appearance (250). So influential was *Radical Tragedy*, and so great the continuing demand for it in university English courses, that a third edition has recently been published. In a foreword to the new edition Terry Eagleton tells us that the book is essential reading for the modern student: it's one of the "necessary" critical works of our time (xiii). Frank Kermode has recently claimed that in the world of Shakespeare studies the discourse of postmodernism is now "virtually omnipotent" (Introduction 5)—perhaps only a slight exaggeration. By the end of the twentieth century the consensus view in what had by then become mainstream Shakespeare criticism was that to read this period through "the grid of an essentialist humanism" (155), as Dollimore put it, is to give a false picture of the age. Shakespeare was in effect a postmodernist "avant la lettre" (Belsey, "Name" 141, 133). Students were warned against the folly of supposing that Shakespeare's plays might have anything to do with human nature (Hawkes, *Alternative Shakespeares* 9–10). What was true of Shakespeare was of course true of literature in general. In *Critical Practice*, Belsey set out to dispel the myth that "literary texts . . . tell us truths . . . about the world in general or about human nature" (2). Students were taught that in this period the human "subject" was thought to be inherently unstable and fragmented (Berry and Trudeau-Clayton 4); that it wouldn't have occurred to people that they might have an inner self (Belsey, *Subject* 18); that the idea of creative originality was an entirely alien concept (Hawkes, *Shakespeherian Rag* 75); and that "in the Renaissance our modern concept of the genius simply did not exist" (Battersby 28). As for gender, that was so indeterminate and had so little connection with biological nature that Elizabethans thought the mere act of putting on an actor's costume could literally turn a man into a woman (Levine 4). Homosexuality hadn't yet been invented. Bruce Smith writes: "there was no such thing in early modern England as a 'homosexual'" (*Homosexual Desire* 12). Following Foucault (*History* 43), Smith sees homosexuality, like any other kind of sexuality, as a cultural artefact: "No one in England during the sixteenth or seventeenth centuries would have thought of himself as 'gay' or 'homosexual' for the simple reason that those categories of self-definition did not exist" (11–12).

The belief that Shakespeare and his contemporaries were radical antiessentialists is not supported by historical evidence. On the contrary, wherever you look in Elizabethan England you find the same insistence on the importance of understanding human nature. As the political historian Janet Coleman reminds us, "for all medieval and Renaissance thinkers, man's nature does not change over time. . . . In all societies throughout history men can be observed to have demonstrated through their actions the same kind of nature, a nature that is specific to humans" (50). For Elizabethan humanists—

the word "humanist" comes via Italian from the Latin *humanitas*, whose primary meaning was "human nature"—the proper study of mankind was man. In his recent *Shakespeare and Renaissance Politics*, Andrew Hadfield implies that belief in a universal human nature is incompatible with an interest in politics (vii). Renaissance historians and political theorists would have found this a puzzling idea. Actually, knowledge of human nature was generally regarded in this period as the key to an informed understanding of history and politics.

Humanist historiographers believed that the study of history was useful because, human nature being much the same in all ages, it could give the politician a valuable key to human action. Literary theorists defended poetry on the grounds that it gives you a much better insight into the way human beings behave than any scholastic treatise could do: one of the main justifications for reading literature was the belief that dramatic poetry could, as Hamlet puts it, hold the mirror up to nature and show us our characteristic human vices and virtues. People naturally argued about what human nature was like, but no one doubted that it existed. That it was important to understand human nature is something that seems to have been accepted by even the most unconventional thinkers. Montaigne's friend Pierre Charron summed up a commonplace of this period when he said that "the first lesson and instruction unto wisdom . . . is the knowledge of our selves and our human condition" (223).

Humanist philosophers from Cicero to A. C. Grayling have argued that any attempt to realise the ideal of a just society must begin with the facts of human nature. In *What Is Good?* Grayling writes: "Before one can get far with thinking about the good for humankind, one has to have a view about human nature, for the very simple and obvious reason that a theory about the human good that drew only on considerations about what, say, dogs and horses are like (and what dogs and horses like) would be of exceedingly little use" (210). Renaissance thinkers shared that belief. However, their intellectual world probably had more in common with Chaucer's than it does with our own. To emphasise the paramount importance that Renaissance thinkers accorded the study of human nature is not to suggest that their educational principles are relevant to the problems of the modern world (Elizabethan humanists showed no interest in the inductive approach to knowledge that was so soon to transform science). Nor is it to endorse Renaissance theories of civilisation (though there was a strong republican element in Elizabethan humanism, much humanist thought was unashamedly elitist). Rather it's an attempt to reconstruct unfamiliar ways of looking at things in the hope that this may correct certain misconceptions about Shakespeare's intellectual world that have become commonplaces in modern criticism. Dr. Johnson said that the task of criticism was to improve opinion into knowledge (4:120). As playgoers and readers we all have opinions about Shakespeare. But it's not until you have established the mental framework within which intellectual debate was conducted and meanings generated in the past that you can begin to judge a writer's response to "the very age and body of the time his form and pressure" (*Hamlet* 3.2.23–24), or evaluate critically the worth of that response from a modern perspective.

Shakespeare's Humanism is about the centrality of human nature in Shakespeare's mental universe. Although in reasserting the importance of *humanitas* in the plays, it runs counter to the general tenor of mainstream, establishment Shakespeare criticism, it's not an argument for returning to the critical past. By listening to what other disciplines have to say about human nature, criticism can move on from an outdated anti-humanism that has its intellectual roots in the early decades of the last century to a more informed modern understanding of the human universals that literature has, in Ian McEwan's words, "always, knowingly and helplessly, given voice to" (3).

Rediscovering Universals

So powerful is postmodernism's anti-humanist rhetoric that many left-wing intellectuals now find it embarrassing to talk of universals: to admit to a belief in something called human nature is to confess to the crassest kind of intellectual naïveté. Dryden defined a play as "a just and lively image of human nature, representing its passions and humours, and the changes of fortune to which it is subject" (25). But first-year university students are now taught that human nature is one of the bourgeois myths that Theory has exploded (Barry 18), and it's left to other disciplines to remind us that great writers have always been "the voice of the species" (Pinker, *Blank Slate* 421). "Works of art that prove enduring are intensely humanistic," writes Edward O. Wilson. "Born in the imagination of individuals, they nevertheless touch upon what was universally endowed by human evolution" (*Consilience* 243). If there were no universal passions and humours, we would have no means of evaluating literature from another age or another culture: a text would have value only for the community in which it was produced. "There is no such thing as a literary work or tradition which is valuable in itself, regardless of what anyone might have said or come to say about it," writes Terry Eagleton. "'Value' is a transitive term: it means whatever is valued by certain people in specific situations, according to particular criteria and in the light of given purposes" (*Literary Theory* 11). That's why, as Peter Lamarque and Stein Olsen point out in *Truth, Fiction, and Literature*, reception theory is increasingly taking the place of literary criticism (441). In a world where "everything exists and nothing has value" (Forster 147), all criticism can do is record the history of a text's reception.

But there's no philosophic or scientific justification for our embarrassment about universals. Wilson claims that "the greatest enterprise of the mind has always been and always will be the attempted linkage of the sciences and humanities" (*Consilience* 6) (for a recent attempt to bring modern psychological theory to bear on the study of literature, see Carroll, *Evolution*). One of the most puzzling aspects of postmodern criticism is the fact that, despite its proclaimed interdisciplinarity, it has chosen to isolate itself from so many other disciplines. Concerned though it is with the history of "man," postmodernism has ignored the rapidly growing body of work in archaeo-anthropology, evolutionary psychology and neurobiology that has transformed modern thinking on social behaviour, the mind and the mystery of human creativity. Catherine Belsey speaks for many postmodernists when she says she finds sociobiology "deeply distasteful" (*Desire* 14). Just as fundamentalist creationists prefer to disregard the scientific evidence of the fossil record because it conflicts with their religious beliefs, so postmodernists would rather ignore the extensive scientific literature on selfhood, gender and consciousness because they believe that it's incompatible with their own political ideals.

Literary postmodernism prides itself on being the *avant-garde* of the modern intellectual world. It's ironic, therefore, that its intellectual models are thinkers who have long been regarded as obsolete in their own fields. Joseph Carroll writes:

> The conceptual shift that takes place when moving from the Darwinian social sciences to the humanities can be likened to the technological shift that takes place when travelling from the United States or Europe to a country in the Third World. . . . It is as if one were to visit a country in which the hosts happily believed themselves on the cutting edge of technological innovation and, in support of this belief, proudly displayed a rotary dial-phone, a manual typewriter, and a mimeograph machine. (*Literary Darwinism* 10)

While postmodernists in Britain and America were popularising outdated Continental theories of "man" based largely on intuition rather than empirical investigation, a revolution was taking place in the biological and social sciences (Malik, chap. 7). In the decades immediately following the Second World War the application of evolutionary theory to the study of human nature was unthinkable. Though groundbreaking advances were being made in molecular biology—Francis Crick and James Watson cracked the riddle of DNA in 1953—memories of the Holocaust, and the eugenicist theories that were used to justify it, were too close to contemplate studying the biological basis of human nature. In her autobiography Margaret Mead recalled how the danger of misinterpretation meant that any study of innate human characteristics would have to wait for "less troubled times" (222) (about the similar and more recent constraints on research into gender, see Baron-Cohen, *Essential Difference* 11). However, by the 1960s sociobiologists had begun to challenge behaviourism's blank-sheet theory of human nature and to argue that universal patterns of human behaviour must owe something to our biological nature. Though sociobiology was well established by the mid-1970s—Edward Wilson's controversial *Sociobiology* was published in 1975—it continued to face hostility from those who believed that it gave support to racist theories of humanity. But as the pioneering evolutionary psychologists John Tooby and Leda Cosmides point out, it's difficult to see how an attempt to provide empirical evidence for the existence of a universal nature common to all humanity can be accused of being racist ("Psychological Foundations" 19–36).

In his influential *The Interpretation of Cultures* the American anthropologist Clifford Geertz wrote: "our ideas, our values, our acts, even our emotions, are, like our nervous system itself, cultural products" (50). In a simile popular in constructionist theory, he compared the infant human with a computer before any software has been loaded. What provides us with the programmes we need for survival is culture. Without those programmes our life would be "virtually ungovernable; a mere chaos of pointless acts and exploding emotions" (46). But sociobiologists and evolutionary psychologists argue that if diverse cultures result in endlessly varied forms of social behaviour, that's not because the infant mind is like a computer before any software has been loaded. Rather it's because we come into the world with a pre-installed operating system specifically "designed" to enable us to assimilate and interact with the culture that surrounds us from birth (Tooby and Cosmides, "Psychological Foundations"). Evolutionary psychology has been criticised for substituting one form of determinism for another and allegedly insisting that all human behaviour can be explained in terms of inherited drives (Malik, chap. 10). But this is to misrepresent evolutionary psychology. Far from denying the influence of culture in shaping human behaviour, evolutionary psychologists try to show how we are innately and uniquely equipped to interact with the cultural world and operate as cultural beings. As Edward Wilson explains, our minds are indeed shaped by culture. But it's our genetic inheritance that determines which parts of the cultural world we absorb. Those inherited predilections in turn shape culture. Wilson calls this process "gene–culture coevolution" (*Consilience* 139).

The term that some evolutionary psychologists use to describe the universal human ability to integrate with the social world and become flexible and interactive participants in its discourses is "metaculture." In "The Psychological Foundations of Culture," John Tooby and Leda Cosmides explain how

> [t]he variable features of a culture can be learned solely because of the existence of an encompassing universal human metaculture. The ability to imitate the relevant parts of others' actions, the ability to reconstruct the representations in their minds, the ability to interpret the conduct of others correctly, and the ability to co-ordinate one's behavior with others all depend on the existence of a human metaculture. (92)

Informing and co-ordinating all these specialised skills is the sense that we have both of our own subjectivity and that of others. We are born with an inbuilt theory of mind, a sense that other people too have inner selves (Baron-Cohen, *Mindblindness*; Pinker, *Blank Slate* 61–62, 223–24). Without that innate sense of subjectivity, discourse couldn't exist; we couldn't operate as social beings. In exceptional cases this inbuilt subjectivity "module" is impaired or apparently non-existent, with devastating consequences.

Could a classically autistic person who seemed to lack a sense of inwardness really be an artist in the conventional sense of the term? Oliver Sacks's impression that autistic people appear to lack an inner self was confirmed by his meeting with a well-known autistic person, the American biologist Temple Grandin. Grandin is unusual in having written, with the help of a journalist, an autobiography. She is interested in her own condition, recognising that in subtle ways she is different from other people. "She surmises," writes Sacks, "that her mind is lacking some of the 'subjectivity,' the inwardness, that others seem to have" (*Anthropologist* 275). Looking for words to describe the unusual phenomenon of people who seem to lack that sense of self which most of us take for granted, Sacks uses the phrase "*no living centre*" (*Man* 188). Autistic people are in effect de-centred human beings.

Sacks uses the same metaphor of an absent centre in his description of hebephrenia, a condition characterised by incessant facetiousness, punning and wisecracks:

> "funny" and often ingenious as they appear—the world is taken apart, undermined, reduced to anarchy and chaos. There ceases to be any "centre" to the mind, though its formal intellectual powers may be perfectly preserved. The end point of such states is an unfathomable "silliness," an abyss of superficiality, in which all is ungrounded and afloat and comes apart. (*Man* 113)

Wits might be tempted to suggest that Sacks's description of hebephrenia sounds like an uncannily accurate description of classic deconstruction of the kind that used to be practised by critics like Paul de Man and Geoffrey Hartman. But in truth it's not a joking matter. If we really want to know what it means to be a de-centred person with no inner self, "frail, precarious, dispersed across a range of discourses," we cannot do better than read Sacks's poignant accounts of autists, hebephrenics and patients suffering from other neurological disorders.

It's a strange irony that postmodernism should have adopted as the basis for its theory of mind a pathological condition (promulgated in particular by a psychotic who ended his days in an institution for the insane) so disabling that it would appear to prevent the possibility of any true creativity. Indeed it's on the question of selfhood and agency that postmodernist arguments are most puzzling. Postmodernism says that culture is everything. Yet as Cosmides and Tooby argue, the development of culture depends on our ability to communicate with other people, which in turn depends on our ability to recognise that other people are intentional agents with minds of their own. As Tooby and Cosmides point out, "humans everywhere include as part of their conceptual equipment the idea that the behavior of others is guided by invisible internal entities, such as 'beliefs' and 'desires'" ("Psychological Foundations" 89; see also Ridley, *Nature* 210–11). If our ancestors had all been autists there would be no advanced civilisations. In denying the reality of the self, postmodernists deny the very thing that makes culture possible. However, modern neuroscience appears to confirm what common sense has always told us about the inner self. Though we may play many roles over a lifetime, it's the sense of a constant inner core of our being that gives meaning and coherence to our lives: we may change our minds and even our most fundamental beliefs; we may put on antic dispositions; we may express our passions in riddles and dreams (as Webster's Duchess of Malfi puts it [1.1.445–46]). But we don't usually forget who we are. When we do forget as a result of some neurological disorder such as Alzheimer's disease or Korsakov's syndrome, the result is a catastrophic destruction of personality. For the gerontologist Raymond Tallis, senile dementia is "an indirect reminder that to be human is to be explicitly extended in and across time" ("Dark Mirror" 15). Modern neuroscientists believe that it's the brain's synapses—the interfaces between its neurons—that allow us to store memories of who we are and who we used to be. Without those memories "personality would be merely an empty, impoverished expression of our genetic constitution," writes the neurobiologist Joseph LeDoux; "we wouldn't know if the person we are today jibes with the one we were yesterday or the one we expect to be tomorrow" (9).

If we didn't have a sense of a coherent inner self we would be like autistic people, who seem to have, as Sacks puts it, "no living centre."

No body of work that claims to offer insights into the mystery of human subjectivity can expect to be taken seriously beyond its own limited coterie if it systematically ignores developments in other disciplines that concern themselves with the study of mind and of social behaviour, especially when that new research has produced such a remarkable consensus in a wide range of subject areas. Biologists, from Darwin and Wallace to Dawkins and Gould, have always disagreed over details of the evolutionary process (Cronin, *Ant*), but with only fundamentalist creationists and postmodern anti-humanists rejecting modern Darwinism, natural selection is now accepted in the scientific community as the only known way of accounting for what looks like complex "design" in living organisms. As John Barrow puts it, "wherever we find interwoven complexities, we find the hand of time, slowly fashioning adaptations" (82).

If our bodies are the product of millions of years of selection, so too are our brains; they bear "the stamp of 400 million years of trial and error, traceable by fossils and molecular homology in nearly unbroken sequence from fish to amphibian to reptile to primitive mammal to our immediate primate forerunners" (Wilson, *Consilience* 116). Postmodernism claims that there's no such thing as a universal core of essential humanity. Yet as Steven Pinker reminds us,

> In all cultures, people tell stories and recite poetry. They joke, laugh, and tease. They sing and dance. They decorate surfaces. They perform rituals. They wonder about the causes of fortune and misfortune, and hold beliefs about the supernatural that contradict everything else they know about the world. They concoct theories of the universe and their place within it.
>
> As if that weren't enough of a puzzle, the more biologically frivolous and vain the activity, the more people exalt it. Art, literature, music, wit, religion, and philosophy are thought to be not just pleasurable but noble. They are the mind's best work, what makes life worth living. (*How* 521)

To Pinker's list of human universals we can add the propensity to devise complex systems of moral values. Opinion on human nature has traditionally been divided: pessimists tend to agree with Machiavelli that people never do good unless necessity drives them to it; optimists share the belief of the eighteenth-century *philosophes* in humanity's natural benevolence. Many so-called materialists are also optimists, believing that, if only you can get the economic conditions right, class will disappear and people will live naturally in egalitarian harmony. Evolutionary psychologists argue that both the potential for brutality *and* the benevolence are intrinsic to our nature. History shows that we seem to have an innate capacity for collective cruelty. Evelyn

Waugh didn't exaggerate when he said "Barbarism is never finally conquered; given propitious circumstances, men and women who seem quite orderly will commit every conceivable atrocity" (quoted in Wheatcroft 14). Anyone who doubts the truth of this should read Jared Diamond's *The Third Chimpanzee*. But at the same time, a sense of right and wrong is a feature of all known human societies. Evolutionary psychologists may differ among themselves on the question of whether our sense of morality originally served a survival function or evolved through sexual selection (Miller, *Mating Mind*; Ridley, *Origins*). But they agree that, insofar as human value systems have arisen out of the basic drives of human nature, they have a biological rather than a transcendental origin, and are therefore just as truly universal as if they had been god-given. That's not to say that the laws of every society throughout history are fundamentally the same; not even Renaissance proponents of natural law argued that. What *is* universal is the tendency for both civilised and pre-civilised societies to frame laws or codes of conduct that seek to regulate the ruthlessness, violence, cunning and powers of deception that we share with our closest and most Machiavellian simian relations (de Waal, *Chimpanzee Politics*). Those laws or codes are an expression of our equally innate desire for truth, justice and fair play, and are unique to our species. Matt Ridley writes: "The conspicuously virtuous things we all praise— cooperation, altruism, generosity, sympathy, kindness, selflessness—are all unambiguously concerned with the welfare of others. This is not some parochial Western tradition. It's a bias shared by the whole species" (*Origins* 38). Moral and political values will inevitably vary from one writer to another and one culture to another, but the problems are always recognisably human.

There will always be conflict between different sides of our evolved nature, with some societies privileging one side and some another (*Hamlet* dramatises this conflict in its contrast between the archaic world of heroic values represented by Hamlet's father on the one hand, and Horatio's philo-sophic stoicism on the other). Any enlightened system of human justice is bound to involve compromise as it seeks to allow freedom to the individual while protecting others from the consequences of unlicensed behaviour. But the basic conflicts will always be there and can never be completely resolved. That may be one reason why great literature typically concerns itself with conundrums rather than the "eternal verities" that postmodernists allege are the stock-in-trade of the Western canon.

Of course not all literature deals with those high moral and political conundrums. Despite the solemn tone of the typical humanist treatise on poetics, much of the best Renaissance poetry and drama was anything but serious. Ask a class of English literature students to name their favourite poem from this period and they are more likely to say "The Flea" than "An

Anatomy of the World." Geoffrey Miller has an interesting theory about our fascination with ingeniously frivolous poetic word games. It's all to do with sex. According to Miller our creative intelligence has evolved through sexual selection. While peacocks appeal to their mates' visual sense, we appeal to the mind: "We load our courtship displays with meaning, to reach deeper into the minds of those receiving the signals" ("Looking" 14). Just as the peacock's tail serves no practical function, so evolution has added "baroquely ornamental towers of creativity to our plainly utilitarian foundations of perception, memory, motor control and social intelligence" (14). But that doesn't mean that creativity continues to serve an exclusively sexual function. Once our brains have been wired up for creative thought, the same hardware can be used to run a seemingly endless variety of programmes. The result is "a human creative intelligence that can flood the planet with fictions, but that judges such displays more often for their capacity to excite, intrigue, entertain, and distract, than for their capacity to remind us of the stark, mortal lonely truths of human life" (14; see also *Mating Mind*).

Neo-Darwinian theory is worth listening to, not just because it provides us with an amusingly plausible explanation for our human predilection for the fantastic and the useless (including postmodern literary theory), but because it can give us the confidence to return to those human universals without which the very notion of literature is, as postmodernism itself insists, meaningless.

Literature's record of bringing about social change is not impressive. But with its unique ability to give us the illusion of getting inside another person's mind, one thing it *can* do is tell us about how people think in cultures that are remote, either geographically or chronologically, from our own. In his great book on Freud, Richard Webster writes: "one of the satisfactions afforded by literature is to be found in the way it allows readers to recognise as a part of common humanity feelings which they had previously regarded as individual or private" (480). This is one of the chief justifications for the study of literature: imaginative contact with other minds in periods or cultures remote from our own helps us to appreciate that our common humanity is more important than the ethnic and religious differences that continue to create so much havoc in the modern world. In short, literature can help to teach us the value of tolerance. But deny that there is such a thing as a common humanity, and one of the most powerful arguments for tolerance immediately vanishes. Combine that with postmodernism's extreme relativism—

which logically means that in the absence of human universals there can be no rational grounds for preferring one set of values to another—and tolerance acquires a quite different meaning: it means that we are obliged to tolerate regimes that are themselves brutal and *in*tolerant (Gellner 84–85). It's time we got over our misplaced embarrassment about human nature and recognised anti-humanism for what it really is.

Conclusion

Confrontational in style, and contemptuously dismissive of all that preceded it, anti-humanist Theory has made many enemies (Bradshaw; Carroll, *Evolution*; Crews; Devaney; Ellis, *Against*; Etlin; Freadman and Miller; Jackson, *Dematerialisation, Poverty*; Lee; Sokal and Bricmont, *Intellectual Impostures*; Tallis, *Enemies, Not Saussure*; Vickers). E. P. Thompson, one of Theory's first opponents, made it clear that Althusserian Marxism was, in his view, one of the great intellectual frauds of the twentieth century. Thompson never had any time for Theory. But when even its former champions acknowledge their disillusionment with literary postmodernism, it's obvious that in its more radical form Theory has reached an impasse. In *The Illusions of Postmodernism* Terry Eagleton asks: "What if the left were suddenly to find itself . . . simply washed up, speaking a discourse so quaintly out of tune with the modern era that, as with the language of Gnosticism or courtly love, nobody bothered any longer to enquire into its truth value? What if the vanguard were to become the remnant, its arguments still dimly intelligible but spinning rapidly off into some metaphysical outer space" (1). He answers his own question: "There is, of course, no need to imagine such a period at all. It is the one we are living in, and its name is postmodernism" (20).

Anti-essentialism, and the cultural relativism that is its corollary, are the core of postmodernist thought, and inevitably lead us into a world where, to quote Forster once more, "everything exists and nothing has value." But we don't have to take this route. An alternative lies in a neo-Darwinian view of human nature. Darwinism has been described as "amongst the most comprehensively successful achievements of the human intellect" (Cronin, *Ant* 431). The evolutionary psychology that Darwin himself hinted at (*Descent* 158–84), but which has only flourished in such a remarkable way over the last twenty years, offers a way out of our present theoretical impasse. Neo-Darwinism may not provide the basis for a new critical practice, and there's no reason for insisting that we import evolutionary psychology into literary criticism (though it certainly provides a sounder basis for thinking about human behaviour than Lacanian psychology does). What neo-Darwinism does offer is a body of evidence powerful enough and sufficiently well established to give us the confidence to challenge the orthodoxies that are seldom questioned in modern primers on Theory, and to re-endorse the human universals without which criticism cannot exist, and history becomes impossible. It may even help us to recognise Shakespeare's humanism for what it is.

21

The Reality of Illusion

JOSEPH ANDERSON

The massive outer world has lost its weight, it has been freed from space, time, and causality, and it has been clothed in the forms of our own consciousness. The mind has triumphed over matter and the pictures roll on with the ease of musical tones. It is a superb enjoyment which no other art can furnish us.

HUGO MUNSTERBERG, *THE PHOTOPLAY: A PSYCHOLOGICAL STUDY*

"Why does a movie seem so real? And why do the spokes of a wheel turn backward?" These are questions that many untutored film viewers ask in one form or another, but my interrogator at the moment was not untutored. He was, perhaps, unschooled in theories of film but well trained in the art of asking questions. He was, after all, an attorney.

As a movie viewer, he had no doubt many times been caught up in the enchantment of the world of the motion picture only to have the spell shattered by the intrusion of a stagecoach or carriage wheel that perversely rotated in the wrong direction. His lawyer's suspicions had been aroused. His sense of reality had been toyed with. He knew that something was not quite right. Now he had before him an "expert witness" from whom he would extract the truth.

I squirmed in my chair and perhaps failed to look him directly in the eye, for I knew that out of either naïveté or, worse yet, practiced lawyer's cunning,

Originally published in the work: *The Reality of Illusion: An Ecological Approach to Cognitive Film Theory* by Joseph Anderson. © 1996 by Joseph D. Anderson; reprinted by permission of the publisher.

he had come upon a major inconsistency, a central paradox underlying the art of the motion picture—its capacity for realism and its denial of reality. A generation of film theorists before me had lined up to argue that the motion picture was *not* entirely realistic and therefore could take its place as a bona fide art. Others had argued that the value of the motion picture lay precisely in its capacity for realism. I knew that if I ventured in either direction, mountains of evidence could be weighed against me. Worse yet, he seemed to be asking about more than realism; he wanted an explanation of his experience of a motion picture. What could I say or do? I was trapped.

I took a sip of coffee and slowly looked my inquisitor in the eye. Fortunately, I was not testifying before a court of law but having a cup of coffee with my learned friend. Nevertheless, I responded as truthfully as I knew how. "They are illusions, both the sense of reality and the wheels that rotate in reverse."

My friend leaned against the back of his chair and lightly brushed his moustache with the tips of his fingers. "You have not answered my question," he said evenly. "You have merely given me the word *illusions*, a name, a category, not an explanation. A category is not an explanation. What is required is a reason, a cause, at least a relationship, perhaps a mechanism. If you answer 'illusions' to my question of why movies look so real and yet spoked wheels turn backward, then you must explain what illusions are and what they have to do with motion pictures."

I began searching through my pockets for some change for the waiter. "We had better adjourn this session," I said emphatically, fearing the answers to his questions would fill a book.

What makes the lawyer's tale worth telling is that although we live in a time when crime flourishes, minorities are oppressed, women are victimized, and evil abounds in the world (and these are things with which he is professionally familiar), he chooses to focus not on the overt content of particular motion pictures, but on the source of the power of the motion picture generically, power that he has no doubt witnessed by effect in the world at large, but more to the point, a power he has personally felt while sitting in the theater watching a movie. He wants to know why a motion picture gives him compelling reality in one moment and takes it back in the next. Is it something in the picture or something in him that provides this push-pull in and out of the world of the motion picture? In his unguided and perhaps naïve search for an understanding of the great attractive force he feels tugging at him from the screen, he had focused upon the obvious, the interface between the film and the viewer, an area that professional film scholars have phobically avoided for almost a century. By a strange progression of events, the longer scholars have studied film, the farther they have moved from the interface between the film and the viewer and, it might be argued, from an understanding of the source of cinematic power.

There was no way my barrister friend could have known that he had innocently invaded the domain of film theory, which has existed almost as long as the motion picture itself and has for some time been distinguishable from its sister disciplines, film history and film criticism. If these two disciplines have concerned themselves with the events of film's emergence, with placing its development in the context of other events, with finding causes, and with interpreting and evaluating specific films and groups of films respectively, film theory has endeavored to answer the most fundamental questions as to the nature of film, questions relating to what film is and how it works. Over the last century, several film theories have been offered, usually setting forth one or more principles along with supporting arguments and proposed implications.

For the benefit of my learned friend and anyone else who may not have closely followed the progress of film theory throughout this century, I offer a thumbnail sketch with commentary of the history of film theory. The first film theorist, Hugo Munsterberg, came to Harvard in 1892 at the invitation of William James, not to be a film theorist of course, but to help set up a psychological laboratory. It was as one of the founders of American psychology that he wrote *The Photoplay: A Psychological Study*, published in 1916.

Munsterberg was familiar with the empirical research being done across a broad spectrum of psychological areas. Much of it was being carried out in his own laboratory at Harvard. The tools of what would become the social sciences were being discovered, and they were being put to every conceivable task. The research of the day was a pragmatic enterprise dedicated to finding out how things work. Munsterberg brought this interest in practical problem solving along with his specific knowledge of the results of scientific experiments in psychology to the study of film. In doing so, he demonstrated that empirical investigation can illuminate our understanding of the motion picture. He concluded from his short but intense study of the interface between the film and the viewer that the motion picture is structured in a way that is analogous to the structuring processes of the mind. From such a principle, he was able to elaborate a complex and inclusive theory of film.

It was a stroke of luck for the field of film theory to have as its founder one of the most brilliant and educated men of the day. Specifically, the good fortune lay in that one of the very few people capable of analyzing the interface between the motion picture and the mind of the viewer chose to do so. He set film theory clearly on a path that would have confronted the basic questions about the nature and function of film in a direct and systematic way. Unfortunately, his was a path no one chose to follow.

Those film theorists writing during the first half of the century beginning with Munsterberg and concluding with André Bazin (including Soviet theorists Lev Kuleshov, V. I. Pudovkin, and Sergei Eisenstein, along with Rudolf Arnheim and Siegfried Kracauer) are usually considered together, and their work collectively referred to as classical film theory. Though they diverge greatly in their theories of film, these theorists are united both by their proximity in time and by a common interest in defining film as an art equal in status to the more traditional fine arts such as painting, dance, and theater.

The first wave of film theory was devoted to the defense of film as art, and once film was successfully established as an art meriting the same attention as the other arts, the direction of film theory was pretty well set. Enthusiastic young filmmakers and cineasts in several countries sought to do for the cinema what had been done for the other fine arts: to define the medium's essential components (as art) and to analyze their manifestation in particular works in that medium.

Theorists as divergent in geography and viewpoint as Bazin in France and Eisenstein in the Soviet Union chose to dwell upon issues related to how a motion picture might best be structured in order to maximize those aspects of the medium they each believed best revealed its unique character. Eisenstein set forth a theory of montage informed by Hegel's principle of dialectics. He described the process of montage as the collision of two shots (or particular aspects of shots) as a dialectical clash of two material concepts, resulting in a third and totally new abstract concept. And André Bazin offered a theory of film consistent with the philosophical movement of phenomenology, which denied a split between mind and matter, between the self and the world. For Bazin, the film was neither a product of the mind nor a clash of concepts but rather a photochemical record of reality. Upon this central premise he built an intricate and subtle realist theory of film.

The *auteur theory* that grew out of the publication *Cahiers du cinéma* in the fifties also accentuated the focus on film as an art, asserting that if film is an art then the director is the film artist. By virtue of this logical construct, it became possible to analyze not only an individual work of film art, but to look at the whole of an author's *oeuvre* as in the study of literature, drama, and painting. The way was thus paved for film to enter the academy in the full humanist tradition.

In the decades of the sixties and seventies, a number of film scholars, most notably Christian Metz, held that film was more than an art, that it was a language, and they based their film theory upon the linguistic theory of Ferdinand de Saussure. From such a perspective, they set out to develop a semiotics of film, to see film as a language or at least language-like and to identify its components. In semiotics, there was hope of answering the question of what film is (a system of signification) and to come to know how it works in a thorough and disciplined way.

But the optimism of classical and early semiotic film theory was soon supplanted by what must seem to the noninitiate as a most bizarre program derived from a marshaling of Freudian psychoanalysis in support of an academic strain of Marxism. The theory was an admixture of the neo-Marxism of Louis Althusser and the neo-Freudianism of Jacques Lacan in which concepts from psychoanalysis were fused with Marxism and applied to film (Carroll, *Mystifying*). From this perspective, film in general, as well as a specific motion picture, was seen as a covert and perhaps unwitting instructor of political ideology. For example, movies from America in particular were believed to propagate the concealed assumptions of capitalism (which carried a negative value in the psychoanalytic/Marxist system). Film in general, and indeed any particular film, came to be seen as a patient symptomatic of a sick society. Desires systematically repressed by the ruling elite could be brought to the surface and revealed, thus exposing the social disease. The film became the patient, and the film theorist took on the role of psychoanalyst. In time, feminist film theory came to occupy a section of the larger purchase of psychoanalytic/Marxist theory and tended to see film as an instrument of oppression and victimization. The role of the film theorist became that of exposing hidden agendas of power embodied in the films themselves.

In recent years, film studies, along with the arts and humanities collectively, has fallen into an attitude characterized as postmodernism, an attitude that has tended to revel in the eclectic, to advocate a revisionist view of history, and to retain a fascination with the revealing of things hidden. I use the word *attitude* rather than *theory* because while postmodernism appears to encompass many positions it is not itself a theory; it does not set forth a general principle; it is not a related body of facts. It is not even a hypothesis. It can be defined only by what it is not: it is not modernism, it is not science, and though it embraces eclecticism, it is none of the historical entities from which it pilfers. The attitude is one of self-absorption, and the perspective is elitist. There is the not-so-hidden assumption of a kind of worldliness, of a sophisticated cynicism on the part of initiates. Any view that is nonreflexive or nonironic is characterized as naïve.

An apparent advantage to writing from this perspective is that film writers may feel free to appropriate the language of psychoanalysis, Marxism, or any pop culture movement or special interest cause, without assuming responsibility for their theoretical imperatives. Such writers may claim adherence to no theory at all. From such a perspective, film is seen merely as a vehicle for revealing problems of social conflict and authority. Postmodernists look *through* the medium of film and discuss its overt content (or perhaps what they see as its *covert* content). To the extent that they concern themselves with the medium, they are interested in it only as a purveyor of ideology. In the postmodernist era, there is no need to ask what film is or how it works. There is no need to pursue an understanding of the nature of film *qua* film. There is no need for film theory.

In its inherent nihilism, its self-conscious quoting of the recent past, and its position in time at the end of both a century and an era, today's postmodernism has much in common with the mannerism that followed the Italian Renaissance and saw the sixteenth century to a close. If the analogy holds, there is hope for the future, for as the sixteenth century ended, the disillusionment of mannerism gave way to a new surge of the human spirit expressed in the art of Bernini and Rembrandt and the science of Galileo and Newton.

Actually, a life-affirming, reality-embracing revolution is already under way that offers a refreshing alternative to the effete cynicism of the postmodernist era. Scholars from such diverse fields as perceptual and cognitive psychology, linguistics, artificial intelligence, neurophysiology, and anthropology, who have confidence in the scientific method and an interest in understanding the workings of the human mind, are sharing information in pursuit of their common goal. They have spawned what has been called the cognitive revolution.

By the mid-eighties, a few courageous film theorists suggested that cognitive science might be a more productive path than the then-pervasive, psychoanalytic/Marxist approach to film study. Notably, David Bordwell in *Narration in the Fiction Film* suggested, among other things, that a film spectator might be *cued* by a film rather than *positioned* by it. And Noël Carroll in *Mystifying Movies* argued in detail that existing film theories served more to mystify than to explain the workings of motion pictures.

In spite of science's unparalleled success in explaining the workings of the universe, the efforts of film theorists to apply the methods and findings of cognitive science to the issues of film studies have not met with unanimous approval from film scholars. Many have rejected not only the special thrust of cognitive science but science outright. Indeed, some are apparently fixed in the position that science is but another set of conventions, that its claim to special status is no greater than could be made by any culture for its religion or institutions, that science is no more than a tool of cultural imperialism with which Western culture attempts to maintain its dominance over the rest of the world. Though such a view apparently has a certain surface appeal, in order to maintain such notions advocates must somehow be willing to deny either the existence of reality itself or the possibility of knowing it.

Admittedly, the course of science has not been a straight path of progress. It has taken twists and turns and has sometimes backtracked. As Karl Popper has observed, it is, after all, a human endeavor like all human endeavors, albeit a special one.

> The history of science, like the history of all human ideas, is a history of irresponsible dreams, of obstinacy, and of error. But science is one of the very few human activities—perhaps the only one—in which errors are systematically criticized and fairly often, in time, corrected. This is why we can say that, in science, we often learn from our mistakes, and why we can speak clearly and sensibly about making progress there. In most other fields of human endeavor there is change, but rarely progress. ("Truth" 78)

That science *can* be said to progress is part of its claim to a special category among human enterprises. Science is built upon the assumption that there is a physical world and that it can be known by observation. Science proceeds by the formation of hypotheses about the world that are then tested. The tests are required to be open and repeatable, and results are continually questioned and reevaluated. It is the specialness of science, its uniqueness among human endeavors, however, that some film scholars have refused to grant. This is not the place for a full exposition of the argument, but the fact that film scholars took such a position for ten to fifteen years left film study in the predicament in which we found it at the beginning of this decade. In the absence of any criteria for establishing the relative accuracy of a given theory, we were left with what E. H. Gombrich has called *conventionalism*, an attitude that counts all theories as equally valid, all signs as conventional, all expectations as solely culturally determined, a "manifestly absurd relativism" (Krieger).

It is against such a backdrop of almost a century of film theory ranging from the brilliantly creative to the totally absurd that I must attempt to explain to my friend why movies seem so real. But my task is made much easier by the work of cognitive film theorists and cognitive scientists. I shall not repeat the work of my colleagues Bordwell and Carroll, who freed film theory from the chokehold of the psychoanalytic/Marxist paradigm in the eighties and replaced it with the perspective of cognitive science, which though not yet universally accepted by film scholars is now firmly in place. In the following pages, I shall take for granted the assumptions and methods of science in general and cognitive science in particular. My efforts will be directed toward placing cognitive science in a relationship to the film medium, and I shall describe a metatheory that can encompass both. I call this metatheory *ecological* because it attempts to place film production and spectatorship in a natural context. That is, the perception and comprehension of motion pictures is regarded as a subset of perception and comprehension in general, and the workings of the perceptual systems and the mind of the spectator are viewed in the context of their evolutionary development.

I shall simply begin at the beginning. For the motion picture, the beginning is the interface between a spectator and sounds and images on a screen. No detail of this interface can be responsibly ignored. When viewers sit before a theater or video screen to watch a movie, they face a sequence of images and sounds. The precise nature of the sequence is neither arbitrary nor random, but of a most carefully crafted order. The makers of the movie have often spent many months and millions of dollars to achieve perfection of individual elements and overall form.

The particular way a motion picture is crafted, those elements often referred to collectively as *style* have, particularly in the United States, developed in the direction of *accessibility*. Hugo Munsterberg in 1915 predicted that film would become the domain of the psychologist ("Why" 31). It has instead become the province of the entrepreneur. From the beginning, American films were subject to the contingencies of the free enterprise system; capital was raised and the picture was produced and then exhibited to viewers for the price of a ticket. For each picture, the success or failure of the entire venture was in the hands of the thousands of individual consumers who either purchased a ticket or did not. If a particular motion picture failed to sell enough tickets to return a profit to the investors, its producer and/or director was not likely to get the opportunity to make another picture. That is the way the system worked as the so-called *Hollywood style* developed in the second, third, and fourth decades of the twentieth century, and that is the way it works today. Such a system caused producers of motion pictures to make movies that appealed to a wider and wider audience. And that appeal is largely measured by the film's accessibility. That is to say that individual moviegoers are more apt to buy tickets to movies that are accessible to them, accessible in the most fundamental ways, such as whether it is possible to comprehend the fictional events that occur on the screen and to follow the basic story line. Apart from the obvious problem of language differences, which can be largely overcome with subtitles or dubbing, problems of accessibility are problems of perception. And though few if any of them had training in perceptual psychology, filmmakers in Hollywood proceeded to discover how to make their products accessible to individuals across economic and class boundaries and across national and cultural boundaries as well. A number of rules of thumb developed: that every shot should advance the story line, that actors should avoid broad gestures and never look directly into the lens, that there should be a change of camera angle and image size from one shot to the next, that the camera should be kept on one side of the action (the 180-degree rule), that the fictional world of the movie should never be intruded upon by the workers and equipment involved in its construction, and that the story should be told in action rather than words whenever possible. These rules were followed because they tended to make the events of the picture understandable to the individual spectator. For reasons both obvious and profound, following these rules made the picture accessible to a wider audience. Motion pictures made by these rules were generally thought to be more realistic, and indeed they were. Purely by trial and error, the moneymen, the technicians, and the artists who made up the American film industry succeeded in developing a style of filmmaking that was potentially accessible to every human being on earth. Whatever its shortcomings, the classical Hollywood style became more universally accessible than any of its competitors, and it remains so today.

Let me note that accessibility is of theoretical and not political interest here. My purpose is not to argue for a privileged status for the classical Hollywood style, but to point out that the problem of accessibility in a motion picture is not merely a matter of culture. It is more fundamentally a matter of perception.

It is readily apparent that the motion picture/viewer interface is not equally balanced, for while a motion picture is created specifically for the viewer, the viewer was not created for watching motion pictures. The implications of the imbalance, however, are not so easy to grasp. It may be helpful to draw an analogy. The motion picture can be thought of as a program. And it is more precisely a program than either a language or a mere set of stimuli. It is a very complex set of instructions utilizing images, actions, and sounds, a string of commands to attend to *this* now, in this light, from this angle, at this distance, and so forth, and to recall earlier sequences and anticipate future ones. The program cannot be run on a projector or a videotape machine. These devices have no capacity to interact with the instructions. The program can "run" only in the mind of the viewer.

The viewer can be thought of as a standard biological audio/video processor. The central processing unit, the brain along with its sensory modules, is standard. The same model with only minor variations is issued to everyone. The basic operating system is also standard and universal, for both the brain and its functions were created over 150 million years of mammalian evolution.

Filmmakers can be seen as programmers who develop programs to run on a computer that they do not understand and whose operating systems were designed for another purpose. Since the filmmakers/programmers do not understand the operating system, they are never sure exactly what will happen with any frame or sequence of their programs. They therefore proceed by trial and error. They follow certain filmic conventions and then go beyond them by guessing. They test the outcome (that is, how their programs will be handled in the minds of viewers) by becoming viewers themselves and running the program in their own minds. This might be considered a fairly risky procedure, because each human mind is a little different, different subroutines may be initiated by the same instruction, and different meanings may result in each mind. But the filmmakers-turned-viewers are not proceeding completely recklessly and irresponsibly, because the "hardware" of the mind and most of the "software" is standard and universal.

Such an analogy between the human mind and a computer is useful to the extent that it serves to help us grasp the relationships between the cinematic apparatus, the filmmaker, and our own processing systems. It also points out just how different our mind is from present-day computers. Both are capable of computing functions of considerable complexity, but the computer that I now employ in writing this text is a serial, digital, highly programmable device; the mind is none of these things. It is, first of all, much more complex. The brain itself is very likely the most complex structure in the universe. (The structure and function of the brain may be so integrated that to speak of hardware and software even by analogy may be unjustified.) It processes in parallel as well as sequentially, and it frustrates most attempts at reprogramming. Yet the brain is not as good as the simplest commercial computer at carrying out tasks like keeping track of expenses or balancing checkbooks, and there is little one can do to change its capacity or procedures for calculation. But it is very good at guiding our movements in three-dimensional space, so we do not bump into trees, fall into chasms, or lose our way in the countryside. And it is very good at precisely locating an object of prey in space and guiding our arm in the hurling of a stone or spear with deadly accuracy. The point is that by whatever device or analogy, we must understand that our brains and sensory systems, indeed our very consciousness, our sense of self, our mind in all its implications, is the present result of past evolution, and for most of the time during this evolution, when our capacities were being cruelly sorted by the processes of natural selection, the contingencies of existence were quite different from what they are today.

For example, the origins of the human visual system vastly predate the emergence of humans. To move about with speed and agility, and to hunt successfully, an animal needed accurate information about the location of things in space. By the time humans emerged, the visual system in mammals was pretty well defined, and its central organizing principle was *veridicality*. This is to say that an individual's perception of the world needed to be a very close approximation of that world. It had to be accurate enough to act upon because the consequences of error were severe. If a creature could not detect the presence of a potential predator, it could not take evasive action, and its chances of surviving long enough to reproduce were greatly diminished. If a predator were not correct about the location of prey in space, it would not be successful as a hunter. It would starve, its young offspring would starve, and its genes would never be passed on to succeeding generations. The cold, indifferent process of evolution selected for veridicality in the visual system, not through purpose, but through contingency.

An interesting paradox of human perception is that although perceptual development has tended toward veridicality, we at times perceive illusions. The simplest definition of an illusion is that it is a nonveridical perception. But it is a *wide-awake* nonveridical perception. If we are asleep we may dream, if we are drugged or deranged we may hallucinate, and if we fast and meditate we may have a vision, but illusions occur even when we are awake, sane, and skeptical. For example, in the three-dimensional illusion, we perceive depth, when in reality, there are but two slightly offset images projected simultaneously upon a flat screen. Such illusions are particularly revealing about our perceptual systems. Visual illusions, like 3-D, result when the visual system, following its own internal instructions, arrives at a percept that is in error if compared to physical reality. That is, illusions occur *because* the system follows its own internal rules even though the resulting percept is in error. To perceptual psychologists, and by extension to those of us who would understand motion picture viewing, illusions are of special interest because they reveal the rules according to which the perceptual system functions, rules that are ordinarily invisible. By studying the system when it makes an "error" we can see the rules exposed; by studying the rules we gain a greater understanding of how the human mind interacts with a motion picture.

The motion picture, or the phenomenon of cinema, can best be understood by utilizing the methods of science within an ecological context. We are and always have been part of a larger ecology. In this interlocking relationship with the larger ecological setting, we developed, through eons of evolution, elaborate and sophisticated capacities to gain information. Today, we interact with the synthesized images and sounds of a motion picture, but we have no new capacities for gaining information from them. We have only the systems developed in another time, in another context, for another purpose. We must process the images from the glass-beaded screen and sounds from the metal speakers with the same anatomical structures and the same physiological processes with which we process scenes and sounds from the natural world. To ask how we process continuity and character and narrative in motion pictures is to ask how the forces of evolution equipped us to know where we are in space and time, to make rapid judgments of character, and to narratize the events of our existence. When we delve into research on these questions, we quickly realize that our capacities exist within boundaries, and seldom if ever can those boundaries be overridden by transitory cultural fad or clever linguistic fabrication.

22

Darwin and the Directors

FILM, EMOTION, AND THE FACE IN THE AGE
OF EVOLUTION

Murray Smith

It has been commonplace to think of film as "the art of the twentieth century"—not only because the cinema emerged during the first twenty years of the last century as an aesthetically novel and economically powerful institution, but also because it had a major impact on the practice and self-definition of most of the traditional arts. But the acceptance of film as an authentic art, with all of the formal, expressive and symbolic power of literature, drama and painting, was a fraught affair. Most serious critical writing about film up to the 1950s concerned itself with justifying film as an art, and with teasing out those aspects of film which made it distinct from earlier forms. And although the battle is in many ways long since won, even today one often encounters a casual condescension to film on the part of many influential commentators, suggesting that it is still regarded in some quarters as a glitzy but insubstantial upstart, a glamorous newcomer whose seductive appearance flatters to deceive.

Our own time is also one in which Darwinian theory has come of age. But "the age of evolution" also refers to that epoch of human (pre-) history

"Darwin and the Directors: Film, Emotion, and the Face in the Age of Evolution." *Times Literary Supplement*, February 7, 2003, 13–15.

when, biologically speaking, the species *Homo sapiens* took shape. This is the special concern of evolutionary psychologists, who ask how the evolutionary history of the human species has created a legacy of psychological features, functions and preferences which (directly or indirectly) still affects us today, millennia after the environment in which these features were adaptively formed has disappeared. And part of that inquiry has concerned the ways in which our evolutionary inheritance might bear upon such things as morality, beauty, art and fiction. At first sight fiction looks like one of those things that couldn't possibly have an evolutionary explanation, because it is hard to see how stories about non-existent people and places could have been much help in terms of survival. But since we have fiction, the evolutionist would insist that it must have descended to us either because it was selected as something that enhanced survival (that is, it is an adaptation), or, perhaps more plausibly, because it is a by-product of something else that was an adaptation. That "something else" is our capacity to imagine: one thing that sets us apart from other species is our ability to simulate, in our minds, circumstances which we might encounter, or indeed which we have encountered in the past. And in doing so, we are able to rehearse how things might go in circumstances we have not actually experienced. The imagination, in other words, enhances our foresight and supercharges our ability to plan; and it is not hard to see how this improves our fitness in the environment of human action.

Fictions, in turn, engage our emotions, and emotion too is a subject about which evolutionary thinkers have had quite a bit to say. Most of us would accept that our emotions are part of our basic biology, but would equally quickly agree that emotions are a crucial part of social and cultural life, and that our experience of (most) art is emotionally coloured. So emotions span the apparent gulf between our biological, evolutionary inheritance, and the sophisticated, endlessly varied phenomena of our modern cultural existence.

The dominant tradition in Western thought has regarded emotion as a burden to human existence, an impediment to reason—a view manifest in thinkers as varied as Descartes, Kant and the playwright Bertolt Brecht. And if anything, this is a view of emotion even more entrenched in popular culture than it is in the realms of philosophy and art theory—think of those models of supreme intelligence, the emotion-free Spock and Data from *Star Trek* (Gene Roddenberry, 1966–). One doesn't need to listen to news coverage for very long before encountering some manifestation of the idea that someone is being led astray by their emotions, or not taking a sufficiently detached (unemotional) view of a situation.

To be sure, there have always been thinkers representing a different view of the emotions, one which grants them an important, even revered, place in human existence. But evolutionary theory poses the sharpest challenge to the anti-emotion tradition by asking the question: Why would emotions—these subverters of rationality—have been naturally selected at all? Evolutionary theory offers two compelling arguments. First, emotions provide us with a kind of motivational gravity, allowing us to grasp the world and act decisively in it, rather than drifting among an array of equally weighted options. Second, emotions provide a rapidity and intensity of response to a changing environment which reasoning alone cannot provide. Given that we live in a changing and sometimes hostile environment, our chances of survival are enhanced if we have a kind of in-built "rapid reaction" force alongside our more precise, but much slower, mechanisms of reasoning. Whether it's a wild animal or a car suddenly bearing down on us, it is mighty handy that we have an instinctive fear reaction to unexpected loud noises and fast movements—that we leap out of the way immediately, rather than calmly trying to assess the nature of the moving object, its size, speed and intentions. In short, emotions are characterized by what Dylan Evans terms an *evolutionary rationality*—given certain environmental or ecological conditions, particular emotions will serve you far more effectively than would reason divested of all emotional attachments.

The evolutionary view of emotions enables us, indeed exhorts us, then, to take emotions seriously—rather than regarding them as an irrelevance, a distraction or an embarrassment. This is as true for art as it is for life. And it is surely true for film, which not only names many of its genres after emotions (thrillers, weepies, horror movies), but depends to a greater extent than any preceding art on the interplay among emotions as these are expressed in the human face and voice (as well as in posture and gesture). Take a look at any mainstream feature film—and a great many other films as well—and it will be obvious that the visual landscape of these films is dominated by shots in which facial expression is legible (note that this does not mean simply close-ups), while their corresponding soundtracks resonate with the cadences and intonation of emotionally expressive human voices.

Another kind of argument for the foundational significance of facial expression in film is offered by films which depict characters who have lost the power of facial expression—take, for example, the burnt airman's face in *The English Patient* (Anthony Mingella, 1997); or the facial prosthesis masking the disfigured face in the recent Spanish film *Abre los ojos* (Alejandro Amenábar, 1997); or again in the classic French horror film, *Les Yeux sans visage* (Georges Franju, 1959) (figure 22.1). Faces without the capacity to express emotion are striking and troubling, because they deprive us of such a basic means of social interaction. Expressions of emotion point in two directions: inwards, towards the agent's felt state of being, and outwards, to others perceiving and interacting with the agent. The consequences of an inability to

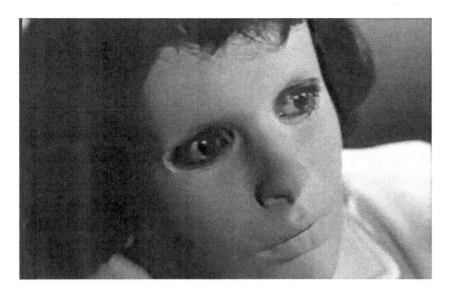

FIGURE 22.1 The inexpressive face in Georges Franju's *Les Yeux sans visage* (1959).

express emotion facially—whether caused by disfigurement or neural disorders, such as Parkinson's disease—can be as severe as the complete destruction of emotional orientation experienced by some victims of brain injury. It is here that evolutionary theory makes a more specific contribution to our understanding of film as an art. Inspired by Darwin's *The Expression of the Emotions in Man and Animals* (1872), there is now a large body of literature on the nature and functions of facial and vocal expression, which discriminates among types of emotion, their degree of cultural variability, their qualities and functions.

When filmmakers, critics and philosophers first began to reflect on the nature of film—something which really took off in the 1920s—they saw as their basic task the need to specify what made film a distinctive art form, and neither merely a passive technology of recording nor a mere parasite on older forms of art (theatre, painting, or literature). And according to one very influential view, associated especially with the great Soviet filmmakers of the 1920s such as Sergei Eisenstein, editing—or montage, as it was known at the time—was the aspect of film that made it distinctive. That is to say, it was the capacity of film to take diverse shots, often derived from entirely disparate or unrelated locations, and weave them together into a new whole, that seemed to the Soviet theorists to be the most original power of film, and one unique to it. And they can hardly be faulted for this insistence; the power of film editing is indeed astonishing. But while montage theory captured a certain truth, it was also a thesis with flaws. It turned out to be one of those truths whose very light obscures other, adjacent truths, whose existence can only re-emerge over time, as the novelty of the original insight wanes.

The most pertinent item of montage film theory for us to consider here is the so-called Kuleshov effect, named after one of the very earliest of the Soviet filmmaker-theorists, Lev Kuleshov. In a demonstration of the potency of editing, Kuleshov edited together a shot showing a man with a neutral expression, glancing off-screen, with shots of several different objects—respectively, a bowl of soup, a child in a coffin, and a sunny landscape. According to Kuleshov, audiences who viewed these variations not only made the assumption that the man was, in each case, looking at the object, but ascribed to him different affective states—hunger, sadness and happiness—and marvelled at his performative skill (even though it was the same shot of the actor on each occasion) (51–55, 200; see also Pudovkin, *Cinema Writings* 168). Thus the power of editing: what we see in each shot is less significant, on this account, than the meanings and emotions which arise out of the editing together of shots. Our interpretation of the man's situation and state of mind changes according to the shots that are juxtaposed with the shot of him.

Evolutionary studies of facial expression, however, give the lie to the notion that shots of faces are entirely plastic, and can be moulded by contextual shots to suggest any kind of emotional state. On the contrary, the work of Paul Ekman—the most well-known contemporary evolutionary scientist of facial expression, and the editor of the recent edition of Darwin's *Expression of the Emotions*—provides extensive evidence that a range of basic emotional expressions are easily and quickly recognized cross-culturally (happiness, sadness, disgust, anger, fear and surprise) (*Emotion*; see also Izard). A shot of a smile, a grimace, a frown, or an angry sneer (figure 22.2)—to take some key examples—cannot just be overridden by adjacent shots which might suggest some other emotional state; the facial expression is itself a powerful determinant in evoking specific emotions. Such facial expressions don't, obviously, tell us everything about the dramatic situation of which they are a part; they have to be interpreted in context, and that context will include many culture-specific things. The key point is that basic facial expressions play a crucial role in giving us an initial orientation towards the scene, on the basis of which we then apply our contextual knowledge.

Basic expressions, then, while very important, are not the whole story. The use of a neutral facial expression in the Kuleshov experiments draws our attention to all those types of expression which exist alongside the basic-emotion expressions. These would include expressions associated with the so-called higher emotions, like love, which admit of more cultural variation, and expressions

FIGURE 22.2 The sneer as a basic emotion expression in the character Begbie in Danny Boyle's *Trainspotting* (1996) and in soccer manager Kevin Keegan.

modified by, in Ekman's words, culture-specific display rules—that is, cultural conventions which govern who can express what sort of emotion in which social contexts (DeStano and Salovey 151–52; Parrott 18–19). The classic example here concerns the different rules evident in American and Japanese society, but let me take one closer to home: the gender-related display rules governing the facial expression of fear in Hollywood films. Lead male characters get frightened in American movies, but they are far less apt to express this facially than either their female compatriots (or minor male characters); think of such "facially laconic" icons as Robert Mitchum or Clint Eastwood.

So there is plenty of room within a Darwinian account of emotions for neutral, suppressed, culture-specific or otherwise ambiguous expressions. And this is a good thing, for such expressions form an essential complement to more transparent basic-emotion expressions in the dramatic fiction film. Indeed directors such as Otto Preminger and Robert Bresson have developed styles in which oblique or ambiguous facial expression plays a key role. Moreover, the emphasis here on facial *expressions* of emotion—visible, physiological manifestations of inner states—should not be taken to overlook the role of less straightforward forms of these same expressions in social communication (including deliberate miscommunication or deception).

The sceptic, however, might still ask: Did we need these evolutionary studies of emotions to know that some emotions are plainly and forcefully legible? When Ekman started out in the 1960s, he began with the—at the time—widespread assumption that all facial expressions were culture-specific and none was universally recognizable. Darwin's thesis had fallen completely out of favour. (And such relativism, it must be said, has been dogmatically assumed for much of the history of academic film studies.) What's more, the fact that basic-emotion expressions are universal lends support to another idea that was important to early film theorists—namely, the notion that film was a peculiarly internationalist medium, that it constituted a kind of universal language. Writing in 1924, the Hungarian film theorist Béla Balázs extolled the power of film to capture facial expression and gesture, which he characterized as the "aboriginal mother tongue of the human race" (42). And in the late 1920s, many filmmakers and critics became agitated about the loss of this internationalism, as the talkies displaced silent cinema and the "universal language" of facial expression was overlaid and obscured by the babble of particular verbal languages.

As we watch a fiction film, then, our ordinary perceptual, cognitive and emotional capacities are engaged, including our ability both to recognize and respond to facial expressions. (In this sense, fiction films were creating "virtual realities" for spectators long before the phrase was invented; virtual in the sense that they are at once imagined realities and yet sensuously apprehended.) Basic-emotion expressions play a vital role here as a bridgehead between actual and fictional experience, and between universal and culture-specific experience. Small wonder that the face was so important to some early film theorists; facial expressions underwrote both the special realism and the internationalism that seemed to be the peculiar attributes of cinema.

But this is still not the end of the story with regard to the significance of the Darwinian account of facial expression. In addition to the universal expression and recognition of a certain range of emotions, there is also evidence that when we see such emotions expressed in others, we mimic them in ourselves. In other words, as we move through social space, we don't only perceive and categorize the emotions of others, we feel them, albeit in attenuated form—a phenomenon which has been discussed by Elaine Hatfield and her colleagues as, among other concepts, *emotional contagion* (a notion which relates historically to the concept of empathy). The adaptive origins of such contagion can be traced to an alarm-and-rescue function – the evolutionary ancestor of altruism—in which individual members of social species would be sensitive to, and feel as if their own, the emotional states of their conspecifics.

A striking instance of emotional contagion occurs at the climax of Alfred Hitchcock's wartime espionage thriller, *Saboteur* (1942). The saboteur has been cornered at the top of the Statue of Liberty, and in desperation climbs out from the viewing deck onto the torch. There is nothing very subtle about the symbolism here: an enemy of democracy splayed helplessly on one of its most famous icons. But Hitchcock peppers the sequence with close-ups of the saboteur's expression of terror—shots which are apt to elicit mimicry on our part, so that we begin to feel the saboteur's fear (figure 22.3). Up to this point, the saboteur has been an object purely of hatred and disgust—unsurprisingly, given the film's wartime theme and context. This contagion creates an unexpected emotional cross-current, however, as for a few moments his fear, and

FIGURE 22.3 Terror and emotional contagion in Alfred Hitchcock's *Saboteur* (1942).

his vulnerability, become palpable for us. The imagined satisfaction of running the traitor to ground and vanquishing him is undercut by these shots.

Perhaps Hitchcock was trying to suggest how we—the defenders of liberal democracy against the Nazi threat—are capable of complex, sympathetic reactions even to our enemies; and that even in victory we might regret that the battle had to be fought at all. What is certainly true is that in a later wartime thriller, *Lifeboat* (1944), which comes to a very similar climax in which a German U-boat captain is pushed overboard from the lifeboat when his malicious intentions become apparent, Hitchcock avoided any shot of the captain at all—let alone a close-up making his expression maximally legible and inviting emotional mimicry. The kind of play with our sympathy and antipathy for villainous characters that we find in *Saboteur* can be traced throughout Hitchcock's career, but it seems that he came to feel (or was made to feel) that such play was out of place in a propagandistic film, in which our allegiances should remain clear and unambiguous throughout, and the creation of sympathy for the Nazi figure, at least at this climactic point in the story, politically and morally inappropriate. Now all of this is to argue that an aspect of the biology of the emotions is enlisted in a cultural and political cause. But the biology of emotional mimicry doesn't determine anything about the politics of the sequence; rather, it demonstrates what Stephen Jay Gould calls *biological potentiality*, that is, the way in which our biological capacities can be exploited in various ways and to various ends ("Biological Potentiality"). Neither is the meaning of the sequence being "reduced" to the biological level. Hitchcock's exploitation of emotional contagion in the sequence and the argument about it is but one link in a chain. We do not have to make a choice between an evolutionary, biological explanation, and a purely cultural one; and if we do make that choice then we weaken our explanation.

I have been happily running together discussion of emotional expression in film with emotional expression in everyday life. It would be naïve, however, to suggest that facial expression, as an aspect of film acting, can be understood wholly as a matter of the imitation of facial expression in life. Any film art is going to modify, shape and redirect facial expression to some degree, if only by bracketing, clarifying and intensifying expressions which strike us as otherwise quite ordinary. There will always be a process of aesthetic reshaping in play—that is, depending on the particular artistic goals of the work in question, facial expression will be moulded in ways which support those goals. I want to outline, very briefly, the strategies of three directors, all of whom develop performative styles of facial expression in strong contrast to the bold expressions we see in Hitchcock: the French director Robert Bresson, and the contemporary filmmakers Takeshi Kitano (from Japan) and Wong Kar-Wai (from Hong Kong).

Kitano has achieved renown as both an actor and a director, and often appears in the lead role in his own films. He has developed a style of performance which presents the face as a blank, expressionless mask, broken only by the occasional smile and facial twitch. Given that Kitano works mostly in the gangster genre, we might regard his approach to facial expression as an extreme stylization of the macho culture of inexpressiveness (the same culture as that of Hollywood's tough guys). Kitano himself cites the influence of Japanese theatrical traditions, and has said that his aim is to make the intentions and feelings of his character enigmatic. We might think, therefore, of Kitano as pursuing an aesthetic of emotional reticence (figure 22.4).

Robert Bresson is from a much older generation, his career running from the 1940s to the 1980s. Bresson developed a still more consistently austere style of performance, in which facial expression as such is displaced, in two ways: his performers modulate their expressions only very subtly, and Bresson will often cut to parts of the body other than the face at moments of dramatic intensity. The effect is a kind of ritualization of human action, by which Bresson aims to remove his characters from the sphere of ordinary human psychology, asking us to regard them instead as governed by a kind of spiritual destiny (figure 22.5).

The films of Wong Kar-Wai depend on a more recognizable, naturalistic style of performance, but this is overlaid by a cinematic stylization which often presents characters obliquely: in profile or from behind, partially hidden behind other objects, sometimes obscured by layers of translucent

FIGURE 22.4 Emotional reticence in Takeshi Kitano's *Sonatine* (1993).

FIGURE 22.5 Ritualized bodily gesture in Robert Bresson's *L'Argent* (1983).

material (curtains, rain) or made hazy by special effects (figure 22.6). David Bordwell has suggested that Wong's work is defined by an aesthetic of the glimpse (*Planet Hong Kong* 285), in which characters' emotional states are often at one remove from us, echoing the way that the romantic desires of his characters are perpetually just beyond their grasp.

Hitchcock, Bresson, Kitano, Wong: each of these directors, I have argued, has developed very different artistic strategies in relation to facial expression. The fact that Hitchcock's approach is more "naturalistic," however, in the sense that his performers adopt expressions which are manifestly similar to those we encounter in life, while in the cases of Bresson and Kitano we find an attenuation of emotional expression which we would find disturbing or even pathological in life, does not render the evolutionary perspective on emotional expression irrelevant. The latter kind of directors provide a more culture-specific form of "negative evidence" for its relevance: understanding how facial expression ordinarily functions sharpens our appreciation of the aesthetic sculpting of expression by particular artists.

Contemporary opposition to Darwinian theory, at least as it is applied to social and cultural matters, can be boiled down to three groups: creationists (of course), humanists (of a certain sort), and many of those on the political Left. Creationists I shall leave aside; the criticisms levelled by humanists and leftists, however, concern me more. Traditional humanists and leftists often argue that an appeal to natural science, and especially evolutionary theory, necessarily results in explanations that are reductive, deterministic

FIGURE 22.6 Oblique and elliptical depiction of emotion in Wong Kar-Wai's *In the Mood for Love* (2000).

and politically reactionary. I hope that, in arguing for the pertinence of natural science to the theoretical understanding of cinema art, I have at least suggested the inadequacy of the conventional wisdom. I hope too it is clear that I am not arguing that the concerns of artists or critics can be reduced to those of scientists. Artists, whatever else they are, are craftspeople, who use the techniques of a given medium to fashion objects or performances that are designed to elicit particular responses. Good artists possess a practical knowledge of how the medium that they work with works *on* audiences; and one of the roles of the critic and theorist of art is to elucidate this knowledge with an explicitness which we do not expect from practitioners. An insular humanism that disdains scientific insight, in the misguided belief in its complete autonomy from the stuff of the natural sciences, is a much impoverished one. For this reason, the humanities cannot simply turn away from the hypotheses and discoveries of science, including evolutionary science; and a human science confident of its own grounds and legitimacy will not feel the need to do so.

23

What Snakes, Eagles, and Rhesus Macaques Can Teach Us

David Bordwell

I once projected a kung fu film, *Snake in the Eagle's Shadow* (Yuen Woo-ping, 1978), in a Cantonese-language print lacking both English dubbing and subtitles. The question was, How much of the film could the audience grasp without knowing its native language?

Actually, quite a lot. It might seem too obvious to mention, but we in the audience perceived the film. In *Snake in the Eagle's Shadow*, we saw patches of color, patterns of light and dark, trajectories of movement, and changing shot displays. But our perception wasn't really of abstract configurations. Humans evolved to detect objects and actions in a three-dimensional world, and in watching *Snake* we definitely *recognized* things. We saw young and old men and women, all going about activities in a voluminous space. We saw a youth involved in social interactions, in locales—a village, a clearing—that we could recognize, at least generically. We heard speech, and if it had been in English, we could have grasped it as quickly and involuntarily as we

From *Poetics of Cinema*, pp. 43–53, by David Bordwell. Copyright 2008. Routledge and Taylor and Francis Group; and *Moving Image Theory: Ecological Considerations*, edited by Joseph D. Anderson and Barbara Fisher Anderson. Published by Southern Illinois University Press.

grasped the sight of a human face. We heard noises, such as the blow of fists on flesh, or labored breathing, or the sound of a cobra hissing. We heard music, mostly in a tradition we recognized, and it registered as such.

But also, and more interestingly, we viewers of *Snake in the Eagle's Shadow* understood a lot of the story. We understood that the protagonist was a young servant in a kung fu school. He wants to learn martial arts and meets an old man who, despite his shabby appearance, is a master fighter. The youth undergoes arduous training and eventually comes to defeat a villainous master. These features aren't simply given in perception; we had to bring in large domains of knowledge to arrive at this story. Viewers who were familiar with the kung fu genre could structure the film along familiar lines, but even those who weren't martial arts fans understood a good deal of the action because they had skills in understanding any type of story. At one point, when the protagonist sees a cat fight a cobra, all of us realized that he was inspired to model his kung fu technique on the cat's attack. Call our activity of this sort *comprehension*, a grasp of the concrete significance of the perceptual material as patterns of social action. In this case, the patterns are presented in the form of a story.

Finally, spectators used the film in various ways. *Snake in the Eagle's Shadow* wasn't intended to be shown in a college classroom, but I drew it into my own agenda. Some of the students took the film as an occasion to celebrate the prowess of Jackie Chan. Others took it as proof of the artistic bankruptcy of Hong Kong cinema. Some students from Hong Kong read it as a statement of local pride in the face of adversity. Those who practiced martial arts themselves spotted techniques that they could try themselves. Let's say that all of us *appropriated* the film, in however disparate ways.

These types of activity suggest how poetics can address what I'm calling the *processing* of films by viewers. If poetics is concerned with how filmmakers use the film medium to achieve effects on spectators, we ought to have some idea of how those effects might be registered. Film researchers aren't psychologists or sociologists, but we can draw upon the best scientific findings we have to mount a plausible framework for considering effects. The poetics I propose is thus *mentalistic*: It assumes that we can characterize the spectator's embodied mind as engaging with the film. It's also *naturalistic*, presuming that scientific investigation of mental life is likely to deliver the most reliable knowledge. I'd also propose that the best mentalistic and naturalistic framework we have available is that provided by what we can broadly call the *cognitive* approach to mental life.

Adopting this perspective makes some researchers worried. Some object that it neglects the influence of society, ideology, or culture on viewers. But this is to assume that a mentalistic and naturalistic framework focuses wholly on individuals. It doesn't. Cultural activities are mental in an important sense: They're learned, recalled, rethought, and so on by the embodied minds of social agents. The framework presumes some intersubjective regularities of mental activity across individuals, but cultural theorists do the same thing when they discuss how members of a subculture come up with a resisting reading. Other critics have argued that conceiving of the spectator in the way I propose neglects the differences of race, gender, ethnicity, and other markers of identity. Yet clearly there are common effects across such groups; people of all sorts feel suspense in a thriller and sadness in a melodrama. Studying such commonalities isn't on the face of it unreasonable or uninteresting. Moreover, there's always a degree of idealization in discussing spectatorship. Just as linguists create the idealization of "the native speaker" in order to understand grammatical principles, virtually all researchers are obliged to idealize the spectator, even the female or African American spectator, to some degree. Finally, although not every conceptual framework fits well with every cluster of research questions we might want to float, I think that some identity differences can be understood from the standpoint of poetics.

We can start to understand the effects of films by borrowing a distinction from classic cognitive psychology, that of top-down and bottom-up mental processing. Top-down processing is concept driven; bottom-up processing is data driven. A classic instance of top-down processing is problem solving. Given a crossword puzzle, you draw upon your stored knowledge about language and the world (including the stratagems of crossword puzzle designers) to fill in the blanks properly. By contrast, bottom-up processing arises from a moment-by-moment encounter with the world. As you enter an unfamiliar room, for instance, your visual system picks up information about edges, brightness differences, and a host of other features that coalesce into a spatial whole.

Our brains can process information in both "directions" at the same time, so any particular experience will be a mixture. While searching your memory for the right word, you get bottom-up information about the crossword puzzle from the written clues and the array of empty spaces and black ziggurats on the page. Upon entering an unfamiliar room you quickly rework the perceptual input in the light of knowledge. Identifying a chair in the shape of a beanbag and a lamp bubbling like lava leads you to make a higher-level inference about the tastes of the people living there. On the whole, bottom-up processes are fast, involuntary, cheap in cognitive resources, and fairly consistent across observers (Pylyshyn, chaps. 2 and 3). In an important sense, all TV viewers watching the horrendous crash of the airliners into the World Trade Center saw and heard the event in the same way. Top-down processes are slower, more voluntary, more expensive in cognitive resources, and more variable across observers. Having seen the Trade Towers assault, viewers interpreted its significance in different ways—as an act of war, as a response to globalization, and/or as a counterthrust to U.S. imperial ambitions.

Perceptual uptake occurs in milliseconds, and for good reasons. Our brains evolved in situations in which survival demanded reasonably accurate information about spatial layout and other agents. Consequently, the activity of our perceptual mechanisms is hidden from us; we can't watch our retinal image or our neuronal firings. And although experimental films like James Benning's *Ten Skies* (2004) create noticeable visual effects, like illusory movement, we can't really probe the mental hardware yielding the experience. Nearly as fast are intuitive judgments, as when we sense that a person is arrogant or kindly, or when we just know we'll like a class after hearing just a little of the teacher's opening lecture. Although we can make these judgments in a few seconds, they draw on stored knowledge and are thus to some degree top-down (Ambady and Rosenthal; Gladwell; Hassin, Uleman, and Bargh; Myers; T. Wilson). Yet even these remain fairly impervious to introspection.

The top-down/bottom-up distinction drastically simplifies a complex process that would probably be best modeled along several dimensions rather than a single vertical one. Doubtless neurological research will eventually show that any experiential process involves complicated feedback and input-output among many mental systems. Take mirror neurons, which can be found in various areas of the brain. Watching someone lift a heavy weight, either in front of you or on a movie screen, stimulates some of the neurons in your brain that would fire if you lifted a weight yourself. Many of these mirror neurons are linked to intentional action on your part, so that when they fire, you can spontaneously understand the actions of others as products of *their* intentions. It seems that we have a powerful, dedicated system moving swiftly from the perception of action to empathetic mind-reading (Ramachandran).

This is only one instance of how contemporary research asks us to consider that many of what we take to be learned or culturally guided mental activities will turn out to be packed into our biological equipment. Psychological research in the cognitive paradigm has steadily diminished claims for a blank-slate conception of the human mind and belief in the unlimited plasticity of human capacities. More and more activities (e.g., language, recognition of emotional signals, and attribution of intentions) seem traceable to humans' supersensitive natural endowment. Many specialized faculties need only triggers from the regularities of our world to lock in and function at high levels quite quickly. As research goes on, many "higher-order" activities will probably be revealed as grounded in a rich perceptual system present at birth but awaiting activation and tuning from the environment. In the long run, the discovery of mirror neurons is likely to refine the

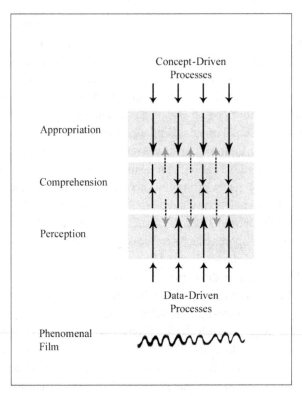

FIGURE 23.1 A schematic model of the spectator's activities. Continuous arrows indicate the primary direction of processing, with dotted arrows indicating a degree of feedback among processes.

cognitive perspective. Instead of treating thought as higher or central processes that rework sensory stimulations, we may discover that some garden-variety thought, including mental representation, takes place at the level of cells and cell networks.

So I grant the schematic quality of my distinctions. As a first approximation, however, they can clarify how our minds interact with movies. I suggest that we can characterize viewers' interactions with films along a continuum of activities: perception, comprehension, and appropriation (figure 23.1; for similar formulations, see Anderson; Persson 16–24). Sensory input drives perceptual processing; perceptual processing feeds into comprehension and appropriation, in the "bottom-up" direction. Appropriation drives comprehension to some degree and perception to a lesser degree. There are secondary feedback effects, too, as when the manner of appropriation can recast perception or comprehension. For example, a decision to interpret the film a certain way can lead us to look more closely at the film and notice or comprehend aspects that might otherwise be missed. I'd argue that such feedback systems can't go all the way down or all the way up, because perception can't in every respect determine appropriation, and appropriation can't completely reshape perception or comprehension. Wishing that Thelma and Louise don't die won't make it so.

Perception evolved in large part to give us reliable information about the three-dimensional world in which we live. Representational films solicit this activity straightforwardly: We involuntarily see the world depicted on the screen. We recognize our conspecifics and their surroundings. We hear noises, music, and language. We also see movement where there is only a stream of rapidly projected still pictures.

The perception of film as a representation of the world emerges very early in human development. Many of the experiments testing babies' reactions to facial expressions use televised images of the mother, which indicates that children spontaneously identify an audiovisual representation of the caregiver (for instance, see Trevarthen, "Communication"). Paul Messaris (chaps. 3 and 4) has shown convincingly that people in cultures without images recognize films and photographs as presenting persons, places, and things. The perceptual mechanisms that film engages seem to be shared with other primates. Experimenters routinely use videos to test chimps, monkeys, and their cousins, and the results indicate that these creatures identify their counterparts on the screen. Other experiments suggest that perception is also attuned to displays of emotion. When rhesus macaques who are unafraid of snakes watch a film of other monkeys shrinking from a snake, they begin to show fear themselves (Mineka and Cook).

This isn't to say that the processes of filmic perception are innate, as if a newborn could enjoy *Snake in the Eagle's Shadow*. Perceptual development unfolds in response to the environment. Humans and other primates have evolved to be ready for first encounters with the world's regularities. By the time people watch movies with any degree of perceptual understanding, they have developed the capacities to negotiate the three-dimensional world as well. And perception isn't terribly plastic. Given a stable input, perceptual capacities will develop along well-marked pathways. No one learns to see in infrared, or to move as if the world were two-dimensional, or to differentiate the separate frames of film flickering past. We are not so made.

The activities of filmic perception tend to be neglected by scholars today, but there's a long tradition of film aesthetics that places importance on the moment-by-moment effects of composition, lighting, cutting, and the like. From the Russian montagists through Rudolf Arnheim, André Bazin, and Noël Burch, theorists have paid attention to fine-grained creative choices that structure the viewer's perceptual uptake. Although strict experiments on filmic perception are welcome, there's a lot to be said by those of us not wearing lab coats. We can be sensitive to how patterns and practices of the medium shape such apparently simple strategies as directing the viewer's attention.

In comprehending a film, we construe the outputs of filmic perception as representing a hierarchical pattern of actions, a conception, or simply a train of sensuous elements (as in an abstract film). The viewer applies a wide range of knowledge to make sense of film, segment by segment or as a whole, and to give it some literal meaning. Narrative comprehension is the clearest instance. In my *Snake* experiment, spectators were able to build out of a perceived world a story about an ambitious young man who wants to master kung fu. Comprehension also comes into play when we're asked to grasp a cinematic argument or lyric. Comprehension is evidently a matter of degree; some viewers get more, some get less. In *Snake*, the viewers who knew Jackie Chan recognized him as the star and probably hypothesized that he would triumph through vigorous abuse of his body. Those who didn't recognize Jackie were probably surprised by the punishment he inflicted on himself. But the fact that comprehension varies in degree only indicates the extent to which it's a top-down process. Not everyone has the same set of conceptual schemes.

Again, poetics has a lot to contribute to understanding comprehension. The technical choices made by filmmakers organize perception in ways designed to enhance comprehension. Filmmakers design their shots and scenes so that spectators can follow the movie's large-scale form. Focusing on certain traditions or particular films, we can study how principles of style, narrative, and the like aim to provide a distinct experience for the viewer. For instance, since the 1920s most films in most countries have organized their perceptual surface according to some basic principles. Commercial storytelling cinema has long followed the conventions of analytical editing: master shot, followed by a two shot or over-the-shoulder shots, followed by singles highlighting each participant in shot/reverse-shot fashion. In fact, seldom do we find in any art a style with such pervasive presence and 100-year longevity. These norms have provided easy, comprehensible ways for narrative action to be understood.

Beyond stylistic patterning, it seems clear that comprehension also rides upon action schemas that the spectator can activate. When one man slaps another, and the second responds with a punch, there's not much doubt about what is going on in the story world at this point: insult and physical conflict. Likewise with a theft, an abduction, or glances that suggest attraction between man and woman. The large-scale form of the film is designed to create a flow of cues that ask viewers to apply schemas for typical situations and human actions and reactions, locking them into place quickly.

Indeed, there's good reason to believe that these action schemas enable us to learn the stylistic schemas that present them. We know that people tend to face one another when they converse, so this regularity of social interaction makes comprehensible the stylistic option of shot/reverse-shot editing. As Gombrich puts it, "It is the meaning which leads us to the convention and not the convention which leads us to the meaning" ("Image" 289; I develop this point in relation to cinema in *Figures* 258–60).

The perceptual surface can also be so roughened that holistic action patterns become difficult to grasp. In *The Two Minutes to Zero Trilogy* (2003–2004), Lewis Klahr presents shaky fragments drawn from comic books, showing only bits of words and imagery of bank holdups and police chases (figure 23.2). The images are so broken up and they shudder past us so quickly that we never get time to figure out character relations or construct a complete story. The slender cues summon up action-based schemas but also frustrate our efforts to absorb them into scenes and larger narrative patterns.

Comprehension occupies an intermediary place in my framework, balancing between data-driven and concept-driven features. Appropriation is much more top-down. Here the viewer uses the film in a more or less deliberate way, drawing it into her personal projects, and she may stray far from the phenomenal film. I appropriated *Snake in the Eagle's Shadow* as a classroom example, but fanboys have appropriated it as a cult object. Films are appropriated by individuals and communities for all manner of purposes. People employ favorite films for mood management, watching *Die*

FIGURE 23.2 A bank robbery is evoked in jittery comic book imagery in *The Two Minutes to Zero Trilogy* (Lewis Klahr, 2003–2004).

Hard (John McTiernan, 1988) to pump themselves up or *Sleepless in Seattle* (Nora Ephron, 1993) to have a good cry. Bloggers may use films to flaunt their tastes or strike a posture, whereas academics interpret films to validate a theory. Social groups appropriate films to a multitude of ends, treating some as praiseworthy representations of political positions and castigating others as harmful. Filmmakers are always surprised by the range of ways in which people take their movies.

Accordingly, much of what interests cultural critics are acts of appropriation. Some Asian Americans have attacked Charlie Chan movies as exemplifying Hollywood racism, not only in their plots and characterizations but also because the Chinese Hawaiian detective is played by a Westerner in "yellowface." They have perceived the films and comprehended them; but they have appropriated them in a way very different than the makers intended, or could probably have imagined. Much of what Janet Staiger attributes to "perverse spectators" consists of unusual forms of appropriation. She writes,

> Knowing that fictional narratives are produced permits many viewers to concentrate on narrational issues related to the production of the text. A study of some 1950s gay male viewers of *A Star Is Born* (1954) revealed that they were much more interested in constructing the story of the production of the film (when did Judy Garland shoot which scene) than in the film's plot—which at any rate was already "known." (37)

It's an interesting fact about films that groups (and individuals) can build unforeseen inferences out of particular aspects of a film that interest them. Nonetheless, what Staiger calls "uncooperative spectators" tend to perceive and comprehend the film in quite convergent ways, as she indicates by saying that the gay audience already grasped the film's plot.

Staiger's example typifies the tendency of cultural critics to stress divergence of response among groups. We know as well as that there's likely to be considerable differences among individuals in any group we pick out. Both sorts of divergence are explicable in the light of top-down appropriation, for in this domain an indefinitely large range of conceptual schemes can be brought to bear on any phenomenon. A balanced account will also note the high amount of convergence at all levels. Academic interpretations display great agreement at the levels of perception and comprehension, as well as a surprising degree of overlap in interpretations too, as when one critic revises the reading of her or his predecessors (Bordwell, *Making* 224–48). More importantly, there are many convergences among spectators at the level of comprehension. All storytelling traditions evidently deploy such concepts as protagonist, goals, personal agency, conflict, and causal change—all concepts relevant to comprehension. Patrick Hogan has shown that some prototypical narrative patterns, such as romantic tragicomedy, are to be found throughout the world's literatures (*Mind* 101–9).

This three-stage framework helps us understand the range of control available to filmmakers and viewers. Critics often ask, How much does "the text" control its "readings"? This framework lets us give some focused answers. As we move up the ladder, from bottom-up to top-down processing, the filmmaker's control diminishes and the spectator's power increases.

By constructing the phenomenal film, the filmmakers control very strongly, though not absolutely, the viewer's perception of it. It's impossible for a viewer to perceive the hero of *Snake in the Eagle's Shadow* as, say, tall and blond. In addition, all the options of film style and structure can be mobilized to guide the viewer's notice to certain material. Framing can center key information, whereas cutting can highlight a detail. Style operates at all levels, but among its basic tasks is organizing the stimulus for uptake—even if that uptake is made difficult by an oblique technical choice or a problematic narrative.

At the level of comprehension, the filmmaker still has a lot of control, because features of theme and subject, style, and large-scale form are mobilized to guide the spectator's overarching understanding of the material. Still, no formal pattern can anticipate every question that can be asked about it. In grasping narrative form, for instance, the spectator contributes a lot—picking up the cues planted by the filmmakers, as well as inferring, extrapolating, filling in gaps, and the like. Most of this inferential elaboration is foreseen and governed by the filmmaker, but not all of it is. Shakespeare famously leaves us with the question of whether Macbeth and Lady Macbeth ever had children. She claims she has "given suck" to a babe, but Macduff says, apparently of Macbeth, "He has no children." Did the couple have a child who died? Did Lady Macbeth suckle another woman's baby? Spectators will differ in the ways that they deal with such zones of indeterminacy. Most will ignore them, but others will use them as occasions for appropriation, as when fans write fiction filling in gaps in *Star Trek* or *The Lord of the Rings*. No narrative can avoid leaving some openings for inferential elaboration of this sort. Louisa May Alcott couldn't have anticipated Geraldine Brooks's novel *March*, a fictional biography of the father depicted in *Little Women*. Of course, some artworks deliberately introduce gaps that they decline to fill, as with ambiguous endings.

Clearly, filmmakers have the least control over the activities I've gathered under the rubric of appropriation. Having perceived and comprehended (to a greater or lesser degree) *Snake in the Eagle's Shadow*, you're free to do with it as you will. Viewers sympathetic to gay rights can take *Brokeback Mountain* (2005) as a plea for tolerance, whereas those opposing gay rights can treat it as Hollywood propaganda for alternative lifestyles. Director Ang Lee and his colleagues can seek to shape the film's reception through publicity campaigns, but they can't anticipate every way the movie can be appropriated. The wave of mashup trailers that swept the Web (*Brokeback to the Future*, *The Empire Brokeback*) weren't foreseen by the filmmakers, however much they may have welcomed them. In sum, as we move up my chart, filmmakers' freedom wanes and spectators' power increases.

I haven't mentioned a prime component of film's effects: emotions. In my past work, the historical poetics I've proposed has slighted emotions, leading some people to think that a cognitive perspective can't tackle such matters. In part, leaving emotion out of the picture is a simple piecemeal idealization of the phenomenon; studying the grammar of a joke may not yield insights about what makes it funny. In addition, in the early 1980s, when the cognitive perspective was hitting critical mass in several disciplines and when I imported some of its observations into film studies, emotional matters were set to one side. But they weren't legislated out of existence, and the 1990s saw vigorous efforts to incorporate emotional life into the cognitive framework. This was also reflected in film studies, in the work of Murray Smith (*Engaging Characters*), Ed Tan, Torben Grodal, Greg Smith, Plantinga and Smith, and many other researchers since. It's not an area of specialization for me, but I think that the processing framework I've proposed can accommodate emotion as an integral part of a film's effects.

Emotion is part of our evolutionary heritage, and it has largely served in tandem with cognition. That is, rather than being the foe of emotion, reason has used emotion and emotions have exploited reason. Certain sorts of reasoning would be maladaptive without some emotional upsurge that halts thinking and forces action. The hominids who lingered to investigate whether the stripes glimpsed in the underbrush belonged to a predator didn't leave as many offspring as those who, driven by fear, simply fled at first glimpse. Emotions offer quick and dirty solutions to problems that make thinking risky. Alternatively, so-called commitment emotions may have evolved to strengthen group bonds, even if they work against self-centered rationality. Fathers have no rational reason to hang around after a woman is impregnated, but it seems likely that men who had romantic attachments to the mother had more children who survived, and so love helps unite father and mother across the lengthy period in which children grow to self-sufficiency (Frank). Within specific cultural contexts, of course, people learn to judge the proper moments to express feelings, to mask them with other feelings, or to send emotional signals.

In cinema, I'd suggest that emotion operates at all three of the levels I've sketched out. Most obviously, acts of appropriation are shot through with emotion. Fans cherish their favorite movies, critics get worked up in attacking a film they loathe, and unhappy viewers can wax indignant about a film's moral shortcomings. Less apparent are the ways in which emotions function in perception. A controversial case would be our startle response, which can be triggered quite automatically, as when you jump at a sudden burst of sound in a horror film. Startle isn't a prime candidate for being an emotion—it seems to prepare the way for the emotion of surprise—but it does lead to physiological arousal of a sort that primes affect.

More common and central is our sensitivity to emotional signals sent by other humans. Just as the rhesus macaques recognize signs of distress in their mates in a movie, we are prepared to grasp many facial expressions. Newborn babies can reliably read their mothers' smiles, eye movements, and eyebrow play. A film's soundtrack can arouse us quite directly by cries, bellows, and other signals, just as infants respond to the mother's coos and baby talk (Dissanayake, *Art* 26–42). The weight of the evidence shows that evolution has primed us to engage in encounters with others by making us sensitive to the slightest signs of their emotional states. And these "affect programs" seem cross-cultural in large part, although we should expect them to vary extensively from person to person within any given culture (Griffiths 122–25). There is some evidence that in-group familiarity leads to faster recognition of facial expressions; Asians, whether living in Asia or the United States, recognize emotions on Chinese faces somewhat more readily than non-Asians do (Elfenbein and Ambady).

More obvious are the emotions that fund comprehension. As we come to understand a narrative, we begin to run scenarios that require "emotional intelligence"—good guesses about how characters will react to the story situations. At the same time, we gauge a character's personality or current attitude on the basis of their emotional responses. Our inferential elaboration of the cues we're given is guided by the emotions that characters register. At the same time, the emotions we feel shape our sense of the film's macro-action. If we feel that a character has been wronged, we may mimic, in weakened form, her anger and self-righteousness. Screenwriters provide strong prompts for sympathy (Hauge), such as making sure that the protagonist is treated unfairly, and many screenplay manuals argue that the skillful filmmaker evokes both hope (for the character's success) and fear (of the character's failure) (D. Howard 52–56; Keating 4–15). Again, there may be considerable cross-cultural regularities in these emotions, most of which depend upon recurrent social situations that people in most cultures encounter—sympathy for children, anger at being wronged, and a sense of fairness or justice.

In comprehension, emotion and thought mesh. Greg Smith argues that narrative films tend to sustain moods and then punctuate them with bursts of emotion proper (*Film Structure*, chaps. 1–4). These in turn can focus our attention on story developments. In *Rope* (Alfred Hitchcock, 1948), our knowing where the corpse is hidden generates suspense, a mood that in turn makes us hyperattentive to every movement toward the chest. When Rupert lifts the lid, the mood prepares us to concentrate on his face, which betrays his shock and distress. Emotion also affects memory; in real life, a traumatic event becomes sharply etched into our minds. Films exploit this tendency by making the most vividly emotional scenes crucial for the plot—a death, a separation, a reunion. Ben Singer has proposed a catalogue of prototype scenarios, drawn from intense emotional experiences in ordinary human life, that melodramas draw upon. For example, the pathos we feel when seeing people degraded by misfortune forms the basis of a scene in *Mother India* (Mehboob Khan, 1957), when the mother, in the aftermath of a flood, digs desperately through the mud for something that will feed her dying children (Singer). Patrick Hogan's survey of transcultural story patterns traces their constant features to the way they make salient certain emotion-based proto-types of happiness (*Mind* 86–101).

There's a great deal to be studied about how emotion works within our cinematic experience, including the bonding effects of watching a film with others. I propose that narrative films often model social intelligence, and that modeling in turn rests upon sensitivity to emotional signals. Here all I want to indicate is that the ways audiences process films can be investigated through systematic studies of the range of emotional effects that a film can have.

What processes enable us to perceive, comprehend, and respond emo-tionally to moving pictures? Here, in gross outline, is one answer. As humans we have evolved certain capacities and predispositions, ranging from per-ceptual ones (biological mechanisms for delivering information about the world we live in) to social ones (e.g., affinities with and curiosity about other humans). Out of these capacities and predispositions, and by bonding with our conspecifics, we have built a staggeringly sophisticated array of cultural practices—skills, technologies, arts, and institutions.

Moving pictures are such a practice. We designed them to mesh with our perceptual and cognitive capacities. What hammers are to hands, movies are to minds: a tool exquisitely shaped to the powers and purposes of human activity.

A great deal of movies' effects—more than many contemporary film theo-ries allow—stem from their impact on our sensory systems. We are prompted to detect movement, shape, color, and sounds, and this is surely one of the transcultural capacities that movies tap. Similarly, films from all nations and times draw upon more "cognitive" skills, such as categorizing an object as living or nonliving, or seeing a face as furious—abilities that, it's reasonable to think, are part of our evolutionary heritage. And because affective states and counterfactual speculation are of adaptive advantage, it is likely that an artistic medium that permits emotional and imaginative expression would have appeal across cultural boundaries.

If we consider culture to be an elaboration of evolutionary processes, there's no inherent gulf between "biology" and "society" in this explanatory framework. True, these elaborations vary historically, yielding (among other things) what we usually call *conventions*—local practices that seem "artificial" and that differ from one society to another. Yet some conventions are less artificial than others. Writing a verbal language, mastering chess, and solving differential equations take years to learn, largely because all rest upon hard-core conventionality. Other conventions can be picked up fast because they are functionally similar across cultures. Some countries require you to drive on the right side of the road, and others on the left, but the idea of ordering the traffic flow is congruent across both. Still other conventions require only the slightest adjustments of our natural proclivities. In a picture, if the most important element occupies the center of the format, viewers from any culture will probably not be surprised. Centering (manifesting the principle of symmetry) is in some sense a convention of pictorial composition, but it seems to run with the grain of our visual predispositions, taking the "line of least resistance." Strategic decentering, on the other hand, may be a convention that requires a little more tutoring.

Films use conventions. In most movies, characters face each other in an odd way: Their bodies and faces are conveniently tilted in 3/4 view for the camera. Scenes are cut according to the tactics of continuity editing. We may hear music that does not issue from the locale of the scene, and a dissolve or fade may convey a passage of time. Still, such conventions are mostly of the quickly learned variety. Many of them piggyback on our natural predispositions; others require only slight adjustments. Several amplify and streamline regularities of human interaction, as when movie characters talking to one another stare more fixedly and blink far less than they would in real life. We understand movies fairly easily because in many respects their conventions are easy to learn: They are simplifications of things we already know.

Of course, a particular filmmaker may wish to block that easy understanding—to be, as we say, unconventional—but very often, she will have to tap into other capacities and proclivities we have. If the story is told out of order, then we will need some redundant cues to that design as well, such as *Pulp Fiction*'s (Quentin Tarantino, 1994) replay of the opening dialogue when the action returns to the diner for its climactic scene. Nevertheless, a great deal of what is conveyed in a movie is conveyed naturally—through those perceptual-cognitive-affective universals that are part of our biological inheritance.

Most scholars in the humanities tend to doubt the existence (or the importance) of empirical universals. Further, the framework hypothesizes causal and functional explanations for social practices. Most humanists, though, prefer interpretation to explanation. When they do seek explanations, they rule out biological causes or functions as too deterministic and prefer some form of social-learning theory. The framework I traced also takes rational-empirical inquiry, of which science is our most successful exemplar, as the most promising way to explain cultural practices. But academic humanists on the whole mistrust science and, sometimes, rational empirical inquiry more generally.

Film academics are on the whole even more suspicious of this framework than their peers in other disciplines, I believe. This is largely because film studies, entering university humanities departments in the late 1960s, became rather quickly attached to certain doctrines. Most of these, such as semiotics and psychoanalytic theory, were deeply antinaturalistic (at least in the versions that became influential). Although these particular doctrines have lost their grip, an extreme version of cultural constructivism is at the base of most film studies. Consider just a few premises.

- All personal experiences—identity, concepts, feelings, and even perceptions—are socially constructed.

(Constructed out of what? That matter is not addressed.)

- Because everything is socially constructed, there is no such thing as a more or less realistic representation; every sign is equally arbitrary.

(Can the concept of the arbitrary sign be intelligible without a concept of the *non*arbitrary sign? Shouldn't one then consider the possibility that there might *be* nonarbitrary signs? And why should all signs be equally arbitrary? These questions are not asked.)

- Realism is a myth because no representational system provides total access to some "reality out there" (if, indeed, such a thing exists).

(Doesn't this set the bar unreasonably high? A realistic representation need not preserve all aspects of its referent in order to be reliable, as we see in architectural design and forensic photography. But these objections are caricatured as "naïve realism.")

- Every culture creates its own web of meaning. There may be "hybridity" when cultures come in contact, but there is no universal human culture.

(If every culture is sui generis, how could theorists have grasped enough features of alien cultures to arrive at this generalization? This self-refutation isn't considered.)

Despite some claims that the discipline has become more pluralistic since the 1980s, premises like these, invoked ritualistically in the literature and taught by rote and exemplar in courses, have become operational assumptions of most academic film writing.

Film studies also got off on the wrong foot methodologically. Instead of framing *questions*, to which competing theories might have responded in a common concern for enlightenment, film academics embraced a *doctrine-driven* conception of research. They embraced a scholastic conception of their work, holding that certain theorists had revealed core truths and that their gospel could be applied, in a more or less mechanical fashion, to particular movies. First came Mulvey's "gaze" theory, then postmodernism, then versions of identity politics, multiculturalism, and "modernity theory"—none weighed as candidate answers to a puzzle or problem but were accepted unskeptically, then used to churn out interpretations of film after film. Film studies remains, in a word, dogmatic.

Contemporary theory assumes that cinematic representation is almost wholly conventional (and the conventions come from culture); what is not conventional is very little (often called "physiology"!) and not very important. According to the ecological view, cinematic representation relies on a great many nonconventional capacities and processes, and the conventions are correspondingly small in number and easy to learn—riding as many of them do upon just those ecologically constrained processes.

Part IV

Interpretations

24

Homeric Women

RE-IMAGINING THE FITNESS LANDSCAPE

JONATHAN GOTTSCHALL

The Rape of Troy is about men—about their battles, real and ritualized, for dominance, wealth, women and, ultimately, for reproductive success. This emphasis is almost inevitable owing to the subject matter of Homeric poetry: men—men competing, venturing, arguing, killing, and being killed. The women in Homer's poetry play mostly supporting roles or appear as mere extras; even Penelope is only the third most important character in the *Odyssey*, and no female character is so prominent in the *Iliad*. The women featured so far in *The Rape of Troy* have, with some exceptions, behaved passively. They have waited to be acted upon by men who hash out rights to them in exuberant competitions: men compete for women in contests, in combat, and by offering the biggest bride prices and the brightest wooing gifts. But were Homeric women really so passive, really so content to wait until "taken" by victorious men?

No. While women's strategies for surviving and reproducing are different from men's—most strikingly in that they are less likely to entail violent

From Jonathan Gottschall. *The Rape of Troy: Evolution, Violence, and the World of Homer*, pp. 100–118. Cambridge, Cambridge University Press, 2008. Copyright © Jonathan Gottschall.

aggression—they are no less complex, effective, or potentially Machiavellian (on female competitive strategies, see A. Campbell; Cummins). While mammalian biology dictates that women vary less in reproductive success than do men, there is still substantial variance. Thus women have much incentive to compete—for status, for the best mates, and for access to resources to provision their young. Women who were easy dupes for male exploitation could never pass on genes as successfully as women equipped to avoid exploitation, and even to act as exploiters themselves, in pursuit of their own goals. As Sarah Hrdy has written, the passive, compliant, apolitical, exclusively nurturing woman of the stereotypes surely never evolved (*Mother Nature, Woman*).

The Homeric epics, emerging from a muscular patriarchal culture, were composed by a man (or men, depending on how you answer the Homeric question), mainly about men, and largely for men. (Samuel Butler's argument that the *Odyssey* was composed by a woman is without foundation and has been universally discredited.) Women are pushed to the poetic margins. For instance, since Homer rarely allows us to glimpse interactions among women, we have the vaguest sense of how and under what conditions they would have competed and cooperated with one another. However, women are not pushed entirely from his page. In Homer we see rich women and slaves, mothers and daughters, maidens and crones, portraits of female propriety and of wantonness; we see women in times of peace and peril, we see them bustle through their daily routines, and we see them confront extraordinary situations.

In order to assess female strategies as depicted in Homer, and how they influence male competitiveness, we must first accomplish a massive shift in perspective. We must break away from the position of predominant identification with men, which the poems force on most male and female readers alike, and try to re-imagine Homer's world from a female standpoint. Since *The Rape of Troy* analyzes the epics not only as works of literary art, but also as precious time-capsules reflecting Greek life sometime around the eighth century B.C.E., we must also appreciate that Homer's women have passed through two distorting prisms: that of a vigorously male perspective and that of epic exaggeration and romanticization. However, drawing on multiple streams of evidence we can "see through" the distortions to produce an anthropologically coherent reconstruction of some aspects of the lives of Homeric women.

The fitness landscape, as I am employing the term, can be envisioned as a foreboding and inhospitable expanse, pocked with hazards, traced by deep canyons, fast rivers, and jagged mountains. On the horizon, for the person hearty enough to swim the rapids, scale the cliffs, and find and negotiate the snowy passes, one can just make out a land of gentle hills and fertile plains, representing the reward of reproductive success—the only thing that really matters in evolutionary terms. To understand the strategies of Homeric women we must first understand the fitness landscape they picked their way across. We must appreciate how different it was, how much more menacing, than those traversed by most women in modern societies. We must also appreciate how much different it was from those traversed by men. While Homeric women faced radically different challenges from those faced by men, the challenges were not less severe and the potential consequences for wrong turns were not less dire.

In short, the principal hazards facing Homeric women stemmed from the darker aspects of Homeric men: their propensities for violence, and for promiscuous and coercive sexual activity. Women sought to marry and raise families in a world where, virtually without warning, their husbands and kinsmen could be massacred, their children murdered or enslaved and raped, and they themselves could be raped and enslaved. Such scenarios were familiar enough to be the material of harrowing similes:

> As a woman weeps and flings herself down on her beloved husband, who has fallen fighting in front of his town and his people, trying to ward off the day of doom from his children and his town—seeing the man fall and convulse, she clings to him and pours out lamentations in a loud clear voice, while the enemy from behind prod her in the back and shoulders with their spears, leading her into bondage, to work and woe, while the most pitiful grief wastes her cheeks—so from under Odysseus' brows the pitiful tears fell. (*Odyssey* 8.523–31)*

If they escaped the worst-case scenario—and probably most of them did— Homeric women were obliged to pursue their goals within the tight and potentially dangerous strictures of intense male possessiveness and an exaggerated sexual double standard (for suggestions of this, see *Odyssey* 5.118–29, 6.285–88). Running foul of male possessiveness or violating the strict sexual morality governing their behavior could bring severe consequences (Fulkerson), ranging from reputational damage and abandonment, all the way up to physical abuse and even murder (*Odyssey* 6.273–88, 22.419–73; *Iliad* 1.565– 67, 15.16–22). Finally, while strict sexual fidelity was demanded of Homeric wives, their husbands faced no social costs for open pursuits of promiscuous liaisons—if anything, promiscuity among the men was a mark of status and

---

*All translations are by the author, and all references are to the Loeb Classical Library editions of the *Iliad* and the *Odyssey*.

pride. While many Homeric women resisted (*Iliad* 9.444–53; *Odyssey* 1.429–33), they were generally expected to endure the infidelities of their husbands, to face the real possibility of being supplanted by younger and/or more beautiful slaves (*Iliad* 9.444–53, 1.113–15), and to accept the likelihood that a significant proportion of family resources would be diverted away from their own children in support of their husbands' bastards (D. Ogden, chap. 4).

How did Homeric women negotiate these features of their fitness landscapes given that the only roles they were allowed to play were as wives, mothers, and household managers? Lacking political muscle in the community and physical muscle in the home, isolated by patrilocal marriage patterns from the protection of blood kin, confined by the nature of their work mainly to the household, and able to enhance their social and material capital *only* through marriage to the right type of man, how did Homeric women maneuver to give themselves the best opportunity to traverse their fitness landscapes: to survive, reproduce, and live to see their children reproduce?

Beauty Is Power

When Odysseus takes leave of Nausicaa, finally going home to a reunion with his own wife, he says, "May the gods give you all your heart desires, a husband and a house, and may they also give you oneness of heart in all its excellence" (*Odyssey* 6.180–82). The words reflect something important about the limited scope available for female ambition in the Homeric world. Men are likewise strongly motivated by concerns of marriage and family. But they also strive intensely for social status, power, and riches. Odysseus' words suggest that women were content simply to make a good, stable marriage and raise a family. He would be wrong. Homeric women were also interested in status, power, and resources. However, women could gain these ends only by attaching themselves to a high-quality mate. Wives of high-status males gained privileged access to scarce resources and some degree of community standing. While women had little direct power, the wives (and lovers) of leading men were treated with great deference, and could enjoy considerable political influence. Hera is a good example of such a woman (e.g., *Iliad* 18.364–67), as is Arete. (Also *Odyssey* 6.298–315; *Iliad* 21.498–501.)

When it came to attracting desirable suitors, beauty was a woman's greatest asset. This is not to slight the role that social position played in marriage. Marriages in Homer serve as linkages for whole kin networks. Homeric men consider it useful and desirable to marry into rich and powerful families. Thus, when the son of Castor mentions his bride he brags that she was the daughter "of rich people" (*Odyssey* 14.211–13), not that she was beautiful.

Furthermore, even the highest-status Homeric women, the wives of chiefs and gods, do work that is utterly indispensable to the functioning and survival of the household. These slaving "queens," "princesses," and goddesses are, for me, a particularly eloquent indicator of the simplicity of Homeric society. It is as though Homer is incapable of even imagining a social stratum where women's lives are *not* taken up with work. Consider the depiction of "Princess" Nausicaa doing laundry (*Odyssey* 6.85–96) and the Laestrygonian "princess" hauling heavy buckets of water from the spring back to town (*Odyssey* 10.105–8).

However, while a woman's desirability is affected by her social rank, wealth, and skill at work, nothing about Homeric women is stressed as much as their beauty, and no female attribute is more important to men in choosing potential mates. The tremendous importance of beauty to female life outcomes is compellingly reflected in the famous story of the Judgment of Paris, which is identified in the *Iliad* (24.25–30) as the cause of the Trojan War. In this episode, which is described in detail in a preserved summary of a lost epic (the *Cypria*), Paris is chosen as judge of a beauty contest among Hera, Athena, and Aphrodite. To boost their chances, each goddess offers Paris a secret bribe. Paris' preference for Aphrodite's bribe of the love of the most beautiful woman in the world, over Athena and Hera's offers of power and glory, sets off a chain reaction that culminates in the Trojan holocaust and trails off in subsequent bloody years. The young Trojan's choice of Aphrodite inspires the unreasoning and everlasting hatred of Hera and Athena. Troy burns to satisfy their vain pique.

The myth symbolizes two interlocked truths about Homeric men and women. First, the goddesses' shameless maneuverings to be chosen by Paris, and their spiteful fury when passed over, indicate how important beauty was to Homeric women. It suggests that beauty formed a road to dominance of female status hierarchies in much the same way that physical size and strength led to dominance of male hierarchies. A similar inference can be drawn from Odysseus' warning to Melantho, the most beautiful of his female slaves. Still disguised as a beggar, he uses his own riches-to-rags story to illustrate how quickly Zeus can dash a person's fortunes. The moral of his tale: "Therefore, woman, beware that you too some day do not lose all the beauty wherein you now are preeminent among the handmaids" (*Odyssey* 19.81–82). Odysseus' analogy implies that loss of beauty is as devastating for a woman as a man's loss of wealth and status.

Beauty was the most potent weapon a woman wielded, not only for competing against other women, but also for competing on a more level footing with politically and muscularly powerful men (for review of research of the role of physical attractiveness in the formation of female hierarchies, see

A. Campbell). Second, the myth emphatically defines what is most important to men: finally every enticement pales before the allure of a desirable woman. Paris chooses Aphrodite to further his "grievous lustfulness" (*Iliad* 24.30), a weakness consuming mortal men and immortals alike.

These truths feed on one another. Beauty is important to women because, as Paris' choice shows, beauty is so important to men. And this is not only because beautiful women marry better than the less beautiful. Rather, Paris' choice shows that desirable women possess something that men covet above all other things—and this is a source of endless power.

Hera fully deploys this power in her marriage to Zeus. In the *Iliad* we hear much of Zeus' overwhelming strength. He is "by far the strongest"; even a coalition of all of the other gods could not rival him in strength (*Iliad* 8.18–27; *Odyssey* 15.105–8). Zeus is the alpha male of all the Homeric alpha males that strut and preen and bellow their power—a portrait of puissance that effeminizes the most hirsute hero by contrast. Yet even Zeus is no match for a woman, provided she is not only beautiful but guileful (*Iliad* 19.95–99). Successfully competing against men requires equal measures of beauty and brains.

As Hephaestus says, "The Olympian [Zeus] is a hard foe to meet in strife" (*Iliad* 1.588–89), so most of the male gods do not even bother to try; and if they do, pain teaches them the error of their ways. However, goddesses have more success because they harbor no illusions about their capacity to overcome Zeus through strength. Unlike male gods, goddesses can compete with Zeus by exploiting sexual vulnerabilities he shares with all men. Aphrodite, in her symbolic capacity of the sex drive, is the only force capable of defeating Zeus, an otherwise omnipotent being. While the other gods strain and cower under his sometimes capricious rule, Aphrodite always has her way with him. Zeus' sexual vulnerabilities are consistently depicted in the Homeric epics, Greek mythology, and the Homeric Hymn to Aphrodite:

> Muse, tell me the deeds of golden Aphrodite the Cyprian, who stirs up sweet passion in the gods and subdues the tribes of mortal men and birds that fly in the air and all the many creatures that the dry land rears, and all that the sea. . . . There is nothing among the blessed gods or among mortal men that has escaped Aphrodite. Even the heart of Zeus, who delights in thunder, is led astray by her; though he is greatest of all and has the lot of highest majesty, she beguiles even his heart whenever she pleases and mates him with mortal women. (Hesiod 1–39)

Hera exploits Zeus' sexual weakness with great skill, confidence, and intelligence. In a scene intriguingly reminiscent of the ritualistic arming scenes

of the epics, where warriors carefully strap on all of their battle gear, Hera girds herself with female weapons and armor: she cleans herself with ambrosia, smears her gorgeous body with redolent olive oil and other perfumes, combs and primps her hair into sleek braids tousled about her shoulders, and decks herself in smooth robes, jewelry, and veil. Finally, between her breasts she tucks Aphrodite's embroidered breast band. Thus armed "with all this loveliness" (*Iliad* 14.187) she sallies forth to meet Zeus and she quickly subdues him. At a glance he is seized with "sweet desire" (*Iliad* 14.328); "lust closely enwrapped his mind, just as when they had first mingled in love" (*Iliad* 14.294–95). She expertly seduces him, enervates him, and puts to sleep his "wise mind" (*Iliad* 14.165) while she enacts her own agenda, turning the tide of battle against the Trojans she hates.

Other scenes suggest the ability of women to use their sexual appeal to manipulate powerful males and advance their own interests. For instance, Hera's stratagem succeeds only through the collusion of the god Sleep. Sleep's debilitating fear of crossing Zeus, and his steadfast refusal to entertain any of Hera's bribes, is instantly quashed by Hera's offer of a beautiful Grace, one of the younger ones, for a wife (*Iliad* 14.263–76). Similarly, Penelope cannily turns the lust of her suitors to her advantage. At one juncture in the poem, she arms herself as Hera does above, accentuating her beauty with oils, perfumes, and alluring dress and jewelry. She then displays herself to her suitors with the express purpose of wringing precious gifts out of them. Odysseus observes the whole performance, and rejoices "because she drew gifts from them, bewitching their souls with winning words, but her mind was set on other things" (*Iliad* 18.281–83).

In short, Homeric women use the few tools their society cannot deny them—sharp intelligence, manipulative stratagems, and sexual allure—to ably defend and promote their interests in spite of their muscular and political disadvantages. Hera interferes with Zeus's plans at every turn: she hectors and badgers, flatters and wheedles, and she constantly maneuvers within her encumbrances to pursue her own agendas. Homer's portrait of Hera shows that a subtle and courageous woman could win substantial benefits in challenging her husband, but that she also faces dire risks. For her various transgressions, Zeus has physically tortured Hera in the past (*Iliad* 15.16–22), and when she pushes him too hard he warns her to yield lest he come for her with his "irresistible hands" (*Iliad* 1.565–67). But Hera has learned exactly how far she can push Zeus. In the *Iliad*, she plays him with ease, like a virtuoso playing children's music. She wins their wars by pretending to capitulate in their smaller battles, convincingly acknowledging Zeus's strength and authority, and making him feel that he has won the day. But when the battle of wills between Zeus and Hera winds to a close, Hera

has successfully wrought her vengeance for the Judgment of Paris. Zeus's favorite city, Troy, is a smoldering ruin, his favorite people, the Trojans, are, depending on their sex, either dead or degraded, and his most beloved mortal son, Sarpedon, is dead. Thus in sharp contrast to Homer's invocation to the muse (*Iliad* 1.5), it is *not* the will of Zeus that is accomplished in the Rape of Troy, it is the will of Hera.

In emphasizing the way that Homeric women exploit male sexual vulnerabilities, I do not mean to imply that all women in the epics are portrayed as calculating temptresses. Homeric women are not only the agents of Aphrodite, they are also, like Helen, her hapless victims; Aphrodite is, in fact, pointedly labeled as a "deceiver of women" (*Iliad* 5.349). Moreover, Homer celebrates romantic bonds as among the deepest and most rewarding in human experience (e.g., *Odyssey* 6.180–85). He shows that sex can be a political lever, but he also describes the desperation of Hector and Andromache on the wall, the tender pillow talk of Odysseus and Penelope, the sweet anticipation of young lovers, and interludes of comfortable warmth even in the long-fraying marriages of Zeus and Hera, Menelaus and Helen. But this chapter is less about describing the strategies of Homeric women and more about exploring how these strategies shape the qualities of Homeric men, especially their propensities for violence.

Female Choice

Whoever her father commands; whoever pleases her.

ODYSSEY 2.113–14

Making a good marriage was by far the most important challenge a Homeric woman faced in negotiating her fitness landscape. This may jar modern sensibilities, where we all grin a little at the feminist aphorism, "A woman needs a man like a fish needs a bicycle." While it may be more challenging, modern women can, and regularly do, negotiate their fitness landscapes without male protection, support, or provisioning of offspring. However, we too readily forget how unusual this situation is; in the tens of thousands of years of human history predating the novel conditions that engendered the feminist revolution, in order to survive, thrive, and see children reach maturity, a woman needed a man *like a fish needs water*. Lacking an investing mate, women had difficulty providing sufficient calories for their children, especially the vital fat and protein calories provided by male hunters; lacking protection, they faced the constant specter of sexual coercion and harassment (Good and Chanoff; Smuts, "Apes," "Male Aggression").

Moreover, in preindustrial societies fatherless children had lower rates of survival due to deficient nutrition, smaller funds of kinship support, and, in the worst case, the violence of potential stepfathers (on threats from human stepfathers, see Daly and Wilson, *Homicide*, *Truth*; Hrdy, *Mother Nature*). As Sarah Pomeroy writes, in the Homeric age and most of the rest of Ancient Greek history, "the lives of women lacking the protection of men were truly pitiful" (44).

But this dependence was mutual. In Homeric society, the ideal marriage was envisioned as an unshakeable partnership (*Odyssey* 6.180–85); men were also vitally dependent on their wives. Family survival in non-state societies could be a difficult and dangerous affair, and it required the participation of competent male and female investors. As in all preindustrial societies, Homer's had a clear division of labor into male and female categories (on the sexual division of labor, see Eagly and Wood; Good and Chanoff 73; Wood and Eagly). While we often see the men engaged in sport, leisure, and idle debate, we almost never see women—whether free or slave, mortal or divine—doing anything but working. Young Nausicaa and her maids do play girlishly at ball, but only after they have spent the day scrubbing their knuckles and laundry in the river. Women weave and sew, prepare and cook meals, thresh and grind grain, pick grapes, haul heavy buckets of water, clean house, wash clothes, and raise children. The wives of important men have many maids, but they do not sit idly and watch them work. Rather they are more like forewomen of bustling production facilities, working alongside and supervising a female workforce as it churns out the broad variety of products needed to sustain a large household.

In short, Homeric men require mates who are much more than beautiful brood mares. In addition to beauty, men value loyalty, endurance, sense, decorum, and skill and industriousness in work. In short, they require—and therefore desire—extremely competent partners to help them negotiate the substantial challenges of subsistence in a world with ever-present dangers, small margins, and no social safety net beyond kinsmen and close friends. (This sense of partnership is amply conveyed at *Odyssey* 6.180–85.) Thus, while Homeric men place a high premium on physical attractiveness, the ideal mate is very far from an empty-headed beauty. This fact is embodied in Penelope—an ideal wife combining diverse and numerous virtues—and it is continually reinforced in female epithets that stress not only physical allure, but wisdom, good sense, and proficiency at work.

Of course, Homeric women were as constrained in their mate choice decisions as they were in making other important choices. In fact, while fathers, brothers, and potential husbands all had roles in arranging marriages, many references to marriage lack information about what role a woman played, if

any, in choosing her own husband. But does this mean that Homeric women were at the mercy of their kinsmen when it came to mate choice?

A close look at the evidence strongly suggests that they were not. First, there is sustained attention to the details of only one potential marriage in all of Homer—the lead up to the (eventually averted) marriage of Penelope to one of her suitors. If Penelope's example is at all representative, marriageable women negotiated their mate preferences with those of their kinsmen, especially their fathers. The preferences of fathers and daughters appear to have been substantially aligned. For instance, Alcinous and Nausicaa both consider Odysseus to be an ideal mate/son-in-law, and suitors court potential brides and fathers-in-law in exactly the same way: by giving large bride prices to the father and expensive wooing gifts to the woman.

But their interests were not necessarily identical: a woman's family could realize great social and economic gains from her marriage. When brides "earn" oxen from the families of their grooms (unmarried young women are described as "earners of oxen" [*Iliad* 18.593]), it is the woman's family who principally benefits. In the *Odyssey*, for instance, all of Penelope's closest kinsmen—her father, her brothers, her son—pressure her to choose a new husband, in spite of her reluctance (*Odyssey* 15.16–18, 19.157–60, 19.530–34). We should assume that this pressure is not all self-interested: after twenty years, the men think it is time for Penelope to accept that Odysseus is obviously dead, for her to dry her tears and enjoy the pleasures of life and marriage while time remains. But they also stand to benefit economically. Penelope's family will receive a great infusion of wealth, and the suitors' steady consumption of Telemachus' patrimony will come to an end. Penelope's ability to resist this pressure for almost four years suggests that women, while not free agents in the marriage market, did wield significant influence. The tension between the preferences of a prospective bride and the wishes of her family is intimated in a suitor's comment that Penelope should marry "whoever her father commands; whoever pleases her" (*Odyssey* 2.113–14). And, while the risks were immense, women who were unhappy with their mates had infidelity as an option. Helen, Clytemnestra, and Aphrodite (not to mention Odysseus' slave women) all exercise this option, and Homer builds much tension around the possibility of Penelope straying. In short, the epics give the impression that women's mate preferences—hemmed in as they were by familial constraints—did have non-negligible scope for expression.

One could reasonably object that this picture, based as it is primarily on the example of Penelope, is merely a side effect of an unusual situation. Penelope is not a maiden; she is a grown woman, ostensibly widowed, and she is not surrounded by strong kinsmen. Was the situation markedly different

for young women at the age of first marriage? Were they at the mercy of their kinsmen? For several reasons, I think not. First, the notion of powerless young women is manifestly at odds with the ubiquitous and clearly described Homeric custom of wooing gifts (e.g., *Odyssey* 6.158–59, 8.269, 11.116–17, 11.281–82, 15.16–18, 15.366–67, 19.528–29, 20.334–35). When a man seeks to win a bride he approaches *her* with gifts, not her father; negotiating a payment to the father appears to be a separate transaction. Such was the case, in fact, during Penelope's youth: Odysseus wooed her "with many gifts" (*Odyssey* 24.294). And Hector wooed Andromache in the same way (*Iliad* 6.394, 22.88), before paying her father a huge bride price and leading her to Troy (*Iliad* 22.471–72). If young women lack significant influence, then why do suitors shower them with expensive gifts? Why if not to win favor? And why seek to win favor if women lack influence?

Furthermore, while there are situations in Homer where women are married off apparently without input (e.g., *Iliad* 9.141–48, 18.429–34; *Odyssey* 11.281–91), there are widely scattered indications that they usually had more freedom to follow their mating preferences. For instance, there are allusions to women actively instigating illicit affairs (e.g., *Odyssey* 11.235–47; *Iliad* 6.160–62). And there are familiar images of youthful courtship—of young men and women displaying for each other at a dance (*Iliad* 18.593–606), privately trading soft words (*Iliad* 22.126–28), and sneaking off to consummate their love in secret from their parents (*Iliad* 14.295–96).

Finally, there is the evidence of comparative anthropology. Across cultures, even in those with arranged marriage, young women usually exert nonnegligible influence over mate selection (Broude and Greene; Geary, *Male, Female* 124–26; Small). The alternative to the hypothesis that women had a significant role in choosing their mates is to deny the generality of Penelope's example, to ignore the clear implications of wooing gifts and other aspects of youthful courtship, to downplay myriad instances of illicit affairs, and to posit that Homeric society was, in this respect, anthropologically unusual. As the examples of Aphrodite, Helen, Nausicaa, Circe, Calypso, Penelope, Anteia, Tyro, Melantho, and others illustrate, Homeric women would have obviously had pronounced mate preferences. It is unlikely, based on the information in the poems and anthropological inference, that these preferences could have been entirely controlled and contravened by men.

On the contrary, Homeric marriage customs appear to be of the most common anthropological type: where the interests of daughters and their families are substantially, though not perfectly, aligned, and where the choice of groom is a negotiation between the young woman's preferences and those of her family. I suggest that the Kipsigis, a pastoral group in Kenya, are one of many good models for the Homeric situation:

Choice of marriage partners is technically made by the young woman's kin and is influenced, in part, by the man's bride-price, or bride wealth offer, in addition to his social reputation and his political influence. In some cases, the preferences of the woman and the best interest of her parents strongly conflict, and in these cases female choice is sometimes circumvented by the woman's kin. . . . In most cases, however, the parents' decision is influenced by their daughter's preference. . . . With the Kipsigis, female choice is . . . intertwined with the material and political interests of the woman's kin, but in most cases these interests largely coincide with her preferences. (Geary, *Male, Female* 125–26)

Size Matters

To an extraordinary degree, the predilections of the investing sex—females—potentially determine the direction in which the species will evolve. For it is the female who is the ultimate arbiter of when she mates and how often and with whom.

HRDY, *WOMAN*

The fact that, in most species, females choose and males compete to be chosen means that the female's preferences go a long way toward determining the traits and qualities of the male. Female preferences and male characteristics co-evolve, the one always shaping the other. If men, for instance, tend to love and invest in their children, it is partly because ancestral women rewarded fatherly men with sexual access while denying it to those who lacked paternal feelings and a corresponding predisposition to invest in children. Likewise, if men compete for status and dominance, it is partly because ancestral women rewarded high-status, dominant men with reproductive opportunities. John Hartung was not being frivolous when he said, "males are a breeding experiment run by females" (pers. comm., quoted in Batten 22). Rather, he was colorfully expressing the core of Darwin's theory of sexual selection: the choices of females significantly shape many, though by no means all, male traits (on sexual selection, see Andersson; Cronin, *Ant*; Darwin, *Descent*; Miller, "How").

What qualities do Homeric women seek in their mates, and why? First, it is clear that they look for a man capable of gaining and controlling resources. At a minimum a suitor (and/or his family) must be able to afford a substantial transfer of resources to the bride and her family. As discussed above, the epics recurrently suggest that the best way for a man to woo a woman is to shower her with gifts. The payment of bride price, along with the less formal wooing gifts, indicates a suitor's ability to provide material care for a woman and her children.

While possession of a minimum level of wealth seems an absolute prerequisite for any suit, Homeric women also care about the physical appearance of potential mates. In this, the pragmatism of Homeric women only apparently gives way. As we shall see, their attraction to men with certain physical characteristics is as practical as their attraction to men with appealing socioeconomic profiles.

While Homeric women gravitate toward handsome men—men with glossy hair and beards, taut skin, and handsome faces—they seem particularly attracted to men who are large and strong. Odysseus is as devastating a romantic hero as he is a martial hero; while not as tall as Agamemnon or Menelaus (*Iliad* 3.193–94, 3.210) he is never described as short and he is once described as tall (*Odyssey* 21.334). Moreover, he is so powerfully built that, even amid a swarm of strong warriors, he stands out like a great strapping ram among ewes (*Iliad* 3.197–98). Women move to him as though magnetically compelled. Such is the case, for instance, when Odysseus' charms sweep Circe, Calypso, and Nausicaa, not to mention Penelope, off their sandal-clad feet. Odysseus is a composite representation of all the things, according to folk wisdom and science, women crave in their mates: he is unfeasibly handsome, dominant, intelligent, rich, and vigorous. Odysseus would be as comfortable in the world of the modern-day Harlequin Romance as in the world of the *Odyssey* (for a seminal study of cross-cultural tendencies in women's mate preferences see Buss, "Sex Differences"; for a relatively current review see Buss, *Evolution*).

Homer stresses all of these qualities, not least of all his masculine size, strength, and vigor. Homer and his characters unabashedly marvel at Odysseus' massive arms, his broadly muscled shoulders, and his rippling hero thighs (e.g., *Odyssey* 8.457–59, 11.11–21, 11.336–37, 18.66–72). While men take note of Odysseus' appearance—carefully marking him as a formidable potential foe or ally—so too do women. For instance, after Odysseus has washed away the muck and brine of his ordeal at sea, young Nausicaa sees how handsome he is, how strong and impressive. She gazes on him "in wonder" and sighs to her maids, "If only such a man were called my husband, living here, and it pleased him to stay here forever" (*Odyssey* 6.244–45). Similarly, the scenes with Circe and Calypso, breathtaking goddesses who lust after Odysseus and hope to marry him, confirm Odysseus' status as a masculine ideal.

Moreover, there is some evidence that Penelope does not remarry partly because no man measures up to Odysseus in physical power and fighting spirit. By personally organizing the archery competition to choose a new husband, Penelope forsakes the usual custom of favoring the suitor who offers most. The contest is designed to identify the man with most strength and skill—the man who can, figuratively and literally, wield Odysseus' massive bow. Penelope usually huddles in the women's quarters and wails her fate while the suitors enjoy their drunken sports. But during the contest Penelope joins them, carefully observing as each suitor pitifully fails to string Odysseus' bow. The scene is a mating ritual, reminiscent of bird leks, in which a receptive female sits and observes while flocks of males dance, ruffle, preen, and swell their chests in hopes of being chosen. When Odysseus easily bends and strings the bow, and coolly makes a shot that is virtually impossible, we can see that *here* is a man worthy of the choosiest woman in the world, for not even Helen had more than 100 suitors to choose among.

The preference of Homeric women for large and strong men is not arbitrary. With only a couple of exceptions, the leaders of Homeric groups—whether of war parties or towns—are described as their largest, most powerful, and most formidable members. In Homer, as in many ethnographical societies, the "big men" *are* big men, and they enjoy clear advantages in competition for wealth and prestige (on the advantages tall men enjoy, see Sugiyama 315–21; see also Barrett, Dunbar, and Lycett 106–7).

Demonic Males

As the genocidal destruction of Troy demonstrates, women obviously have much to lose from male competitiveness, perhaps especially when it moves beyond the individual level to competition between groups. Among groups of humans, as among some other social animals, including lions and gorillas, male competitiveness can have disastrous consequences for females. When one group of males vanquishes another, it can mean death for young children (Hrdy, *Mother Nature* 237–44, 413; Valero). Moreover, it can mean rape—the loss of the precious prerogative of joining reproductive fortunes with the best man available. Obviously, these outcomes run counter to female fitness goals. How did ancestral women respond to these selection pressures? Paradoxically, it was probably by choosing men who were more, not less, competitive.

Homeric women played a role in perpetuating male competitiveness and violence by rewarding brave warriors with favor and denying it to the weak and uncourageous. These dynamics are reflected in Helen's disgust at Paris' performance in his fight against Menelaus. Paris is soundly thrashed by Menelaus, who drives him to the ground and is dragging him off the field to execute him. However, Aphrodite intervenes and spirits Paris off the field in a mist, depositing him in Helen's boudoir. When Helen finds him she is not elated that her lover has survived; she is appalled and ashamed. Instead of tending his wounds, and soothing his injured pride, she conducts an emotional assault on the beaten man that is as furious as Menelaus' physical assault. She is so disgusted by his weakness that she cannot look at him; with eyes averted, she taunts him with reminders of his boasts of strength, and coldly expresses her regret that he survived the encounter (*Iliad* 3.428–36).

Crucially, at this moment, Helen is filled with revulsion for Paris, the paragon of male sexual attractiveness. She tells Aphrodite, "To that place [Paris' bed] I will not go. It would be disgraceful to share his bed. All of the women of Troy would heap blame on me" (*Iliad* 3.410–12; see also 6.349–53). Despite her contempt for him, Paris finds her more attractive than ever and wants ardently to take her to bed. Helen has no interest, but consents when Aphrodite threatens her with dire consequences. The whole scene carries the message that Homeric women respect bravery and strength in their men, and are loath to reward the weak and cowardly with favor. These tendencies are reinforced by social pressure from the female community as a whole ("the women of Troy would heap blame on me"). The scene also communicates the strong effect this fact has on the men, since Helen's tirade shames Paris back to the battlefield. After taking leave of Helen, Paris runs into Hector, and tells him: "I only wanted to give myself over to anguish. But my wife convinced me with winning words, urged me back to the fighting" (*Iliad* 6.335–36).

The respect of the Trojan women for men of valor is similarly communicated in Hector's fond—and sadly chimerical—daydream that his doomed son will one day be a fierce and mighty warrior. He envisions his baby as a grown man, returning from battle, laden with bloody trophies stripped from dead foes. Part of the savor of the vision is his anticipation of Andromache's own satisfaction, as an aging mother, in claiming such a brave, capable son (*Iliad* 6.476–81). That warriors may have regularly presented battlefield spoils to women is also suggested by Zeus' statement that he will not allow Hector to survive to present Achilles' armor, stripped from Patroclus, to Andromache (*Iliad* 17.201–8).

These scenes suggest that Trojan women respect and expect valor in their men. This is despite the fact that Andromache futilely begs Hector not to go back to the plain, but to defend Troy from behind the walls. Of course, as a loving wife, Andromache fears for Hector's safety and has strong misgivings about his return to a pitched battle. But, despite her restraining words, Hector insists on venturing back to the field. And his reasons for doing so are instructive. He says that, as much as he would like to remain with his family, he must rejoin the fight or "feel great shame before the Trojans, and the Trojan women with the trailing robes" (*Iliad* 6.441–42). This is not the first or last time that a warrior goads himself to fight, at least in part, by considering his reputation among women. Paradoxically Hector's decision to resist his wife's pleas seems based on the fact that, were he to submit, he would not be the great warrior that she loves and respects.

The evidence suggests that Homeric women find those men most attractive whose bodies signal formidable warrior potential. The ugliest man in Homer is Thersites, whose body—scrawny, stooped, lopsided—highlights his weakness. There is something important behind the fact that the most physically desirable men are usually the greatest warriors. The greatest mortal warrior, Achilles, is also the most beautiful Greek on the field (*Iliad* 2.673–74); the second best warrior, Ajax, is also second in physical beauty (*Iliad* 17.277–80). Perhaps importantly, Achilles and Ajax are also defined as the largest Greek warriors by far—the only ones who could wear each other's armor (*Iliad* 18.188–93). Eurypylus and Memnon—preeminent warriors among the Trojan allies and leading heroes elsewhere in the Trojan Cycle—are identified by Odysseus as the handsomest men he has ever seen (*Odyssey* 11.519–22). And the war god, Ares, is gorgeous, the natural mate for the most desirable goddess, Aphrodite (*Odyssey* 8.309–10).

Thus in Homer, as with much of Greek myth and literature, the war hero is virtually synonymous with the erotic hero. This is true of Odysseus, of the Athenian hero Theseus, and of the greatest Greek hero of all, of whom Plutarch wrote, "It would be a labor of Heracles to enumerate all his [Heracles'] love affairs, so many were they" (*Erotikos*). The one prominent exception to this pattern is Paris, who is breathtakingly handsome but uncourageous. (The only other exception I can recall is Nireus, who is very briefly described in the Catalog of Ships as a strikingly handsome weakling [*Iliad* 2.671–75].) Paris is so weak that Menelaus—far from the strongest Greek—overcomes him barehanded, after his sword has shattered and his spear has been thrown in vain. But Paris actually represents the exception to prove the rule. Because Paris does not *look* like a wimp; he has all the physical appeal of a formidable warrior. When Hector harangues Paris for attempting to shirk the fight with Menelaus, we learn of a powerful correlation between male physical attractiveness and fighting strength: "I think the long-haired Achaians will scoff, saying you are our best champion on account of your fine looks, but there is no might in your heart, no courage" (*Iliad* 3.43–45).

In summary, the epics suggest that women's mating preferences are partially responsible for the violence of individual men and of the culture as a whole. Women in Homer are not only the victims and objects of male violence, though they certainly endure great traumas at the hands of men. Rather, through an active system of sexual and reputational rewards to men with powerful bodies, combative dispositions, and courageous spirits, they reinforce, encourage, and perpetuate male competitiveness. This is in no sense to "blame the victim," because Homeric women have little choice in the matter. They are forced down this path less by the meddling of their kinsmen in their romantic choices than by the behaviors of males in their world. As the primatologists Wrangham and Peterson argue, in their book on the evolution of aggression in humans and other apes, once males develop what they call "demonic" reproductive strategies—characterized by intense, frequently violent competitiveness—females have little choice but to reward them with sexual access. In the course of human evolutionary history, women have consistently chosen mates with the capacity for aggressive behavior for an utterly compelling reason: there are dangerous and aggressive men in the world (Smuts, "Evolutionary Origins," "Male Aggression," Wrangham and Peterson). This is part of Andromache's attraction to Hector: he stands between her and the aggression of other men, protecting their child, and warding off her "day of bondage" (*Iliad* 6.462–63). A viciously cyclical feedback loop takes hold: men can be dangerous so women reward the strong and potentially aggressive with sexual access; as a result males grow, generation by generation, more "demonic," both by genetic predisposition and by sociocultural pressure; this, in turn, places women in the position of having to select for progressively more demonic mates. Demonism breeds demonism, and neither men nor women have much choice in the matter.

25

New Science, Old Myth

AN EVOLUTIONARY CRITIQUE OF THE
OEDIPAL PARADIGM

MICHELLE SCALISE SUGIYAMA

Literary interpretation and theory derive their legitimacy from the tacitly accepted yet largely unexamined premise that characters are representations of human beings and, as such, exhibit the same psychology as their author and audience. To put it another way, literary characters do not exhibit the thought processes of okapis, ostriches, octopi, or any other species. (Even the metamorphosed Gregor Samsa thinks, perceives, and responds primarily as a human.) We assume that literary characters have human beliefs, desires, emotions, and perceptions—for example, that a (mentally competent) character's conceptualization of dog, fetch, and devotion reliably corresponds to our own. By virtue of its subject matter, then, all literary criticism is in one way or another psychological criticism, and, in a fundamental way, literary study is the study of human cognition.

Nicholas K. Humphrey writes that the "novelist is in the most literal sense a 'modeller' of human behavior, someone whose skill as a psycholo-

An earlier version of this essay first appeared in *Mosaic: A Journal for the Interdisciplinary Study of Literature* 34, no. 1, pp. 121–136.

gist is required not simply to comprehend but to invent the things that other people do" (67). The same principle applies to the literary scholar, whose skill as a psychologist is required to understand the things that literary characters and narrators do. Yet, despite the fact that literary scholarship regularly makes assumptions about the operations of the mind, its practitioners customarily receive no training in cognitive design and evolution. This essay addresses a specific—yet widespread—manifestation of this problem: the persistence of the oedipal paradigm. This model, which in various literary permutations is commonly invoked to analyze everything from male sexuality to family dynamics to narrative structure, is founded upon an inaccurate conception of what the mind is designed to do. Freud did not understand that, in order for a psychological feature to evolve, it has to contribute to fitness (a biological term referring to the differences in physical and psychological attributes that cause some individuals within a given population to contribute more genes to subsequent generations than other individuals do). As a result, he posited a highly unlikely phenomenon. Freud's mistake is understandable, but, given what is now known about human cognition, behavior, and biology, the continued use of this model by his intellectual descendants is not.

With few exceptions (Carroll, *Evolution, Literary Darwinism*; Constable; Cooke; Easterlin and Riebling; Nesse; Scalise Sugiyama, "Food," "On the Origins," "What's"; Storey, "I Am," *Mimesis*; Whissell), contemporary psycho-literary criticism gives the impression that human evolution has nothing whatsoever to do with the study of literary activity. Mainstream psycho-literary criticism remains founded on the century-old conceptualization of the mind originally developed by Freud, a model no longer taken seriously in cutting-edge psychological research. Although Freud's model has been "revised" by psychiatrists and literary scholars (perhaps most notably by Jacques Lacan), these amendments are more properly labeled "refinements": because they do not approach human cognitive processes on the level of function (in the biological sense of subserving fitness), they do not fundamentally alter Freud's original conceptualization of the mind. Although the veracity of Freud's work has been seriously challenged in recent years, the implications of these critiques for the oedipal theory's validity as a heuristic tool have been overlooked or ignored. Regardless of whether it is explicitly Freudian, much psycho-literary analysis assumes that the oedipal phase is a normal part of human development, and the oedipal complex is used to explain a wide assortment of conflicts (sexual and non-sexual). Critics unblinkingly speak of "oedipal narrative" (Lupton and Reinhard 2; L. Williams 81) and the "conventional, linear, oedipal plot" (Moon 144), of "oedipal development" and "the oedipal moment" (L. Williams 28), of "the classical oedipal father–son conflict" (Boker 9) and the "oedipal family situation"

(Boker 139), of "oedipal rebellion" and "oedipal hostility" (Griswold 94, 35), of "the pre-oedipal phallic mother, who threatens castration" (Lydenberg 1078), and even of the negative Oedipus complex (Boker). Moreover, this paradigm is not localized to a particular sub-field: a cursory search of oedipal analyses produced over the last decade or so yields studies spanning a wide range of genres, periods, and authors ranging from film (Moon) to American children's classics (Griswold), from Hamlet (Lupton and Reinhard) to Huckleberry Finn (Boker), and from Augustine (Rudnytsky) to Thoreau (Pfitzer) to Hemingway (Boker). That these scholars are not household names is testimony to the model's pervasiveness: the oedipal paradigm is not the hermeneutic domain of an exclusive group of celebrity critics but, rather, common literary property.

The basic premise of the oedipal complex is that all children go through a phase in their sexual development when they are sexually attracted to their opposite-sex parent and consequently desire to kill their chief rival, namely their same-sex parent. As Frederick Crews asks, "Would we allow a less ingratiating rhetorician than Freud to sell us such an idea on his personal guarantee that it was true?" (62). Freud's assertion was based on a handful of what he, himself, referred to as "neurotics"—not, in other words, on a representative sample. Contrary to Freud's hypothesis, there is compelling evidence that sexual apathy or aversion toward family members is activated during the early years of childhood.

As Robin Fox so frankly puts it, to understand incest avoidance, we need to understand how the human breeding system evolved (4). Specifically, we need to bear in mind that the human breeding system is the product of natural selection. Incest—that is, mating with a close genetic relative—tends to increase the chances of producing an unhealthy offspring. (Throughout the vast majority of human evolution, no contraceptive devices were available; thus, every act of heterosexual coitus between fertile, sexually mature persons carried with it the very real possibility of pregnancy.) Unhealthy offspring are unlikely to survive to reproductive maturity. Incestuous mating, then, tends to decrease an individual's chances of passing his or her genes on to subsequent generations. Conversely, non-incestuous mating tends to produce healthier offspring, which in turn increases an individual's chances of passing his or her genes on to subsequent generations. In the process, the cognitive capacity by means of which this individual avoids incest gets passed along as well. As we can see, then, a mechanism inhibiting incestuous mating is much more likely to have evolved than a mechanism motivating incestuous mating (Daly and Wilson, *Sex*).

The notion that our aversion to incest is culturally conditioned rather than part of our evolved psychological architecture was based in part on the erroneous assumption that other animals commonly mate incestuously (Degler 245–69; Erickson). Data from a variety of studies, however, show that incest in animals is rare (Bischof; Degler; Erickson; Kevles; Reynolds; Sade). Also, it is now known that incest is biologically hazardous: one of its most disastrous effects is "an increased expression of deleterious recessive genes . . . the net result [of which] is a marked increase in early mortality" (Erickson 412). Evidence suggests that early humans were aware of this correlation: in a survey of incest myths in the Human Relations Area Files, Melvin Ember found that one-third of the stories mentioned deformed offspring as the product of incestuous matings.

What, then, is the mechanism by means of which the vast majority of humans avoid incestuous mating? In 1891, Edward Westermarck proposed that children who are raised together, even if unrelated, rarely develop sexual feelings for one another. Earlier in the century, Jeremy Bentham made a similar observation, commenting on the absence of sexual feeling between family members and, conversely, the "novelty," or lack of familiarity upon which sexual attraction is dependent (cited in Degler 247). Westermarck's idea is logically compelling. A prerequisite to avoiding incest is recognizing individuals who are closely related to you. Cross-culturally, close relatives typically inhabit the same household, and we have no reason to believe that our Pleistocene ancestors did any differently. Thus, throughout human evolutionary history, a reliable cue of relatedness would have been continuous proximity in early childhood. Because persons living closely together were highly likely to be near relatives, a mechanism that squelched sexual desire between cohabitants would have tended to prevent incest under familial living conditions.

Westermarck's argument is strongly supported by both ethological and ethnographic data. Close contact with immediate kin in early life is crucial to the onset of incest-avoidance mechanisms in a wide variety of animals, from prairie voles to baboons to chimpanzees (Erickson; Gavish, Hofmann, and Getz; Pusey; Smuts, *Sex*). By far the most compelling evidence, however, is a body of data collected by several independent researchers from three culturally distinct human populations. These data overwhelmingly indicate that people do not develop sexual feelings toward their childhood familiars. The first of these studies was conducted by Yonina Talmon, who found that, of 125 couples who had grown up in Israeli kibbutzim, no two mates were raised from birth in the same house or peer group. Nor did she find any love affairs between same-group members (*sabras*). Subsequent studies of kibbutzim yielded similar results: no marriages and very few extramarital sexual relationships between *sabras* (Parker). The most ambitious of these later studies was conducted by Joseph Shepher, who examined the marriage records of over 2,769 kibbutz-raised individuals. His survey revealed an interesting pattern: he found no marriages between persons who had been members of the same kibbutz throughout early childhood (ages 0 to 6), only eight marriages between persons who had been members of the same kibbutz throughout late childhood (ages 6 to 12), and only nine marriages between persons who had been in the same kibbutz throughout most of their adolescence (ages 12 to 18). Shepher's research indicates that the mechanisms that inhibit sexual desire between familiars are most sensitive from birth to age six. Shepher did not stop with marriage records, however. He also investigated the premarital sexual preferences of adolescents raised in kibbutzim: the young people reported that marrying a person from one's own house would be like marrying one's brother or sister. Tellingly, Shepher found only one instance of heterosexual activity between adolescents who had been members of the same kibbutz, and that case involved a male who had not entered the kibbutz until he was ten years old. These findings are especially compelling given the kibbutzim policy of openness about sexual matters and given parental encouragement of marriage between *sabras* (Shepher).

Another "natural experiment" in support of Westermarck's hypothesis was reported by Arthur Wolf, who, with his colleague Chieh-shan Huang, examined *simpua* marriages in Taiwan ("Childhood"; *Marriage*). In *simpua* marriage, a couple is betrothed in infancy, and the bride-to-be is raised in her future husband's household. Wolf and Huang found that *simpua* couples commonly have an aversion toward consummating their union, and that infidelity and divorce rates are higher among *simpua* couples than among other couples. Additionally, they found that the birth rate was about 30 percent less for *simpua* couples than for other couples (due presumably to lower frequencies of intercourse).

Further evidence that childhood familiarity breeds sexual apathy or aversion comes from the Lebanese practice of marrying the son of one brother to the daughter of another. Since brothers tend to live near each other, their children grow up in continual contact with one another. Justine McCabe found that these marriages produced 23 percent fewer children and were four times more likely to end in divorce than marriages between unrelated persons. Furthermore, marriages between cousins who were not raised near each other exhibited a profile similar to marriages between unrelated persons, indicating that childhood familiarity and not degree of relatedness was the cause of diminished sexual desire between cousins who grew up together.

Pierre van den Berghe takes Westermarck's idea one highly plausible step further: in addition to siblings, he notes, parents (especially mothers) are typically intimate associates of children in their early years; it thus stands to reason that children will not develop sexual feelings for their primary caregiver(s). Mark Erickson develops this idea more fully in his "familial bond" hypothesis, in which he proposes that "an innate bonding process in early childhood" (411) between the infant and its immediate kin functions to establish both incest avoidance and kin selection. He doesn't specify what this innate bonding process might be, but one possibility is suggested by Fox: "No one is more familiar with a son than a mother who suckles him and some anthropologists have seen in the suckling experience precisely one of those universal, early imprinting traits that might account for mother–son incest aversion" (53). The plausibility of this hypothesis is increased by the fact that, in known hunter-gatherer cultures (whose living conditions are evocative of Pleistocene conditions), children are commonly nursed for three years (Chagnon; Shostak).

In light of this research, it is important to note that Oedipus does not fall in love with Merope, the woman who raised him and—significantly—the woman he believes to be his biological mother. (Nor does he murder Polybus, the man who raised him and the man he takes to be his biological father.) As Jean-Pierre Vernant noted over twenty-five years ago, technically speaking, Oedipus has no Oedipus complex (107–9). If he did, his "oedipal" feelings would be directed toward Merope, not Jocasta. Van den Berghe describes the phenomenon observed in the kibbutzim as "culture fooling Mother-Nature" (quoted in Degler 260). In the Oedipus tale, we have a similar situation except that the story deals with mother–son incest avoidance instead of brother–sister incest avoidance. In Oedipus's case, the cultural institution that "fools" his evolved incest-avoidance mechanisms is not the kibbutz but, rather, adoption. His feelings toward the female caregiver with whom he has been familiar since birth are precisely the feelings evolutionary theory and van den Berghe's extension of the Westermarck hypothesis would predict: lack of sexual interest. His feelings towards Jocasta are similarly predictable given that he was not in contact with her during the crucial period identified by Shepher (ages 0 to 6): absence of sexual revulsion. The myth thus supports Westermarck's proposal that childhood familiarity, not relatedness, is the mechanism that triggers incest-avoidance mechanisms. The myth also supports the corollary that incest tends to occur under atypical familial circumstances, circumstances that thwart the psychological mechanism evolved to trigger sexual antipathy among close kin.

From a biological perspective, childhood feelings of sexual desire for one parent and murderous jealousy toward the other are implausible: there is no need driving their emergence. The evolution of oedipal desires would be a waste of energy because a child is incapable of carrying them out. Even if this were possible, such desires would be counterproductive: on the one hand, copulation before sexual maturity is fruitless, and, on the other, killing one of its providers would at best sharply decrease a child's quality of life and at worst drastically reduce its chances of survival. The inefficiency and implausibility of this model become glaringly apparent when viewed from a design standpoint: the proposed mechanism requires (1) sexual desire before sexual maturity (2) for a biologically unsuitable partner that (3) must then be redirected. In other words, the proposed mechanism provokes the very problem it seeks to prevent. Given the high energy costs of a complex brain (Pinker, *How*), a design that expended so much energy on unnecessary operations would not survive for long. In sum, the oedipal complex postulates the existence of psychological mechanisms that not only are unnecessary but also would be deleterious to the individual possessing them.

It may be that Freud named the oedipal complex after the infamous king of Thebes not because Oedipus's childhood experience mirrored the developmental phase he described but simply because Oedipus was readily recognizable as a man who killed his father and had sex with his mother. Or perhaps Freud unconsciously misconstrued the myth to suit his purposes. Regardless of his motives, the myth has come to be closely identified with the complex: Freud's model is regularly referred to or treated as an interpretation of the Oedipus myth (e.g., Boker 138; Nussbaum, "Oedipus Rex" 43; Rudnytsky 129). As a handful of scholars have pointed out, however, it is actually a misinterpretation. Stith Thompson, for example, noted nearly fifty years ago that the psychoanalytical version of the myth overlooks the value system and world view of the culture that produced it:

> Certainly the Oedipus story of Sophocles involves only quite accidental events and, so far as I can see, has no bearing whatsoever upon the so-called "Oedipus complex." The only enmity Oedipus has is toward an unknown man who tries to drive him off the road. From the Greek point of view, the fact that this man happened to be Oedipus' father came entirely from the workings of Fate and not from any psychological law. The same is exactly true of the unwitting marriage with his mother. (quoted in Lessa 74)

Moreover, as Martha Nussbaum notes, erotic desire is not a major theme of Sophocles's play, even in the marriage between Oedipus and Jocasta: "The marriage is a political one, and is never described as being motivated by eros. Eros is mentioned frequently in Sophocles—but not in this play. . . . And it is crucial to its construction that the collocation of circumstances that strikes Oedipus down is not regarded, by him or by anyone else in the play, as the product of his sexual intentions, whether conscious or unconscious" ("Oedipus Rex" 43–44). As these critiques illustrate, the psychoanalytical version

of the myth glosses over some important details: (1) Oedipus does not have sexual feelings for the woman he knows as his mother, Merope; and (2) Oedipus does not harbour feelings of sexual rivalry toward the man he knows as his father, Polybus.

The Freudian interpretation of the Oedipus myth also overlooks the important biological fact that infants are incapable of reproduction. Sexual desires at this age are a waste of time, energy, and resources that could be—and are—directed toward more immediately useful development such as motor coordination, language acquisition (Pinker, *Language*), social cognition (Baron-Cohen, *Mindblindness*), and understanding the physical and psychological properties of objects in their environment (Leslie, "Spatiotemporal Continuity"; Premack; Spelke). Tellingly, in "The Oedipus Rex and the Ancient Unconscious," Nussbaum notes "the complete absence of any belief in infantile, or even childhood, sexuality" (48) in ancient theories about child development. She adds that, in pre-Freudian Western culture, sexual desire was assumed to awaken during puberty, not in infancy or childhood (49). Freud put the cart before the horse, assuming that the relationship between parent and child is sexual in nature when in fact the opposite is true: adult sexual relationships borrow from infantile and parental behavior (Eibl-Eibesfeldt, *Love*). As Robert Storey explains, "The bonding rites of human sexual behavior have originated in infantile or parental gestures. . . . The kiss, for example, has its probable source in the caretaker's mouth-to-mouth feeding; the 'kiss with the nose' in the sniffing of kin; the 'love bite' in the friendly grooming nip; the nuzzle in the infant's search for the breast" ("I Am" 54). It can still be argued that the Oedipus myth is evocative of universal human psycho-sexual experience; however, the common denominator of this experience is incest aversion, not incestuous desire.

The Oedipus myth also indirectly evokes the adaptive problem of parent–offspring conflict. Robert Trivers defines parental investment as "anything done by the parent for the offspring that increases the offspring's chance of surviving while decreasing the parent's ability to invest in other offspring" ("Parent–Offspring" 249). As Daly and Wilson explain, in his interpretation of the Oedipus myth, Freud conflated two separate sources of potential father–son conflict, both of which involve parental investment, but only one of which involves the mother (*Homicide* 115). In the conflict involving the mother, father and son are indeed competing for the mother's attention. However, each wants a different kind of attention: the father wants sex, the child wants care. A sibling—the likely result of mother's sexual attention to father—diminishes the amount of care and resources mother can expend on her older child. Thus, "it is not implausible that young children have evolved specific adaptive strategies to delay that event [the birth of a younger

sibling] by diminishing mother's sexual interest and thwarting father's access to her" (Daly and Wilson, *Homicide* 115). For example, it is commonly known that, upon the birth of a younger sibling, children often engage in regressive behavior (Dunn and Kendrick). Daly and Wilson interpret this as an exaggeration of dependency aimed at regaining parental investment now being directed at the younger offspring (*Homicide* 97).

At the heart of Freud's primal horde theory is the notion of a power struggle between older and younger males over sexual access to women (Fox 76), and this is the essence of the second parent–offspring conflict obfuscated by the oedipal paradigm. Here Freud simply misunderstands which women the father and son are competing for. Once a male child reaches sexual maturity, he and his father may become sexual rivals not for the child's mother but for other women (in cultures that allow polygyny, for example). Daly and Wilson explain this conflict as follows: "Human fathers frequently command resources (including status and titles as well as the obvious material resources) that are familially held and patrilineally transmitted. Such resources are limited, and hence impose limitations upon reproductive and other ambitions. A still robust father can therefore be an impediment to the aspirations of a young man" (*Homicide* 117). In other words, a father may choose to apply his resources toward acquiring a mate for himself or for his son. In the former case, parent–offspring conflict is almost sure to erupt. Robert LeVine repeatedly observed such tensions in sub-Saharan Africa:

> Here the tension often centers about the father's fear that the son, particularly the eldest, will wish to hasten his death in order to gain independence from the old man or replace him as head of the family. In societies where the father is expected to transfer goods to his son for brideprice payment, conflict may arise over the father's delaying the transfer and/or the son's appropriating the goods in advance of formal permission from the father. The sexual rights of father and son also present problems: where the father exercises authoritarian domination over his sons, the danger exists that he will commit adultery with their wives, and this danger is recognized in many African groups. On the other hand, when the father is a senescent polygynist, wealthy enough to acquire numerous young wives but not virile enough to keep them sexually satisfied, the danger exists that the sons will commit adultery with his younger wives. In some groups—where the sons inherit these wives upon the father's death—this may be overlooked when the father is very old, but in others it is a real source of conflict. (193)

Thus, a son may indeed harbor murderous feelings for his father, and the wellspring of these feelings may be sexual desire, but the object of this desire

is not the mother. Conversely, a father and son may compete for the mother's attention, but it is only the father who seeks sexual attention.

According to Fox, "Freud was a great reader of the anthropology of his day. He had read Darwin and Westermarck (of course)" (53). Indeed, Freud discusses Westermarck's hypothesis in *Totem and Taboo* (122–23). It is surprising, then, that a man of such powerful intellect did not see the relevance of these men's ideas to his own. More surprising is the tenacity with which literary criticism has held on to such a counterintuitive paradigm. Although Freud probably did not have access to such information, it has been known for decades that incest among humans is statistically rare and mother–son incest the rarest of all (Fox). This trend is echoed in the world folklore record. In their exhaustive review of oedipal myths spanning five continents and Oceania, Allen Johnson and Douglass Price-Williams found stories in which a young man kills his father and marries his mother to be uncommon (3). Thus, through its continued use of the oedipal model, and its disregard of a century's worth of advances in psychological and anthropological science, literary scholarship presents a picture of the human psyche that is both inaccurate and incomplete.

26

The Wheel of Fire and the Mating Game

EXPLAINING THE ORIGINS OF TRAGEDY AND COMEDY

DANIEL NETTLE

One of the most remarkable attributes of human beings is their propensity to engage with worlds that are imaginary. If other species attend to nonveridical narratives, and we have no evidence that they do, those narratives are private and internal to one individual. In the human case, by contrast, all cultures create stories that are shared, are attended to jointly by several people, and continue to hold attention despite being universally known to be false. These are not peripheral activities; the world of fiction and myth engages a great deal of time and energy in societies both affluent and under the most basic of conditions. There are two key questions concerning the origins of this attention to the imaginary. First, why should human beings be predisposed or vulnerable to attending to non-veridical events in the first place? Second, why do particular forms of shared representation persist and hold attention, rather than other logically possible ones?

The two questions may in fact be best tackled together, since it is not nonveridical representations *in general* that humans are predisposed to attend

Daniel Nettle. "The Wheel of Fire and the Mating Game: Explaining the Origins of Tragedy and Comedy." *Journal of Cultural and Evolutionary Psychology* 3, no. 1 (2005): 39–56.

to, but rather, particular classes of representations that vary from culture to culture but might turn out to have some universal similarities or tendencies. Thus the central question is simply: Why do we engage in the fictional cognition that we do? By fictional cognition, I mean cognition about non-veridical scenarios, represented verbally, visually or kinetically, where knowledge of the non-veridical nature does not lead to a cessation of attention.

This paper concerns the evolution of the dramatic mode of fictional cognition, and in particular, the comic and tragic genres (section 3). However, I will begin by discussing what we know of the adaptive problems that human minds are designed to solve, which allows some predictions to be generated about the types of fictional representations which might be most effective at getting themselves noticed. As a result of this, I will argue that the deep appeal of the dramatic mode, and in particular comedy and tragedy, is easy to explain. Thus, once such a form had arisen, it is no surprise that it rose to cultural pre-eminence and endured for so long. Analyses of one comedy (section 4) and one tragedy (section 5) are given in support of these contentions.

Psychological Foundations of Drama

He who understands baboon would do more toward metaphysics than Locke.

CHARLES DARWIN

Human beings belong to the order of primates. They are often characterised as the exceptional primate, which obviously they are in terms of language, culture and so on. However, human beings are not aberrations; in many ways, they are merely the most clear or extreme example of tendencies that characterise the whole order, and especially the monkeys and apes. Primates, for example, have relatively large brains.

This tendency is increasingly marked from prosimians up through New and Old World monkeys to apes. We stand not as an exception, but as the summit of a tree. Another essential characteristic of most primates is their sociality (Dunbar, *Primate*; Smuts et al.), with most species living almost all their lives in groups of a handful to a few dozen individuals. Again, the tendency becomes more marked up through the order and culminates in us.

Primate social groups have special characteristics that differ from, say, the social groups of bees and ants. Bees and ants are locked into hereditary castes dictated genetically and regulated physiologically. Primate groups, by contrast, are much more dynamic. Individual animals have places within a status hierarchy that determines mating and resource holding, but this hierarchy is constantly renegotiated through behaviour. Status may be achieved by direct physical dominance, through the formation of coalitions based on reciprocity, or through piggybacking on the dominance of kin. Status is desirable to achieve as it determines access to food and, most importantly, the likelihood of reproduction. Low-ranking individuals can suffer considerable physiological stress (Keverne, Martensz, and Tuite).

Within primate societies, "cliques" and "groups" can be distinguished (Kudo and Dunbar). The group is the total number of individuals that travel, feed or sleep together. Cliques are smaller sets of individuals within this who have stronger than average direct relationships with each other. The total group is tight; any two individuals in a group can usually be linked either directly or by one remove; either they are in a clique with each other, or they are in a clique with someone who is in a clique with the other. The "social glue" that holds this intricate arrangement together is grooming. Wild primates spend up to 20 percent of their time going through each other's fur, removing parasites (Dunbar, "Coevolution"). Subordinates groom superiors, alliance partners groom each other, kin are tended, potential mates flirt, outcasts go ungroomed. One consequence of the flexibility of primate social systems is that you need to keep track of a lot of social knowledge. A bee can *smell* who is a worker and who a drone; a monkey has to *remember* who is dominant and who subordinate by close observation of who has been grooming who. A bee knows who has the right to reproduce and who not, because only the queen has, and her physiology is an indelible mark; a monkey has to work out, negotiate, remember. Since primate rank is not entirely hereditary, life contains vicissitudes, including the possibility of rising to dominance, and the risk of falling from it, all contingent, all dependent on one's own behavioural decisions and the reading and remembering of the behaviour of others. It follows from this that the larger the social group a primate lives in, the more complex the psychological task it needs to perform to behave in an adaptive way. This is because as the group size increases, the number of relationships one needs to observe and keep track of rises exponentially.

Robin Dunbar has shown that, across primate species, the larger the social group, the larger the relative size of the neocortex, the "higher cognitive" part of the brain ("Coevolution"; Kudo and Dunbar). More social information to track demands more computing power to do it, even at the evolutionary cost of laying down more cerebral tissue, one of the most metabolically expensive tissue types we have in our bodies (Aiello and Wheeler). And humans have the largest neocortices of all, suggesting what common observation shows to be true, that our nature is to live in large and complex social formations. One consequence of this fact is that observing social interaction within a group (seeing who is with whom and what they are doing together) is inherently attention-grabbing, well-remembered and salient. As Dunbar has cogently

argued, social groups as large and complex as ours would be very difficult to keep track of by grooming relationships alone. To directly observe all the pairwise grooming relationships amongst even a modest hunter-gatherer band would take a lot of time, and to groom or be groomed by every other individual even more so. But human beings have language. We can *share* social information. "X is trying to curry favour with Y, so Y will cooperate in supplanting your father Z," we can say. This has the double advantage of quickly disseminating useful information, and of creating a reciprocal bond between speaker and hearer, which is itself a kind of coalition that may come in handy later. Language binds us together as grooming binds monkeys and apes (Dunbar, "Coevolution").

Place a group of monkeys or apes in a room together, and if they do not fight or mate they will groom each other. Repeat the same experiment with a group of people, and if they do not fight or mate, they will talk. And Dunbar's observational studies suggest that they will talk overwhelmingly about the social worlds they inhabit (Dunbar, Marriott, and Duncan): people they know in common, the behaviour of other individuals, their own relationships. Technological, institutional, philosophical or aesthetic matters may intrude, but most natural conversations will soon return to the social network within which all of these other activities are embedded. For monkeys and apes, grooming releases the body's endogenous opiates, which is why they find it so rewarding (Keverne, Martensz, and Tuite). In humans, it seems that language may have sequestered this mechanism; we like nothing better than a good conversation. Conversations, then, are activities we are inherently motivated to seek, for good reasons to do with our evolutionary past. Any activity which could mimic the relevant features of a really good conversation would be extremely potent at capturing our attention. Good conversations are proto-typically concerned with exchanging information about the vicissitudes of relationships within a small social group. The foregoing paragraphs lead to very simple predictions. Evolution will have provided us with cognitive mechanisms or biases towards observing and tracking the behaviour of others within a social context, and attending strongly to information about such behaviour that might be given to us by third parties. We should find people-watching and gossip intrinsically rewarding. Thus any fictional cognitive activity which involved (1) direct observation of key interpersonal behaviours (for what "key" might mean, see below), and/or (2) shared information about social group motivations, deceptions and coalitions should strongly and persistently engage attentional mechanisms. Conversations are only interesting to the extent that you know about the individuals involved, and your social world is bound into theirs; as their distance from you increases, the interest level declines. Fictional social

cognition, which is based on invented individuals, has to compensate for this effect somehow. One way is for it to be not just a simulacrum of ordinary social cognition, but an *intensified* simulacrum. That is, the content of the cognition has to be at the maximum interest level possible in order to hold onto receiver attention, equalling or even exceeding the most attention-worthy aspects of real life. This is like the "supernormal" stimulus effect in animal behaviour. An egg elicits nesting behaviour from a female gull; a football elicits an abnormally strong nesting reaction (Tinbergen), even though eggs as big as footballs do not actually exist.

What does this "supernormal conversation" hypothesis predict more particularly? We have stated that human beings evolved in small, tight-knit social groups in which one person's opportunity to maximise fitness was closely bound up with the attempts to do so by all the others. Much conversation thus typically concerns behaviour by other people in our circle in relation to their attempted maximisation of their biological fitness; their illness, health, mating, rises and falls in status and coalitions and so on. The more extreme the fitness stakes, the greater the interest level. Betty going shopping is a lot less interesting than Betty leaving her husband for another man, because its effect on her fitness is much more significant, and, in a small-scale society, it would have a likely effect on *our* fitness, because there weren't many other mates to choose from, and the individuals involved were likely to be our allies or kin.

We would thus predict that as the fitness-change stakes go up, the attention-grabbing power of a story would increase. In particular, we should be especially interested in attempts by others to sequester scarce social resources. The key social resources in any primate society are status and mates. Members of both sexes, but males in particular, seek to maximise status with a combination of coalitional behaviour, deception, and direct physical confrontation. The way others around us are doing this is so important to us because status is inherently zero sum, and thus someone else's rise may be our fall. Potential mates are also limited in number, and so we should attend to who is looking like pairing off with whom. Thus, a supernormally interesting conversation would be one about status competition so severe it led to the death of one or more of the parties, and/or the process of pairing mates together in our immediate social group.

In a small-scale society, the attempts at fitness maximisation by one individual would be bound to impinge heavily on others, and this will bring about conflict. What is more, there can be conflict within one individual between the possible pathways for fitness maximisation. In primate societies, the ultimate goal of fitness maximisation decomposes into several distinct more proximate goals. At the most basic level, we can identify the following fundamental sub-goals:

1. *Self-preservation.* An animal should stay healthy, which means being well fed and keeping away from fights and predators.
2. *Mating.* An animal should seek to mate when opportune. Males need to ensure fidelity from their females, in order to be sure of paternity, whereas in altricial species females need to secure post-reproductive investment from their males.
3. *Status.* An animal should maximise its position on the status hierarchy, as this dictates access to mates, and resources in times of scarcity.
4. *Coalition formation.* An animal should maintain its coalitions.
5. *Kin.* An animal should protect the interests of its kin. (For a very similar list, see Carroll, "Deep Structure.")

All organic life involves trade-offs, and in primates these five goals are often in conflict, suggesting different strategies in the game of life. Should I seek the mate I want (2) even at the risk of physical danger (1) or the wrath of my friends (4)? Should I seek the alpha status position (3), even at the expense of my kin (5)? There is no definitive answer to these questions. The future can only be guessed at, and besides, the outcome of my choosing one course will depend on what everyone else decides to do. Thus the dilemma resonates ceaselessly; solutions can be guessed at but never known. These are the universal problematics of our order. Any primate should be interested in how other individuals solve them, for two reasons. First, it will have to solve them too, and seeing the result of someone else's negotiation of them might provide a partial model. Second, in tight groups, the way one individual solves them will have direct effects on the fitness prospects of all the others too. Given all these considerations, we can derive a design specification for a fictional form that would have an extremely strong intrinsic capture potential for human minds.

i. It should essentially involve the observation and conveyal of social information about relationships within small tight groups similar to those typical of our species' natural behaviour.
ii. These *groups* should interact in smaller units with especially strong relationships (*cliques*).
iii. The individuals involved should make attempts to maximise their biological fitness, with reference to goals 1–5 above.
iv. The more extreme the fitness stakes, the greater the intrinsic interest. The most significant domains are mate choice and status competition, and the extreme outcomes in these domains are mating and death.
v. The attempts by the protagonists to maximise their fitness will bring about conflict, either between different individuals, or between different sub-goals

for one individual. The richer and more complex these conflicts, the more captivating the product will be.

In the next section, we search for a fictional form in Western culture that fulfils these criteria.

The Dramatic Mode

All tragedies are finished with a death
All comedies are ended with a marriage

BYRON, DON JUAN

The dramatic mode is perhaps the dominant mode of fictional representation in Western culture. This was not always the case. As Aristotle argues in chapter 4 of the *Poetics,* the dramatic mode was developed in the fifth and fourth centuries B.C.E. in Greece, when the epic (story-telling) mode of performance was expanded to include more actors, who directly represented interactions in dialogue, rather than describing them. Once established, however, the mode has dominated the whole history of Western theatre, and subsequently film, television, and now interactive video gaming and other new media experiences. The essence of the dramatic mode is that characters in a story are directly impersonated (by actors), rather than being talked about by a narrator. Thus, drama is a mimetic art.

Dramatic modes have developed apparently independently in diverse cultures at diverse times. I make no argument that the dramatic mode is cross-culturally universal, or historically inevitable, or the only effective way of telling a story. Rather, I wish to argue that drama, as it has evolved in the West, has developed a set of structural features that make it enduringly powerful, and these features can be understood in terms of the adaptation of the genre to the evolved minds of the audience. Overlaid over these general tendencies, of course, there will be specific historical factors that may explain the development of particular types or interpretations of dramas at particular times. But underlying these is a general attention-grabbing power of the dramatic mode that is probably universal.

Drama consists of the creation of a (fictional) tight-knit social group. The audience has the chance to directly observe social interaction, but also—usually—to be part of a conversation between the characters about what is going on within the group. The groups depicted, like real human social networks, typically consist of blood relatives, and coalitional cliques, and sometimes a few strangers. Characters have wants or objectives, and these belong

ultimately to the set of basic motivations 1–5 above. Often those wants conflict with those of the other characters (and other wants of the same characters), and the work of the play is to resolve those conflicts. These characteristics already mean that dramatic presentations should have high attention-grabbing power for our evolved social cognitive mechanisms. Other features make this even more strongly the case. First there is the size and structure of the group. The number of characters in a drama rarely exceeds two or three dozen (see Stiller, Nettle, and Dunbar for Shakespeare, for example). This is a number of the same order as an average person's close social network (Hill and Dunbar), so well within the number of relationships we are attuned to keeping close active track on. Within the group portrayed, characters either interact directly or are separated by at most one intermediate link in a chain of interaction. Thus, everyone is closely enough connected for their behaviour to influence everyone else's, and in particular for their fitness striving wants to cause zero-sum conflict with those of others. This re-creates what surely must have been the situation of ancestral human societies. Most interestingly, the audience, necessarily, is directly connected to every character that appears on the stage. The audience often knows more about what is going to happen than any particular character, and, through devices such as asides and soliloquies, is the most informationally privileged member of the social group.

Second, the content of the conflicts depicted is such as to make them especially attention-grabbing. Right from classical times, dramas have mainly been discussed in terms of two categories: comedies and tragedies. These categories apply fairly accurately for most dramas for the entire intervening 2,500 years. Tragedies involve serious, often political, conflicts, usually leading to a negative outcome for the protagonist. Comedies typically involve conflicts that tend to the ridiculous and which are positively resolved. Thus, Byron's heuristic (quoted above) separates the two classes reasonably well. If there is marriage after the final death, it is usually a comedy, and if a death after the final marriage, usually a tragedy.

I will argue that tragedy and comedy quintessentially represent explorations of the domains of status competition (tragedy), and mate choice (comedy). Thus, they are the dramatic forms that fulfil feature iv above. Tragedies generally end with a death because that is the logical extreme fitness change that can arise from status competition. Comedies generally end with a marriage, because that is the key fitness-change event that can arise from mate choice procedures. These generalisations hold quite widely. However, the best way to demonstrate them is to take a concrete analysis of some representative plays. In the next two sections, I present such an analysis, taking two Shakespearean dramas, *Twelfth Night* (a comedy) and *Richard III* (a tragedy, for present purposes). Shakespeare has been chosen for several reasons. I will argue that the main lines of tragedy and comedy run all the way from classical theatre to contemporary cinema, and Shakespeare stands chronologically at the centre of this cultural history. He is also culturally at its centre too, as the dramatist whose works have been most performed and reinterpreted, not just in the land of his origin, but all over the world (Bate). Almost any Shakespearean (or other) play could be analysed in the manner presented here, but these are two particularly interesting examples.

The technique of analysis is to show how the play can be understood in terms of (1) desires by different characters to maximise their biological fitness by either mating, status enhancement, coalition building, or kin nepotism; (2) the conflict engendered by different characters' conflicting fitness desires; (3) some kind of structural perturbation of the matrix of fitness desires that allows the conflict to be overcome; and (4) extreme changes in biological fitness as the outcome of the action.

TWELFTH NIGHT

Twelfth Night (1600–1602) is the story of two twins, Sebastian and Viola, who are washed up in a strange country by a shipwreck. In that country is a noble lady, Olivia, who is wooed by the local Duke Orsino, and also by a clownish knight, Sir Andrew Aguecheek. Aguecheek is a close friend of Olivia's cousin Sir Toby Belch. One of Olivia's servants, Malvolio, is also in love with her. Viola, alone in a strange country, adopts male disguise and, passing herself off as a boy called Cesario, is employed by Orsino as a servant. In the course of her duties carrying messages between Orsino and Viola, both Viola (thinking she is a young man) and Orsino (detecting her feminine qualities) are smitten with her. The love knot is untied when Sebastian, her male twin, turns up, since she can now reveal her true sex and marry Orsino, whilst Sebastian is only too happy to marry Olivia. In a sub-plot, Belch and Malvolio vie for status within the household and Malvolio goes away, having been shamed in a plot devised by Sir Toby Belch and the servant girl Maria. Maria impersonates Olivia's hand-writing to trick Malvolio, and Belch is so impressed that he marries her.

We can express the structure of *Twelfth Night* in terms of a number of heavily interlocked fitness volitions. Orsino wants to marry Olivia. Olivia wants to marry Viola (as Cesario), and Viola wants to marry Orsino. This is the interlocked fitness dilemma at the heart of the main plot. The characters in a small group have nonsymmetrical mate choice preferences, as a result of which nobody can mate despite the availability of potential mates of both sexes. Meanwhile, in the sub-plot of lower-ranking individuals, both Sir Andrew Aguecheek and Malvolio also want to marry Olivia. In Malvolio's case it is one strand in his attempt to rise in status, vis-à-vis Sir Toby Belch. Sir Andrew Aguecheek is a natural coalitionary partner for Belch in his battle against Malvolio, for he is spurred on by the thought of access to Olivia, and the Belch–Aguecheek coalition is naturally inimical to Malvolio's status and mating interests. Maria for her part wants to marry Belch, but is of lower status than him and thus needs to prove herself to do so.

This interlocked structure is completely irresolvable without some perturbation or transformation of the matrix. A list of the main fitness volitions is shown in table 26.1. For each, I have specified the agent, the action (in terms of the fundamental motivations, 1–5, above), and the other individuals implicated. I have also specified which of the other actions provide an obstacle to each other. The ten listed would seem to be the primary ones; secondary issues follow, such as Olivia's conflict over whether to throw Belch out; on the one hand he is an obstacle to her match with Viola as Cesario, on the other he is kin.

The work of the plot is to transform the matrix structurally, so that each of the ten actions is either achieved or becomes definitively impossible, resulting in change in status and fitness for the relevant characters. In fact, with remarkable economy, only two transformations are needed to do this. The first is the substitution of Sebastian for Cesario. The offstage narrative of the

TABLE 26.1 Structure of Interlocking Fitness Volitions in *Twelfth Night*

	Agent	Action	Target	Obstacle or counterpoint
1	Orsino	Mating	Olivia	2
2	Olivia	Mating	Viola as Cesario	3
3	Viola	Mating	Orsino	1
4	Aguecheek	Mating	Olivia	2
5	Malvolio	Mating	Olivia	2
6	Malvolio	Status	Belch et al.	7
7	Belch	Status	Malvolio	6
8	Belch	Coalition	Aguecheek	4.6
9	Maria	Mating	Belch	Low status
10	Sebastian	Mating	Olivia	2

twins and the shipwreck allows this exchange of targets at the heart of the resolution: one person (Viola) who can logically only be of one sex and marry one spouse, is replaced by a pair of twins, one of each sex, who can take two spouses. Olivia's marriage to Cesario/Sebastian is now unblocked, and Aguecheek's, Malviolo's and Orsino's suits made impossible (Belch's coalition with Aguecheek is now destroyed too). Once Sebastian replaces Cesario, Viola can reveal her true identity and actions 1 and 3 can be fulfilled. The third transformation is the successful impersonation of Olivia's handwriting by Maria. This allows Belch to triumph over Malvolio; Maria thus rises in Belch's esteem to the point where he marries her.

Thus, in a very simple way, all ten volitions are resolved. The play ends with three weddings as the characters whose fitness has increased cement the bond, and the characters whose fitness has fallen disperse (Malvolio, Aguecheek). The play's particular strength is cramming a maximal richness of conflicting actions into a small group, and resolving them by a remarkably simple set of contrivances.

Shakespeare's comedies all have essentially similar structures; a central set of potential couples (two in *A Midsummer Night's Dream* and *Much Ado About Nothing*, four in *Love's Labour's Lost*, and so on) and related mating motivations, obstacles or confusions in the path of these, and structural transformations of the network which unblock the matrix of desire. The result is always a marriage, be it double, triple or quadruple. And this pattern continues into contemporary culture, with romantic movies such as *When Harry Met Sally* and *Four Weddings and a Funeral* concerning the playing out of the dynamic of mate choice, finally resolving in a wedding or mating.

RICHARD III

Richard III (1592) is the story of the rise and fall of Richard, Duke of Gloucester, who was King Richard III of England from 1483 until his death in 1485. It is the fourth and final part of a series of plays that tells the story of the Wars of the Roses and civil strife of England beginning in 1485. Since the play is based on historical events and non-fictional sources, it is usually classified as a history, distinct from the tragedies such as *Othello* and *Hamlet*. However, *Richard III* and several other of the history plays exemplify well the main features of tragedies, and so we will consider it as such for present purposes.

The play is dominated by one of Shakespeare's most memorable central characters, and equally dominated at a structural level by a single fitness volition: Richard, Duke of Gloucester's desire for status (exemplified, as so often in Elizabethan drama, by the crown of the kingdom). The play can be seen as the tale of four high-status men. Henry VI, the former king whose clan was deposed by Richard's, is already dead by the time the action of the play begins, but his presence lives on, as we shall see. Edward IV, Richard's older brother, is on the throne at the beginning of the play. Richard becomes king by the middle of the play, and Henry of Richmond becomes King Henry VII in the denouement.

Richard faces a trade off amongst competing motivations. His fitness could be enhanced by becoming king. He has already played a key role in the destruction of the Lancastrian clan, and the placing of a Yorkist (his own brother) onto the throne. Now there are few individuals ahead of him in succession. The problem is that they are his own older brothers, Edward IV and the Duke of Clarence, and their children. Thus, a kin protection motivation is finely balanced by status enhancement one.

There is no universal solution to such dilemmas, and drama represents to some extent the playing out of different possible strategies. The theory of kin selection does not predict that individuals will never harm kin; only that they will only do so if the benefit is very great, which in the case of the crown, it is. One could envisage a continuum of responses from giving the kinship motivation greatest weight, to giving it least. Richard's strategy represents the latter extreme, perhaps, as often suggested by characters' own understandings, pathologically so. Richard's strength of will is the central motor of what happens.

Richard's rise and fall can be plotted in terms of support networks. Across the primate world, a subordinate will usually require coalition partners to displace a dominant male. The extent to which others can be recruited to such a coalition depends on what is on offer; if they can be made to believe their fitness will be enhanced in the process, they will join. At the beginning of the play, Edward IV has an impressive support network in his new role of king; not only his own brothers, Clarence and (as he thinks) Richard, but his wife, Elizabeth, and her extensive kin (Rivers, Vaughan, Grey, Dorset), as well as the high-ranking professional politicians (Buckingham, Stanley, Hastings). The only dissenters are represented by the young widow Anne and the former Lancastrian queen Margaret. They are the members of the former king Henry VI's support coalition who have not been successfully drawn on board by Edward.

Richard's action is twofold; he mops up Henry VI's residual coalition by marrying the deceased king's widowed daughter-in-law, Anne. Anne can be little enthused by the prospect, but since the fall of the Lancastrians she is left utterly without a support network. Richard's offer may be her least bad option. Richard also sets out to destroy Edward's coalition by building up his own. He first uses machinations to engineer between Edward and Clarence a mistrust that leads to Clarence's execution without Richard seeming to be implicated. Edward's subsequent shock at having so violated his kin protection drives without fully meaning to leads to his death. The natural succession would pass to Edward's children, but they are still too young to maintain dominance. In the vacuum, Richard picks apart their father's coalition. Buckingham is lured by the promise of an estate under the new order. Hastings is sounded out for a similar defection, but, seeming unwilling, is hastily executed. Rivers, Vaughan and Grey (but fatefully for Richard not Dorset) are executed on trumped-up charges.

Thus by the middle of the play, the young sons of Edward IV, too insubstantial to maintain dominance without a support network, are completely lacking a coalition, and are marginalised and eventually killed. Richard has assembled Anne and Buckingham and a host of lower-ranking nobles and political groups, and has removed opponents like Hastings from the picture. However, Richard's one-sidedness in pursuit of his status enhancement, which has been his strength in ascent, becomes his flaw. Coalitions can only be maintained by reciprocity. Despite becoming king, Richard fails to deliver Buckingham's bait of material enhancement. Buckingham flies to the young Duke of Richmond, as does Dorset, whose own social network has been massacred by Richard. Richmond's support network grows. Stanley would defect, but Richard holds his son hostage, thus binding Stanley unwillingly into the coalition by his own kin-protection motivation. Richard relies on lower- and lower-ranking allies to carry out his intentions, and in desperation now tries to recruit Elizabeth, who holds him off whilst actually supporting Richmond.

The final battle at Bosworth Field is essentially a formality; the structural work of the drama has been done by the changing composition of the different support networks through time (table 26.2). Richard III has often been seen as a study of the tyrannical individual, informing, for example, Brecht's study of Hitler in *The Resistible Rise of Arturo Ui*, but it can equally be seen as an exploration of the dynamics of social coalitions. An individual can rise in status to the extent he is adept at creating a coalition, by playing on the fitness, kinship and self-protection motivations of others; but his status cannot be maintained unless he is also adept at keeping the coalition going, through reciprocal reward of his partners. Betrayals and failed promises come, quite literally, back to haunt Richard, and in the end, the quality and weight of the social coalition is always the determinant of the outcome. As before, the changes of state explored in the play are the logical maxima on the dimension of the action; kingship, which is maximal conceivable status, or death.

TABLE 26.2 Support Coalitions in *Richard III*

Act one

Henry VI[†]	Edward IV[§]	Richard	Richmond
Margaret	Clarence	Hired thugs	Margaret
Anne	Buckingham		
	Hastings		
	Stanley		
	Queen Elizabeth		
	Rivers		
	Vaughan		
	Dorset		
	Grey		
	Young princes		
	Minor nobles		

Acts two–four

Henry VI[†]	Edward IV[†]	Richard[§]	Richmond
		Anne	Margaret
		Buckingham	
		Stanley	
		Mayor and citizens	
		Minor nobles and hired thugs	

Act five

Henry VI[†]	Edward IV[†]	Richard[†]	Richmond[§]
		Minor nobles and hired thugs	
		Dorset	
			Buckingham
			Elizabeth
			Stanley
			Margaret

[†] Dead

[§] King

The path of Richard III up, and then down, the status-fitness hierarchy is an exemplar of a clear model that runs through all of Shakespeare's history plays, and many of the tragedies (indeed some of the comedies too, such as *As You Like It* and *The Tempest*). It is what Jan Kott calls the "grand mechanism" of history. One man struggles to reach the top; he does so by a combination of direct power and coalition; in turn he is displaced by the next generation coming up behind. We hope that a more peaceable social order will descend on us; that coalitions and consensus will hold ambition at bay, but there is a gap between what we might hope for, the fragile and cherished good of cooperation, and what happens. It is never long before the next challenge, be it battle or murder or rebellion, comes along.

From Dionysus to *Die Hard*: Underneath New Culture Lies Old Psychology

Lechery . . . still wars and lechery. Nothing else holds fashion.

SHAKESPEARE, *TROILUS AND CRESSIDA*

We have seen how the evolutionary principles expounded in general terms compellingly apply to a representative comedy and a tragedy. Comedy centrally concerns the procedure of pairing up sexually eligible individuals within a small group to everyone's satisfaction. Its denouement is therefore marriage. Tragedy essentially involves competition for status within a social group; it may involve the attainment of dominance, perhaps temporarily, but its logical outcome is death. Both forms have high intrinsic attention-grabbing power because they are intensified conversations in the social domains that, because they affect our fitness most, we are most interested in. Independent evidence for this cognitive bias comes from the newspapers, which are full of stories of bitter rivalry, and the mating game.

It is beyond the scope of this paper to show that similar analyses can be applied to tragedy and comedy of different historical periods, or different cultures, though these are important researches to begin. Suffice it to say that the modern romantic comedy and action film are astonishingly directly related to *Twelfth Night* and *Richard III*. The classification should perhaps be a little more complex than I have presented here (Nettle, "What Happens"). In fact, there would seem to be four key types of dramatic plot (table 26.3). As well as the question of whether the central fitness action is the mating game or status competition, there is the question of whether the ultimate change in fitness for the characters with whom the audience is most allied

TABLE 26.3 The Fourfold Classification of Dramas

	Resolution	
	Negative	Positive
Central conflict		
Status	Tragedy	Heroic
	e.g., *Richard III*	e.g., *Henry V*
	Taxi Driver	*Die Hard*
Mating	Love tragedy	Comedy
	e.g., *Romeo and Juliet*	e.g., *Twelfth Night*
	Hedda Gabler	*When Harry Met Sally*

Source: Nettle, "What Happens."

is positive or negative. Thus, a mating game with positive fitness outcomes is a comedy, like *Twelfth Night* or *Four Weddings and a Funeral,* whereas a mating game with negative fitness outcome is a love tragedy, like *Romeo and Juliet* or *Hedda Gabler.* A status game with a positive fitness outcome for the central character is heroic, like *Die Hard* or *Henry V.* A status game with a negative fitness outcome is a tragedy, like *Othello* or *Taxi Driver.*

It is remarkable how many dramas, from any historical era, can be fitted easily into one of the four cells in this matrix. Of course, some have elements of several cells, and great art often exhibits a constantly shifting perspective (for example, back and forth between positive or negative as the audience's allegiances change). Nonetheless, the typology appears compelling.

The purpose of this paper has not been to argue that the dramatic mode, or comedy and tragedy, are in any way innate or direct products of our evolved psychology. They are social constructions with a particular historical origin and cultural history. Instead, the argument has been that fictional representations must compete to garner human attention, and this influences the way that cultural traditions, drama in this case, evolve. It follows from what we know about the human mind that social information will have high attention-grabbing potential, and in particular information about how others

in the social group around us are trying to maximise their fitness. Dramas appear well designed by cultural evolution to exploit this underlying psychology. Love and status in those around us are two enduringly interesting features of human interaction, and, because of this, perennial persistence of the comic and tragic forms is no surprise.

27

Jealousy in *Othello*

MARCUS NORDLUND

In the third act of *Othello*, Desdemona complains to Emilia that she has given her husband no cause for jealousy. Emilia's response is mysterious:

> But jealous souls will not be answered so:
> They are not ever jealous for the cause,
> But jealous for they're jealous. It is a monster,
> Begot upon itself, born on itself.
>
> (3.4.159–62)*

What shall we make of this explosive emotion that flies so completely in the face of common sense? The inordinate force of jealousy has puzzled many people, including Shakespeare's critics, for instance, Mark Breitenberg, Maurice Charney (55), and Allan Bloom (109–10); near-contemporaries such as Robert Burton (273), Edmund Tilney (cited in Vaughan 76), and Michel

de Montaigne (3:88); and Shakespeare's influential predecessors, most notably Ovid (104). One thing we must accept, with A. P. Rossiter, is that what "we say about *Othello* will necessarily depend greatly on our attitudes toward jealousy" (189).

Unfortunately, this important point is too often conceived in the relativistic sense that all readers are doomed to colonize the literary texts they read with their own Trojan hobbyhorses. What must be resisted is the widespread but incoherent assumption that *Othello* will gladly mirror all our theoretical assumptions back at us, provided they are good enough. No theory of human nature or historical context, no matter how sophisticated, should ever be more than a dialogical framework or reference point for the analysis of a fictional text (Wachterhauser 6).

Anyone who reads up on modern accounts of sexuality or sexual jealousy in English Renaissance literature is soon struck by the predominance of a single mode of explanation, which we have come to know as social constructivism. Most recent books or articles on the subject begin with an assertion that gender and sexuality are "constructions" (as if this were an interpretive axiom) and then spend the rest of the discussion demonstrating the truth of the assertion.

A representative example is Mark Breitenberg's *Anxious Masculinity*, where it is assumed that men's anxieties in this period are "historically rather than essentially constructed" (7). Since he has defined early modern men as products of their historical situation, Breitenberg explains their worries as symptoms of "specific social tensions that are endemic to the early modern sex-gender system, the very tensions that produce the masculine subject in the first place" (13). Furthermore, since the act of worrying reproduces the existing sex-gender system, the result is a circular situation where male anxiety is "both cause and effect" (5). Culture generates culture, which generates culture. But what generates that first instance of culture? The answer can only be more culture, since there is no essential self, no human nature, from which it might emerge. Like Emilia's view of jealousy, Breitenberg's patriarchal culture becomes "a monster begot upon itself, born on itself."

Our first step must be to improve on this constructivist framework by combining historical explanation with the fact that humanity is an identifiable species and not just a random assortment of cultural or historical fashions. In this discussion I will focus mainly on male jealousy because we cannot take it for granted that jealousy will be experienced in exactly the same way, or even for the same reasons, by men and women (Wiederman and Kendall).

When we turn to the literary text, I hope to show that a combination of the biocultural perspective on human nature with some hermeneutical awareness can serve as a strong corrective to excessive literary reductionism

(either biological, cultural, or otherwise). My ulterior purpose will thus be essentially negative: to show that Shakespeare's text is far more enigmatic, mysterious, and interesting than some previous critics have claimed.

The Nature of Jealousy

The most serious problem with the constructivist account of early modern jealousy is that it advances a particularistic explanation of a phenomenon that seems to be universal to human societies. Jealousy has been recorded in most societies all over the world, from tribes in the Amazon region to the Tiwi islanders of Australia, and it even exists in cultures whose worldviews seem diametrically opposed to the idea of jealousy (Brown, *Human Universals*; Buss, *Dangerous Passion*; H. Fisher 270–71).

The most obvious biological cause of human jealousy is that we are sexual and affectionate mammals, endowed with a capacity for active choice that we usually prefer to exercise if we can. When we take a broad sweep of the natural world, it also becomes clear that the males of many different species go to great lengths to prevent their mates from having sex with others. Zoologists call this phenomenon "mate guarding," and it takes a wide variety of shapes in different species. Among mammals, the neuropeptide vasopressin may be a key neurochemical player, since elevated levels of this hormone trigger defensive and aggressive reactions toward potential rivals (Panksepp and Panksepp 121–22; Winslow et al.). Since claustration practices have developed independently among humans on all five continents, and still survive in parts of the Islamic world, it would appear that humans have not been entirely averse to such tactics either. But two other human phenomena—a comparatively high degree of male parental investment and the fact that females have concealed ovulation—provide particularly important clues to the nature of jealousy.

The basic theory of parental investment will not be rehearsed here. A second and more universally familiar issue is that women are automatically assured of their maternity, while men can hardly be equally certain. Compared to close relatives like bonobos and chimps, who could not care less about their individual offspring because they live in promiscuous groups, human males are notable for their high investment in children they cannot safely call their own. By the cold logic of natural selection, which only commemorates genetic fathers and disregards those who actually changed the diapers or brought home the bacon, that spells trouble.

The question arises whether sexual jealousy can be regarded as an evolved psychological trait. The most daring reply to this question comes from some exponents of evolutionary psychology. According to evolutionary psychologists Margo Wilson and Martin Daly, the ubiquity of mate guarding, claustration, and legally codified sexual double standards in human civilizations suggests that we may be dealing with an evolved component of the male psyche: "the repeated convergent invention of claustration practices around the world and the confining and controlling behavior of men even where it is frowned upon . . . reflect the workings of a sexually proprietary male psychology" ("Man" 301; see also Sidanius and Pratto 298).

In all likelihood, human beings of both sexes do have an innate, general disposition for jealousy, which is all too apparent in very small children. But the view of *sexual* jealousy as a specific psychological adaptation—or more precisely, the algorithmic, modular neural structure posited by many evolutionary psychologists—remains conjectural in spite of a wealth of supporting evidence from the human and animal world. There are too many gaps in the explanation, and the "massive modularity hypothesis" that informs it remains controversial (for a particularly controversial example, see Buss, *Dangerous Passion* 207). Most likely, the question is not whether sexual jealousy is rooted in human nature but whether it is separable from the attachment drive and involves specialized neural structures. In the light of the functional independence of love and sex, it is at least noteworthy that the same neuropeptide underpins both *attachment to females* and *aggression toward rivals* in male mammals (Winslow et al.).

The suggestion that jealousy is an evolved psychological adaptation should not make us ignore social factors. Parental investment theory predicts that *investment* of any kind (resources, time, etc.) will be a reliable clue to mating patterns in most species, and for humans, to marry and have children is to make a substantial investment that is at once sexual, emotional, and economic. Drawing on this insight, Mildred Dickemann has shown that wherever claustration practices occur they are status-graded: "the higher the socioeconomic status of the family, the greater the intensity of the practice" (313). There is also a correlation between this practice and social stratification, since it receives its most extreme expression in those societies that exhibit the greatest inequalities between the rich and the poor.

Now that we have provided some evolutionary background to the problem of jealousy, we can zoom forward again to the early modern age and enjoy the benefits of a larger perspective. Historians now believe that the monogamous nuclear family had been the typical base of English social structure since at least the fourteenth century (Macfarlane 147). The system of lifelong monogamy—with its rigid codification of sexual proprietary attitudes and double standards—may well have added to the intensity of sexual jealousy. Yet a widespread model for married life in England was, as Susan Amussen puts it, that of "benevolent patriarchy, not authoritarian government" (39). While the division of labor tied English women more closely to the home, they enjoyed much greater freedom than women in southern European countries like Spain and Italy (149). English plays of the period, like Thomas Middleton's *Women Beware Women* (1622), typically associate claustration of women with (Catholic) southern Europe.

If one should look for the most important social factor that contributed to male jealousy in this period—apart from the universal scourge called syphilis—then the best candidate might be a legal one. English civil law pragmatically defined all children born to a married woman as legitimate, regardless of who had fathered them—indeed, even if "everyone knew that they had been procreated by another father" (Macfarlane 147). A law that automatically defined all children born to a married woman as legitimate may have had grave psychological consequences for husbands. It may well be an important factor, together with the more general importance of honor and reputation in this period, behind the innumerable jokes about cuckoldry in early modern ballads and plays.

"One Not Easily Jealous"?

Let us now turn to Shakespeare's *Othello* and consider the problem with which I began my discussion: Emilia's contention that jealousy has no cause. The gist of my argument so far has been that jealousy is not simply a social construction but a universal phenomenon that goes to the heart of human nature. Even if its precise mechanisms still remain unknown and its proximate causes must be both variable and complex, we have at least learned not to focus too exclusively or single-mindedly on Shakespeare's historical context in the interpretation of his play. It is, however, equally important to avoid rushing in and proclaiming aprioristically that Shakespeare vindicates the biocultural perspective on jealousy, since there is no obvious reason to expect him to do so.

Anyone who wants to explore this problem must confront an interesting objection raised by Virginia Mason Vaughan in her excellent contextual study of the play. Vaughan's argument is that the modern concern with Othello's psychological motivation is really an anachronistic red herring. She supports this contention mainly with references to two sources: Thomas Wright's contemporary discourse *The Passions of the Mind*—where psychological motivation is "not an issue" and "it is not necessary to explain sudden changes in behavior" (78; the formulation is Vaughan's and not Wright's); and Elizabeth Carey's *Tragedy of Mariam*, where Herod's jealousy seems equally unexplained. If Vaughan is right about this, then we have reason to take Emilia's protestation quite literally: we could be looking for something in Shakespeare's play that is not meant to be there. Must our search for the cause(s) of Othello's jealousy be historicized and given up even before it has started?

The thing to remember about such detailed contextual or historical explanations is that they necessarily place selective emphasis on certain texts or historical records at the expense of others. If we turn instead to Robert Burton's *Anatomy of Melancholy*, written two decades after Shakespeare's play, we find that the author lists no less than eight potential causes for sexual jealousy, from melancholy and impotence to a desire for sexual variety. Nor do we need to dig very deeply into Shakespeare's own writings to find attempts at causal explanations of emotional states. In fact, we need go no further than the very remark that prompts Emilia's rejection of causality—Desdemona's contention that she has given Othello no cause for jealousy—which obviously presupposes some sort of connection between identifiable grievances and strong feelings. Human beings have always had a strong incentive to interrogate the underlying causes of each other's emotional responses, especially when these become violent and aggressive, and Shakespeare certainly expected his audience to do the same with his main protagonist.

The thing that really clinches the argument, however, is that *Othello* displays such a pervasive concern with intentions, psychological motivations, and our fragile attempts to uncover them—both in ourselves and in other people. In one of the finest book-length studies to date, Jane Adamson finds it "remarkable how explicitly the play dramatizes and explores the ways and means by which different people 'make sense' of what happens in their lives, including what they merely imagine to be happening" (4). This preoccupation spreads from the intratextual level to include the audience's attempts to make sense of what we see before us.

Since no person on earth has ever had unmediated access to another's intentions or motivations—let alone those of literary characters—the assessment of psychological causes must always contain an element of speculation. We must deal in degrees of probability and ask ourselves, for example, whether it is *reasonable* or *plausible* to suppose that a particular cause has brought about a particular emotional response or behavior. It is worth noting in this context that Shakespeare repeats the question "Is't possible?" no fewer than four times throughout *Othello*, which happens nowhere else in his works (2.3.278, 3.3.361, 3.4.70, 4.1.42). The exact formulation is used by all three main protagonists—Iago, Othello, and Desdemona—and in three out of four cases it is directly concerned with the plausibility of other people's intentions and behaviors.

We can also connect this insistent question with the dramatic parallel that critics have already detected between Othello's hopeless quest for ocular proof and the war council in the Venetian senate. Debating whether or not the apparent impending assault on Rhodes is a tactical ruse, the Duke gradually rifles a "possible" interpretation into something that is held "in all confidence" and finally deemed "certain" (1.3.9, 33, 44); the Turks are indeed heading for Cyprus.

From this discussion we can derive an interpretive lesson that concerns characters, audiences, and literary critics alike. On the intradramatic level, Iago is able to dupe Othello because there will always be a *possibility*, however slight, that Desdemona has actually slept with Cassio (or even with Emilia or Brabantio for that matter). Whether this is a *plausible* proposition is a different question. Like Othello, we too will delude ourselves if we confuse what is merely *possible* with what is *plausible*, *probable*, or *certain* as we explore the mechanisms that underpin his jealousy. Indeed, the play more or less forces such a dialectical approach on us since the attempts of the characters to make sense of their experience are mirrored by our own endeavor to make sense of the tragedy that unfolds before us.

It is surprising, therefore, that so many critics have cast themselves so unreservedly into the hermeneutical trap that Shakespeare holds up to view. They have, in short, rehearsed Othello's fatal mistake by jumping to premature conclusions on the basis of flimsy evidence. As is usually the case in the current critical arena, the main instigating factor has been a strong ideological impulse that automatically privileges certain modes of explanation over others, that does not hesitate to skip past strong evidence to the contrary, and that therefore ends up with a distorted account of Shakespeare's play.

An important reason for the strong recent interest in *Othello* has been the widespread academic concern with race, gender, and politics. In such an environment, a play about a black man who kills his wife must clearly be regarded as something of a godsend. But even though it must be hoped that most readers of Shakespeare are against racism and for sexual equality, these political readings must be subjected to the same rigorous examination that one would apply to any others.

Let us start with the widespread contention that Othello's jealousy derives from the color of his skin (Cowhig; Hogan, "*Othello*"; Kirsch 32–33; Newman 153; Rubinstein). At first sight, this "racial" reading seems to have much to offer. Othello is clearly exposed to racist slander in the first act, and there is also something worrying about his own suggestion later in the play that Desdemona's "name, that was as fresh / As Dian's visage, is now begrimed and black / As mine own face" (3.3.389–91). Not only does Othello seem to recognize that black may not be beautiful, but his words yoke together two metaphors: the blackness of his skin and the blackness of shame. If we leave the play aside for a moment, it does not seem impossible that a man could be driven half insane by racism and end up murdering his wife. But a *possible* reading is not necessarily a *plausible* or *convincing* reading, and we cannot pass judgment until we have also considered its potential limitations.

For example, when Ruth Cowhig turns racism into something of a prime mover in *Othello*, she skips merrily past material that points us in the opposite direction. When Othello has started to grow jealous, he briefly considers three reasons why Desdemona might lose interest in him—that he is not particularly eloquent, that he is black, and that he is old. But Cowhig ends her citation just before Othello evaluates his improvised explanations of Desdemona's supposed adultery: "yet that's not much" (3.3.270). Not only are we given no reason to prefer blackness above the other two explanations, but all three explanations are rejected by Othello himself. We must also remember that the three men that utter some sort of disparaging remarks about Othello's race—Brabantio, Roderigo, and Iago—all have large axes to grind. In the case of Brabantio, the racism certainly cannot be very deep-seated, since he has "loved" and "oft invited" (1.3.129) Othello to his house before the latter stole his daughter. No other character in Shakespeare's play ever discusses Othello's race or color in derogatory terms. There is, in other words, little support for Ania Loomba's recent assertion—in overt opposition to any reading based on human nature or human universals—that "Shakespeare goes out of his way to draw attention to Othello's colour and race" (156).

It is a necessary component of the racial argument that Othello internalizes the racism that surrounds him; he must feel "racial despair" (Hogan, "*Othello*") or "complicitous self-loathing" (Newman) and then channel it into jealous rage. The most damaging blow to the "racial insecurity reading" comes from textual evidence that points emphatically in a different direction. For one thing that Shakespeare takes considerable pains to establish in this play is how *loved* and *admired* his main protagonist is. Most characters seem to flush with admiration when they talk about the Noble Moor. As they wait for Othello to arrive in Cyprus, Montano expresses his admiration (2.1.34–36). Later in the play, when Lodovico is confronted with Othello's horrendous abuse of Desdemona, he cannot reconcile what he has seen with the person who was everyone's darling in Venice:

> Is this the noble Moor whom our full senate
> Call all in all sufficient? This the nature,
> Whom passion could not shake? whose solid virtue
> The shot of accident nor dart of chance,
> Could neither graze nor pierce?

> (4.1.264–68)

These are strong words. Othello's reputation has been no less than that of a complete man, one who fully embodies such central Renaissance ideals as virtue, constancy, and martial prowess. That he is capable of genuine love and affection is also suggested elsewhere. Even Iago, his sworn enemy, admits to himself in private that the Moor has a "free and open nature" (1.3.397), indeed, a "constant, loving, noble nature" which suggests that he will prove a "most dear husband" to Desdemona (2.1.286, 288).

Othello is aware of the love and admiration that surrounds him. When he lands on Cyprus he remarks that "I have found great love amongst them" (2.1.204) and that he therefore expects his wife to receive the same good treatment. The same self-confident awareness of his own worth has enabled him to steal the daughter of one of the most influential men in Venice without any fear of punishment: "Let [Brabantio] do his spite; / My services, which I have done the signiory, / Shall out-tongue his complaints" (1.2.17–19). The play gives us no real evidence that Othello is feeling insecure prior to Iago's onslaught on his mind or that he exhibits any signs of either self-loathing or racial despair. It does, however, give us plenty of reasons to think otherwise.

The most recent trend in criticism on Shakespeare's play has been to combine the aforementioned study of race with that of gender. Karen Newman's 1987 article—where it was argued that blackness and femininity are perceived as equally monstrous in *Othello* (153)—was something of a landmark in this area, and later critics have approached the same ideological alloy with similar conclusions (Barthelemy, "Ethiops" 95; Loomba 165; for more strictly gender-oriented readings, see Drakakis; French 212; Neely 81).

Although there is much to be criticized in these hyper-ideological readings it would be foolish to deny that there are patriarchal norms and values at work in Shakespeare's play. These norms make their appearance immediately when the enraged Brabantio seeks to assume his power as head of the family, and Desdemona soon affirms them too by professing to exchange one "lord" (her father) for another (her husband). Patriarchal norms are also present, but this time presupposed and rejected, when Emilia delivers her famous critique of the sexual double standard.

Shakespeare may not have had recourse to the jargon of gender studies, but it is quite likely that he consciously depicted a Venetian society that was much *more* patriarchal than his native England, regarding both women's freedom and the intensity of male jealousy. In the *Anatomy of Melancholy,* Robert Burton regarded it as a commonplace that Italians were more jealous and possessive of their wives than were Englishmen: "England is a Paradise for women, an hell for horses; Italy a Paradise of horses, hell for women, as the diverbe goes" (3.3.2:282). But as Richard Levin reminds us, even the Italians in *Othello* do not view the murder of Desdemona "as one of your everyday patriarchal events; instead, they consider it a horrifying violation of the norms of their world" ("Feminist Thematics" 127–28). What must be questioned, therefore, is not the existence of patriarchal norms in Shakespeare's play but the idea that these norms are a sufficient cause for Othello's disastrous action.

Having failed to uncover a sociological cause for Othello's jealousy, we must now consider another explanation of his defeat at the hands of the green-eyed monster: that his love for Desdemona is flawed. Many years ago, F. R. Leavis started something of a critical campaign against Shakespeare's main protagonist wherein he denied that the latter's love for Desdemona was really love at all: "[I]t must be much more a matter of self-centred and self-regarding satisfactions—pride, sensual possessiveness, appetite, love of loving—than he suspects" (quoted in Rossiter 201). More recently, James Calderwood has found Othello guilty of self-centered possessiveness toward his wife and defined his love as "flawed from the beginning" (361). David Bevington suggests that Othello does not really love Desdemona as much as he loves her love for him and that his feeling for her is really a form of self-regard. For both Bevington and Calderwood, it is Othello's deep-seated dependence on other people's praise that makes him vulnerable and ultimately destructive because it twists his love into a conception of Desdemona as an extension of himself.

To argue for a "flawed" love presupposes some sort of normative idea of what a true, healthy love should look like, and we can trace a clear pattern in these critical responses. Othello is to blame because he is in love with love rather than with Desdemona; he is too dependent on the capacity of her love to bolster his own self-image; he is possessive; and he is self-centered or even egotistic.

The emotion of romantic love, with its fusion of passion, intimacy, and commitment, is characterized by—among other things—a desire for exclusivity and emotional dependency on the loved one (Harris, "Rethinking" 102–3; Sternberg, *Cupid's*, "Triangulating"; for evidence that romantic love is universal, see Gottschall and Nordlund; Jankowiak and Fischer). Hence we should not be surprised if a romantic lover exhibits at least a mild degree of possessiveness and finds it difficult to imagine a life without the other person. It is difficult to imagine a union between two people that does not somehow turn them into "extensions" of each other. We would have much more reason to be suspicious of Othello's love if he were not in the least possessive or dependent on Desdemona's love—and, say, merely shrugged at the idea of Desdemona and Cassio doing the beast with two backs.

There is one more aspect of the "flawed love" argument that we still have not dealt with: the idea that Othello's love is egotistical or self-centered. Bev-

ington's and Calderwood's chief support for their claim comes from Othello's speech at the Senate, when Brabantio has just accused him of having practiced witchcraft on Desdemona (Calderwood 360). But look at the dramatic context: Othello finds himself in the absurd position of being asked to explain why Desdemona fell in love with him. How could anyone hope to explain such a thing? When Othello retells the story of their courtship and concludes that Desdemona "loved me for the dangers I had passed" (1.3.168), this reads like a modest attempt to steer the attention away from his own personal characteristics. It is quite simply the best thing available to a man who is forced to answer an impossible question and who has recently indicated that he knows his worth but does not want to brag about it (1.2.19–24). In fact, his account of the courtship also gives us an interesting glimpse into the considerable mutuality and balance of power between himself and Desdemona. Contrary to Harold Bloom's view that Othello is "essentially passive" in the courtship (*Shakespeare* 450), the passage actually gives us a delicious example of two sensitive lovers who gradually nudge each other toward a mutual confession of love: *This to hear would Desdemona seriously incline . . . I found good means to draw from her a prayer . . . I did consent . . . she wished that heaven had made her such a man. . . . Upon this hint I spake* (1.3.129–70).

Like the claim for various sociological causes of Othello's jealousy, the notion that his love for Desdemona is flawed or self-regarding does not hold up well under scrutiny. At best, it remains a weak possibility that is plagued by solid evidence to the contrary. It seems much more convincing to suggest that Shakespeare has deliberately painted Othello's and Desdemona's love in the sunniest colors he could find: a love that is mutual, strong, generous, and calmly defiant of the prejudice and resentment that surrounds the lovers early on. The necessary consequence is that the contrast between the two Othellos—the carefree soul prior to the temptation scene in act 3, and the tortured and doubtful being who finally ends up killing his wife—becomes all the more mysterious and unsettling.

The Plague of Great Ones

Another Shakespeare play confronts us with a seemingly inexplicable case of jealousy. Unlike Othello, his counterpart Leontes in *The Winter's Tale* seems to lack any external incentive for his destructive jealousy. Harold Bloom defines him nicely as "an Othello who is his own Iago" (*Shakespeare* 639).

Iago's success derives at least partly from his capacity to maneuver his commander into an intolerable position between hope and despair, so that the latter finally comes to desire the truth—any truth, however painful—

more than he desires a faithful Desdemona: "She's gone, I am abused, and my relief / Must be to loathe her" (3.3.271–72). In this way, Iago gradually destroys a faith so firm that it might easily have been deemed unshakeable. In *The Winter's Tale*, by contrast, Shakespeare diagnoses the opposite but equally disastrous tendency of a single error of judgment to poison all other judgments that derive from it. Camillo realizes that his attempts to cure Leontes of his "diseas'd opinion" (1.2.297) are vain because the latter cannot get around the central assumption that constitutes the foundation of his thinking:

> you may as well
> Forbid the sea for to obey the moon,
> As or by oath remove or counsel shake
> The fabric of his folly, whose foundation
> Is pil'd upon his faith, and will continue
> The standing of his body.
>
> (1.2.426–31)

The realm of "faith" that Camillo points to is concerned with those things that are merely possible, in the sense that they can be neither corroborated nor denied conclusively. Othello can never know for sure that Desdemona is faithful, and so he must have faith in her (in the restricted and nonmystical sense that he must accept a belief that is reasonable but can never be categorically proven). In *The Winter's Tale*, it is equally *possible* that Hermione has actually slept with Polixenes, but Leontes turns this gratuitous possibility into a cornerstone for his thought and feeling.

Why is it that Leontes and Othello, of all people, are so particularly inept at distinguishing what is merely possible from what is likely? And why does their jealousy become so incredibly intense and destructive? The jealousy in both plays concerns a woman of matchless beauty and virtue. But we should be wary of any critical attempt to turn either Leontes or Othello into mate-guarding beasts, proprietary patriarchs, or for that matter intrinsically insecure and self-loathing figures. In spite of their detestable actions, Shakespeare leaves us little doubt that their vulnerability is rooted in love rather than more callous or self-regarding considerations.

Of course, a distinctly normative theorist of love can always chide Othello for the depth of his existential as well as emotional investment in his wife, but such a charge is perhaps better directed toward the emotion rather than the character who feels it. To love *is* to become vulnerable to loss and pain, as Leontes realizes with great distress when he suspects that Hermione and Polixenes may be laughing behind his back. We have no reason to

dispute Othello's final account of himself as a man who "loved not wisely, but too well" and who therefore "threw a pearl away / Richer than all his tribe" (5.2.344, 347–48).

The second characteristic shared by Leontes and Othello is that they are extremely powerful men with almost unrestricted authority. (Othello may not be king of Venice, but he *is* commander of Cyprus, endowed with supreme authority over the Venetian subjects on his island.) This means that there is no larger social or institutional check for their jealousy once it gets out of hand, and the same problem also afflicts their personal relationships, since their own wives are held back by the domestic ideal of obedience and duty.

This exalted position in society—where each is king of his own little dramatic world, metaphorically or otherwise—has another consequence for Othello and Leontes. Since they occupy the apex of a social structure, their fall becomes all the more dramatic and painful. In fact, Othello appears to regard his own elevated social position as an instigating factor for jealousy. After rejecting skin color, lack of eloquence, and age difference as explanations for his wife's supposed transgression, he refers to cuckoldry as "the plague of great ones" (3.3.277). This again suggests that the Noble Moor, at least, does not subscribe to the modern critical commonplace that he is a self-loathing or inherently insecure being. His basic problem is rather that he has been sufficiently attractive to attract a very attractive woman who is likely to attract a substantial number of other attractive men.

We can conclude from this that the jealous man in these Shakespeare plays is not a discursive construct specific to the early modern period. He is a privileged and powerful human being who loves his wife deeply and who therefore has something very precious to lose. This situation will quite naturally be exacerbated in a specific culture like early modern England that places a particularly high premium on honor and that inserts its men and women into a formally unequal relationship where love becomes a matter of duty and obedience. But these are no more than specific variations on a much larger human theme where love or jealousy are neither immaterial substances that "transcend" social reality nor reducible to some crass cost-benefit analysis, social determinist dictate, or "sexual economy." When we say, as I think we should, that *Othello* presents us with a timeless theme in the history of our species, it is important to get all the prepositions right in this formulation. The drama of jealousy has always unfolded *in* history, not outside it.

But let us pause right here, just before we raise the biocultural flag and proceed to celebrate Shakespeare as the world's first literary exponent of gene–culture interactionism. For the question remains whether this string of psychological causes—for all its textual and theoretical support—can really

be regarded as a *sufficient* cause for the singularly violent and destructive jealousies that Shakespeare depicts. Is it convincing to suppose that most powerful men who marry beautiful women whom they love deeply will sooner or later turn jealous to the point of murder? Hardly, for then we would expect our royal palaces or government buildings to be strewn with female bodies, and the homicide statistics in Shakespeare's society and our own would also be much higher (Sharpe 111). It would seem, then, that we still have not addressed the fundamental question that must have been equally disturbing for Shakespeare's contemporaries as it is for us: how can it be that these men are transported so far beyond everyday pangs of jealousy into destructive or even murderous rage?

In my view, there are two reasons why this question must remain unanswered. First, we must not forget—as so many literary critics and critical theorists do today—that we are dealing with intentional artifacts bound by literary conventions. In the case of *Othello*, we are looking at a tragedy. As Robert Storey observes, tragedy educates its readers in three ways: it invites empathetic identification, it creates ambivalence about the emotional allegiance that results, and it enables a vicarious experience of catastrophe (*Mimesis* 138). So while a moralist stands on the outside looking in, or rather, looking *at* the tragic protagonist, a tragedian or tragic reader explores how it feels on the *inside*. This vicarious experience enables us to explore deeply troubling things that we might well have to moralize heavily about in our everyday lives. Now if there were a clear answer to Othello's madness it would probably make the play less unsettling, and *Othello* is a brilliant tragedy because it is *extremely* unsettling. In the words of Hans-Georg Gadamer, "it is part of the reality of a play that it leaves an indefinite space around its real theme. A play in which everything is completely motivated creaks like a machine. It would be a false reality if the action could all be calculated out like an equation" (498).

As human beings, we have an innate desire—a so-called cognitive imperative—to explain atrocities like Desdemona's innocent death so that we may render our experience intelligible and manageable. What makes *Othello* so powerful and captivating is that it plays on this desire but refuses to satisfy it. The play both invites and frustrates our attempts at psychological explanation by presenting us with psychological causes that either prove too many (in the case of Iago's malice) or ultimately fail to add up (in the case of Othello's murder of his loved one). The empathetic identification Storey speaks of would be impossible unless we shared some common ground with Othello, but as the play progresses our desire to make sense of him, to explain him, increases in proportion to our growing sense of estrangement.

To "explain" Othello's murder of his wife in a satisfying manner would be to show that his actions are actually quite plausible or even predictable given a certain conception of human nature or the specific situation he finds himself in. But even in real life, let alone art, we would be hard pressed to find a concept of human nature that could serve such a purpose since this theory would have to deny the extraordinary complexity—both biological and social—that creates human universals as well as individual and cultural variation. Given their specific circumstances and the apparent universality of jealousy in human cultures, we could certainly have expected Othello or Leontes to become a little wary in the presence of admirable men like Polixenes and Cassio. But what Shakespeare treats us to in *Othello* and *The Winter's Tale* is not everyday jealousy or a stage version of the average Elizabethan man. He confronts us with the horrifying individual exception rather than the psychological rule.

The scary and interesting thing about Shakespearean jealousy, then, is that its deepest mystery cannot be explained away. We cannot assign it to a barbaric and unenlightened past, to a pathological insecurity that can be eradicated by equal doses of therapy and social change, or to some hardwired mate-killing module. And therein, I suppose, lies its perennial horror and fascination.

28

Wordsworth, Psychoanalysis, and the "Discipline of Love"

Nancy Easterlin

In the past three decades, psychological approaches to literature, including feminist interpretations, have been overwhelmingly psychoanalytic, and this is still the case even as cognitive psychology emerges as a relevant and fruitful secondary field for literary scholars. The dominance of psychoanalysis holds true for Wordsworth scholarship, an area in which, given the poet's developmental concerns, psychological orientations seem particularly apropos. Unfortunately, Freud's most basic assumptions about infant experience, still credited in various forms by Lacanian and many feminist scholars, are no longer accepted by developmental psychologists, who regard the infant as a self-organizing system engaged in a *fundamentally productive* and social relationship with his primary caregiver, usually his mother. By contrast, psychoanalysis, which opposes *union with the mother* in the state of primary narcissism to *separation and individuation*, envisions the mother–infant relationship as paradigmatically conflicted. Though both psychoanalysts and literary critics have pointed to the methodological weakness of

From *Philosophy and Literature* 24, no. 2, October 2000, pp. 261–279, by Nancy Easterlin.

placing "pathomorphically chosen clinical issues . . . in a central developmental role," the implications of this insight for literary criticism have not been fully examined (Stern, *Interpersonal World* 20; see also Bowlby, *Attachment*; Crews; Scalise Sugiyama, "New Science").

In misconstruing infant psychology and growth along the lines suggested by Freud and his followers, many of Wordsworth's interpreters unintentionally misrepresent and devalue both the poet's conscious understanding of that interaction as well as the unconscious motivations for the poet's attachment to nature. Most especially, the adoption of this conflict model has hampered interpretation of the "infant Babe" passage in Book II of *The Prelude*, a passage whose central importance has not, by all accounts, received its due. In the following pages, I will review the Freudian and neo-Freudian readings of Wordsworth, critique commonly employed psychoanalytic assumptions about infant experience and, demonstrating the correlations between the current research model and Wordsworth's description of infant experience, make a case for the foundational importance of mother–infant interaction in *The Prelude*.

Since much of Wordsworth's poetry is manifestly concerned with the formative character of childhood experience, Freudian and neo-Freudian notions about the stages and nature of infant development have been applied to the poetry with great regularity. In accord with the procedures of depth psychology, these readings seek to uncover latent meanings, yet in addressing the "infant Babe" passage, a manifest statement about infant development, they find it, curiously, consistent with the latent dynamics. Thus a common theme of psychoanalytic readings is that nature and Dorothy are mother-substitutes who reveal the poet's regressive libidinal desire for union; in construing the attachment to the mother or mother-substitute as infantile by definition, all such readings correlate *pathology* with *union with the mother*. In Freud's formulation, "there are regressions of two sorts: a return to the objects first cathected by the libido, which, as we know, are of an incestuous nature, and a return of the sexual organization as a whole to earlier stages" (*Introductory Lectures* 341). Even critics such as Barbara Schapiro (*Romantic Mother*) and James Heffernan, who discern the emotional and imaginative efficacy of mother–infant interaction, are paradoxically and simultaneously obliged by the psychoanalytic paradigm to interpret representations of mothers or presumed mother-figures as informed solely or chiefly by infantile sexual desires.

More recently, feminist readings influenced heavily by the modified psychoanalytic models of Nancy Chodorow and Carol Gilligan, which focus on the hypothesized pre-Oedipal phase and its implications for feminine development, continue to detect the poet's regressive desire for union with the mother—again symbolized as nature and/or Dorothy—but, in a new turn, seem especially to observe a specifically masculine desire for control over the mother, a critical theme lent further support for some critics by Kleinian notions of infant aggression and the Lacanian identification of language with the Symbolic order of the father. Romanticists including Diane Hoeveler Long, Alan Richardson ("Romanticism"), Anne Mellor, Margaret Homans (*Bearing*), Marlon Ross, and Mary Jacobus, among others, emphasize the appropriative and destructive impulse of the masculine toward the feminine, thus employing a theoretically driven gender dichotomy that is the inevitable result of the Freudian agon.

A strong theme, then, in thirty years of psychoanalytic Wordsworth scholarship, is that of the poet torn between the regressive desire for union with the mother on the one hand and the desire for separation and individuation on the other. On the whole the most recent readings emphasize the masculine will-to-power over the mother and all female others, who are interpreted almost exclusively as mother-substitutes. Though Wordsworthians have voiced concern over the lack of nuance in emerging feminist readings and in the theory-driven character of psychoanalytic approaches (Blank; Wolfson), it has not been generally proposed that the psychoanalytic paradigm of development may be in good measure responsible for the progressively negative direction of these readings. And if one of the dubious effects of recent feminist theory has been to rescue female writers from the blame cast upon males by adopting Chodorow's and Gilligan's theories that feminine development, characterized by pre-Oedipal attachment and an ethic of care, is *essentially* different from Oedipal, individuation-oriented masculine development, it cannot be said that mothers are placed in a very positive light. For whether construed as socially produced or innate, pre-Oedipal and Oedipal stages establish a norm of conflict whose initial locus is the mother. Of course, in the pursuit of knowledge, none of this is a problem *if* psychoanalytic speculations about the infant are in fact true to subsequent observations of babies.

As all of these readings attest, the psychoanalytic model of human development is fundamentally agonistic because closeness to the primary caregiver is said to be at odds with selfhood and socialization, the imperative to individuate severing infants from the parent with whom they have existed in supposedly perfect union. Hence, central to the agonistic depiction of the mother–infant relationship is Freud's hypothesis of primary narcissism or primary identification itself. Moreover, for Freud, this conflict between individuation and union with the mother applies equally to female and male infants. Though feminist theorists astutely note masculine bias in psychoanalytical models, they locate that bias in the misapplication of the Oedipus complex to girls and women, never questioning the validity of Freud's notions of sexualized primary attachment and individuation.

Critics working in the Freudian tradition assume that sexual impulses first emerge in a narcissistic mother–infant union, and that, in fact, the "drive" underwriting the mother–infant bond is essentially sexual. As Freud points out, "when children fall asleep after being sated at the breast, they show an expression of blissful satisfaction which will be repeated later in life after the experience of sexual orgasm. This would be too little upon which to base an inference. But we observe how an infant will repeat the action of taking in nourishment without making a demand for further food. . . . We describe this as sensual sucking. . . . It is our belief that [infants] first experience this pleasure in connection with taking nourishment but that they soon learn to separate it from that accompanying condition" (*Introductory Lectures* 313–14)

If in the concept of primary narcissism Freud imagines an idealized unity that one could only resent losing, the additional inference that the pleasure of sucking is identical in kind to sexual pleasure and, therefore, that nursing is the ground of sexual instruction determines that the mother must become, in the psyche of the child, seductress and cheat. Via nursing, sexual jealousy between father and infant is ushered into the household, and the groundwork is laid for the Oedipal stage, when children must learn that the mother is not, alas, theirs to acquire. Freud's conflict model reached its foreseeable conclusion in the theories of Melanie Klein, who held that the process of individuation is so deeply traumatic that sadistic fantasizing constitutes a stage in normal development.

In contrast to the fundamentally agonistic Freudian model whereby the child in the first year of life is torn between two modes of being, present-day psychologists view development as a progressive phenomenon from the time of birth onward—and, indeed, doubt that issues of separation principally characterize this age (Stern, *Interpersonal World* 10). In both their views of an evolving self-concept and those of connection to others, researchers see human infants as engaged in organizing their perceptions and responses; continuity and growth, rather than strict stages, characterize development. These early modes of orienting and relating, moreover, are universally apparent (for cross-cultural comparisons, see Eibl-Eibesfeldt, "Patterns").

As the developmentalist Daniel Stern explains, a self-concept evolves out of, rather than in opposition to, the early relationship with the mother (*First Relationship*). So also the emotional bonds that develop in tandem with mother–infant interaction are not, as psychoanalysis would have it, opposed to an earlier connection or union with the mother but result from it, as attachment theory explains. Attachment behavior, defined as the "seeking and maintaining of proximity to another individual," consisting of crying, calling, smiling, gesturing, and babbling, and presumed to facilitate species survival, is observable in human infants by six months and highly observable in the second year of life (Bowlby, *Attachment* 194). Shortly after attachment behaviors are first observed, they become directed toward other family members. In short, while the Freudian concept of primary narcissism implies that connection with the mother inherently threatens adult psychic health and extensive social relations, Stern and Bowlby claim that a strong initial relationship with the mother provides the *foundation* for a child's growth, individuation, and sociality. When Bowlby tells us that "in the early months of attachment the greater the number of figures to whom a child was attached the more intense was his attachment to mother as his principal figure likely to be," he is noting that, far from encouraging incestuous, infantile isolation, a strong primary bond with the mother provides security and teaches the practical and affective rewards of social interaction (*Attachment* 202).

And whereas Freud posits that the infant's attachment to the mother is a byproduct of nursing, a secondary drive derived from a sexualized primary drive for food, developmentalists have long considered these views in error (Bowlby, *Attachment* 177–78, 210–17, 361–78). First, studies of animal behavior consistently demonstrate across a wide variety of species that there is no causal relationship between food and attachment, the most important research in this respect being Harlow's 1961 experiments with infant rhesus monkeys who were isolated in cages and would cling to a cloth and chickenwire "mother" rather than eat, and who later manifested severe emotional and social disturbances. These findings are supported by Konrad Lorenz's studies in the 1930s of imprinting behavior, which demonstrate that even chicks attach to a parent-figure independent of food availability. Indeed, primates and humans characteristically develop strong attachments to those who do not meet their physiological needs, a fact which psychoanalytic theory is poorly equipped to explain.

Second, Freud's assumption that incestuous desire is natural and normative should have struck a discordant note even in the nineteenth century, for before the theory of natural selection and the discovery of genetics, breeders had long known of the damaging effects of inbreeding depression. This alone makes it improbable that attachment to the mother is mobilized by a sexual drive. An alternative hypothesis known as the Westermarck effect, which holds that those with whom one associates closely in early childhood are avoided as sexual objects, was an early rival to Freudianism and seems more plausible in the face of all subsequent evidence (Brown, *Human Universals* 118–29; Darwin, *Origin*; Wilson, *Consilience* 173–80).

Taken together, these developmental studies delineate the productive character of mother–infant interaction, and it is, I believe, this *essentially* productive formative relationship that Wordsworth acknowledges in Book II of *The Prelude*:

                    Bless'd the infant Babe,
(For with my best conjectures I would trace
The progress of our being) blest the Babe,
Nurs'd in his Mother's arms, the Babe who sleeps
Upon his Mother's breast, who, when his soul
Claims manifest kindred with an earthly soul,
Doth gather passion from his Mother's eye!
Such feelings pass into his torpid life
Like an awakening breeze, and hence his mind
Even [in the first trial of its powers]
Is prompt and watchful, eager to combine
In one appearance, all the elements
And parts of the same object, else detach'd
And loth to coalesce. Thus, day by day,
Subjected to the discipline of love,
His organs and recipient faculties
Are quicken'd, are more vigorous, his mind spreads,
Tenacious of the forms which it receives.
In one beloved presence, nay and more,
In that most apprehensive habitude
And those sensations which have been deriv'd
From this beloved Presence, there exists
A virtue which irradiates and exalts
All objects through all intercourse of sense.
No outcast he, bewilder'd and depress'd:
Along his infant veins are interfus'd
The gravitation and the filial bond
Of nature, that connect him with the world.
Emphatically such a Being lives,
An inmate of this *active* universe;
From nature largely he receives; nor so
Is satisfied, but largely gives again,
For feeling has to him imparted strength,
And powerful in all sentiments of grief,
Of exultation, fear, and joy, his mind,
Even as an agent of the one great mind,
Creates, creator and receiver both,
Working but in alliance with the works
Which it beholds.

                                        (II.238–79)

Because Freudians make much of the mother's breast as a site of sexual excitation and thus an object arousing possessiveness and concomitant rage, literary readings that follow Freud, Klein, and Chodorow discern a conflictual relationship between mother and child in Wordsworth's passage. Combining psychoanalysis with historicism, for instance, Richardson claims that the 1850 version of this passage, the basic substance of which is the same as the 1805 version quoted here, corresponds to other images of nursing in Romantic literature, "graphically [representing] the male child's absorption of his mother's sympathetic faculty even as his primary affective bond is established" ("Romanticism" 17). But it is worth noting that, although Wordsworth employs the words "nursed" and "breast" in the passage, neither he nor the child he describes is preoccupied with breasts and breastfeeding. Instead, the poet's focus on the *place* of nursing ("in his Mother's arms") and of continued holding ("Upon his Mother's breast"), as the cradled child is lulled to sleep, highlights physical contact—holding and the warmth, protection, and security that such proximity confers. Furthermore, the repetition of "his Mother's" in conjunction with the specific descriptions of the infant's closeness to his parent conveys not only the importance of continued contact but also the value, at a time when wet nurses were commonly employed by the middle and upper classes, of being nurtured and protected by one's *actual* mother. Aware that newborns demand to be held, Wordsworth here is less fixated on the mother's breast than appreciative of the ongoing nurture that originates in bodily contact. Indeed, he makes this point explicitly in the subsequent paragraph when he asserts that as "a Babe, by intercourse of touch / [He] held mute dialogues with [his] Mother's heart" (ll. 282–83), engaging in a conversation-like, turn-taking affective exchange that enables the "infant sensibility" (l. 285) to form the core of adult consciousness.

Whereas Freud sees bodily contact and the early emotional rewards connected to it as a byproduct of feeding, Wordsworth apparently places primary importance on the contact itself. If the Harlow experiments demonstrate that bodily contact and the attachment behavior soon to emerge from it correspond to basic biopsychological needs, Wordsworth seems to have intuited the crucial importance of holding and closeness for infants well before psychology's theoretical formulations.

Wordsworth clearly envisions a causal relationship whereby the physical closeness of mother and infant, possible only through the parent's initiative, stimulates positive affective development whose ultimate outcome is interest in and a sense of connection to the larger world. As Jean Hagstrum notes, "natural and bodily energy inspires and shapes the poetry of both tender familial affection and transcendental vision" (74). The child's enthusiasm is not solely the result of holding, but equally of the other feature of mother–infant interaction on which Wordsworth focuses, eye contact. If touch forms the most basic ground of nurture and affective exchange, it is augmented by early gaze behavior, as the child "[gathers] passion from his Mother's eye," the resulting positive feelings stimulating cognitive development and promoting the organization of sensory experience. More than windows on the soul, the eyes are the conduit of emotion, motivating the child "to combine / . . . all the elements / And parts of the same object, else detach'd / And loth to coalesce" (ll. 247–50)—that is, to create a coherent picture of his environment. While "[sleeping] / Upon his Mother's breast" (ll. 240–41) provides the security for further growth, the progression of the passage attributes cognitive development to the exchange of feelings he experiences as he awakens to and explores not his mother's breast but her eyes and face.

In asserting the primacy of visual exchange, Wordsworth is once again ahead of Freud and his constituents, perceiving a key feature of development recognized by current psychologists. Basing his conclusions on research that indicates the innate preference for facial configuration and/or specific facial features, Daniel Stern tells us that it is the predisposition to attend to the human face that draws an infant's attention to his mother's gaze. As he puts it, "From the very beginning, then, the infant is 'designed' to find the human face fascinating, and the mother is led to attract as much interest as possible to her already 'interesting' face" (*First Relationship* 37; for confirmation that this process is reciprocal, see Csermely and Mainardi; Eibl-Eibesfeldt, "Patterns"; Trevarthen, "Interpersonal Abilities"). It is no accident that the normal focusing distance for the human neonate is about eight inches, the same as the approximate distance between the eyes of the mother and those of the infant during feeding.

Rather than the breast forming the locus of visual and emotional development, as Freud and Klein theorized, our more recent knowledge of the non-symbolic centrality of gaze behaviors and the maturation of the visual system provides a physiological basis for understanding that psychic and physiological growth mirror one another, and that both are primarily progressive rather than conflicted. Mutual gaze, a nonsymbolic interaction, provides the emotional stimulation for further exploration of the environment; given the prolonged course of human development, the human visual system matures remarkably early. By three months of age, a human infant can track, fixate upon, or bring an object into focus as well as an adult. Remarks Stern, "This developmental landmark is extraordinary when contrasted with the immaturity of most of his other systems of communication and the regulation of interpersonal contact, for instance, speech, gesture, locomotion, manipulation of objects" (*First Relationship* 38–39). Of the two other motor systems to mature early, sucking and head movement, the latter serves primarily to support and expand the range of the visual system.

Apparently sensitive to the efficacy of eye contact between mother and child, Wordsworth glorifies without sentimentalizing gaze behavior, identifying its centrality in affective development, creative inspiration, and cognitive coherence. Securely ensconced in his mother's arms, the child looks into her eyes and senses her love, "[gathering] passion" that at once catalyses his responsiveness and unequivocally identifies this moment, through the metaphor of the "awakening breeze" that harks back to the beginning of the poem and betokens creative inspiration, as the origin of poetic power. (On the primacy of emotion, see Storey, *Mimesis* 7–15. On the breeze as a metaphor for poetic inspiration, see Abrams, "Correspondent Breeze.") For the time being, the motivating force of powerful feelings encourages the "prompt and watchful" child to direct his nascent synthesizing ability to the world beyond himself—first and foremost, to the "object" offering these affective rewards, his mother's face.

This remarkably accurate depiction of the emotional and cognitive results of maternal nurture justifies Wordsworth's subsequent deification of the mother through the exalted terms "Being" and "Presence" and through the introduction into the passage of light imagery associated with her. Far from drawing the child back into an infantile and hermetic relationship, the mother "irradiates and exalts / All objects through all intercourse of sense" (ll. 258–60)—metaphorically shedding light on the full range of their interactions and on the child's early excitement about the world. Indeed, the chronological progression in the passage from interest in the mother, expressed through mutual gaze, to interest in the surrounding environment also corresponds to developmental research, which indicates that by three to four months infants prefer object exploration and hand games over face-to-face interactions (Trevarthen, "Interpersonal Abilities"). At the same time, as David Miall notes in his commentary on the dynamics of feeling in this passage, "the infant's feelings about an object will be governed by the feelings it senses in the adult" ("Wordsworth" 234). As the mother's disciple, her apprentice in the "discipline of love," the child is united with the world through the feeling it now reciprocates, a light which, far from following an outmoded Freudian economy of drives and energy, is augmented through positive emotional exchange. Whereas the Freudian model suggests that we all experience an infantile desire to return to a preconscious sanctuary of primary union and primitive emotions (pre-emotions?) embodied in the mother, and thus, as Homans puts it, "look for her representation in the later objects of [our] love" (*Bearing* 49), Wordsworth envisions goal-directed living in the world (i.e., active *being*) as the outgrowth of the initial mother–infant bond, which renders the child, *like his mother*, a deified being who now shares with her and the "one great mind" the status of "creator and receiver."

Understood within the larger context of *The Prelude*, this passage represents an epiphanic moment of primary importance, because it identifies the source of the moral imperative to write. Though its structure and content has been far less frequently analyzed than the other visionary epiphanies in *The Prelude*, the "infant Babe" passage has priority over them inasmuch as neither the experiences of vision nor the epiphanic recognitions are possible without the foundational mother–child relationship and grounding epiphany of the poem (on the organization of epiphanic moments in the poem, see Abrams, *Natural Supernaturalism*; Easterlin, *Wordsworth*; Johnston; Lindenberger; Miall, "Alps"; Nichols; Ogden).

The "infant Babe" passage is the culminating moment in Wordsworth's struggle to find an appropriate topic for a long poem and fortify his resolve to write it. Wordsworth's self-mocking, introductory "glad preamble" (I.1–54) locates easy claims for poetic inspiration and productivity squarely in the past, and the poet, tempted alternately to luxuriate passively in nature and to dwell on unfulfilled ambition and guilt (I.55–272), generates his forward course out of the guilt and despair of feeling "[l]ike a false Steward who hath much receiv'd / And renders nothing back" (I.270–72). But the pronounced tonal shift from this despair to the exuberant rhetorical question—"Was it for this / That one, the fairest of all Rivers, lov'd / To blend his murmurs with my Nurse's song" (I.272–74)—does not explain why "rendering back" is so necessary. Alternating between depictions of the soul's "fair seed-time," in which, alongside his playfellows, the boy is tutored "by beauty and by fear" (I.306–7), and general commentary asserting the spiritual and developmental value of those experiences, the remainder of Book I "[fixes] the wavering balance of [Wordsworth's] mind" (I.651) and implies the poem's subject—that mind's development—but it does not explain how poetry represents a reciprocal gift to nature. Rivers, after all, cannot read.

Coming several hundred lines into Book II after Wordsworth describes his growing love of the peace and solitude to be felt in nature, the "infant Babe" passage explains the moral necessity of writing even as it corrects—by anticipation—the quietistic direction of the adolescent boy's new awareness of nature. Neither the "vulgar joy" nor dreamy solitude are possible without the relationship that founds, simultaneously, emotional attachments and interest in the world, and this most basic social relationship, too, is a part of nature. It is the specifically human dimension of natural life that obligates the poet to supersede contemplative ease and that dictates the subject of his poem, the growth of the creative mind in nature whose source is mother-love.

Within the immediate context in Book II, Wordsworth signals the importance of the passage by preceding it with a verse paragraph depicting a moment of cognitive hesitation in the analysis of his own experience, thus aligning the passage with the many like juxtapositions, wherein despair, disappointment, doubt, or relaxation are reversed by a overflow of emotional and/or intellectual confidence. Beginning with the claim that he had reached a new stage in his development, Wordsworth immediately questions his compartmentalization of experience, asking,

> Who knows the individual hour in which
> His habits were first sown, even as a seed,
> Who that shall point, as with a wand, and say,
> "This portion of the river of my mind
> Came from yon fountain"?
>
> (ll. 211–15)

Even if the adolescent's contemplative appreciation of nature is remembered by the writing poet as a newfound joy, Wordsworth nevertheless implies that its origin cannot be known. Indeed, the final lines of the passage seem to consolidate this nascent epistemological nihilism:

> Hard task to analyse a soul, in which,
> Not only general habits and desires,
> But each most obvious and particular thought,
> Not in a mystical and idle sense,
> But in the words of reason deeply weigh'd,
> Hath no beginning.
>
> (ll. 232–37)

Taken in isolation, this passage suggests the logical endpoint of the poem, announcing as it does a disinclination to divide life into stages that feel to the poet like artificial constructs. Instead, in suddenly exhorting us with the blessed life of the neonate, Wordsworth tells us, more powerfully than he would if he stated it directly, that at least the river of "general habits" can be traced to its fountain, identifying the source of human well-being as the mother–infant relationship (Heffernan 266). All else is "best conjecture," the passage seems to say, but *this* we know. Behind or above human life may reside some obscurely intimated divine entity, some transcendent source or cause, the "Presences of Nature" (I.490) or the "Wisdom and Spirit of the Universe" (I.429), but in our earthly existence it is the nurturing mother who is confidently identifiable as the source of the poet's being.

In accord with this reading, Wordsworth's claim at the end of the poem that the successful poet's heart will "be tender as a nursing Mother's heart"

(XIII.207) is not a denial of his own mother's death or a co-optation of female sensibility, but a final confirmation and acknowledgment of all that is due to mother-love. Even while the emergent dualism of *The Prelude*'s final book suggests that "love more intellectual" (XIII.166)—spiritual love—comes from a separate, divine source than human love, the logic of the poem, which has asserted that feeling and creativity have their source in the mother–infant relationship, indicates that spiritual love, which "cannot be / Without Imagination" (XIII.166–67), arises from and depends upon human love. But more than this, Wordsworth has taken care to point out that divine love does not supplant human love in the poet's affective constitution; rather, it is his nurturing sensibility that makes possible common pleasures, sublime poetry, and a network of loving relationships exactly *because* sublime poetry reciprocates nurturing affection, rendering back many times over—to Coleridge, Dorothy, Mary, and, according to the poem's logic, Wordsworth's readers—a love that begins with the mother (Easterlin, *Wordsworth*; Haney; Schapiro, *Romantic Mother*).

The stunted creature who yearns for the infantile state of primary narcissism, a sort of intercourse without effort, little resembles the poet who celebrates "the discipline of love" as the source of social feeling and poetry. Surely the durability of the psychoanalytic paradigm in literary studies attests to an abiding sexism in our general culture which replaces the positive fact of nurture with a primarily destructive dynamics. For as Wordsworth perceived and subsequent research in developmental psychology bears out, the primary caregiver, almost always the mother, holds enormous power, central as she is to the lifelong well-being of the child, and it is this positive power, so long culturally devalued, that we have failed to see as so justly central to Wordsworth's vision. Certainly, conflict exists between parents/caregivers and children, but this conflict has little to do with a lost primordial union or sexual competition for the mother: one thinks of the exhausted new mother, her sleep interrupted every few hours by a hungry infant as yet incapable of rendering back love and feeling or, a few years later, of the preschooler who cannot understand the annoyance of napkins soaked in milk and bran flakes arranged in casual piles on the kitchen table. (I'm sure the Wordsworths experienced their age's equivalent.) By the same token, adults are psychically blind, however intellectually schooled and imaginatively gifted, to the significance of these activities for the child. Occupying such different places in the life cycle and motivated by different kinds of commitments to our own cohorts and to other generations, parents and children will inevitably have disagreements. But those differences emerge against the background of an efficacious first relationship and its outgrowth, and it is within the lives resulting from that productive, formative time that conflict must be negotiated.

29

Vindication and Vindictiveness

OLIVER TWIST

WILLIAM FLESCH

My charity is outrage.

MARGARET IN SHAKESPEARE'S *RICHARD III*

The Pleasures of Vindication

Recent insights of evolutionary biology and behavioral game theory permit two related and converging arguments. The argument that has conceptual priority is the one for humans' prosocial disposition. Even when given the opportunity to be free-riders, people tend not to be. We usually cooperate instead of maximizing the net gain we can get out of the cooperation of others. We do this because of an interestingly complex attitude toward others: we have a native inclination toward altruism; we expect others to have such an inclination as well; these inclinations are self-reinforcing since we expect others to pay to punish us if we don't share enough of what we have with them; and therefore our prosocial behavior is partly predicated on an uncalculated expectation that others will behave in accordance with how well we meet or defy their uncalculated expectations for how we behave.

Reprinted by permission of the publisher from "Vindication and Vindictiveness": in *Comeuppance: Costly Signaling, Altruistic Punishment, and Other Biological Components of Fiction* by William Flesch, pp. 155–156, 164–168, 170, Cambridge, Mass.: Harvard University Press, Copyright © 2007 by the President and Fellows of Harvard College.

The second argument is just as important, though conceptually it comes after the first. We tend to reward others who engage in genuinely altruistic behavior, and to approve of anyone else who rewards them, and we tend to approve as well of others who punish antisocial or socially noxious behavior. This is a special element in the prosociality the first argument describes. Some such self-monitoring procedure must have evolved to prevent free-riders from outcompeting altruists, and so eventually causing altruism to die out. The more widespread this self-monitoring procedure, the less it costs to be an altruist.

This second argument is more or less where we can locate our interest in narrative. We care about the narrative report of what some people do to other people because we care about whether they treat them altruistically or self-ishly. *Altruistically* needn't mean *generously*. They treat others altruistically by punishing them for defecting as well as by rewarding them for their own altruism. They treat them selfishly by ignoring an obligation to punish, as well as by defecting themselves. We like characters who engage in effective altruistic behavior and we dislike their opposites. Parties to narrative (narrator and narratee, author and reader) are then in at least a third-order relation to the behavior represented in narratives. We could say, in very general terms, that in narrative at least one character undertakes prosocial behavior either on behalf of or against another character (it needn't be but can be both), and that we approve of the prosocial character. Our relation is therefore one of approval of a genuine altruist. We ourselves can't reward or punish the character we want to see rewarded or punished, but we can cheer on the altruistic character who does—and the storyteller who arranges these things as well.

Oliver, Trusted and Reviled

Oliver Twist (1838) provides a good example of this play of judgment, in particular people's judgments as to Oliver's trustworthiness. After Oliver is saved from the false accusation of picking his pocket, the kindly Mr. Brown-low takes him home, gives him a new suit of clothes, and asks him to take, unsupervised, a five-pound note as well as books worth the same amount of money back to a poor bookseller. His spirit is aroused as he gives Oliver this errand, since he wishes to prove Oliver's honesty to his friend Mr. Grim-wig and so to show Mr. Grimwig up. Mr. Grimwig doubts that Oliver will return, such a return being contrary to what Mr. Grimwig thinks Oliver will perceive as his own interests, which would be to abscond with the clothes Mr. Brownlow has furnished him, with the money, and with the valuable books. Mr. Brownlow is sure he will return, and that return will vindicate

Oliver to Mr. Grimwig and vindicate Mr. Brownlow as well: Mr. Brownlow and Mr. Grimwig are the audience of Oliver's action.

Oliver, luckless boy, is on his way to fulfill his errand scrupulously when he is waylaid by Bill Sikes and Nancy, who force him to come with them. They do this by accusing him in the public street of having deserted his family—Nancy claims to be his sister—so that his denials and pleas infuriate the abduction's spectators, who think that Oliver is lying. They condone Nancy and Bill's apprehension of Oliver. Meanwhile, Mr. Brownlow and Mr. Grimwig wait to see whether Oliver will return, and who will win the contest as to accurate prediction of Oliver's behavior and character.

The scene is a highly effective one, and belongs to a highly effective genre, that of a frame-up. (Hitchcock is the great cinematic exponent of this situation.) We're aware of the frame, and in our own response we'll typically anticipate vindication while resenting the vindictiveness of those who don't recognize it. As we've noted, the two things are related. Mr. Brownlow and Mr. Grimwig bait each other: Mr. Grimwig with a "provoking smile," Mr. Brownlow with a "confident" one. The different tonalities of the phrases—*provoking* versus *confident smile*—suggest a difference in our sense of the two old men, but the similarities matter also: each is confident of being right and provoked by the other's confidence. Likewise, vindication and vindictiveness name differences in degree and in the moral judgment of whoever uses one term or the other. Vindictiveness is spiteful, whereas vindication demonstrates the reasonableness of the vindicated person's altruism or trust or risk. We might add a further distinction by saying that vindication is something like a demonstration that someone has been wronged or judged unfairly or even vindictively, that demonstration being a triumphant end in itself, while vindictiveness has a far greater component of overweening revenge in it (Bacon's wild justice), so that the vindictive seek not only to be vindicated but to harm the adversaries they conceive have harmed them. They imagine that they are entitled to cause great harm—that they would be acquitted for doing so—because their very willingness to show their spite to a world that disapproves of spite would be a sign of how great was the wrong they suffered: Look how spiteful you've made me; look how spiteful she's made me! Spite is perhaps a sign of how wronged they do, in fact, feel. To risk looking vindictive is itself a kind of emotional anticipation of vindication (one almost always disappointed, and sometimes spitefully proud of being disappointed).

Spite isn't particularly evident in this scene (although it's to be found passim in the novel), but the vector by which vindication threatens to turn into vindictiveness is. Mr. Brownlow anticipates being vindicated. He does it by anticipating Oliver's vindication. Here, vindication is manifestly a vicarious attitude: he thinks Oliver will vindicate his trust in him. And Mr. Grimwig also seeks to have his judgment vindicated. But Dickens couches this in a more selfish and therefore somewhat more spiteful language:

> The spirit of contradiction was strong in Mr. Grimwig's breast, at the moment; and it was rendered stronger by his friend's confident smile. . . . It is worthy of remark, as illustrating the importance we attach to our own judgments, and the pride with which we put forth our most rash and hasty conclusions, that, although Mr. Grimwig was not by any means a bad-hearted man, and though he would have been unfeignedly sorry to see his respected friend duped and deceived, he really did most earnestly and strongly hope at that moment, that Oliver Twist might not come back. (137)

The scene is exemplary because the reader too looks to see Oliver vindicated and Mr. Grimwig thwarted. But there is more than one person we are anxious about: we want to see Oliver vindicated, especially in the face of Mr. Grimwig's provoking certainty—that'll show him! We want to see Oliver vindicated before Mr. Brownlow too, but mainly perhaps we want to see Mr. Brownlow's trust in Oliver vindicated. We resent Mr. Grimwig's selfish preference for his own judgment to Mr. Brownlow's humane sense of Oliver. (Remember that what we feel is Mr. Grimwig's selfishness is selfish in a psychological sense, in no way a rational or biological one, and indeed derives from what Hume calls a general propensity of benevolence toward mankind.)

Mr. Grimwig's provoking skepticism has consequences: had he not doubted Oliver Mr. Brownlow would not have attempted to vindicate him by ostentatiously entrusting the errand to him, and Nancy and Bill would not have found him on the way to the bookseller. Our anger with them is exacerbated into fury by our frustration that Mr. Grimwig will think himself vindicated, that Mr. Brownlow will be disappointed, and that both will wrongly conclude that Oliver is depraved, Mr. Grimwig triumphantly, Mr. Brownlow sorrowfully. Dickens expertly elicits this frustration in the next scene when he has the bystanders disbelieve Oliver as Nancy and Bill abduct him:

> "What's the matter, ma'am?" inquired one of the women.
> "Oh, ma'am," replied the young woman, "he ran away, near a month ago, from his parents, who are hard-working and respectable people; and went and joined a set of thieves and bad characters; and almost broke his mother's heart."
> "Young wretch!" said one woman.
> "Go home, do, you little brute," said the other.
> "I am not," replied Oliver, greatly alarmed. "I don't know her. I haven't any sister, or father and mother either. I'm an orphan; I live at Pentonville."
> "Only hear him, how he braves it out!" cried the young woman. (143)

The manifest injustice of the two women stands for the general injustice of those who judge Oliver throughout the book, and that injustice is palpable to readers. And yet what are they but mistaken altruistic punishers, cooperating with what they take to be the imperatives of the urchin's family? And here we can judge how vexatious it is to see altruistic punishment gone wrong: how mistaken or unjust altruistic punishment becomes itself something that we want to see altruistically punished.

It's important to note that it's we who feel this way, not Oliver. He doesn't resent the injustice. We resent it for him. He feels only alarm, and it is a tribute to our sense of his essential innocence that, to use Philip Fisher's phrase, we volunteer the affect for him. Oliver himself is not vindictive, nor does he even think to seek vindication. But we are close to vindictiveness ourselves by this point, and while we will find some satisfaction in the fate of Bill Sikes, it's a good thing that Dickens doesn't cherish a grudge against these bystanders, since they will never learn the truth and repent.

I want to call attention to the following elements of this scene. We are concerned about several interconnected things at once. We worry about Oliver's fate. Part of that worry is about his relation to Mr. Brownlow. We worry therefore about whether Oliver will vindicate Mr. Brownlow's trust. But there has to be an audience for that vindication within the world where it takes place. Our knowledge of Oliver's innocence is not enough. That audience is often primarily the person whose faith we hope to see vindicated (from Odysseus to Gilbert Redman), but there is more pleasure to be anticipated in seeing a vexatious skeptic like Mr. Grimwig get his comeuppance. Mr. Grimwig would be the witness of his own confutation. He would also be the necessary witness—not of Oliver's vindication, but of the vindication of Mr. Brownlow's faith in Oliver.

So we sympathize, or volunteer, or hope to volunteer affect not only for Oliver, and not only for the sympathizing Mr. Brownlow, but—oddly enough—for Mr. Grimwig. We care about what he feels: we want him to see that he has been wrong. We are outraged that he doesn't know it. If we want Mr. Brownlow's faith vindicated, we also anticipate some vindictive pleasure in making Mr. Grimwig eat his head (as he says he will), or his words. Notice that Dickens says of Mr. Grimwig that he is not a bad-hearted man and would be sorry to see his friend duped. Dickens has plenty of bad-hearted men, but Mr. Grimwig has a trickier function: we care about what he feels rather than just wishing him away, as we have wished away the awful magistrate Mr. Fang. Dickens has to balance him delicately between selfish wrongheadedness and a capacity to enter into relations of sympathy. (Scrooge, of course, is another such figure.) When Mr. Grimwig returns later on, now to help punish the tormentors of Oliver, enough time has gone by, a sufficient number of actually selfish characters have appeared, that we feel indulgently disposed toward him and happy to see him absorbed into the general good fellowship of prosocial characters.

Note too that such a capacity for sympathy belongs as well even to the bystanders who excoriate Oliver. Their anger is volunteered on behalf of the mother whose love for Oliver they understand without sharing. Of course they take pleasure in excoriating him, and of course this pleasure is selfish or spiteful. But the alibi for that pleasure is also its real motive: the sense of punishing someone who has done a real wrong (as they think) to his mother.

As a final observation about this sequence, it's worth remarking the narrative's own tender observation of Oliver's ignorance of all this. We sympathize with Oliver, although he is ignorant of the extent of the misfortune we ascribe to him. We resent the wrong done him behind his back, the wrong done him through the misjudgments of Mr. Grimwig and, as we fear, of Mr. Brownlow as well. After retelling Mr. Brownlow's disappointment in Oliver (and Mrs. Bedwin's continued faith in him) upon the testimony of the egregious Mr. Bumble, the narrative insists on Oliver's ignorance of the disappointment he has been the occasion for, and the misconstructions put upon him, in a single-sentence allusion to his feelings: "Oliver's heart sank within him, when he thought of his good kind friends; it was well for him that he could not know what they had heard, or it might have broken outright" (164). His heart would have broken in sympathy for them and for their disappointment, and not in self-pity. Vicarious experience outweighs his own situation.

Likewise, the subjunctive mood of this statement ("it might have") nevertheless elicits our sympathy for Oliver (despite the fact that his heart is not broken) since, in fact, he could not have known what they felt. We sympathize with his capacity to sympathize, with the fact that he might have had his heart broken by the heartbreak he causes, and these reciprocal vicarious feelings, their tendency to volunteer affect for each other, is just what gets us to volunteer affect for them.

One interesting aspect of *Oliver Twist* is the virtual disappearance of Oliver from the second half of the book. He has become a Maguffin, the catalyst for a plot where his disparagers, his vindicators, and the witnesses of his vindication become the central characters. We could say much the same about Cordelia in *King Lear*; she is vindicated offstage, and her absence from the plot functions much like Oliver's. So too does Hermione's in *The Winter's Tale*; the oracle vindicates her but the vindication isn't real until Leontes spends the next sixteen years ratifying it. In stories of the vindication of the innocent, from Cordelia to Hermione to Oliver to *Dial M for Murder* (Hitchcock, 1954), much of the anxiety that we feel is about whether the altruistic punishers—the police or patrons or fathers—will come to see that they have mistrusted wrongly, have punished wrongly. Will they now make it right? Doing so will, in its turn, vindicate them before those they have wronged. This may be so even if the wronged are dead ("I killed the slave that was ahanging thee," Lear assures the dead Cordelia). In *Hamlet* we may say that the revenger attempts to vindicate the ghost, the nearly absent Maguffin who watches the story of his son's vengeance. But to be absent (especially in Act V) he has to be first present in the plot.

30

The Cuckoo's History

HUMAN NATURE IN *WUTHERING HEIGHTS*

JOSEPH CARROLL

Wuthering Heights occupies a singular position in the canon of English fiction. It is widely regarded as a masterpiece of an imaginative order superior to that of most novels—more powerful, more in touch with elemental forces of nature and society, and deeper in symbolic value. Nonetheless, it has proved exceptionally elusive to interpretation. There are two generations of protagonists, and the different phases of the story take divergent generic forms that subserve radically incompatible emotional impulses. Humanist readings from the middle decades of the previous century tended to resolve such conflicts by subordinating the novel's themes and affects to some superordinate set of norms, but the norms varied from critic to critic, and each new interpretive solution left out so much of Brontë's story that subsequent criticism could gather up the surplus and announce it as the basis for yet another solution. Postmodern critics have been more receptive to the idea of unresolved conflicts, but they have tended to translate elemental passions into semiotic abstractions or have subordinated the concerns of the novel to current political and social preoccupations. As a result, they have lost touch

From *Philosophy and Literature* 32, no. 2, pp. 241–57, by Joseph Carroll. © Joseph Carroll.

with the aesthetic qualities of the novel. Moreover, the interpretive solutions offered by the postmodern critics have varied with the idioms of the various schools. Surveying the criticism written up through the 1960s, Miriam Allott speaks of "the riddle of *Wuthering Heights*" (*Introduction* 12). Taking account both of humanist criticism and of seminal postmodern readings, Harold Fromm declares that *Wuthering Heights* is "one of the most inscrutable works in the standard repertoire" (*Academic Capitalism* 128) (for surveys of postmodern criticism on *Wuthering Heights*, see G. Frith; Stoneman, *Emily Brontë*).

Brontë's novel need not be relegated permanently to the category of impenetrable mysteries. The critical tradition has produced a good deal of consensus on the affects and themes in *Wuthering Heights*. Most of the variation in critical response occurs at the level at which affects and themes are organized into a total structure of meaning. In the efforts to conceptualize a total structure, one chief element has been missing—the idea of "human nature." By foregrounding the idea of human nature, Darwinian literary theory provides a framework within which we can assimilate previous insights about *Wuthering Heights*, delineate the norms Brontë shares with her projected audience, analyze her divided impulses, and explain the generic forms in which those impulses manifest themselves. Brontë herself presupposes a folk understanding of human nature in her audience. Evolutionary psychology converges with that folk understanding but provides explanations that are broader and deeper. In addition to its explanatory power, a Darwinian approach has a naturalistic aesthetic dimension that is particularly important for interpreting *Wuthering Heights*. Brontë's emphasis on the primacy of physical bodies in a physical world—what I am calling her naturalism—is a chief source of her imaginative power. By uniting naturalism with supernatural fantasy, she invests her symbolic figurations with strangeness and mystery. From the perspective of evolutionary psychology, the supernaturalism can itself be traced to natural sources in Brontë's imagination.

An evolutionary account of human nature locates itself within the wider biological concept of "life history." Species vary in gestation and speed of growth, length of life, forms of mating, number and pacing of offspring, and kind and amount of effort expended on parental care. For any given species, the relations among these basic biological characteristics form an integrated structure that biologists designate the "life history" of that species. Human life history, as described by evolutionary biologists, includes mammalian bonding between mothers and offspring, dual-parenting and the concordant pair-bonding between sexually differentiated adults, and extended childhood development. Like their closest primate cousins, humans are highly social and display strong dispositions for building coalitions and organizing social groups hierarchically. All these characteristics are part of "human nature" (Barrett, Dunbar, and Lycett; Buss, *Evolutionary Psychology*; Dunbar and Barrett, *Oxford Handbook*; Gangestad and Simpson). Humans have also evolved unique representational powers, especially those of language, through which they convey information in non-genetic ways. That kind of informational transmission is what we call "culture": arts, technologies, literature, myths, religions, ideologies, philosophies, and science. From the Darwinian perspective, culture does not stand apart from the genetically transmitted dispositions of human nature. It is, rather, the medium through which we organize those dispositions into systems that regulate public behavior and inform private thoughts (Boyd, "Literature," *On*; Carroll, "Evolutionary Paradigm," *Literary Darwinsim*; Dissanayake, *Art*; Dutton; Gottschall and Wilson; Wilson, *Consilience*). In writing and reading fabricated accounts of human behavior, novelists and their readers help to produce and sustain cultural norms. Novelists select and organize their material for the purpose of generating emotionally charged evaluative responses, and readers become emotionally involved in stories, participate vicariously in the experiences depicted, and form personal opinions about the characters.

Beneath all variation in the details of organization, the life history of every species forms a reproductive cycle. In the case of *Homo sapiens*, successful parental care produces children capable, when grown, of forming adult pair bonds, becoming functional members of a community, and caring for children of their own. With respect to its adaptively functional character, human life history has a normative structure. In this context, the word "normative" signifies successful development in becoming a socially and sexually healthy adult. The plot of *Wuthering Heights* indicates that Brontë shares a normative model of human life history with her projected audience, but most readers have felt that the resolution of the plot does not wholly contain the emotional force of the story. Brontë is evidently attracted to the values vested in the normative model, but her figurations also embody impulses of emotional violence that reflect disturbed forms of social and sexual development.

The elements of conflict in *Wuthering Heights* localize themselves in the contrast between two houses: on the one side Thrushcross Grange, situated in a pleasant, sheltered valley and inhabited by the Lintons, who are civilized and cultivated but also weak and soft; and on the other side Wuthering Heights, rough and bleak, exposed to violent winds, and inhabited by the Earnshaws, who are harsh and crude but also strong and passionate. Conflict and resolution extend across two generations of marriages between these houses. In the first generation, childhoods are disrupted, families are dysfunctional, and marriages fail. The destructive forces are embodied chiefly in Catherine Earnshaw and Heathcliff, and Brontë depicts their passions with extraordinary empathic power. In the second generation, the surviving children, Catherine Linton and Hareton Earnshaw, bridge the divisions between the two families, and the reader can reasonably anticipate that they will form a successful marital bond. Through this movement toward resolution, Brontë implicitly appeals to a model of human life history in which children develop into socially and sexually healthy adults. Nevertheless, the majority of readers have always been much more strongly impressed by Catherine and Heathcliff than by the younger protagonists.

The differences between the two generations can be formulated in terms of genre, and genre, in turn, can be analyzed in terms of human life history. The species-typical needs of an evolved and adapted human nature center on sexual and familial bonds within a community—bonds that constitute the core elements of romantic comedy and tragedy. Romantic comedy typically concludes in a marriage and thus affirms and celebrates the social organization of reproductive interests within a given culture. In tragedy, sexual and familial bonds become pathological, and social bonds disintegrate. (On the structure of romantic comedy and tragedy, Frye, after more than half a century, remains the most authoritative source.) *Wuthering Heights* contains the seeds of tragedy in the first generation, and the second generation concludes in a romantic comedy, but the potential for tragedy takes an unusual turn. In most romantic comedies, threats to family and community are contained or suppressed within the resolution. In *Wuthering Heights*, the conflicts activated in the first generation are not fully contained within the second. Instead, the passions of Catherine and Heathcliff form themselves into an independent system of emotional fulfillment, and the novel concludes with two separate spheres of existence: the merely human and the mythic. The human sphere, inhabited by Hareton Earnshaw and the younger Cathy, is that of romantic comedy. In the mythic sphere, emotional violence fuses with the elemental forces of nature and transmutes itself into supernatural agency. Romantic comedy and pathological supernaturalism are, however, incompatible forms of emotional organization, and that incompatibility reflects itself in the history of divided and ambivalent responses to the novel.

Brontë would of course have had no access to the concept of adaptation by means of natural selection, but she did have access to a folk concept of human nature. To register this concept's importance as a central point of reference in the story, consider three specific invocations of the term "human nature." The older Catherine reacts with irritated surprise when her commendation of Heathcliff upsets her husband. Nelly Dean explains that enemies do not enjoy hearing one another praised. "It's human nature" (77). Reflecting on the malevolent mood that prevails under Heathcliff's ascendancy at Wuthering Heights, Isabella observes how difficult it is in such an environment "to preserve the common sympathies of human nature" (106). The younger Cathy is sheltered and nurtured at the Grange, and when she first learns of Heathcliff's monomaniacal passion for revenge, she is "deeply impressed and shocked at this new view of human nature—excluded from all her studies and all her ideas till now" (172). In Heathcliff, human nature has been stunted and deformed. Apart from his passional bond with Catherine, his relations with other characters are almost exclusively antagonistic. The capacity for hatred is part of human nature, but so is positive sociality. No other character in the novel accepts antagonism as a legitimately predominating principle of social life. Brontë shares with her projected audience a need to affirm the common sympathies that propel the novel toward a resolution in romantic comedy.

Folk appeals to human nature provide a basis for comparing an adaptationist perspective on *Wuthering Heights* with humanist and postmodern perspectives. Humanist critics do not overtly repudiate the idea of human nature, but they do not typically seek explanatory reductions in evolutionary theory, either. Instead, they make appeal to some metaphysical, moral, or formal norm—for instance, cosmic equilibrium, charity, passion, or the integration of form and content—and they typically represent this preferred norm as a culminating extrapolation of the common understanding. Postmodern critics, in contrast, subordinate folk concepts to explicit theoretical formulations—deconstructive, Marxist, Freudian, feminist, and the rest—and they present the characters in the story as allegorical embodiments of the matrix terms within these theories. In their postmodern form, all these component theories emphasize the exclusively cultural character of symbolic constructs. "Nature" and "human nature," in this conception, are themselves cultural artifacts. Because they are contained and produced by culture, they can exercise no constraining force on culture. Hence Fredric Jameson's dictum that "postmodernism is what you have when the modernization process is complete and nature is gone for good" (ix). From the postmodern perspective, any appeal to "human nature" would necessarily appear as a delusory reification of a specific cultural formation. By self-consciously distancing itself from the folk understanding of human nature, postmodern criticism distances itself also from biological reality and from the imaginative structures that Brontë shares with her projected audience. In both the biological and folk understanding, as in the humanist, there is a world outside the text. An adaptationist approach to *Wuthering Heights* shares with the humanist a respect for the common understanding, and it shares with the postmodern a drive to explicit theoretical reduction. From the adaptationist perspective, folk perceptions offer insight into important features of human nature, and evolutionary theory makes it possible to situate those features within the larger theoretical system of human life history analysis. (For representative humanist readings of *Wuthering Heights*, see Allott, "Rejection"; Cecil; Leavis; Mathison; Nussbaum, "*Wuthering Heights*"; Van Ghent. For representative postmodern readings, see Armstrong, "Emily Brontë"; Eagleton, *Myths*; Gilbert and Gubar; Homans, "Name"; Jacobs; J. Miller.)

A Darwinian approach to fiction involves no necessary commitment to a metaphysical ideal or to an ideal of formal aesthetic integration. Identifying human nature as a central point of reference does not require the critic to postulate any ultimate resolution of conflict in a novel. Quite the contrary. Darwinians regard conflicting interests as an endemic and ineradicable feature of human social interaction (Bjorklund and Pellegrini; Geary, "Evolution"; Geary and Flinn). Male and female sexual relations have compelling positive affects, but they are also fraught with suspicion and jealousy. Even when they work reasonably well, these relations inevitably involve compromise, and all compromise is inherently unstable. Parents have a reproductive investment in their children, but children have still more of an investment in themselves, and siblings must compete for parental attention and resources. Each human organism is driven by its own particular needs, with the result that all affiliative behavior consists in temporary arrangements of interdependent interests. Nelly Dean understands this principle. Reflecting on the ending of the brief period of happiness in the marriage between Catherine Earnshaw and Edward Linton, she explains, "Well, we *must* be for ourselves in the long run; the mild and generous are only more justly selfish than the domineering—and it ended when circumstances caused each to feel that the one's interest was not the chief consideration in the other's thoughts" (72). The prospective marriage of Hareton and Cathy invokes a romantic comedy norm in which individual interests fuse into a cooperative and reciprocally advantageous bond, but no such bond is perfect or permanent, and many are radically faulty. The conclusion of *Wuthering Heights* juxtaposes images of domestic harmony with images of emotional violence that reflect deep disruptions in the phases of human life history.

In modern evolutionary theory, the ultimate regulative principle that has shaped all life on earth is the principle of "inclusive fitness"—that is, of kinship, the sharing of genes among reproductively related individuals. Kinship takes different forms in different cultures, but the perception of kinship is not merely an artifact of culture. Kinship is a physical, biological reality that makes itself visible in human bodies. The species-typical human cognitive system contains mechanisms for recognizing and favoring kin, and perceptions of kin relations bulk large in folk psychology (Barrett, Dunbar, and Lycett 45–66; Kurland and Gaulin; Salmon and Shackelford).

As one might anticipate, then, kinship forms a major theme in the literature of all cultures and all periods. In *Wuthering Heights*, that common theme articulates itself with exceptional force and specificity. Kinship among the characters manifests itself in genetically transmitted features of anatomy, nervous systems, and temperament. The interweaving of those heritable characteristics across the generations forms the main structure in the thematic organization of the plot.

Heathcliff and Catherine are physically strong and robust, active, aggressive, domineering. Edgar Linton is physically weak, pallid and languid, tender but emotionally dependent and lacking in personal force. Even Nelly Dean, fond of him as she is, remarks that "he wanted spirit in general" (52). Isabella Linton, in contrast, is vigorous and active. She defends herself physically against Heathcliff, and when she escapes from him she runs four miles over rough ground through deep snow to make her way to the Grange. Her son Linton, weak in both body and character, represents an extreme version of the debility that afflicts his uncle Edgar. Linton Heathcliff is "a pale, delicate, effeminate boy, who might have been taken for my master's younger brother, so strong was the resemblance; but there was a sickly peevishness in his aspect that Edgar Linton never had" (155). Isabella's son has "large, languid eyes—his mother's eyes, save that, unless a morbid touchiness kindled them a moment, they had not a vestige of her sparkling spirit" (159). Despite his inanition, Linton Heathcliff can be kindled to an impotent rage that recalls his father's viciousness of temper. Witnessing an episode of the boy's "frantic, powerless fury," the old servant Joseph cries in malicious glee, "Thear, that's t' father! . . . That's father! We've allas summut uh orther side in us" (192). With respect to Linton Heathcliff, Nelly Dean participates in the brutal physical naturalism of her creator's vision. She observes that Linton is "the worst-tempered bit of a sickly slip that ever struggled into its teens! Happily, as Mr. Heathcliff conjectured, he'll not win twenty!" (186). He lives into his mid-teens but in manner remains infantile—self-absorbed and querulous. The younger Cathy is as physically robust and active as her mother and her aunt Isabella. She also has her mother's dark eyes and her vivacity, but she has her father's blond hair, delicate features, and tenderness of feeling. "Her spirit was high, though not rough, and qualified by a heart sensitive and lively to excess in its affections. That capacity for intense attachments reminded me of her mother; still she did not resemble her, for she could be soft and mild as a dove, and she had a gentle voice, and pensive expression: her anger was never furious; her love never fierce; it was deep and tender" (146). The younger Cathy has not inherited her mother's emotional instability. Nor does she display her mother's antagonistic delight in teasing and tormenting others. Her cousin Hareton Earnshaw is athletically built, has fine, handsome features, and his mind, though untutored, is strong and clear. He has evidently not inherited the fatal addictive weakness in his father's character. The inscription over the door at Wuthering Heights bears his own name, Hareton Earnshaw, and the date 1500. In his person, the finest innate qualities in the Earnshaw lineage come into flower.

The interweaving of heritable characteristics across the generations progresses with an almost mechanical regularity, but the meaning invested in that progression ultimately resolves itself into no single dominant perspective. The narrative is all delivered in the first person; it is spoken by participant narrators who say "I saw" and "I said" and "I felt." Through these first-person narrators, Brontë positions her prospective audience in relation to the story while she herself remains at one remove. The story is told chiefly by two narrators—Lockwood and Nelly Dean. Lockwood, a cultivated but vain and affected young man, holds the place of a conventional common reader who is shocked at the brutal manners of the world depicted in the novel. Nelly is closer to the scene, sympathetic to the inhabitants, and tolerant of the manners of the place—characteristics that enable her to mediate between Lockwood and the primary actors in the story. She provides a perspective from which the local cultural peculiarities can be seen as particular manifestations of human universals. When Lockwood exclaims that people in Yorkshire "*do* live more in earnest," she responds, "Oh! here we are the same as anywhere else, when you get to know us" (49). Both Nelly and Lockwood express opinions and make judgments, but neither achieves an authoritative command over the meaning of the story. Nelly is lucid, sensible, humane, and moderate, and she sees more deeply than Lockwood, but her perspective is still partial and limited. She has her likes and dislikes—she particularly dislikes Catherine Earnshaw and Linton Heathcliff—and some of her most comprehensive interpretive reflections fade into conventional Christian pieties that are patently inadequate to the forces unleashed in the story she tells. She holds the place of a reader for whom the impending romantic comedy conclusion offers the most complete satisfaction.

Behind the first-person narrators, the implied author, Emily Brontë herself, remains suspended over the divergent forces at play in the two generations of protagonists. The resolution devised for the plot is presented in an intentionally equivocal way. In its moment of resolution, the novel functions like an ordinary romantic comedy, but the pathological passions in the earlier generation are too powerful to be set at rest within a romantic comedy resolution. The narrative offers evidence that the earth containing the bodies of Catherine and Heathcliff is still troubled, and always will be, by their demonic spirits. In the last sentence of the novel, Lockwood wonders "how any one could ever imagine unquiet slumbers for the sleepers in that quiet earth." But it is Lockwood himself, during the night he spends at Wuthering Heights, to whom the ghost of Cathy appears in a dream, crying at the window. Nelly Dean deprecates the rumor that the spirits of Catherine and Heathcliff walk the moors, but she also reports that the sheep will not pass where the boy saw the ghosts of Heathcliff and Catherine, and she herself is afraid to walk abroad at night.

Nelly introduces her story of Heathcliff by saying it is a "cuckoo's" history (28). It is, in other words, a story about a parasitic appropriation of resources that belong to the offspring of another organism. That appropriation is the central source of conflict in the novel. The biological metaphor incisively identifies a fundamental disruption in the reproductive cycle based on the family. Heathcliff is an ethnically alien child plucked off the streets of Liverpool by the father of Catherine and Hindley, and then, almost unaccountably, cherished and favored over his own son Hindley. When the father dies, Hindley takes his revenge by degrading and abusing Heathcliff. When Heathcliff returns from his travels, he gains possession of Hindley's property by gambling, and after Isabella's death, he uses torture and terror to acquire possession of the Grange. He abducts the younger Cathy, physically abuses her, and compels her to marry his terrorized son Linton Heathcliff. From the normative perspective implied in the romantic comedy conclusion of the novel, Heathcliff is an alien force who has entered into a domestic world of family and property, disrupted it with criminal violence, usurped its authority, and destroyed its civil comity. In the romantic comedy resolution, historical continuity is restored, property reverts to inherited ownership, and family is re-established as the main organizing principle of social life. The inheritance of landed property is a specific form of socio-economic organization, but that specific form is only the local cultural currency that mediates a biologically grounded relationship between parents and children. The preferential distribution of resources to one's own offspring is not a local cultural phenomenon. It is not even an exclusively human phenomenon. It is a condition of life that humans share with all other species in which parents invest heavily in offspring (Figueredo et al.; Trivers, "Parental Investment and Sexual Selection"). The cuckoo's history is a history in which a fundamental biological relationship has been radically disrupted.

Brontë assigns to the second generation the thematic task of restoring the genealogical and social order that has been disrupted in the previous generation. The younger Cathy serves as the chief protagonist for this phase of the story. During the brief period of her relations with Linton Heathcliff and Hareton Earnshaw, she meets moral challenges through which she symbolically redeems the failures of her elders. Despite the ill usage she has received from Heathcliff's son, she nurses and comforts him in his final illness, and after his death, she establishes a wholesome bond with Hareton Earnshaw. Linton Heathcliff is wretched and repugnant in his self-absorbed physical misery. By nursing him in his final illness, the younger Cathy introduces a new element into the emotional economy of the novel—an element of redemptive charity. Her attitude to Hareton, at first, reflects the class snobbery that had distorted her mother's marital history. Cathy feels degraded by her cousinage with Hareton, and she mocks him as a lout and a boor. By rising above that snobbery and forming a beneficent bond with him, she resolves the conflict between social ambition and personal attachment that had riven the previous generation. Linton Heathcliff had embodied the worst personal qualities of the older generation—the viciousness of Heathcliff and the weakness of the Lintons—and Hareton and the younger Cathy together embody the best qualities: generosity and strength combined with fineness and delicacy. Even Heathcliff participates in the romantic comedy resolution, though in a merely negative way. The eyes of both Hareton and the younger Cathy closely resemble the eyes of Catherine Earnshaw. Seeing them suddenly look up from a book, side by side, Heathcliff is startled by this visible sign of their kinship with Catherine, and his perception of that kinship dissipates his lust for revenge. By dying when he does, he leaves the young people free to achieve their own resolution.

It seems likely that one of the strongest feelings most readers have when Heathcliff dies is a feeling of sheer relief. In this respect, both Lockwood and Nelly Dean serve a perspectival function as common readers. Lockwood leaves the Grange just before the final crisis in the story. All around him, he sees nothing except boorish behavior, sneering brutality, and vindictive spite. The mood is pervasively sullen, angry, bitter, contemptuous, and resentful. The physical condition of the house at Wuthering Heights, where Heathcliff, Hareton, and the younger Cathy are living, is sordid and neglected. When Lockwood returns, Heathcliff is dead; there are flowers growing in the yard; the two attractive young people are happy and in love; and Nelly Dean is contented. Very few readers can feel that all of this is a change for the worse. It is something like the clearing of weather after a storm, but it is even more like returning to a prison for the criminally insane and finding it transformed into a pleasant home. (In one of the best and most influential humanist interpretations of the novel, Cecil identifies images of storm and calm as symbols for an ultimate metaphysical equilibrium—"a cosmic harmony" [174].)

Readers have often expressed feelings of pity for Catherine and Heathcliff, but few readers have liked them or found them morally attractive. The history of readers' responses to the two characters nonetheless gives incontrovertible evidence that they exercise a fascination peculiar to themselves. In the mode of commonplace realism, they are characters animated by the ordinary motives of romantic attraction and social ambition, and in the mode of supernatural fantasy, they are demonic spirits, but neither of these designations fully captures their symbolic force. At the core of their relationship, a Romantic identification with the elemental forces of nature serves as the medium for an intense and abnormal psychological bond between two children. Describing her connection with the earth, Catherine tells Nelly that she once dreamed she was in heaven, but "heaven did not seem to be my home; and I broke my heart with weeping to come back to earth; and the angels were so angry that they flung me out, into the middle of the heath on the top of Wuthering Heights; where I woke sobbing for joy" (63). Catherine plans to marry Edgar Linton because he is of a higher class than Heathcliff, but she herself recognizes that class is for her a relatively superficial distinction of personal identity. "My love for Linton is like the foliage in the woods. Time will change it, I'm well aware, as winter changes the trees—my love for Heathcliff resembles the eternal rocks beneath—a source of little visible delight, but necessary. Nelly, I *am* Heathcliff" (64). As children, Heathcliff and Catherine have entered into a passional identification in which each is a visible manifestation of the personal identity of the other. Each identifies the other as his or her own "soul." Each is a living embodiment of the sense of the other's self. This is a very peculiar kind of bond—a bond that paradoxically combines attachment to another with the narcissistic love of one's self. Self-love and affiliative sociality have fused into a single motive that transforms the unique integrity of the individual identity into a dyadic relation. Dorothy Van Ghent astutely characterizes the sexually dysfunctional character of this dyadic bond. The relationship is not one of "sexual love, naturalistically considered," for "one does not 'mate' with one's self" (158). In normally developing human organisms, a true fusion between two individual human identities occurs not at the level of the separate organisms but only at the genetic level, in the fertilized egg and the consequent creation of a new organism that shares the genes of both its parents.

The unique integrity of the individual identity is a psychological phenomenon grounded in biological reality. Individual human beings are bodies wrapped in skin with nervous systems sending signals to brains that are soaked in blood and encased in bone. Individual bodies engage in perpetual chemical interchange with the substances of the environment—air, water, and food—but they nonetheless constitute self-perpetuating physiological systems that can be radically disrupted only by death. Brontë's imagination dwells insistently on the reality and primacy of bodies, and that naturalistic physicality extends into the depiction of the peculiar psychological fusion of individual identity in Heathcliff and Catherine. At the time of Edgar Linton's funeral, eighteen years after that of Catherine, Heathcliff describes to Nelly his necrophiliac excursion to Catherine's grave. He has the sexton uncover her coffin, knocks out the side next to his own anticipated grave, and bribes the sexton to remove that side of his own coffin when he is buried. Catherine and Heathcliff achieve consummation not in a reproductively successful sexual union but in the commingling of rotted flesh. If necrophilia can reasonably be characterized as a pathological disposition, the empathetic emotional force that Brontë invests in the relationship between Catherine and Heathcliff can also reasonably be characterized as pathological.

The pathology that culminates in necrophilia disrupts the reproductive cycle, and it arises from disruptions in an earlier phase in that cycle—in childhood development. After Catherine has had the first attack of the hysterical passion that ultimately leads to her death, she tells Nelly that she came out of a trance-like condition, and "most strangely, the whole last seven years of my life grew a blank! I did not recall that they had been at all. I was a child" (98). She feels she is "the wife of a stranger," and she yearns passionately to return to her childhood. "I wish I were a girl again, half savage, and hardy, and free." The fixation on childhood has a seductive romantic appeal, but the passion behind the romance derives much of its psychological force from the traumatic disruptions in the family relations of the two children. Heathcliff is an orphan or an abandoned child. Catherine's mother—like Emily Brontë's own mother—dies when she is a child, and her father is emotionally estranged from her. Both children display a hypertrophic need for personal dominance, and their capacity for affectional bonding channels itself exclusively into their relation with one another. Neither Heathcliff nor Catherine ever becomes a socially and sexually healthy adult. Heathcliff's social relations, including his relation to his own son, are all destructive, and he finally also destroys himself. Catherine is torn apart by the unresolvable conflict between her childhood fixation and her adult marital relation. She remains estranged from her husband until her death, and she dies two hours after giving birth. The violence of feeling through which she destroys herself is a symptom of a psychological stress sufficiently strong to shatter a robust physical constitution.

Psychosexual symbolism is clearly a main organizing principle in the novel, and it has evoked a steady stream of interpretive efforts. Most have been distorted by appeal to Freudian theory as the most readily available form of "depth psychology." Despite this severe handicap, psychoanalytic readings of *Wuthering Heights* have made real progress in getting at the imaginative significance of the story. The juvenile character of the adult romantic relation between Catherine and Heathcliff has caught the attention of both humanist and postmodern critics (Cecil 167; Mendelson 47–55; Spacks 138; Stoneman, "Brontë Myth" 234–35; Van Ghent 158–59, 169). Gilbert and

Gubar observe that "all the Brontë novels betray intense feelings of mother-lessness, orphanhood, destitution" (251). Bersani notes that "the emotional register of the novel is that of hysterical children" (203). Subsuming Freudian developmental theory within an ethological perspective, John Bowlby illuminated the lasting traumatic effects of maternal separation. Chitham describes Emily Brontë's own traumatized response to motherlessness (205, 210, 213–14). Berman reads the novel psycho-biographically in the light of Bowlby's concepts (78–112). Wion, using a psychoanalytic framework, characterizes Heathcliff as a maternal surrogate for Catherine Earnshaw (146). Massé gives a Freudian account of Catherine's narcissism. Moglen (398) and Schapiro (*Literature* 49) both contrast the narcissistic disorders of the first generation with the norm of maturity in the second. Bersani, in contrast, though using a similar Freudian vocabulary, valorizes the disintegrative emotional violence of the older generation (214–15, 221–22). Looking at the novel from a Marxist angle, Terry Eagleton seeks resolution in a utopian social norm alien to Brontë's own perspective, but he nonetheless powerfully register's psycho-social stress in the novel (*Myths*).

In the folk understanding of human nature, the needs for self-preservation and for preserving one's kin have a primal urgency. From a Darwinian perspective, those needs are basic adaptive constraints through which inclusive fitness has shaped the species-typical human motivational system. In *Wuthering Heights*, the movement of the plot toward the resolution of the second generation demonstrates that Brontë herself feels the powerful gravitational force of that system. Her empathic evocation of the feelings of Heathcliff and Catherine nonetheless indicates that her own emotional energies, like theirs, seek a release from the constraints of human life history. Some of the most intense moments of imaginative realization in the novel are those in which violent emotions assert themselves as autonomous and transcendent forms of force—moments like that in which Catherine's ghost cries to be let in at the window and like that in which she haunts Heathcliff and lures him into the other world. For both Catherine and Heathcliff, dying is a form of spiritual triumph. The transmutation of violent passion into supernatural agency enables them to escape from the world of social interaction and sexual reproduction. In the sphere occupied by Hareton and the younger Cathy, males and females successfully negotiate their competing interests, form a dyadic sexual bond, and take their place within the reproductive cycle. In the separate sphere occupied by Heathcliff and Catherine, the difference of sex dissolves into a single individual identity, and that individual identity is absorbed into an animistic natural world (on the supernatural and animistic aspects of Brontë's imagination, see Cecil 162–70; Geerken; Maynard 204–9; Traversi; Van Ghent 164–65).

The fascination Heathcliff and Catherine exercise over readers has multiple sources: a nostalgia for childhood, sympathy with the anguish of childhood griefs, a heightened sensation of the bonding specific to siblings, the attraction of an exclusive passional bond that doubles as a narcissistic fixation on the self, an appetite for violent self-assertion, the lust of domination, the gratification of impulses of vindictive hatred and revenge, the sense of release from conventional social constraints, the pleasure of naturalistic physicality, the animistic excitement of an identification with nature, and the appeal of supernatural fantasies of survival after death. All these elements combined produce sensations of passional force and personal power. In the prospective marriage of the second generation, those sensations subside into the ordinary satisfactions of romantic comedy, but Brontë's own emotional investments are not fully contained in that resolution. The ghosts that walk the heath are manifestations of impulses that have never been fully subdued. In becoming absorbed in the figurations of Catherine and Heathcliff, readers follow Brontë in the seductions of an emotional intensity that derives much of its force from deep disturbances in sexual and social development. They thus follow her also into a restless discontent with the common satisfactions available to ordinary human life.

Wuthering Heights operates at a high level of tension between the motives that organize human life into an adaptively functional system and impulses of revolt against that system. In Brontë's imagination, revolt flames out with the greater intensity and leaves the more vivid impression. Even so, by allowing the norms of romantic comedy to shape her plot, she tacitly acknowledges her own dependence on the structure of human life history. She envisions her characters in the trajectory of their whole lives. The characters are passionate and highly individualized, but life passes quickly, death is frequent, and individuals are rapidly re-absorbed within the reproductive cycle. Catherine and Heathcliff seem to break out of that cycle, but in the end, they are only ghosts—elegiac shadows cast by pain and grief. Investing those shadows with autonomous life enables Brontë to gratify the impulse of revolt while also satisfying a need to sacralize the objects of elegy. That improvised resolution points toward no ultimate metaphysical reconciliation, no ethical norm, no transcendent aesthetic integration, and no utopian ideal. Brontë's figurations resonate with readers because she so powerfully evokes unresolved discords within the adaptively functional system in which we live.

31

Human Nature, Utopia, and Dystopia

ZAMYATIN'S *WE*

BRETT COOKE

In literary dystopias we rebalance ourselves. Ostensibly utopian fictions push us too far, too fast. Recoiling from future shock, we feel drawn to older modes of life. The severe environmental disorientation these fictions create typically impels us to acknowledge our enduring human nature. No other narrative mode so effectively points out behavioral universals, if only by so directly confronting them. This, in a nutshell, is the thematic structure of dystopian novels like Yevgeny Zamyatin's masterpiece, *We* (1920).* As in the other great dystopias, including Aldous Huxley's *Brave New World* (1930) and George Orwell's *1984* (1949), social engineers tinker with traditional modes of sexual reproduction, the rearing of children, and other vital elements of daily life, creating a world not made perfect but rather rendered unfit for human habitation. The mishandling of core human concerns in *We* and other dystopias triggers a powerful aversive reaction on the part of fictional characters and most sensitive readers. This shared response can help us understand how our

---

From *Human Nature in Utopia: Zamyatin's "We"* by Brett Cooke, pp. 3–10, 13–15, 16–20, 22.
Copyright © 2002 by Northwestern University Press.
*In quoting from Ginsburg's translation of *We*, I have made some slight alterations.

common evolutionary heritage underlies both our daily behavior and our aesthetic preferences.

The primary rationale for building a utopian society has always been the promise of reconstructing its inhabitants. This goal follows from the Marxist dictum that environment conditions consciousness—a belief also shared by many non-Marxists. The predominant received opinion in the social sciences still remains that culture largely shapes human nature. A child constitutes at birth a "tabula rasa": its personality is plastic, amenable to any mold to which it is subjected. Such "social construction" holds out the hope that if we can control the culture in which we live, we can reshape ourselves and especially our children, and thus achieve a utopian state. This view necessarily repudiates the idea that our behavior has a genetic core. If natural selection influences not just our physiology but also our actions and attitudes, then human nature is not wholly amenable to social engineering. Utopia is not only a dream but quite possibly a dangerous, inhuman fantasy.

Whether human behavior and consciousness are socially constructed or shaped according to their evolutionary heritage is a question scholars and scientists are likely to debate for years. But the issue was much more than academic in 1918–1920, when Zamyatin wrote *We*. Plans then were afoot in the nascent Soviet Union to put social construction into action so as to mold its citizens into "the New Soviet Man" and establish a social utopia within a few decades. Bolsheviks were optimistic at the outset that many of their countrymen could be easily reshaped with the application of a few new policies. Sadly, the subsequent history of the USSR may be read as an instructive example of what happens when an aggressive social policy is based on false science.

Zamyatin's *We* constitutes a remarkable case of aesthetic precognition. As many readers have noted, the novel, written before the institution of most Soviet policies, anticipates many unfortunate developments in the Soviet Union, like the controlled press with its official optimism, political constraints on the arts, the one-party system, the cult of personality, fixed elections, secret police, and show trials. Some of these were incipient during the years of War Communism, while Zamyatin was writing. The Proletkult poets who pointedly preferred the pronoun "we" to "I" in all likelihood provided an ironic title for a novel in which the first word is "I" (1). But in *We* the former Bolshevik took Communist thought to its logical extreme. Presumably he intended what he later called his "most jesting and serious work" to hold Communist ideas up for ridicule and elicit horror at the same time, the better to set in motion a series of unending revolutions against any and all status quos (*Soviet Heretic* 4). No wonder the novel was the first ever to be banned in the USSR. Yet *We* counts as much more than mere anti-Soviet dissidence, as it was received by its first readers, both in Russia and in the West. It may give us a glimpse of the intended future for Communism, but it is really about us, what we are everywhere simply because we have a shared human nature. Though written as a satire on Bolshevik dreams, Zamyatin's novel has outlived its immediate cultural context because it engages so strongly with evolved features of our psyche.

We depicts an almost wholly controlled walled city of the twenty-ninth or thirtieth century. Its "numbers" (people) live lives so standardized as to resemble cogs in a smoothly running machine. As the chief builder of the first rocket ship, D-503 writes a journal to inform inhabitants of other planets of the glories of the Single State. He boasts that utopia is on the verge of achievement and that the end of social development is in sight. He argues that their way of life is far superior to any other known society, but sensitive readers see through his naïveté and suspect that the reverse is true: the regime constitutes not a utopia but a dystopia. We soon learn that eight hundred years and a massive governmental apparatus will not suffice to alter humankind's essential character. D-503 is seduced by the Mata Hari–like I-330, a leader of the rebellious Mephis, who wish to seize the rocket ship in order to stage a coup d'état. This encounter causes age-old human qualities to surface. Privately, D-503 recovers emotions familiar to us, including love and sexual jealousy; publicly, open revolt breaks out. Although D-503 continues to waver between loyalty to the State and renascent humanity, his development is cut short when the authorities seize him and subject him to a brain operation. As he writes his fortieth and last entry, the battle for utopia still rages, but by now Zamyatin's narrative has aroused in most readers an aversive reaction to the Single State and, hopefully, all such regimes.

The more that the rulers of the Single State try to force its numbers into regimented, antiseptic garb, the more the hairy arms of D-503 remind him of his innate atavism—and the more we readers are reminded of our primeval selves. Such is the message of D-503's brief sojourn beyond the Green Wall that protects the city from the outside world. He has been so conditioned by life in his city of glass that he does not recognize trees. Everything in the natural realm outside the Single State is novel and strange to him, but he has no trouble perceiving that the band of hunter-gatherers he encounters in the outside world are his own kind. Although they are naked and hairy, millennia distant from him in terms of cultural development, and they do not speak his language, he quickly establishes a communion of sorts with them. Obviously they speak the same body language. The people beyond the Wall offer him food and drink, and D-503 joins them in spirit by giving a short, irrational speech at their ritual ceremony. This passage resembles "first contact" encounters between Western anthropologists and, for example, the natives of New Guinea. D-503's momentary atavism serves as a pointed reminder to him and to us that integration of personality requires acknowledging our archaic features.

D-503's sojourn beyond the Wall captures the essential message of evolutionary psychology: our psychological unity with the rest of the human race derives from our common heritage of hundreds of thousands of years as foragers. This long stage in our evolutionary history profoundly shaped our mental architecture and persists in our genome. Our appreciation for the arts in general and for utopian fiction in particular reflects this legacy.

Conventional statements about fictional utopias and about *We* in particular often pay lip service to this notion of human nature. The critical literature on Zamyatin's *We* has long acknowledged that the core struggle of the novel sets planned social engineering against what passes for human nature. As Gorman Beauchamp put it, the basic tenet of utopia is that mankind has no essential nature (167). Were that true, we could readily be socially engineered, and utopia could be a real prospect, as the Communist Party of the Soviet Union envisaged. But Marxism and other ideologies of social perfectibility ignore or dismiss human nature and our manifold inherited needs.

Zamyatin himself suggests this perspective by contrasting the effete citizens of the Single State with the hairy people living beyond the Wall. Like the largely Native American inhabitants of the Reservation in Huxley's *Brave New World* and the Proles in Orwell's *1984*, these people unshaped by recent civilization represent universals of "human nature." But the rough hints found both in the novel and in the critical literature only gain their true significance in the light of evolutionary psychology.

Although evolutionists regard human nature as relatively universal, "cross-cultural," they do not envision it as being fixed. Rather, human psychological propensities, like other biological phenomena, continue to change in line with Darwinian natural selection, albeit slowly when compared to the rate of cultural change. One troubling consequence of Darwinist thought is that "natural" biological adaptation has been greatly outpaced by cultural change. As a result, we still possess many behavioral patterns that resemble those of our hominid ancestors and are adapted to hunter-gatherer conditions but anachronistic in our advanced cultural context. Our conditioning often predisposes us to archaic behaviors often hard to reconcile with present standards of rational social justice, some of which are exaggerated in Zamyatin's Single State. Our "biogram" is innate; it can be neglected, "reformed," or extirpated by a would-be utopian society, but only at a massive cost to our psychological integration. Edward O. Wilson cites Abraham Maslow's concept of an ideal society as being that which "fosters the fullest development of human potentials, of the fullest degree of humanness" (*Sociobiology* 550

[twenty-fifth ann. ed.]). Juan López-Morillas seems to refer to the same notion when he says, "[T]he authors of imaginary societies (utopias and anti-utopias) are obeying an imperative of human wholeness, no matter what particular meaning they attribute to such a concept" (59). To some degree, utopian schemes like the Single State exemplify the profound human drive toward social justice. But they still are not good enough for us, and we need to consider why.

Our behavioral choices are affected by our emotional responses to alternative actions. Love, hate, disgust, pleasure, desire, fear, etc. may be regarded as "enabling mechanisms" in that they help to incline an individual to a particular range of behavioral alternatives. As Charles Lumsden and Edward O. Wilson note, for instance, "sex does not merely serve the purpose of giving pleasure. The exact reverse is the case: the feelings of pleasure in the brain make the performance of sex more likely and allow the packets of genes to be taken apart and put together again" (27–28). We spend relatively little time performing such reproductive strategies as seeking mates. Yet our interest in love never flags. Our obsession with gratifying our genetically encoded psychological propensities manifests itself in the arts, where our emotional responses are hypertrophic and we can indulge our natural dispositions through vicarious experiences of love and other emotions. Most narrative works provide some form of "love interest." This usually involves not so much the depiction of sexual intercourse as the process of selecting a long-term mate. Consider the amount of attention—and space—D-503 gives in *We* to his pursuit by, and later *of*, I-330. Similarly for reproduction and child rearing: family relationships, often in terms of filial struggles, occupy much of our narrative literature. D-503 notes the role sex played for "the ancients [as] the source of innumerable stupid tragedies" (22). In contrast, the literature of the Single State is much more closely concerned with the daily tasks that physically occupy its citizens, especially work. But we would hardly queue to buy such future Single State classics as "the awesome red *Flowers of Court Sentences* . . . the immortal tragedy *He Who Was Late to Work* [or] *Stanzas on Sexual Hygiene*" (68). D-503's description of his blissful love triangle in sharing O-90 as a sexual partner with R-13 may be read as a parody of traditional sexual conflicts. His affair with the *femme fatale*, I-330, however, brings him much closer to "ancient" experiences and to what our human nature inclines us to read.

There are few stories that deal so overtly with our evolutionary destiny as utopian narratives and science fiction like *We*. Nevertheless, ethical aspects of fictional narratives typically exert an indirect but nevertheless extremely powerful influence on our values and thus our behavior by conditioning our emotions and interests. Even in a non-biological context, we appreciate what

is "natural" and recoil from the "misbegotten." In *We*, our genetically derived aesthetic sensibility tells us not only that utopias like the Single State are dehumanizing but also that utopian literature which tries to dispense with human nature is emotionally unsatisfying. We also realize that dissident works like Zamyatin's subversive novel appeal much more to the full range of our evolved sentiments.

"Survival of the fittest" entails not just competition for limited resources of food, reproductive opportunity, and safe space but, often, also intraspecific competition, that is, conflict with our fellow humans. We have been plagued by famines, predation, violence, other forms of human oppression, and, yes, plagues. All of this is reflected in our recurring nightmares. Depictions of an idealized society answer the deep cravings of our human nature. That the same tale of plenitude should be told over and over again reveals our fixation on finding a solution for all of our problems—meeting all our material needs and establishing a modus vivendi with all our conspecifics. It is all the more striking that we reject most utopian dreams.

In *We*, disease, hunger, material want, loneliness, and social disharmony have been banished. Humans have complete security. Technology and reason have reshaped the world into a veritable worker's paradise. Virtually every aspect of life is subjected to purportedly rational central planning. "Numbers" live a meticulously scheduled existence from cradle to grave, waking, washing, working, recreating, and retiring largely in unison. Manifold plenitude is assured to all. All basic material needs are met and guaranteed, including regular sexual encounters. The conditions of the Single State must have seemed alluring in comparison with those in Russia, when Zamyatin read selections of the manuscript during the chaotic and difficult years of War Communism, 1918–1920.

But it is not enough. Zamyatin's narrator, D-503, a seemingly perfectly adjusted citizen of the Single State, guides us on a tour of his city state, extolling social engineering on a truly massive scale, more like the dreams than the actual plans of the early Soviet Union. But because he allows many troubling details and subversive thoughts to slip past his guard, Zamyatin can thoroughly subvert D-503's positive portrayal of the Single State. Paradise is not about to be regained, for cracks in the social order are everywhere apparent—provided we look beyond D-503's naive commentary. And even he begins to sense natural misgivings. He is hardly alone in doubts about the regime; he is contacted and partially recruited by an underground conspiracy, the Mephis. As the novel ends, it is unclear which side will prevail. What only days before had seemed to be an indestructible edifice is on the verge of collapse. The regime's only effective recourse is to surgically excise fantasy from its own citizens, often forcibly. What amounts to a lobotomy has devastating

consequences on their personality and behavior. D-503 describes them as humanoid tractors (189). Like dreams, fantasy seems somehow essential for the maintenance of sanity, if not life itself. Much of our activity is occupied by fictions like narratives, speculative and abstract thinking, games, and dreams (Kernan, Brooks, and Holquist 4–6). Excise the fantasy, that core of human nature, and human social engineering could race ahead. Zamyatin seems to be saying that if utopia is to persist, it will not be populated with people like ourselves.

In the Single State, social organization has been *too* tightly achieved, and aiming at utopia, it has slid into *dys*topia. The problem with utopia, as Zamyatin depicts it, is less a matter of material resources or political controls than of human normalcy. For example, D-503 begins to lose his patriotism and incline toward the revolutionaries when he discerns in himself symptoms of what he senses as a psychological disorder. He begins to have dreams, to think in a disconnected, associative, emotive, *and* rational manner. In the Medical Bureau he is dismayed to learn that he possesses a "soul," an incurable malady. (In the Russian original, *dusha* has the double entendre of "psyche," in effect, what we consciously recognize as our own inner world.) Still worse, he soon starts to perceive the same malady in others. A doctor tells him that there is a "soul" epidemic now raging in the city. D-503 comes to the not extraordinary conclusion that virtually all numbers are so afflicted (206). The point, of course, is that such people are merely what we would recognize as healthy.

Utopian narratives wonderfully illustrate Alexander J. Argyros' statement that art is a society's means of selecting amongst possible futures. Our aesthetic sensibilities enable us to design and visit an idealized social order without having to invest the time and expense of actually building it, let alone making the greater investment of committing the rest of our lives to residing there. Art permits us not only to imagine what it would *look* like, but, via the vicarious experiences fiction generates, to sense what it would *feel* like.

In the modern world it can seem that whichever future we select may actually be attainable. As many commentators on utopian fiction have noted, the whole practice of speculating about possible and desired futures was revived by the Renaissance and the ensuing development of science, technology, and industry. As a result of a new capacity to reshape our future, our prospects require serious attention. Utopian fictions have appeared at an accelerating rate that reflects our growing anxieties over the future we might actually create.

And yet, utopia has failed, at least for the time being. The Communist threat—once so pressing or, for some of us, tantalizing—has faded into history. In the wake of the end of the Cold War and the massive disruption of

Eastern Europe, it is hard to recall that people once found the whole enterprise likely, let alone desirable. Meanwhile, utopian (or, better, *eu*topian) fiction has been almost entirely displaced by *dys-* or *anti*utopian narratives. After the Bolshevik Revolution, with the founding of the Soviet Union and, elsewhere, the development of the modern technological state, the tide turned against a future that now seemed all too possible. Thanks to works like the great dystopian classics, *We, Brave New World*, and *1984*, as well as many, many others, we could visit a future that might have seemed attractive in someone's philosophy but that in literary art *feels* as if it is not conducive to human habitation. These texts have helped us shape our future, as Zamyatin, Huxley, and especially Orwell steeled our resistance to the spread of Communism and other forms of social engineering. Just think of how often the name of "Big Brother" is invoked in American political parlance today in ways that have little to do with its actual usage in *1984*.

Modern dystopias have had such an impact because the futures we project are now *negotiable* in the sense that we can actively do something about them. This was the situation Zamyatin faced when he wrote the novel. Lenin himself admitted in 1918 that he could not envision what form the new utopian state would take; evidently, plans were in flux. Nevertheless, as we noted, Zamyatin somehow managed to anticipate features of the future Soviet Union that only appeared subsequently, such as the "war on the [family] kitchen" by means of state-run cafeterias, and the exclusion of the external world. The promulgation of secret "thought" police, state-sponsored propaganda, and the attempted standardization of individual lifestyles were already under way. *We* was probably the best opportunity Zamyatin had to shape the course of events. Because ideas of utopia influence our fate they can also hold our attention. Indeed, Zamyatin's *We* is the second most studied Russian novel of the twentieth century among Western scholars.

Utopias typically appeal by envisioning societies that, despite their relatively modern technology or social organization, resemble the groups our distant forebears lived in as hunter-gatherers for a period of evolutionary shaping far longer than the few millennia since the development of writing. Karl Marx, one of the primary utopian thinkers, apparently had early tribal groups in mind when he envisioned spontaneous communism in primitive societies. In most positive depictions of utopia, communities are small in size, often amounting to only hundreds or thousands of inhabitants, and set in the countryside. Both factors make us feel at home—although hunter-gatherers lived in even smaller groups, probably of about fifty (Kelly 213, 258; Maryanski and Turner 78). Utopian social organization is relatively spontaneous and tends toward egalitarianism. Marx proposed that once a classless society was achieved, the state would "wither away." Communards were to live in a

state of relative anarchy, policing themselves, and voluntarily sharing their resources with each other. But that picture resembles not so much the Single State as the primitive people Zamyatin depicts living beyond the Green Wall. When D-503 ventures there, he encounters a group of three or four hundred foragers (156). Tellingly, he cannot discern who their leaders are.

Dystopian fictions, in contrast, disorient and unsettle us with massive environmental change. The Single State described in *We* is entirely urban. Constructed of glass, it is cocooned from nature by the Green Wall. Its inhabitants all live dormitory-style in tiny one-person, transparent cubicles, where they are simultaneously denied both privacy and personal identity. Instead of being able to wander over a wide territory, as many of our distant forebears did, citizens find their movements limited to little more than a daily promenade, taken four-step to the March of the Single State. Of course forager lifestyles can vary widely. But although Robert Kelly argues that "there is no one original human society, there is no one original selective environment" (338), his survey of the various extant hunter-gatherer societies associates mobility with equality and personal freedom. Conversely, a sedentary existence, such as we find in the Single State, is associated with social hierarchies and inequality (148).

To be sure, an equality of sorts exists in Zamyatin's futuristic metropolis; except for the Guardians and the almighty Benefactor, the "numbers" seem to have achieved a classless society—but a huge one, ten million inhabitants in a single city. Instead of fluctuating relationships, a greatly relaxed hierarchy, if any, and much conviviality, as our primitive ancestors appear to have enjoyed, Zamyatin's "numbers" face relentless regimentation, their actions prescribed down to the number of mastications they make at meals. In contrast, like many foragers who seem to enjoy what Irenäus Eibl-Eibesfeldt calls a "leisure-intensive" lifestyle, Zamyatin's people beyond the Wall seem to have a lot of time on their hands (cited in Maryanski and Turner 79). D-503 finds them quite at ease, while citizens inside the Single State work virtually all their waking hours: according to the "Table of Hours," loyal numbers only have two slightly less regimented "Personal Hours" at their disposal. And the Single State literally has shut out the world with the Green Wall. The apparent spontaneity, freedom, and peace of hunter-gatherer lifestyles that so attract contemporary anthropologists are nowhere to be found within Zamyatin's dystopia.

Issues of scale and selfhood are also manifest in the common use of nonnatural names in dystopia. Whereas traditional societies were small enough for a single name, plus some reference to a parent, to identify an individual, dystopian societies like the Single State affront our sense of individuality by denoting people in an overly rational fashion, often with numerical tags.

To treat people as if they really were statistics is dehumanizing. Each of us desires, in some respects, to be seen as special, even unique. With characters like D-503, R-13, I-330 and S-4711, Zamyatin reminds us of our own resistance to the application of Social Security numbers and other means of serializing human populations. Of course such a practice makes bureaucratic sense in a large society where everyone cannot possibly know everyone else and it has become probable that some people will have the same names. But who likes bureaucrats?

The key to writing an antiutopian story is clear: determine what human nature is and affront it with behaviors we are likely to regard as unnatural and dehumanizing. Dystopian societies, such as we find in *We, Brave New World*, and *1984*, provide a nearly perfect *reverse* image of traditional forager lifestyles. The object, after all, is to convince us we do not want to live there. It rarely is difficult. Although humans have extraordinary behavioral flexibility, we can live virtually everywhere only because we can also continually re-create the environment for which evolution shaped us. So, for instance, we carry nature into our domiciles, often in the form of flowers. Even artificial plants, painted *nature mortes*, and flowered wallpaper will sometimes suffice. Flowers of any kind are alien to the Single State.

Our developing understanding of human nature helps explain not only why life in utopia is dull, but also why positive utopias make boring fiction. Our emotions serve what at least used to be valuable ends. As Frederick Turner argues, our attraction to beauty generally correlates with biological viability, and this tends to draws us to healthier lifestyles. The forager's love of wandering over a wide territory may be linked to our aversion to boredom, which seems to prod hunters and gatherers into maintaining their grasp on proven resources and seeking new ones. Hence the sense of having *arrived* in utopia and therefore of not being able to venture outside the enclosing Green Wall or to divert from its prescribed routines makes the Single State stifling and tedious. In our imperfect but freer modern urban worlds, by contrast, we can satisfy the forager within us by traveling over space and through texts in search of diversion. The avoidance of tedium, significantly, is absolutely central to art. The fictional mode of utopia has almost entirely yielded its place to the restlessness of dystopian narrative over the past century, simply because the latter engages us so much more. *We*'s swift plot of political intrigue and sexual rivalry grips us all the tighter as utopia gives way to dystopia. This is the literary "hedonic" reward we get for defending our human nature.

Zamyatin's *We* does not merely interact with our shared human nature but also manifests it. At the beginning of the novel, Zamyatin presents us with an almost unbelievably awkward narrator in D-503. "Could D-503 possibly mean what he says?" is a question that surely must occur to every reader. As the protagonist becomes aware of his inner self, he very distinctly becomes more human, life-like, and sympathetic. In effect, the novel works by attracting and holding our attention, partly because it manages to project a plausible personality. We come to feel that we know D-503, that we could recognize him, like other great literary characters, were we to encounter him outside of the book, even in contemporary dress. We start to empathize with him. He turns out to be one of us: he shares our common human nature. As the novel shows, we will always be driven to discover what it is to be human, to be "us."

32

Paternal Confidence in Zora Neale Hurston's "The Gilded Six-Bits"

JUDITH P. SAUNDERS

Highlighting male mating behavior, Zora Neale Hurston's short story "The Gilded Six-Bits" (1933) implicitly calls for an evolutionary analysis. Hurston focuses on a husband's reactions to his wife's infidelity, emphasizing the importance of paternal confidence in long-term commitment. The possibility that an unfaithful wife will give birth to a child sired by her extramarital partner represents an evolutionarily grave risk for her husband: if he provides care and resources for a child to whom he is not biologically related, he helps to perpetuate another man's genes rather than his own (Buss, *Evolution* 10, 67; Dawkins, *Selfish Gene* 148; Trivers, "Parental Investment and Reproductive Success" 76). Existing commentary on the story has not focused on the Darwinian implications of female adultery, however. Readers instead have emphasized tensions the story explores between variously defined oppositions: between "real" and "false" values (Peters), for example, between appearance and reality (German), between country and city (Chinn and Dunn; E. Jones), between material and nonmaterial wealth (Gates and Lenke; Wall), and between Caucasian and African American systems of valuation (Hoeller; Wall). Although the presentation of such themes likely forms part of Hurston's auctorial intention, she clearly signals paternity as the central concern in her narrative. Insights from evolutionary biological theory enable readers to explore that concern, together with its implications.

As David Buss points out, human males are confronted "with a unique paternity problem not faced by other primate males" (*Evolution* 66); concealed ovulation in the human female means there is no overtly recognizable cause-and-effect connection between copulation and pregnancy. Lacking observable proof that his mating efforts, rather than another man's, have caused impregnation, the individual male possesses no definitive evidence that he is the biological father of a partner's offspring. (Since gestation occurs within their own bodies, women, obviously, harbor no equivalent doubts about their genetic relationship to offspring.) To reduce the degree of paternal uncertainty inherent in their situation, and to maximize the evolutionary appropriateness of their parental investment, human males demand female sexual fidelity, since it offers the best possible guarantee that a woman's acknowledged mate has in fact sired her children (Buss, *Evolution* 67; Trivers, "Parental Investment and Reproductive Success" 170). Consequently, evidence that a woman has allowed sexual access to other men tends to dilute or end her partner's mating commitment (Buss, *Evolution* 173; Daly and Wilson, "Evolutionary Psychology" 16). Exploring the psychology of the betrayed husband, Hurston's narrative offers insight into the workings of adaptive mechanisms designed to counteract the threat to male fitness posed by female adultery.

The featured couple is working class. Joe Banks brings home weekly wages from the G and G Fertilizer company, where he works the night shift. Small details sufficiently indicate that his job involves hard manual labor in filthy conditions, for Joe is tired when he returns home and in need of a bath; periodically, he complains of pains in his back. Resources are not plentiful, clearly, and this is an important factor in the development of character and plot. The interaction that Hurston selects to demonstrate the vitality of this young marriage (Joe and Missie May have been married for just over a year) is a ritual in which Joe hands over all his pay to his wife. He first flings silver dollars through the doorway, "for her to pick up and pile beside her plate" (87). She responds with mock reproach, which ends in a joyful, erotically charged scuffle as she searches his pockets for small gifts that he pretends to withhold from her, "things he had hidden there for her to find" (88). "A furious mass of male and female energy," they engage in "tussling" and "tickling," a "friendly battle" that bears witness to the couple's energetic sexual relationship (87, 88).

Introducing readers to Joe and Missie May, this lovingly described encounter illustrates with great clarity the evolutionary psychology of long-term mating, from the point of view of both sexes. The man invests his resources in his mate, indicating the intensity and enthusiasm of his commitment by the manner in which he transfers wealth to her. This is no cut-and-dried, "here's your housekeeping money for the week" exchange. Joe's behavior emphasizes that there are no limits to his willingness to invest in Missie May. Playfully prodigal, he in effect announces to her, "here is all my money; I throw it at your feet; I reserve nothing; I put it entirely at your disposal." By supplying her with resources in the form of gifts, small luxuries such as chewing gum and scented soap, he further underlines his desire to please her. The most important luxury item he purchases for her is a bag of "candy kisses," which metaphorically links the proffered resources with the sexual satisfactions the relationship so obviously provides (88). Joe supplements material generosity with verbal expressions of his devotion. "So long as Ah be yo' husband, Ah don't keer 'bout nothin' else," he assures her, for example (91). His strategy for preserving his marriage combines the "provision of resources" with the expression of "love and kindness," features that Buss identifies as key ingredients in the enterprise of long-term mating (*Evolution* 132).

The plot proper begins with a new outlay of expense intended to please Missie May, when Joe proposes an outing to a "real swell" new ice cream parlor (89). Its proprietor, Otis D. Slemmons, has impressed Joe and other local men with his fine clothes, gold teeth, and prosperous, cosmopolitan air. A combination of conspicuous wealth and success with women makes him the object of general male envy. He wears gold coins (five- and ten-dollar pieces) as personal adornment, on his stickpin and watch chain, at the same time letting it be known that "all de womens is crazy 'bout him" and that this is the source of his wealth: "[W]omens give it all to 'im" (90). Slemmons's self-presentation appeals very obviously to male fantasies, for it blatantly reverses the usual relationship between male wealth and female willingness. Instead of having to invest resources in order to enjoy sexual intimacy with women, Slemmons avers that women supply the resources in compensation for sexual access to *him*. The picture he paints compels his listeners' belief, if only temporarily, because it represents male wish fulfillment. *Midnight Cowboy* offers another variation on this fantasy, as readers may recall, and that film's plot hinges on the unrealistic nature of the protagonist's expectations. Not participating in the collective desire of Slemmons's male audience to give credence to his boasts, Missie May responds with telling skepticism to the tales her husband has accepted so uncritically. "How you know dat, Joe?" she demands, arguing that Slemmons's mere word "don't make it so. . . . He kin lie jes' lak anybody else" (90).

It is difficult—if not, indeed, impossible—to think of any human culture in which fertile young women offer material inducements in order to enjoy erotic encounters with men. (Dowry systems are an exception [not relevant here, of course], and Slemmons seems in any case to be referring to short-term mating opportunities rather than to the long-term, contractual relationships regulated by female dowries [Daly and Wilson, *Sex* 289–90, 322].) Because of their biological role in reproduction, including critical facts about egg size, gestation period, and lifetime reproductive potential, females command uniquely "valuable resources" (Buss, *Evolution* 20). Except in unusual instances, therefore, they do not have to offer men any inducement beyond sexual opportunity itself, which is in and of itself precious (Buss, *Evolution* 20, 86). Joe and his fellows nonetheless are hoodwinked by Slemmons's improbable claims. Overwhelmed by his show of wealth to a degree that prevents them from criticizing its manifest boastfulness, they are disposed to take the impressive stranger at his own value. They are further dazzled by the irresistible appeal of the fantasy he represents, a scenario in which women eagerly offer themselves—and their resources—to a uniquely attractive man. Each one of them would like to be that man and enjoy such advantages—to be freed, in short, from the usual rules of the Darwinian game and find himself holding all the cards.

Hildegard Hoeller argues that Slemmons has been receiving money from white women, that he has in effect sold himself in a humiliating, perhaps parodistic, reenactment of slavery (772, 775). There is no textual support for such a reading, however, since Slemmons provides no racially identifying descriptions of his alleged "womens." A more important obstacle to Hoeller's interpretation is the fraudulence, on two different counts, of Slemmons's claims. First, if he had been successful in selling his sexual services to wealthy women, he would not be so quick to alter his tactics with Missie May—to whom he promises "gold" in return for *her* favors. Second, if many women had been giving him money, his boast that he has "money 'cumulated" would be true (90); his prosperity would be genuine rather than pretended. Once his wealth is revealed as sham, his claims that women (of any race or ethnicity) have given him money are exposed as false.

Joe escorts Missie May to the new ice cream parlor with more than one motive. In addition to wishing to give her pleasure, he desires to show her off to the shop's apparently high-status proprietor. Presenting his attractive wife to Slemmons is an act of competitive male display on Joe's part, proof that he has been able to attract an enviable mate of his own (Hoeller 771): "He talkin' 'bout his pretty women—Ah wants 'im to see *mine*" (90–91). He appears to achieve that goal, for Slemmons expresses admiration for Missie May and, implicitly, for the man who can claim long-term access to her: "Ah have to hand it to you, Joe" (91). Joe's triumph is tinged with irony, however, for he has drawn the interest of a womanizer to his wife's attractions. Slemmons begins an intensive pursuit of Missie May, which he conducts along more ordinary lines than those he has boasted of to the men in town. Indeed, the tactics Slemmons employs in his seduction of Missie May effectively give the lie to his earlier bragging. It is evident that he does not expect her to supplement her personal charms with anything of material value; rather, he frankly offers her his money in return for sexual favors. "He said he wuz gointer give me dat gold money," she later explains, "and he jes' kep on after me—" (94). In this brief affair, readers observe evolved adaptations apparently at work: courted by a man who appears to offer her vastly more resources than her current mate, a woman decides that it is in her interest to accept his attentions. As Buss points out, "immediate extraction of resources is a key adaptive benefit that women secure from affairs" (*Evolution* 87).

Newcomers to evolutionary biological theory may be disconcerted by the idea of women exchanging sexual favors for resources. Initially, certainly, the notion can seem retrograde and antifeminist, presenting an unpleasantly acquisitive picture of women. A number of readers have expressed discomfort with Missie May's apparent "prostitution" of herself, either within her marriage (as witnessed by the silver coins that Joe tosses in her direction every week) or in the context of her affair, undertaken for the sake of "gold money" (94; see, for instance, Chinn and Dunn 3; German 5, 11–12; Hoeller 772–73). Even a cursory consideration of women's situation in the ancestral environment, however, suggests why a woman had to consider a potential mate's resources before engaging in sexual activity. A single sexual encounter might lead to conception, and a woman left to survive pregnancy, lactation, and child rearing on her own—without assistance, provisions, shelter, and the like—would be unlikely to succeed in the reproductive enterprise. Throughout most of human history, the connection between available resources and offspring survival has been incontrovertible. If the politically and economically pernicious effects of evolved sexual strategies are to be effectively counteracted in the contemporary, postindustrial environment, moreover, they must be recognized and acknowledged. As Buss explains, "an evolutionary perspective on sexual strategies provides valuable insights into the origins and maintenance of men's control of resources and men's attempts to control women's sexuality" (*Evolution* 212). Elsewhere, Buss describes "possible points of congruence between feminist and evolutionary perspectives" ("Sexual Conflict" 296), and Anne Campbell has undertaken a careful, point-by-point comparative analysis of feminist and evolutionary theory (12–33). Barbara Smuts explores intellectual and political tensions between the two theoretical systems ("Male Aggression"), as do Douglas T. Kenrick, Melanie R. Trost, and Virgil L. Sheets. Griet Vandermassen's detailed consideration of feminism in the context of evolutionary biology is also illuminating in this context.

Because readers do not witness the progress of the courtship but are presented, like Joe, with a fait accompli when Slemmons and Missie May are discovered in bed, it is difficult to assess her motives step by step. From the outset, she expresses doubts about the newcomer's self-proclaimed prowess with women—"Whyn't he stay up ere where dey so crazy 'bout im?" (90)—and hints at vague plans for procuring some of his wealth: "Us might find some [gold] goin' long de road some time" (91). Such remarks leave readers with the impression that her affair may be motivated by the desire to transfer Slemmons's wealth to her husband. This explanation of Missie May's behavior is supported by her grief-stricken tears when Joe catches her in the act of adultery and by her subsequent remorseful conduct. Insisting that she loves Joe "so hard," she appears genuinely devastated by the loss of his trust and affection (94). There is evidence, in sum, that she has not been infatuated by Slemmons's apparent wealth and status, that she feels no love for him, and that she never aspired to become his long-term partner. Instead, she may be exchanging her sexual favors quite deliberately for "dat gold money," intending to present it to her husband; at one point, she avers that the gold coins would "look a whole heap better" on Joe (91). Thus she perhaps justifies her adulterous behavior, in her own mind, as an act intended to enrich her marriage rather than imperil it—"a sin committed out of her love for Joe" (Peters 93). Robert Trivers's analysis of self-deception is useful here, as readers attempt to sort out the tangled layers of Missie May's conscious and unconscious intentions ("Self-Deception" 271–86).

There is no getting around the fact that if she grants secret sexual access to a short-term partner in order to bring new resources to her marriage, Missie May risks lowering her husband's lifetime reproductive success. If Slemmons fathers her first child, the total number of children Joe might conceive with her will have been reduced by one. No matter how generously her motives are interpreted, the consequences of Missie May's infidelity—in terms of fitness—are not nearly so damaging for her as for her husband. The logic of her intentions, very likely not consciously articulated, seems clear when examined from a Darwinian perspective. If she becomes pregnant by Slemmons, she risks giving birth to a lower-quality child than she could have conceived with Joe, perhaps, but the child will still be hers. Thus self-interest might suggest to her that the risk of impregnation by Slemmons is outweighed by the advantage of access to his money. If her affair is not discovered, Joe is likely to acknowledge and support the child without suspecting that he might not be its father—a course of action obviously "contrary to [his] own interest" but highly "adaptive" from Missie May's point of view (Trivers, "Parental Investment and Reproductive Success" 76). Meanwhile, that child, along with any others later sired by Joe, will enjoy the benefits of the unexpected resources that Missie May has extracted from a short-term affair. If Slemmons's wealth had been real, rather than feigned, her decision to seize this opportunity to ensure a more financially secure future for her offspring might have made a positive difference in her fitness and that of her progeny.

As long as an affair remains undetected, and as long as her short-term partner gives a woman otherwise unavailable resources, she stands to reap benefits in fitness. The poorer the woman, the truer this is: a larger percentage of her children are likely to survive and thrive if more food, better shelter, enhanced medical care, or improved vocational opportunities flow to her family through a wealthy extramarital partner. Missie May and Joe belong to the class of the working poor, after all; a woman compelled to dry herself after bathing with a "meal sack" might with reason prove susceptible to the blandishments of a prosperous suitor (87). Disclosure of the affair threatens the stability of the marriage (and this is the principal risk Missie May assumes in accepting Slemmons's attentions) precisely because a wife's adultery stands to decrease her husband's fitness. Missie May's cost-benefit analysis, as she weighs the pros and cons of an affair with a supposedly rich suitor, scarcely will coincide with Joe's. Even a small risk that his wife might conceive a child with another man would be unacceptable from a husband's point of view (Buss, *Evolution* 266). He is unlikely to regard resource extraction from her partner in adultery as adequate compensation for the loss he will suffer in the number of copies of genes passed on to the next generation.

In terms of plot development, Missie May's strong desire to preserve her marriage is crucial. Once detected in an adulterous liaison, she assumes a posture of dignified remorse, cooking and cleaning with energetic dedication while hoping that her husband will accept her contrition and renew his trust in her. After the discovery scene, in which Joe strikes Slemmons and wrests the gold piece from his watch chain, the omniscient narrator shifts the focus of attention to Missie May. In consequence, readers observe the operation of male jealousy, defined by Buss as the psychological mechanism "our ancestors evolved . . . for solving the paternity problem" (*Evolution* 126), through the eyes of the woman at whom it is directed. Joe's initial reaction to the sight of another man in bed with his wife is a stunned incredulity ("the great belt in the wheel of Time slipped"), followed closely by "fury" (93): "he had both chance and time to kill . . . but he was too weak to take action" (93). He lands a couple of punches and sends Slemmons on his way, caught in a kind of emotional paralysis, "feeling so much and not knowing what to do with all his feelings" (94). The powerful emotions that Joe experiences in this moment illustrate the essential elements of male sexual jealousy, including rage and aggression (Buss, *Evolutionary Psychology* 294–95).

In the weeks and months following this confrontation with Slemmons, Joe says nothing to his wife about his feelings or plans. Consistently "polite," but "aloof," he refrains from questioning her about the affair, just as he reveals nothing about his future intentions (95). He maintains the "outside show" of their marriage, going through the motions of ordinary chores and activities with two significant exceptions (96): he ceases their sexual relations, and he stops handing over resources. "There were no more Saturday romps. No ringing silver dollars to stack beside her plate" (95). With the passage of time, he finds himself unable to maintain his sexual reserve: a back rub after "three months" of abstinence leads to more, and "youth triumphed" (95). Missie May anticipates that sexual contact will facilitate reconciliation but finds herself mistaken: Joe maintains emotional distance. He makes his position painfully clear by leaving the gold coin he has yanked off his rival's watch chain underneath her pillow, as if in payment for the sexual encounter.

Only at this point does Missie May learn what Joe has known ever since he acquired the trophy: it is not a real ten-dollar gold coin, but merely a gilded fifty-cent piece. Given the story's composition in 1933, some readers are inclined to interpret the gilded coin in the context of the Great Depression (Chinn and Dunn) and the gold-standard debate (Hoeller). "Making money and the desire for money became national concerns," Chinn and Dunn argue (3). "Popular songs, fiction, and movies throughout the 1930's celebrated the lives of the rich and the famous," they point out, suggesting that Hurston used "gilded money and Otis T. Slemmons to explore the misguided belief that material goods would bring happiness" (3). Hoeller explores the possibility that Hurston forged the emphatic silver–gold contrast in the story in order to voice her critique of the "gold standard and its insistence on the supremacy and universality of white values (and white civilization)" (780). These readers make a well-wrought case that Hurston is taking on broad-based sociopolitical and economic issues in her story. The presence of such concerns does not, of course, alter the biosocial implications of the gilded coins as a male resource. Any political or economic statement is secondary to the biologically fundamental issues that so clearly dominate the story's content and shape its plot.

Since Missie May's affair with Slemmons was motivated, on her own admission, by the resources he promised her—"he wuz gointer give me dat gold money" (94)—it is humiliating for her to learn that those resources were all sham. There never was any benefit to be gained from an involvement with Slemmons, and Missie May has jeopardized her marriage to Joe for nothing. Slemmons's success in impressing her with his supposedly magnificent resources illustrates the "evolutionary arms race between deception perpetrated by one sex and detection accomplished by the other" (Buss, *Evolution*

155). In this instance, a woman has been insufficiently alert to deceptive male tactics. Readers infer that Joe takes some satisfaction in passing on this ironic information to Missie May; he assuages his anger with his wife by demonstrating how she has been duped. Presenting her with the counterfeit gold coin also enables him to express his resentment of her disloyalty through insult, since by offering payment for sex he implies that he regards her as a promiscuous woman whose favors are for sale: "He had come home to buy from her as if she were any woman in the long house. Fifty cents for her love. As if to say that he could pay as well as Slemmons" (96). Joe's unflattering message is that Missie May has ceased to belong to the Madonna-like category of women, characterized by premarital chastity and postmarital sexual fidelity, to whom men make long-term commitments. Her adulterous behavior positions her, instead, in the ranks of promiscuous women with whom men seek only short-term liaisons (Smuts, "Male Aggression" 252; Trivers, "Parental Investment and Reproductive Success" 74).

Missie May returns the counterfeit gold coin as wordlessly as it was given, by placing it in Joe's clothing. Her message to him is as clear as his to her: she indicates that she is providing him with sexual intimacy out of marital love rather than for pay. The coin looms large in her imagination, an object inspiring fear and loathing; it is "a monster hiding in the cave of his pockets to destroy her" (95). She interprets Joe's use of it to torment her as "her punishment" (96). An important effect of Joe's punitive action, clearly, is to convince his wife that he will not tolerate sexual transgressions. If he were to offer her quick or easy forgiveness, he would risk licensing future extramarital escapades on her part. In any cooperative alliance, as Richard Dawkins points out in his analysis of tit-for-tat strategies, it is necessary that individuals be "punished for defection" or cheating will become rampant (*Selfish Gene* 227). Another effect of Joe's behavior is to test the degree of Missie May's commitment to him. Without resorting to either rudeness or violence, he nevertheless succeeds in making his wife extremely uncomfortable. She has no idea when, if ever, his withholding behavior and silent rebukes will end. By putting up with an extended period of coldness and by suffering the insults represented by the coin, she acknowledges fault, communicates remorse, and affirms loyalty. The longer Joe tests her, the more convincingly she proves that the marriage is valuable to her and that she is willing endure discomfort to win back his trust. Both partners in the marriage are engaged in a waiting game. Joe is waiting to see whether Missie May will offer adequate proof of ongoing commitment, while she in turn is waiting to see when and if his resentment of her fault will be healed. Psychologically, this waiting makes sense on both sides: the rift caused by Joe's mistrust can be repaired only gradually, as Missie May's "displays of fidelity" over time provide persuasive evidence of her renewed commitment to sexual exclusiveness (Buss, *Evolution* 114).

The plot takes another turn at this point, as Hurston introduces an evolutionarily critical complication: Joe observes that his wife is showing signs of pregnancy. Before the incident with Slemmons, he had been wishing for exactly this state of affairs: "He thought about children. . . . A little boy child would be about right" (92). Now, of course, his wife's pregnancy is a source of great ambivalence for him: Whose child is she carrying? Neither Joe nor the reader knows for certain whether Missie May's affair with Slemmons involved more than the single sexual encounter that Joe interrupted. There is no evidence, certainly, that they were together often or long. It is possible, of course, that conception occurred on the one night readers know she spent with Slemmons. His wife's pregnancy therefore poses a fitness-related dilemma for Joe. If the baby is his, he longs to nurture it and its mother; if the baby is Slemmons's, he has no such wishes. Joe takes over the heavy chores— "you ain't got no business choppin' wood, and you know it," he avers (96)—a precautionary move to safeguard the health of a fetus that may well be his. At the same time, however, he lets his wife know that he remains wary of investing in this pregnancy. Questioning her assertion that the baby will be sure to resemble him ("You reckon?"), he fingers the gilded coin he still keeps in his pocket. This serves as a deliberate reminder of her infidelity, making his reservations, and their cause, unmistakably plain.

From readers' point of view, it should be noted, there is a frustrating element of imprecision in the time line of Missie May's pregnancy. Joe notices the pregnancy at about the three-and-a-half-month mark, since she gives birth "almost six months later" (96). A vague reference to the passage of time between the resumption of their conjugal relations and his observation of her condition introduces a slight question about whether more than nine months elapse between her affair and the birth of the child: "the sun swept around the horizon, trailing its robes of weeks and days" (95–96). Joe's openly expressed doubts about the child's paternity offer evidence that readers are expected to interpret those "trailing . . . robes" of time as a relatively brief period—that is, less than a month. Joe can do the arithmetic for himself, obviously, and his continued worries indicate that numerical calculations alone will not suffice to eliminate Slemmons from the running as father. Quite apart from any nine-month countdown, moreover, the suspicions awakened by his wife's infidelity work to create a generalized distrust on Joe's part: a wife guilty on one occasion of sexual disloyalty may prove so again. For the best of reasons, Joe's anxieties about paternity loom large.

Norman German reads the "trailing . . . robes of weeks and days" as a fairly extensive period of time, and he concludes that Joe is sure that

Slemmons *cannot* be the baby's father (10). Such a reading is undermined by Joe's openly articulated concerns about the child's paternity; it introduces a further temporal complication, moreover. If the vaguely denoted "weeks and days" represent any amount of time between four and ten weeks, then Joe absolutely cannot be the father, since conception would have occurred during the three-month period of conjugal abstinence. It seems likely, on balance, that the confusion generated by the narrator's reference to "weeks and days" is accidental and that the more definitely noted time periods (for example, three months, six months) are those to which readers are expected to attend.

Hurston assigns to Joe's mother the central role in relieving his doubts. Because a man's relatives also stand to lose if he invests in offspring not his own, it is adaptive for them to maintain a watchful, even suspicious, attitude in response to the problem of paternal uncertainty. Because they share genes with Joe, his relatives will suffer decreased inclusive fitness if he spends years supporting a child conceived by Slemmons and sires fewer children as a result of having been cuckolded. Any assistance to that child given by grandparents, aunts, or uncles on Joe's side of the family (and such assistance from members of an extended family is, of course, common in human societies) similarly would be misplaced, if Joe accepts as his own another man's child (Buss, *Evolutionary Psychology* 236, 249; Dawkins, *Selfish Gene* 186). One commonplace response to the adaptive problem faced by paternal relatives is the tendency of maternal relatives to suggest that an infant resembles its father, or some member of the father's family, presumably with the unconscious hope of allaying fears that might short-circuit paternal investment (McLain et al. 21–22).

Just as it is in the interest of relatives on the maternal side to insist that there can be no possible question about the identity of the father, it is in the interest of relatives on the paternal side to remain vigilant to the possibility of cuckoldry. Because Joe's mother belongs to the naturally suspicious set of paternal relatives, her announcement to him that the newborn baby is "de spittin' image of yuh, son," carries weight (97). She underlines her conviction with notable insistence, telling him that "if you never git another one, dat un is yourn" (97). Similar statements from Missie May's mother would lack the persuasive force of this testimony from Joe's. There is no reason for Joe's mother to imagine a resemblance that is not there; indeed, she has a genetic stake in remaining objective in her assessment of the baby's appearance. Hurston also has provided evidence in her story that Missie May's mother-in-law was displeased from the start by her son's choice of a wife, judging her to be potentially promiscuous—for example, to "fan her foot around" and "get misput on her road" (97). Very evidently, her statements about the child are not motivated by any personal affection for her daughter-in-law or by more generalized loyalty to her sex. Rather, the doubts she harbors about Missie May's character make her particularly apt to question the paternity of a putative grandchild. Her bias against her daughter-in-law renders her assurances to Joe all the more convincing.

Gayl Jones's comment that "the story is perhaps resolved too simply at this point, the '*baby chile*' being a kind of *deus ex machina*" (41), misses the point. The birth of a baby resembling Joe resolves the plot for good reason: it settles, as, indeed, nothing else can, the paternal uncertainty that constitutes the major source of conflict in the story. Joe is in conflict with Slemmons, who has diverted his wife's sexual attentions and reproductive potential away from her marriage. Joe also is in conflict with Missie May, who has been sexually disloyal and who now may be trying to trick him into supporting another man's child. Finally, Joe is in conflict with himself: desperately ambivalent, he yearns to forgive but seethes with mistrust. The child is not a narrative contrivance but an essential means of relieving Joe's doubts and enabling him to continue his long-term commitment to Missie May. If there is a hint of artifice in the story's conclusion, it is Joe's mother, rather than the baby, who plays the role of deus ex machina. Her assurances of Joe's paternity make it possible for him to accept his wife's child as his own and renew his marital commitment. His mother's unsought testimony, offered at just the right moment to allay Joe's painful suspicions, may strike some readers as a little too convenient.

Confirmation of paternity thus serves as the resolution of the story's plot. In the starkest biological terms, what is at stake, more than the happiness of individuals or the sanctity of marriage, is the passing on of genes. Readers can only speculate about what might have happened to Joe and Missie May's relationship if his mother had not volunteered such a strong conviction that Joe is the father of his wife's baby. In the absence of certainty on that point, his commitment to the marriage might be expected to waver. The happy ending that Hurston depicts is possible only because the husband is confident that he will be investing in a child carrying his genes. Hoeller argues that Joe has no "essential proof [of paternity] that goes beyond the surface appearance" (777), dismissing the testimony of his mother. She goes on to voice admiration for Joe's willingness to accept the baby "as his own"; he "makes the baby his own currency" (777). While it is true that his mother's pronouncement does not constitute absolute proof of Joe's paternity, he accepts it as such. The movement of the plot strongly supports her assurances, moreover, reaching its climax in the moment when she avers that the baby is "de spittin' image" of Joe. Only then does he effect a reconciliation with Missie May. Hurston provides ample evidence of Joe's conviction that he is the biological father—behavioral indicators as well as verbal testimony. His certainty on this point is necessarily subjective, but it is unmistakable. Story line and plot development lose their meaning if readers interpret Joe's parental pride at the story's conclusion as altruism rather than as fitness-enhancing behavior.

The importance of this issue is clear: readers must realize, for instance, that Missie May is guilty of adultery no matter who fathered her child. The fact that Joe, rather than Slemmons, impregnated her is irrelevant, ethically speaking. Joe can forgive her, evidently, if her sexual disloyalty will have no negative impact on his genetic legacy. Readers must suspect that if the baby had looked like a tiny Otis, Joe's forgiveness would not have been forthcoming. If Joe is biding his time, as noted earlier, for adequate proof of Missie May's remorse and commitment, he is also awaiting the outcome of the pregnancy. If Missie May were to miscarry, or if the child were to be stillborn, for example, then its paternity would be irrelevant. In such a case, if he were sufficiently convinced of his wife's future loyalty, Joe might continue in his marriage without worrying about Slemmons's threat to his lineage. In the case of a live birth, there is hope, but no guarantee, that the child's appearance will settle the issue of paternity. In an era before genetic testing was available, the most convincing proofs of kinship were provided by signs of physical resemblance.

Once his mother has come forward to settle the question of paternity, Joe resumes investing resources in his wife. In addition to bringing home a large supply of groceries ("all the staples"), he once again flings his pay, fifteen silver dollars, through the doorway (97). He cashes in Slemmons's gilded coin, furthermore, spending the fifty cents on "candy kisses" for Missie May (98). This act signals his wish for a full reconciliation. Disposing of the evidence of Missie May's adultery, he indicates that he will no longer use the coin as an instrument of reproach. Purchasing "gifts of eroticized food" with the same coin, he expresses his desire to renew their sexual intimacy, in all its joyful vitality (Hoeller 774). When the store clerk suggests that fifty cents' worth of molasses kisses is an absurdly large quantity, advising Joe to "take some chocolate bars too," Joe refuses; only an extravagant gesture will do (98). All of Slemmons's money must be spent on "kisses" that illustrate the connection between female reproductive energies and the expenditure of male resources. Like Ado Annie's suitor in *Oklahoma!*, Joe indicates that he is an all-or-nothing man: the prodigality of his outlay communicates the message that he will put *all* available resources at his wife's disposal, with the expectation that she, for her part, will channel *all* her reproductive energy toward him.

Joe's conversation with the store clerk from whom he purchases the candy and other provisions also shows him making an effort to deflect the social humiliation associated with cuckoldry. He describes Slemmons as "a stray nigger" who offended Joe with his boastful manners and pretensions to wealth (97). He even adds that Slemmons was "tryin' to tole off folkses wives from home," a detail he might have suppressed unless he fears that word of Slemmons's womanizing has spread (98). The clerk responds by asking, "[D]id he fool you, too?" raising the possibility that Joe was deceived by Slemmons's mate-poaching activities as well as by his phony gold (98). Joe offers an emphatic denial, claiming that he "knocked 'im down" and seized the gilded coin simply because he was irritated by the stranger's braggadocio, or "smart talk" (98). He makes a point, as well, of affirming paternity of the new baby: "Ah got a lil boy chile home now" (98).

Readers perceive that Joe is misrepresenting his interactions with Slemmons calculatingly, in order to avoid the reputational damage typically suffered by a man who fails to keep his mate's sexual loyalty (Buss, *Evolution* 126). Insisting that he distrusted Slemmons from the outset, Joe hopes to squelch any gossip linking the stranger with Missie May. At the same time, he attempts to present himself as a forceful and aggressive male, well able to fend off challenges from other men. Such masculine "displays of bravado . . . are directed toward other men in an attempt to elevate status and prestige" (Buss, *Evolution* 10). As part of this effort, Joe even adopts some of Slemmons's posturing, telling the store clerk that he has been away in "spots and places" (97). Echoing Slemmons's phrasing to suggest that he is well traveled and sophisticated, Joe imitates the strategy the newcomer used so successfully to impress new acquaintances. The knowledge that Slemmons achieved status and respect (however temporary) by means of fraudulent claims does not deter Joe from making use of similarly deceptive tactics in the hope of bolstering his own image in the community (Chinn and Dunn 8; German 11).

The impact of race on the characters' situation emerges most clearly in the white store clerk's comment after Joe's departure: "these darkies . . . laughin' all the time. Nothin' worries 'em" (98). Interpreting Joe's laughing deprecation of Slemmons as the sign of a carefree nature, the clerk reads Joe as "a type," revealing his conviction that individual psychology is racially determined (J. Lowe 191). To say that Joe never "worries" is to deny his full humanity and, by extension, that of all African Americans. Readers are acquainted firsthand with the suffering that Joe has experienced in previous months, however, and they also understand the attitude of easy confidence that he projects while mocking Slemmons as a function of intrasexual competition (here, the attempt to assert dominance over a rival). Hurston has arranged for readers to realize how utterly mistaken the white man's racially based assumptions are and thus to reject their underlying bigotry. The irony evoked by the clerk's remark is directed, very obviously, toward him. Adaptationist analysis of the story supports Hurston's point fully, as the characters' behavior is shown to be consistent with "universal psychological mechanisms" (Buss, *Evolution* 185).

As Lillie P. Howard states, one of Hurston's recurring messages is "that people, regardless of their color or their peculiar burdens, must inevitably struggle with some of the same life problems" (256). In confronting problematic and conflict-ridden situations, fraught with "infidelity, jealousy, violence, and hatred," her characters express their humanness in all its complexity (257). Further evidence that Hurston regards her characters' feelings and behavior as normative can be found in her chosen narrative strategies. The omniscient narrator keeps readers at some distance from the characters' inner reflections at several critical points in the story. Readers obtain no direct access to Missie May's motives in yielding to Slemmons's seduction, for instance, or to Joe's plans in the months following his wife's adultery. As Gayl Jones points out, "Hurston handles all the emotional reversals and complications in narrative summary rather than in active dramatic scenes" (41). Assuming that readers will understand the two protagonists' motives, reactions, and calculations on the basis of general human experience, she implies that her story is an old one, with universal application. In this way, her narrative method appears to reflect her background as folklorist and ethnographer. She utilizes dialect, rituals, and folkways to locate her characters in an identifiable environment, simultaneously pointing toward psychological mechanisms transcending the local (Chinn and Dunn 4; Hoeller 778).

Certainly men's apprehensions about misplacing parental effort is universal, as countless examples from popular culture and literature bear witness. The final song in Shakespeare's *Love's Labor's Lost*, for instance, declares that the spring call of the cuckoo "mocks married men" (5.2.909). Not only does the bird's call mimic the word "cuckold," warning men of their wives' possible infidelity, but the brood parasitism for which this species is notorious exemplifies reproductive deceit and exploitation at its most extreme. The cuckoo's victims take care of another bird's offspring because they are unable to distinguish the cuckoo's eggs or chicks from their own. In contrast to the bird world, in which both parents are equally prone to deception, the prospect of being so tricked threatens only males in the human realm:

> The cuckoo then, on every tree
> Mocks married men; for thus sings he—
> Cuckoo,
> Cuckoo, cuckoo! O, word of fear
> Unpleasing to a married ear!

(5.2.908–12)

It is highly fitting that such words conclude a play titled *Love's Labor's Lost*. Shakespeare's lyrics remind us that the emotion of "love" is a proximal mechanism driving humans toward reproductive efforts that constitute the central "labor" of most individual lives. To lose the genetic payoff from that labor by lavishing energy on genetic impostors represents an irretrievable loss, indeed.

Like Hurston's story, Shakespeare's song assumes familiarity with the problem that female infidelity poses for men: cuckoldry is presented as a widely understood human concern, not restricted to any one historical moment or social context. Hurston states in her autobiography that her attention as a writer was drawn to commonalities in human nature that underlie surface distinctions:

> My interest lies in what makes a man or a woman do such-and-so, regardless of his color. It seemed to me that the human beings I met reacted pretty much the same to the same stimuli. Different ideas, yes. Circumstances and conditions having power to influence, yes. Inherent difference, no. (*Dust Tracks* 214)

Clearly, her perspective is very like that of an evolutionary psychologist. The store clerk's racist assumptions, like Joe's and Missie May's ritualized games and richly metaphoric verbal exchanges, form part of the particularized cultural context in which the protagonists' evolutionary heritage expresses itself (G. Jones 44–45; Wall 14).

With the restoration of resource provisioning and sexual vitality to Joe and Missie May's relationship, their story comes full circle. The breach in their marriage is healed, and its future prospects are strengthened by a joint parental commitment to a child in whom both partners claim genetic interest. Hurston's plot illustrates with striking clarity the centrality of paternal confidence in male mating decisions: indeed, it may be read as a case study of this particular adaptive problem. It offers an illuminating portrait of male jealousy, identifying fear of misplaced parental investment as a principal source of this powerful emotion. Readers observe that Joe's ability to forgive an adulterous act is tied inextricably to the genetic consequences of the deed. Above all, the adaptationist perspective cuts through any tendency to wrest sentimental or didactic meaning from the story. Often it is read as a record of maturation and forgiveness: marital discord is overcome, false values are rejected, and reconciliation is achieved (see, for example, Baum; Gates and Lenke; Hemenway; L. Howard; E. Jones; G. Jones, J. Lowe; Peters). Neither ethical principles nor romantic ideals dictate Joe's decision to remain in his marriage, however; his chief concern is the safeguarding of his own fitness. If confronted with another instance of infidelity on his wife's part, he can be expected to demonstrate at least as much coldness and anger as he did the first time, very probably more. He has not become more altruistic or more forgiving of human frailty; rather, he has reaped the adaptive benefits of jealousy.

To locate the story's meaning in vague ideas about the power of love or the ethics of reconciliation does great disservice to the tough-mindedness of its statement. "The Gilded Six-Bits" is not a vapid tale of error and forgiveness but an unsparing delineation of Darwinian realities: men practice deceit to gain social status and access to women; happily married women can be tempted to sexual disloyalty if sufficiently impressive resources are on offer; a man can forgive his wife's infidelity if—and only if—he is sure she has not foisted alien genes on him. The author who crafted a fictional situation to test and illustrate these realities is not surprised by the outcomes she depicts, nor does she encourage readers to condemn the portrait of human psychology that emerges. Zora Neale Hurston's narrative quietly accepts genetic self-interest as an inevitable component of our common human nature, a sine qua non that, with luck and a modicum of goodwill, need not be incompatible with tender and lasting relationships. "Tremendous benefits flow to couples who remain committed," after all (Buss, *Evolution* 123). Missie May and Joe have much to gain if they continue their marriage, as long as they can do so without jeopardizing the reproductive success of either partner.

33

Character in *Citizen Kane*

Joseph Anderson

To recognize another person as a distinct individual, to be able to read his emotions, to intuit his intentions, has always been crucial to our survival. In the distant past, our lives depended upon our ability to accurately recognize and judge the moral character of other individuals of our own species. It is no less the case today; we must quickly judge whether we can trust a new acquaintance; we must know whether he will befriend or exploit us. Likewise, we must be able to quickly recognize old acquaintances, and we must continually update our assessments of their characters. The problems of character recognition and attribution are universal. The capacities to cope with these problems were developed through evolution, and the manifestations of those capacities are, as we might expect, similar from culture to culture.

It should not surprise us that a major artifact from our own culture deals with the issue of character. The artifact, a motion picture from 1941, Orson Welles' *Citizen Kane*, is generally considered to be one of the most important

Originally published in the work: *The Reality of Illusion: An Ecological Approach to Cognitive Film Theory* by Joseph Anderson. © 1996 by Joseph Anderson; reprinted by permission of the publisher.

motion pictures, perhaps *the* most important, ever made. It may be more than coincidence that we attribute such importance to a movie that deals with a human problem so central to our survival, that of character attribution. In both form and content *Citizen Kane* is an exploration of character, of a man's character, an exploration of those attributes that make him different from all other human beings, an exploration of those attributes that make him better or worse than other human beings.

Security fences dissolve into wrought-iron gates that enclose a fantasy land of monkeys in a cage, gondolas in a misty lagoon, and reconstructions of classical ruins. A castle dominates this bizarre landscape, and high in its upthrust towers a lighted window goes dark. From inside that same window the first light of dawn appears, the brightness from outside the window revealing a figure stretched out upon a bed and covered up to his waist with a sheet and blanket. In his hand he holds an object that reflects the morning light. Snowflakes fill the screen, and the falling flakes slowly dissolve to a close-up view of a miniature cabin with snow piled high upon its roof. The shot rapidly widens to reveal the glass ball that contains the miniature cabin and the hand that holds the ball. The snowflakes continue falling as the picture cuts to an extreme close-up of a mustached mouth that utters one word: "Rosebud."

Cut to the hand that loses its grip upon the snow covered cabin and its miniature world inside the glass. The ball falls from the hand and rolls down a couple of steps. A cut to the reverse angle, and we see the ball fall from the last step and shatter upon the floor, splashing its contents audibly onto the lens of the camera and into our faces.

Distorted as though through the curved broken glass we see the door to the room open and a nurse enter. Cut to a view near the cabin toppled upon its side looking through the debris of the shattered miniature world, and the nurse crosses the room to the bed. Cut to the nurse folding the hand that held the ball along with the other hand in the crossed-chest self-embrace of death. She pulls the sheet slowly over his hands and face. We return to the long shot of the bed before the window, but now the crystalline globe is gone, and the body is completely covered with the sheet.

Who is this man who lived in such a strange and fanciful place, whose death we witnessed, whose last utterance is burned into our memories? We want to know. And, of course, the remainder of the movie is constructed around the answering of this question. Who was Charles Foster Kane? What kind of a man was he? We have witnessed his departure from the world. This is his judgment day, and we the audience are his judges.

A movie such as *Citizen Kane* is a very complex construction, an intricate, precisely sequenced program designed to interface with human visual and auditory systems, and having gained access in this way, to "run" in the mind of a viewer. And it does "run." We do see and hear, experience emotions, understand causes and consequences, and remember what we have known. We are one with Thompson, who concludes, "You know, all the same, I feel kind of sorry for Mr. Kane," and with Susan, who replies, "Don't you think I do?" But we engage with the movie, even at this level, automatically, without effort on our part, and we would have not the slightest notion of how it all comes about if it were not for researchers laboring for the smallest of gains, reveling in the tiniest insight into how the mind actually works.

Identification

In an ecological model of cognition, one based approximately on Gibson's theory of visual perception, meaning (affordances) cannot simply be perceived abstractly in the events of the world or those of the film we are watching. We, as perceivers, are part of the ecological system, and in the world it is the meaning of events in relation to ourselves that we perceive. To put it another way, I perceive not what something *means* but what it means to me. We are programmed through evolution to perceive meaning in that way, as part of our environment. We cannot therefore just perceive meaning per se in the events of a motion picture. We must perceive meanings in relation to someone, to a character in the movie who inhabits the fictional world of the movie, who is subject to its constraints and affordances. (The methods of science allow us to step outside the ecological envelope to obtain a god's-eye view of the universe. That's why science is extraordinary, special, outside normal everyday perception.)

We usually think of the protagonist of a narrative, whether in a book, on the stage, or in a videotape or a movie, as the person the story is about, the main character of the story. It is an intriguing possibility to consider that privileged character as the one we choose to use as our reference for the interpretation of meaning. We see the events of the fictional world through this person's eyes, as it were. Not that we see only what that character sees or are restricted to what that character knows and feels, but that it is *his* fate, *his* survival or well-being or comfort that we care about, just as it is our own survival or well-being that is our ultimate concern as we make our way in our world.

I have proposed an ecological approach to character identification. But how would such an approach apply to our viewing of *Citizen Kane*? The movie is constructed so that we the viewers must sit in judgment of Kane's character. But a profound indecision engulfs us, and a potentially hazardous hesitation afflicts us. Our own character, our own souls are in grave danger, for as members of the audience we must not only render an intellectual judgment as to the moral fitness of the subject on trial, we must at a most primitive emotional level, decide to identify or not identify with the character Charles Foster Kane. We might choose instead to identify with the reporter, Thompson, who has been assigned to solve the mystery of Rosebud, and in the process the mystery of Kane's character. But he is purposely ill-defined, and we seldom see his face. He remains a shadowy player in a story that is clearly about Kane. Or perhaps we could identify with one of the other narrators; there are five of them, but not one offers a continuing basis for our identification. There is only Kane himself, and to identify with him we run a moral risk. We perhaps identify at first, but then pull back. We totter upon the threshold of our own moral irresolution as we witness Kane, in flashback, living his.

A simple notion of identification with a protagonist or single character does not appear to be adequate. Perhaps the generalized concept of identification needs to be broken down into at least three specific components: *perspective-taking, caring,* and what I will call *role identification.*

To evaluate the affordances in a narrative context (that is, in a diegetic world), one must perceive them in relationship to a character in that world; one must, in other words, perceive them from that character's perspective. The protagonist usually has a problem to solve or a goal to achieve. Whether we are able to share the protagonist's definition of the problem or understand his motivation for pursuing a particular goal, that is, share his perspective, is a factor in our experience of the movie. We may comprehend quite clearly how the protagonist sees things and either disagree or even actively oppose his view, or we may simply fail to care whether he solves his problem, overcomes the obstacles, or achieves his goal.

Not caring results in a disinterest on the part of the viewer, in a lack of sympathy for the protagonist, and in a failure of involvement. A lack of caring concerning the fortunes of the protagonist is quite a different phenomenon from the "malicious joy" (to use Heider's term [242]) that one may experience at the misfortune of the antagonist. Such joy is itself a form of caring and constitutes a major affordance of film viewing. Not caring, in the sense of the present definition, is of the sort that might result in a level of boredom prompting the viewer to switch channels if watching the movie on television. If, as a viewer, one can comprehend the perspective of the protagonist and care whether he achieves his goals, then one will probably not switch channels; one will likely get caught up in the narrative and stick with the movie to the end to see how it comes out. The reference to plot is obvious, but it can be argued that the essential satisfaction one seeks in regard to character is afforded in seeing the protagonist achieve his goal, and in perhaps equal measure seeing the antagonist get his comeuppance.

There is yet another satisfaction associated with a fictional character that is not necessary to one's enjoyment of a film but is a sort of bonus if available, and it will be available to some members of an audience and not others. Such satisfaction is related to our capacity to play, to pretend, and is manifest in role identification. At ten years old it is easy for children to view a movie and pretend that they are the hero, that is, to actively take on the role of the protagonist (even though they know that they are not in fact that person). For adults, such imaginative pretending may not come so easily, but the capacity for all play is not lost, and adults may identify something of themselves in the protagonist. Movies are, after all, a form of adult play, and the marshalling of what one perceives to be his own character traits against overwhelming obstacles and perhaps achieving confirmation of their virtue and effectiveness can be pleasurable and self-affirming. What I have been describing are goods that a movie might afford the viewer firsthand. From moment to moment, however, meanings for the viewer are secondhand, presented as affordances for the characters in relation to their fictional world.

In *Citizen Kane*, we perceive affordances from Thompson's perspective: his task is to solve the puzzle of Kane's character, and in taking his perspective, we ask ourselves the following: What does this or that information tell us about Charles Foster Kane? What does it add to our understanding of the man? Is there a clue here as to who or what Rosebud is? Thompson, however, is not the protagonist we *care* about. The film/script/program is carefully designed to prevent our caring about the person of Thompson. We hardly know him as a character. He is a shadowy presence (but a necessary presence, for we need a character whose perspective we can adopt). It does, however, ensure that what we do care about, along with Thompson, is his quest to better understand the man whose dying word was "Rosebud." We become as obsessed with that search as Thompson himself, and finally the quest takes on a life of its own and outlives even Thompson's presence; we alone as viewers are given the final piece of the puzzle.

Thompson stands in the hallway with a group of other reporters. He is attempting to answer their questions when Katherine, who must have just returned from an assignment in outer Mongolia, asks, "What's Rosebud?" Cut to a shot of Thompson in the foreground facing the group; the camera pulls back on Raymond's line, "That's what he said when he died," and pauses for a moment as Thompson tosses the piece of a puzzle he holds in his hand into the box held by Katherine. The camera resumes its pullback as Thompson takes the box from her, turns, and moving into the foreground, places the box upon a table. The discussion continues and Thompson, admitting defeat in his search for Rosebud, turns back to the group, but the camera, now guided by the hand of an unseen storyteller, continues its retreat all the way to the rafters. From our high perch, we see Thompson put on his coat and lead the others away to catch their train. Dissolve to an even higher vantage point as the reporters make their way between the crates to their exit. Dissolve to a quiet clutter of crates way below. Dissolve and begin a steady and purposeful glide over the tops of the hundreds of crates and downward, veering only slightly to the right, to frame a sled nestled among what appear to be Mary Kane's household effects. Two hands reach in from out of frame and take the sled away. Cut to a longshot of workmen throwing "junk" through the fiery doorway of a furnace. A workman carrying the sled enters the frame and makes straightway for the furnace. The camera follows him to the blazing doorway and rests upon the threshold as he tosses the sled into the flames. Cut to a close-up of the sled in the fire. The varnish bubbles on the surface of the superheated wood and, quickly evaporating, clearly reveals the stencilled white letters R-O-S-E-B-U-D and the rosebud decal beneath. The camera moves in closer to see both the word and the picture consumed in flames. Cut to the exterior with dense smoke rising from the chimney as the sled literally goes up in smoke.

Throughout the film, we have been repeatedly thwarted in our attempts to care about a person. We are held at a distance from Kane. His story is told, not in a continuous chronological narrative, but in bits and pieces, and not just from the perspective of the reporter Thompson, but from his perspective of the stories being told by five different narrators whose own experiences with Kane color their tales, and not infrequently from still another perspective within those tales, that of the implied author of the image we see (by choice of angles, music, lighting, and so forth) (Tomasulo). We are prevented by the multiplicity of narrative voices from getting close to Kane. Moreover, in many of the incidents described by the film's five diegetic narrators (Thatcher, Leland, Bernstein, Alexander, and Raymond) Kane comes across as a not so likable character. The result is an emotional roller coaster. We are drawn to Kane in one scene, repelled by him in the next; we begin to care but then pull back. All of this reminds us that we are still uncertain of the man. We do not know Charles Foster Kane. We have no firm fix on his character. Our emotional response as viewers is thwarted, but we keep trying. Somehow we want to care, and sometimes we come close (as when we, along with Thompson and Susan, feel sorry for Kane) but the real emotional impact of the film is in this way postponed until the end.

As he is leaving Xanadu having to admit defeat in his efforts to learn the identity of the mysterious Rosebud, Thompson quips, "I don't think any word

can explain a man's life. No, I guess Rosebud is just a piece in a jigsaw puzzle, a missing piece." But as he leaves and the camera pans the crates and boxes he leaves behind, we want to hold him by his coattails and say, "Wait, Rosebud is here somewhere! And it *can* define the man. It will tell us who Charles Foster Kane was, what *he* cared about!" As if in response (to *our* feeling), the camera zooms in on the sled in the furnace and unleashes the emotion that has until this time been thwarted. We now know who Kane was and feel the full tragedy of his life.

In most narrative films, the character whose perspective we take and the character we care about are one in the same, and if in addition to that we also identify with the character (that is, see in his personality something of ourselves) the emotional impact is heightened even more. (In all successful characters, of course, there is some element with which we can in some sense identify.) In *Citizen Kane*, however, the emotional impact comes not from identification with the main character, but from the thwarting of that identification, the manipulation of our perspective, the toying with our affections, and the postponing of our accepting as our own his search for moral perfection while tripping over his own vanity. Only when we come to know what he cares about, a sled named Rosebud, a cabin in Colorado, and a mother who loved him, can we *identify* with the character of Charlie Kane, see things from his *perspective*, and *care* what happened to him.

Citizen Kane is perhaps the exception that proves the rule. It is highly unusual for the character with whom one identifies to be someone other than the film's protagonist or for that identification to be deferred until the film's end. But that unusual circumstance in *Citizen Kane* allows us to pull apart and examine the nature of character perception in the motion picture. Because the organizing principle of the film is the quest for the significance of Rosebud, which is Thompson's specific task, we as viewers perceive the affordances of the filmic events through Thompson, asking and finding always partial answers to his question: "Who or what is Rosebud?" As the narrative progresses we, like Thompson, are looking for an answer to that question. What is so unusual (and so intriguing and instructive) about *Citizen Kane* is that only the viewer is provided an answer to that central question, and that critical piece of information allows for the switch to identification with Kane himself. The story becomes not Thompson's quest but Kane's struggle. We go back mentally and reconsider the narrative's events from Kane's point of view in light of the new understanding we have of him. We as film viewers come to feel the tragedy of Kane's life, and we as film theorists come to understand that inside the diegetic envelope of a motion picture things do not just mean, they mean something *to* someone or *for* someone, just as they do inside the ecological envelope of the natural world.

34

Convention, Construction, and Cinematic Vision

David Bordwell

Cinema is partly pictorial representation, and we have come to expect, especially after the dissemination of structuralist and poststructuralist theories, that the most enlightening accounts of pictorial representation will involve a theoretical account of conventions. Yet the humanities have not yet solved the problem of how to understand conventions; indeed, I am not convinced that we know very well what a convention is. (We have, however, made advances [Hjort].)

Shot/Reverse Shot: A Convention?

The problem of convention in filmic representation can be strikingly posed by considering one film technique. What is called "shot/reverse-shot" editing typically involves displaying two figures in face-to-face interaction. The camera shows each one alternately, with either the other character absent or only partly visible. The filmmaker cuts from one shot to another, following the

From *Poetics of Cinema*, pp. 57–82, by David Bordwell. Copyright 2008. Routledge and Taylor & Francis Group.

flow of the conversation and the facial reactions. Sometimes the view is taken from slightly behind each character, putting the other character's shoulder in the foreground in what is called an over-the-shoulder (OTS) shot (figure 34.1).

The shot/reverse-shot device deserves to be called a stylistic invention. It wasn't determined by the technology of the cinema, and I can find no plausible parallels in other nineteenth-century media, such as comic strips, paintings, or lantern slides. It wasn't utilized as a stylistic device in the first 15 years or so of filmmaking; that period was dominated by the so-called tableau style, which showed the entire scene in a single shot. In the early 1910s, some fiction films used the shot/reverse-shot device occasionally, whereas by the end of the teens it was common in American features (on the development of the shot/reverse shot in American cinema, see K. Thompson). Fairly soon after this, shot/reverse-shot cutting was adopted around the world. It continues to be one of the most commonly used techniques in film and television.

FIGURE 34.1 *Metropolis* (Fritz Lang, 1927).

What makes the shot/reverse shot comprehensible? Theorists have offered two fairly distinct answers to this question. The first, and older, view is that the device offers a kind of equivalent for ordinary vision. In an early discussion, Soviet filmmaker V. I. Pudovkin says that editing aims to guide the spectator's attention to important elements of a scene: "The lens of the camera replaces the eye of the observer, and the changes of angle of the camera—directed now on one person, now on another . . . must be subject to the same conditions as those of the eyes of the observer" (*Film Technique* 70). This is rather vaguely put, but the idea that editing simulates the change of glance of an observer makes shot/reverse shot a kind of heightening of our ordinary perception of an event involving participants. More recently, Barry Salt has compared such editing to "what a spectator before the scene would see, standing there and casting his glance from this point to that point within it" (164). For these theorists, then, filmmakers discovered in the shot/reverse shot a correlate to spontaneous perceptual activity. Call this the "naturalist" position.

The naturalist position answers several questions. What enabled shot/reverse shot to be discovered? Presumably, filmmakers seeking to engage audiences hit upon it by trial and error, perhaps guided by their own perceptual intuitions. Why was it so rapidly taken up? Because it achieved the requisite purposes of presenting an intelligible structure of information to the spectator. Why has shot/reverse shot been so enduring and pervasive? Because, as an obvious correlate to perceptual experience outside the movie house, it does not require viewers to have special training in order to understand it.

The chief problem with this account is that shot/reverse shot is in several respects quite unfaithful to perceptual experience. The best equivalent to a viewer moving her or his glance from one character to another would seem to be obtained by simply swiveling ("panning") the camera from speaker to speaker. But this is a very rare stylistic option in mainstream cinema. The instantaneous transfer of attention given by the cut would seem to be a conventional substitute for this swiveling of the imaginary spectator's attention—a substitute that has no exact correlate in ordinary perceptual experience.

The shot/reverse-shot device is also unfaithful to ordinary vision because it changes the camera position so as to favor 3/4 views. When you're a third party to a conversation, you don't typically watch each speaker from an oblique angle, let alone from the changing angles provided by reverse shots. When we watch a face-to-face interaction, we are not perceptually capable of shifting our angle of view as drastically as is normal in shot/reverse-shot cutting. And you certainly don't watch from over each character's shoulder. In the absence of panning from face to face, a profiled shot/reverse shot (such as that in figure 34.2) would provide a closer equivalent to "what a spectator before the scene would see" than does the angled OTS views presented by the majority practice.

FIGURE 34.2 *Class Relations* (Jean-Marie Straub and Danièle Huillet, 1983).

Such difficulties were noticed in Pudovkin's day. He therefore added a proviso: the camera allowed the director to create not an actual observer but an "ideal," omnipresent one. Similarly, as Karel Reisz and Gavin Millar point out, the change of angle within shot/reverse-shot cutting has "no analogous experience in real life" (215). The director aims at creating "a ubiquitous observer, giving the audience at each moment of the action the best possible viewpoint. He selects the images which he considers most telling, irrespective of the fact that no single individual could view a scene in this way in real life." This is justified as artistic selection. But this deviation from the natural-equivalent premise opens the door to quite a different theoretical position. Once shot/reverse-shot cuts depend at least partly upon purely artistic considerations, we can ask if they are not simply conventions. Any artistic device as widely used as shot/reverse shot, if not significantly motivated by perceptual equivalences, is likely to be seen as a stylistic convention.

This presumption, I think, dominates film studies today. From this perspective, shot/reverse-shot cutting is an arbitrary device, having no privileged affinities with natural perception. But what is a convention, on this view?

Minimally, I suppose, most contemporary scholars would say that shot/reverse shot is a convention because it is a piece of artifice, and because it must be learned. Most theorists are content to leave the matter there, but neither point really blocks the naturalist position. The naturalist position does not have to claim that shot/reverse-shot editing is not artificial in some sense. After all, it is an invention; it was not present at the birth of cinema, and people decided to use it. Nor does the naturalist position have to deny the role of learning. Once we have learned to perceive the world, the naturalist might argue, we can learn to grasp artistic devices that provide equivalents to the world. Accordingly, our ability to grasp those devices ought to ride upon the appropriate sorts of perceptual skills.

To this, a contemporary theorist might reply that the typical convention is arbitrary. Here, arbitrariness must mean something like this: in principle, an indefinitely large number of other representations would serve as well; the one chosen is simply assigned that task by the rules of the *langue* or code in force. A dog might as easily have been called a *chien* or a *Hund*.

There are some problems with extending to nonlinguistic phenomena a conception of arbitrariness derived from the lexical items in a language. Consider the turn signals on a car. To an observer on Sirius, the fact that I flash the right signal when I intend to turn right might appear arbitrary. Other options are logically possible: people could signal a right turn by activating the *left* signal. But in fact such mechanisms are designed to fit our propensities to signal rightward movement by something that stands in a rightward relation to our body. Our nonverbal symbol systems, like our technical gadgets, are engineered to our fixed dispositions, including innate ones, and the choice among all possible options is not indifferent.

A similar case can be made about the shot/reverse-shot technique. If the director seeks to represent two people looking at each other, it is less arbitrary to show them looking at each other than to show them, say, looking away from each other, or at the moon. A visual "code" that showed figures looking at each other in order to signify that they are *not* looking at each other would be bizarre in the extreme. We would, I think, be inclined to call *that* alternative code "arbitrary," but not the normal case, which reflects naturalistic assumptions about the image's representation of the state of affairs. For creatures like us, the two options are not equiprobable.

Nevertheless, the naturalist's position on shot/reverse shot remains problematic because of the undeniably "unrealistic" qualities present in orthodox uses of the device. And something theoretically stronger is probably required

to allay the conventionalist's worries. At this point, I want to suggest a middle way between the two positions, one that captures the intuition that such visual devices are constructed and significantly artificial while also preserving the idea that they are not utterly arbitrary.

Primary Theory and a Continuum of Conventions

In contrasting the two views of shot/reverse shot, I followed precedent in distinguishing something called *nature* from something called *convention*. The first step in forging a more comprehensive theory, I believe, is to discard these notions and offer some more flexible concepts in their place.

The term *nature* comes to us fraught with connotations. To most film theorists, it suggests either biologically innate capacities or universal laws operating in the physical world generally. It also suggests the realm of necessity, that which cannot be changed by human will or skill. Such conceptions of the "natural" have been frequently attacked by structuralist and poststructuralist theorists, who insist that all signification is constructed, conventional, and culture bound.

Still, only dogmatists would deny that representation, especially visual representation, relies at least partly on the perceiver's psychophysical capacities. It seems very unlikely that our ability to perceive humans and objects in images owes nothing to our biological heritage. Our understanding of images could hardly be unconnected to our capacities to move through a three-dimensional environment and to recognize conspecifics. The individual's development of language, according to the most powerful theories now available, is as much a biological capacity as the inclination to grow arms rather than wings (Pinker, *Language Instinct*). Certain relevant abilities may not even be species specific: pigeons and monkeys respond to photographs as if recognizing the sorts of things represented (Danto).

Nonetheless, I propose that we can make some progress if we bypass the nature–culture couplet for the moment and concentrate upon some "contingent universals" of human life. They are contingent because they did not, for any metaphysical reasons, have to be the way they are; and they are universal insofar as we can find them to be widely present in human societies. They consist of practices and propensities that arise in and through human activities. The core assumption here is that given certain uniformities in the environment across cultures, humans have in their social activities faced comparable tasks in surviving and creating their ways of life. Neither wholly "natural" nor wholly "cultural," these sorts of contingent universals are good candidates for being at least partly responsible for the "naturalness" of artistic conventions.

Paradigm cases of contingent universals would seem to be practical skills such as the ability to use language for communication, to divide labor tasks, to distinguish between living and nonliving things, and so on. I have stated these rather generally; it is an empirical question as to whether there are not much more specific contingent universals, such as recognizing focal colors or taking turns during conversation (Brown, *Human Universals*).

I have stressed contingent universals as involving behavior, but it seems likely that they constitute a conceptual frame of reference as well. The anthropologist Robin Horton calls such a framework "primary theory" and characterizes it as follows:

> Primary theory gives the world a foreground filled with middle-sized (say between a hundred times as large and a hundred times as small as human beings), enduring, solid objects. These objects are interrelated, indeed, inter-defined, in terms of a "push-pull" conception of causality, in which spatial and temporary contiguity are seen as crucial to the transmission of change. They are related spatially in terms of five dichotomies: "left" / "right"; "above" / "below"; "in-front-of" / "behind"; "inside" / "outside"; "contiguous" / "separate." And temporally in terms of one trichotomy: "before" / "at the same time" / "after." Finally, primary theory makes two major distinctions amongst its objects: first, that between human beings and other objects; and second, among human beings, that between self and others. (228; see also Dawkins, *God Delusion* 367)

Horton suggests that although different communities may emphasize some aspects of primary theory and leave others comparatively undeveloped, as a conceptual framework it does not vary significantly from culture to culture.

Note that no decisive claim need be made that contingent universals, whether practices or "primary theory," are either biologically prewired or culturally acquired. In a trivial sense, the capacity to undertake any action must precede that action, so there must be some "natural" capacities. More strongly, those capacities result from evolution. Like other species, humans have evolved in tandem with their environment, and so we're equipped to detect the sort of primary-theory regularities that Horton points out. We should also remember that our environment includes other humans. As social animals, we're attuned to interact not only with sticks and stones but also with our conspecifics. Cinema, like other narrative arts, relies on displays of social intelligence, some aspects of which are plainly cross-cultural. In sum, many cross-cultural convergences can be traced to our evolutionary heritage. At the same time, a great many aspects of artworks rely on particular

cultural traditions, so our perceptual capacities, our primary theory, our attention to other humans, and so on will still be shaped and fine-tuned by circumstance and culture. We can be quite agnostic about the sources of this or that feature of artworks. As students of visual art, we can assume that, say, the ability to discriminate colors or the skill at working material with tools is a contingent universal of human activity, and leave the detailed story behind that activity to research within the appropriate disciplines.

This perspective casts the concept of "convention" in a fresh light. "Arbitrariness" as a measure of conventionality stems, I think, from a misapplication of Ferdinand de Saussure's claims about the arbitrariness of the linguistic sign (R. Harris 64–69). There is another way to conceive of conventions: as norm-bound practices that coordinate social activities and direct action in order to achieve goals (D. Lewis, chaps. 1 and 2).

If we think of convention in terms of practical action, "arbitrariness" is not a very fruitful way of characterizing it. In one important sense, an action counts as arbitrary if the same goal could have been achieved by an alternative means, with no additional costs or difficulties. If I want a bag of potato chips and I am equidistant between two stores selling the snack, all other things being equal the choice is arbitrary. But most artistic conventions are not arbitrary in this sense. First, for reasons already mentioned, some choices are weighted because human proclivities favor them. It is nonarbitrary that the right rear turn signal on an automobile announces that the driver intends to turn right, not left. Moreover, many artistic conventions are more appropriate to certain ends than others. If I am a film director and I want spectators to study an actor's expression, my choice of a close-up isn't arbitrary, because that's an option more favorable to achieving my purpose than, say, selecting an extreme long shot. The idea that conventions are designed for utility in action echoes Noël Carroll's argument that many "arbitrary conventions" are in fact cultural *inventions* aimed at achieving specific goals (*Mystifying Movies* 142–44). We want an account of convention that accommodates two demands: the "engineering" ought to fit human predispositions, and the means ought to be weighted in relation to ends.

A final piece of brickwork needs to be laid in place. One of the attractions of the concept of culturally conventional "codes" is the premise that works transmit or produce meanings. Meanings are cultural; where there is meaning, so goes the reasoning, there must be codes. Instead, though, we may think of works as producing effects, of which meanings are certain types. If we take the artist's goal to be that of eliciting discriminable effects, we can consider a wider range of theoretical possibilities. Now we can conceive of conventions as part of the artist's means for producing effects of many sorts. And these effects take their place in a fabric of human action; they are

consequences of practical action on the part of artists, and grasping the conventions is bound up with larger activities pursued by perceivers.

Our middle way between sheer naturalism and radical conventionalism, then, is signposted by the notions of contingent universals, conventions as norm-governed patterns of behavior, and artistic goals conceived as effects. The map I propose involves a scale of visual effects, with distinct regions but loose boundaries between them. Here I am picking up on E. H. Gombrich's hint that we could consider "representational method" as ranked on "a continuum between skills which come naturally to us and skills which may be next to impossible for anyone to acquire" ("Image").

At one end lie visual effects, which are dependent upon cross-cultural, even universal factors. Roughly, these would seem to be of two types.

First there are what we can call *sensory triggers*. These are cues that automatically stimulate spectators. In the pictorial arts, contrasts of tonality and texture would seem good candidates. Gombrich's interest in visual illusions has led him to insist particularly on the importance of such triggers. He has frequently drawn analogies to the behavior studied by ethologists, such as the ability of a rigged scrap to draw attack from the stickleback fish. But Gombrich also suggests that such triggering mechanisms need not be in the service of illusion; they can also stimulate a search for meaning, creating perceptual anticipations that run ahead of the evidence. One of Gombrich's great accomplishments is to have discovered that sensory triggers play a much larger role in the visual arts than most theorists had recognized ("Illusion").

All nonlinguistic arts exploit such sensory triggers: scale and volume in sculpture, rhythm and loudness in music, and so on. They are among the best candidates we have for "wired-in" responses. In cinema, we do not have to look far for such triggers. Apparent motion, the basis of cinematic movement, is an obvious one. We still do not know exactly how apparent motion works; it may involve a cluster of specific mechanisms, possibly including motion-detecting cells in the visual system (Hochberg, "Representation of Motion"). Apparent motion is a prime instance of a contingent universal: we did not evolve in order to be able to watch movies, but the inventors of cinema were able to exploit a feature of the design of the human optical system to create a pictorial display that is immediately accessible to all sighted humans. Other sensory triggers available in cinema are the use of extreme contrasts of visual tonality; the startle response evoked by sudden intrusions into the frame; and, if Gombrich is right, the use of lighting to create texture and volume within the shot.

Apart from sensory triggers, there are visual effects that draw upon contingently universal factors. These rely on regularities of experience that are

reasonable candidates for being cross-cultural. Recognizing and reacting to these activities almost certainly require some learning, but their ease of recognition among adult members of all cultures makes them function as contingent universals, instances of Horton's "primary theory."

As in our case of face-to-face interaction in the shot/reverse shot, these contingent universals are so firmly fixed that we can scarcely imagine what arbitrary alternatives would be. For example, we are so used to thinking about the variability of the representation of pictorial space across different periods and places that we often forget that these variations stand out against the background of a remarkable constancy in the portrayal of human beings. If visual representation were truly arbitrary, then we ought to find humans portrayed with four eyes or five legs as frequently as with two of each. Yet in art across the world, the human body is represented in broadly comparable terms: the right number of limbs, the anatomically correct placement of head and feet and hands, approximately similar canons of proportion, and so on (Hochberg, "Representation of Things"). Indeed, deities and monsters are marked as such at least partly by violations of such norms. Just as we can recognize other members of our species in ordinary life, so too can we recognize the human being in art of very distant or ancient societies. Surely cinema draws upon this cross-cultural ability to recognize our conspecifics without any special training.

Returning to our example of shot/reverse-shot cutting, I suggest that face-to-face personal interaction is a solid candidate for a cross-cultural universal. This is probably why a visual code is unlikely to represent shared glances by divergent glances, as noted above. It is also perhaps why the situation portrayed in shot/reverse shot is instantly recognizable across cultures and time periods.

Moving along the continuum, we can turn our attention to visual effects that depend on culturally localized skills but can be learned easily. *Easily* here translates into "quickly, on the basis of comparatively limited exposure, and/or without special training or expert guidance." These are norm-bound practices that can be picked up largely through participating in a culture's life as a whole.

Cinema is full of such easily learned visual effects. Arguably, most transitions, such as dissolves or fades; most acting styles; and most stylistic innovations, such as crosscutting or complex camera movements, instantiate such skills. Moreover, once the viewer has mastered narrative structure to a useful degree, she or he has a sufficiently strong sense of context in which to situate particular cinematic devices. Once the viewer has the working concept of a scene, for instance, she or he can hazard a guess that the darkening and lightening of the screen serve to mark one scene off from another. If cinema does have codes, they are mostly codes of this very easily acquired sort—which makes them significantly different from the codes governing other sign systems, like semaphore or calculus.

At the other end of the continuum are those visual effects that depend on culturally specific skills requiring more learning. Acquiring them is time-consuming and requires wide exposure to exemplars and/or special training and/or expert guidance. In these respects, there is perhaps a genuine analogy to language—not to speech comprehension and production, but to reading and writing. (It's much easier to learn to talk than to learn to spell and to punctuate.)

In film, there may be relatively few visual effects that depend upon such specialized skills. The avant-garde cinema is a plausible place to look for such effects, and it seems likely that the films of Stan Brakhage or Yvonne Rainer require an audience to be conversant with abstract expressionism or contemporary poststructuralist theory. More mainstream art cinema of the 1960s may have cultivated certain comparable devices, such as plays with narrative time or shifts between black and white and color.

The spectrum I've outlined, from sensory triggers to comparatively recondite "expert system" effects, is intended as no more than an initial shot at how to conceive conventions. Many of my examples are speculative and are open to empirical disconfirmation. But the general aim is to produce a frame of reference for theoretical reflection and concrete analysis. Such a continuum lets us avoid the difficulties of the naturalist–conventionalist couplet. In order to achieve certain effects, artists may tap biological propensities and contingent universals; in order to achieve other effects, artists may invoke more localized and recondite skills.

Does all this ascribe too little a role to culture? I think not. For one thing, as Gombrich has often pointed out, it is culture that generates the tasks and interests that shape the ways in which visual effects are manipulated. If conventions are relations of means and ends, the social purposes of a representation necessarily govern how the activity is conducted, how the first region of effects is formed into more complex ones. Moreover, the centrality of artistic schemas—those inherited patterns and formulas through which the artist achieves effects in the medium—assures that culture plays a central role. "Only where there is a way is there also a will," Gombrich notes. "The individual can enrich the ways and means that his culture offers him; he can hardly wish for something that he has never known is possible" (*Art* 86).

Because most of the technical devices we encounter package many sorts of appeals together, it seems plausible to hypothesize that the cues lying closer to the "sensory trigger" end of the scale will specify and constrain those cues that are more culturally specific and more difficult to pick up. That's part of

what we mean by understanding something "because of its context." In the representational package we're offered, the more contingently universal cues lead us to make sense of more esoteric cues in particular ways. This would obviously facilitate learning: not only do we need little exposure to certain effects, but also, in each image, the universal factors reinforce our hypotheses about the proper reaction we should have to the more culturally specific ones.

I suggest that shot/reverse shot is best considered along these lines—as a composite phenomenon, drawing on features from various regions of the continuum. I can't itemize all the relevant cues here, but let me make a start.

In its prototypical form, shot/reverse shot is predicated on a two-person, face-to-face encounter. This phenomenon would seem to be a good candidate for a contingent universal of social intercourse—something that would be intelligible across cultures and periods. This consideration is so rudimentary that neither the naturalist nor the conventionalist position on shot/reverse shot deems it necessary to weigh it, but in my argument it forms a kind of cross-cultural bridgehead. For instance, figure 34.3 from *Yaaba* may present facial or gestural cues specific to rural life in Burkina Faso. Nonetheless, the cutting and camera positions present a face-to-face encounter between the young protagonist and his elder, and they do so through a prototypical shot/reverse-shot construction.

The pattern may capture other contingent universals at work as well. Conversational turn taking, with its interchanging role of speaker and listener, might furnish an approximate structure for the alternation of images we get in shot/reverse shot. Indeed, it would seem likely that historically this alternating editing grew out of an effort to capture the turn-taking phenomenon in cinematic form. Another important cue, at least in the prototypical instance, is the glance of the persons represented on the screen. Noël Carroll has suggested that the informational saliency of eye movements in primates gives filmmakers a powerful opportunity to engage audiences cross-culturally ("Toward"). We don't lack testimony from filmmakers that eyes matter. Here is J. J. Abrams, director of *Mission: Impossible III* (2006): "No matter how many trucks and trailers are at base camp, it's ultimately about those few actors—those eyes, what's being conveyed emotionally" (quoted in Hillner 160). In the terms I have proposed, the direction of the glance would function as a sensory trigger, informing us of the object of the person's attention. It stands as another cross-cultural regularity of human activity that can elicit effects in beholders.

Out of such basic materials—the face-to-face encounter, the marked look, the turn-taking structure of conversation—the cinematic shot/reverse shot elaborates a more complex construction. The immediately intelligible aspects of shot/reverse shot anchor what we might consider to be more

FIGURE 34.3 *Yaaba* (Idrissa Ouedraogo, 1989).

culturally specific sorts of effects. Because it is universally intelligible to people from a very young age, the dyadic face-to-face encounter offers a constant that can contextually guide inferences of more specialized sorts.

Consider the propensity we already noted for shot/reverse-shot images to be 3/4 views. There is some experimental evidence that for human faces in pictures, the 3/4 view may be more easily recognized than other orientations (Bruce, Valentine, and Baddeley). Whereas the straight-on and profile views of police mug shots are aimed at recording measurable facial data and

easing recognition of the real face, the 3/4 view has generally been found to be strongest when *pictures* are to be compared with other pictures. Because a movie viewer doesn't have to pick the actor out of a lineup, the 3/4 view in shot/reverse shot serves the purpose of maximizing rapid recognition (at least in cultures that have pictures). The fact that profiled shot/reverse shot seems to be rare in the world's filmmaking practice suggests that filmmakers have exploited a widespread, easily learned norm of representation.

The saliency of 3/4 views has intriguing implications for over-the-shoulder shots too. An OTS provides relevant and redundant information. We see the face of the favored conversant while also being reminded that her partner is present, and at a certain distance from her. We shouldn't think of this camera position as providing the view of an observer, either realistic or ideal. Rather, the image constitutes a display that makes salient key information about the encounter in a way that permits quick pickup (as, say, a view from steeply above or below wouldn't).

As for the instantaneous change of view that is said to create the "ubiquitous" or "ideal" observer, this would seem to be a special case of the immediate leap in time or space caused by any cut, of any sort. And once spectators, presumably from a very young age, have acquired the skill of taking a cut to signal such a shift in orientation, the other cues present in shot/reverse shot may suffice to motivate the distinct changes of angle (Messaris).

There are doubtless other cues that are ingredient to the shot/reverse-shot device, such as the more localized norm that the figures will be observed from the same side of an imaginary "center line" (Bordwell and Thompson). Nevertheless, these remarks indicate the directions in which my account would move. Against the naturalists, I suggest that we don't have to take the shot/reverse-shot technique as straightforwardly conforming to ordinary perception. It's not necessary to posit the device as creating an invisible observer; it's at least as likely that the shot/reverse shot presents a patterned display organized to highlight certain information. Hence its avoidance of a panning movement to simulate the glance and the physical implausibility of its canonical angles. The shot/reverse shot can best be considered as a bundle of norms, some less stylized than others.

Against the conventionalists, I suggest that this bundle of norms draws upon contingent universals of human culture as well as pervasive, easily learned practices of filmmaking. And it seems likely that the former *constrain* and *specify* the latter: if the viewer knew nothing of face-to-face conversations, eyelines, or turn taking, it would be impossible to grasp the purpose of the camera positions and editing. In a metaphorical sense, the prototype of shot/reverse shot is constructed out of such contingent universals: it is a refined elaboration of them, a piece of artifice serving cultural and aesthetic purposes.

Contingent Universals and Us

If any slogan wins immediate acceptance in contemporary theory in the humanities, it is that a given phenomenon is "culturally constructed." The term gathers its force partly from implicit contrast to alternative positions. The phenomenon is constructed, and thus in some sense artificial; it is the result of human praxis, not natural process. The phenomenon is cultural, and so neither natural nor "individual"; it is broadly social, not narrowly psychic. So far, what I've been sketching out here is consistent with these general implications of the phrase.

What I have been trying to say is not, however, compatible with another implication. Sometimes the phrase "culturally constructed" is used to suggest that the phenomenon is not universal or even widespread; it is assumed to be specific to a particular culture. Yet even if cultural models exercise a local validity, it doesn't follow that all of them are unique to a single society or period. It is perfectly possible for a phenomenon to be culturally constructed and at the same time be very widespread, or even universal, among human societies.

Too often, advocates of radical cultural constructivism have supposed that humans in groups dispersed across time and space never face recurring conditions or problems and that they never develop similar or even identical solutions to these conditions. It is a cardinal error to assume that cross-cultural convergences indicate only a shared "biological" or natural propensity, and that all else must be a matter of divergence and variability, somehow traceable to the vagaries of cultural differences.

Not only perceptual equipment but also the disposition to see the world as a three-dimensional space in which free-standing objects exist independent of the observer; not only language "in general" but also pronouns and proper names, lies and narratives, grammatical redundancy and the greater frequency of short words for familiar objects; not only tool making but also the fashioning of pounders and containers; not only spontaneous smiling but also expressions of skepticism and anger, as well as a fear of snakes and loud noises—all these and many more activities are current candidates for being true cultural universals. Apparently all cultures distinguish between natural and nonnatural objects, between living and nonliving things, and between plants and animals. All societies have created fibers for tying, lacing, and weaving.

The value of recalling such anthropological data is, I hope, to help us get beyond the knee-jerk equation of *cross-cultural* (or even *cross-subcultural*) with *natural* or *biologically determined*. Not even the most hubristic sociobiologist would postulate a genetic basis for proper names, containers, and

twine. It seems likely that regularities of the human body, along with regularities of the physical environment and of interpersonal relations, to which humans are attuned by species-specific propensities, have called forth from social collectivities many similar and even universal practices. If social life requires that humans share information, tacit norms guiding face-to-face interactions and conversational turn taking will assist the process in any circumstance in which humans meet.

At the same time, we ought not to quail at the prospect that these universals frequently have a component rooted in biological predispositions. Academic humanists resist the idea of a human nature, convinced that it leads to reductionist and determinist explanations. But it doesn't, because human capacities and propensities are always reshaped by culture—and those capacities and propensities do as much to create culture as to respond to it. It's clear by now that the nature–nurture split is uninformative, that genes are designed to respond to the environment, and that nature has shaped us to be resourceful enough to adjust behavior in relation to our surroundings. Rather than being the robotic servant of a gene for executing this or that piece of behavior, we are flexible and resourceful. "Nature," writes Matt Ridley, "can act only via nurture. . . . The environment acts as a multiplier of small genetic differences, pushing athletic children toward the sports that reward them and pushing bright children toward the books that reward them" (*Nature* 93).

We ought not, therefore, to balk when the metaphor of construction leads us to recognize that social practices may be "built out" of contingent universals. I've argued elsewhere that a constructivist theory of social convention and mental activity requires some conception of materials out of which a representation is fashioned. These materials need not be raw, nor even material in the strict sense (because *constructivism* is a metaphor to start with). As to the source of these materials, we can be quite agnostic; it's not up to film scholars to do the work of anthropologists, population geneticists, and the like. All we need do is note that some features of the films we study, for whatever reason, are manifested across cultures and may thereby create convergent effects.

Theoretically, the most comprehensive and powerful explanations of conventions in any art would seem to be those that show them to be functional transformations of other representations or practices, some of which may be sensory triggers or contingent universals. Methodologically, the best strategy would seem to be constantly on the alert for the cross-cultural factors that would be part of any representational process. Sometimes, these may go without saying; at other times, examining these may shed light on how familiar formulas achieve their distinctive power.

Something like this position, I think, has the best of both naturalism and conventionalism. This view also points toward ways of understanding how conventions may develop in specific social circumstances. Perhaps most tellingly, a moderate constructivism along these lines points toward an understanding of the cross-cultural powers of visual art.

35

Art and Evolution: The Avant-Garde as Test Case

SPIEGELMAN IN *THE NARRATIVE CORPSE*

BRIAN BOYD

I

Has art evolved, like opposable thumbs and the whites of our eyes? If it has, will knowing so help us understand better not just art in general but particular works, even works of avant-garde art? Over recent decades many have come to accept not only that humans have evolved from other animals but that many features of their minds and behavior can be explained by the deep past of evolution. Yet art remains a puzzle for biocultural analysis. How can we explain art in the hard-nosed terms of biological advantage, especially if it lacks analogies or precursors in other species and seems so pleasurably part of being distinctly human?

Nevertheless we have good reasons to examine whether art might be an adaptation. All cognitively normal individuals in all known societies engage in some form of art, actively or at least passively: music and dance, story, visual art. No one has to be pressured into listening to music on a radio or an iPod. And like language but unlike reading and writing, art develops reliably to a basic level without special training.

"Art and Evolution: The Avant-Garde as Test Case: Spiegelman in *The Narrative Corpse*." *Philosophy and Literature* 32, no. 1 (2008): 31–57. Copyright © Brian Boyd 2008.

Art could be what evolutionary biologists call a byproduct, a mere side-effect of our bigger brains or of culture. Yet that seems unlikely. If art were a byproduct, if it offered on average no advantage in terms of survival and reproduction, many generations of intense evolutionary competition would have eliminated it. If art offered no benefits but exacted the costs it does in time, energy and resources, it would, over the generations, have been selected against through the superior fitness of those less inclined to pay such costs, those who preferred to devote their time and energy to securing material or social advantage or simply to save energy by resting.

We can understand art in all its varieties, I suggest, as *cognitive play with pattern*, with multiple, high-intensity patterns. Play exists throughout much of the animal kingdom. It puzzles and fascinates biologists because it is so widespread and so highly conserved (evolutionary lineages in which play has arisen do not lose it as they continue to evolve) that it must fulfill some function, because it has high costs in energy and risk, and because its functions are not as evident as those of "serious" behavior.

The amount of play in a species correlates highly with the flexibility of its behavior. In all species in which it occurs, play seems remarkably self-rewarding. Animals from pigs to rats play compulsively. In fact this seems the core of play: the self-rewarding nature of play makes animals engage in exuberant, intense behaviors that, under conditions of relative security, allow them to extend their capacities and improve their competence in situations of minimal risk. The skills honed will prove of advantage in real-world situations where it is too late to learn on the job, in situations like fight or flight. Because play is compulsive, animals engage in it again and again, incrementally altering their neural wiring, strengthening and speeding up synaptic pathways, improving their capacity and performance.

Humans uniquely inhabit what has been called "the cognitive niche": we gain most of our advantages from intelligence. We have an appetite for information, and especially for pattern—for information that falls into meaningful arrays from which we can make rich inferences. We therefore like bright, distinct colors, crisp outlines, complex shapes or surface design: think of our special fondness for the look of butterflies and flowers. And we have an appetite for open-ended pattern, new kinds of patterns, not only the patterns we have evolved to detect automatically. Vervet monkeys signal to one another in different ways when they see snakes, or leopards, or eagles, because they need to react in different ways to different threats. Yet they never learn to associate fresh snake or leopard tracks with the likelihood of either in the vicinity. Human trackers in the Kalahari or the Australian desert, on the other hand, over time master the local patterns that enable them to track prey and kin. Or in the distinctly modern world, medical researchers tease out the causal patterns linking particular chemicals to particular cancers.

Until very recently computers have fared dismally at pattern recognition, but living organisms have long been expert pattern detectors. Because the world swarms with patterns at every level, minds evolved as pattern extractors, able to detect the information meaningful to their kind of organism in their kind of environment and therefore to predict and act accordingly. And the more flexibly minds can handle multiple patterns at multiple levels, the more open-ended their mastery of the information around them, the better they can predict and act. "The human mind delights in finding pattern," writes biologist Stephen Jay Gould (*Flamingo's Smile* 199).

Extreme informational chaos, the absence of pattern, as in complete whiteout or dense fog, can cause distress and loss of sensory function. Art offers the opposite of patternlessness. It concentrates and plays with the world's profusion of patterns, with its patterns of interrelated or intersecting patterns. Our perception of pattern and of deviation from pattern produces strong emotional reactions. Art engages us by appealing to our appetite for pattern at multiple levels, in producing or perceiving bodily movement, shapes, surfaces or sounds, words or miniature worlds. Like play it therefore provokes us to continue the activities it offers long enough and to resume them often enough to modify our neural circuitry over time.

If play trains animals in bodily flexibility, art trains us in mental flexibility in the perceptual and cognitive areas that matter most. Sociality, for instance, is the province of story, which always focuses on creatures interacting. Exposure to a single story told once will not transform a mind substantially, any more than a single play-fight will make an animal an accomplished fighter, but as with play, many repetitions, or many different stories, improve capacities for social cognition and scenario construction invaluable in the non-story world.

Art therefore needs to engage our involvement, and it can do so partly through the unique importance of human shared attention. Humans have evolved to have white sclera, contrasting sharply with our colored irises, and wide eyes, because understanding the direction of others' gaze, and therefore where their attention is directed, matters uniquely to our species. From birth our emerging capacities and behavior reinforce attention-sharing. New-born human infants can focus only about eight inches away, the distance from the mother's breast to her eyes, and unlike the young even of other apes, can make and hold eye contact with their mothers while suckling. When human infants reach about eight or nine months, infant–adult interaction attains a new level in what developmental psychologists call *protoconversation*, "more a song than a sentence," "a multimedia performance," as parents around the world spontaneously engage their children in back-and-forth interactions involving movement, sound, rhythm, laughter—a playing with pattern that provides a kick-start for art. Early in their second year all normally developing children begin to engage in pretend play, and continue with it compulsively until old enough to attend to the full-blown stories of the adult world.

The forms taken by our intense sociality from birth and our uniquely prolonged childhood show that we have evolved for open-ended learning from others, in part through play and art. We have a singularly intense predisposition to imitate. Art allows us ready self-stimulation by appealing to our cognitive preferences for pattern, as refined by local traditions that we can easily imitate and adapt.

If we understand art in this way, what functions, what biological benefits, does it have?

Art's benefits start from its being play with pattern. Art improves our production and processing of pattern, especially in the key areas of sight, sound and sociality. And since emotional intensity helps consolidate memory, the more art engages attention and stirs response, the more it can reconfigure our minds. *Second*, for those with unusual talents in art, there will be an additional benefit in their capacity to earn the attention of others, especially since throughout the higher primates the capacity to hold attention correlates with status and hence with reproductive and survival success.

Natural selection can select at multiple levels, at, for instance, the levels of the gene, the individual, and the group. Art's additional functions arise at individual and group levels. *Third*, art, regardless of content, intensifies the advantages and skills of shared attention, the sense of shared purposes, and *fourth*, it often specifically strengthens group allegiance, through processes of tribal identification like scarification, tattooing, or costume, through chants and anthems, through mythical, heroic or prosocial stories. *Fifth*, the sheer stimulus of art offers benefits to offset the rising costs in social tension that result from high population concentrations. *Sixth*, art generates confidence that we can transform the world to suit our own preferences, that we need not accept the given but can work to modify it in ways we choose. *Seventh*, it supplies skills and models we can refine and recombine to ensure our ongoing cumulative creativity.

II

In the parlor game of Consequences each person adds another phrase or line to what others have written or drawn, and passes it on to the next contributor, each time folding the page so that the next person can see only the latest addition. In 1925 the Surrealists adopted this as a mode of composition, which they called *Le Corps exquis*, the Exquisite Corpse, to subvert old notions of art: chance instead of design, caprice instead of rationality, crazy disjunction instead of continuity. Three remarks:

1. The idea itself exemplifies the close relationship between art and play.
2. It typifies what Colin Martindale calls the logic of artistic innovation, the artist's need to counter habituation—the diminished responsiveness of neural tissue to a sustained or repeated stimulus—in order to earn or refresh attention.
3. It also shows that time-honored forms of continuity actually hold attention better, precisely because they are not random, because their persistent patterns allow inferable sense.

The Surrealist *Corps exquis* are more widely referred to than read.

But this Surrealist experiment appeals to experimentalists. From 1980 to 1991 the comics artist Art Spiegelman and his wife, Françoise Mouly, published *Raw*, which became the world's leading avant-garde comics magazine. There Spiegelman published serially his project *Maus*, which developed into a two-volume account in comic form of his parents eluding and then surviving Auschwitz. *Maus*, which presents Jews as mice, Germans as cats, Poles as pigs, won Spiegelman a special Pulitzer Prize for comics and has done more than any other work to make comics respectable in Anglophone high culture.

As a project for *Raw*, Spiegelman conceived a comics Exquisite Corpse. A single comic stick figure with a black circle for a head would be the protagonist; each comic artist would have a week to draw three frames, and send them on to the next artist. *Raw* itself ceased publication before the project finished. In its final form, as a long thin stand-alone comic, *The Narrative Corpse* involved 69 comics artists, world famous within comics, from Will Eisner, who started his career in the 1930s, through Robert Crumb, the leading underground comics artist in the 1960s; Matt Groening, the creator of *The Simpsons*, who started in the 1970s; and Chris Ware, who emerged in the 1980s. To speed things up, Spiegelman had a second sequence started and was left to draw three bridging frames to splice the two sequences together. I will focus especially on his three frames.

III

Before we probe this particular example, what are the consequences in general of an approach to art that incorporates an evolutionary perspective and stresses adaptation, play, pattern and attention? The first advantage of a biocultural approach to art is that it makes possible a multilevel analysis, incorporating the deep perspective of the evolutionary past but also focusing on the fine detail of the artist's moments of choice. It can explain art at multiple levels: at the *global* or species-wide level; at the *local* level, in terms of historical, political, economic, technological, cultural, intellectual and artistic contexts; at the level of the *individual* artist or audience member; and at the *particular* level, the decisions behind *composing* this or that feature of a particular work, or *responding* to it for a particular purpose or in particular circumstances. An evolutionary approach can apply this multilevel analysis both to the artistic process, from the origins of art or of a particular work to its reception, and to the worlds represented in artistic works. It does not ignore the local, but nor does it imbalance art by seeing it as primarily a product of local extra-artistic conditions.

An evolutionary analysis of art will consider the costs and benefits of art as a behavior in general—a kind of biological reckoning now "a core approach within evolutionary biology." It will also assume that creatures act in ways they suppose advantageous in composing and responding to art. Artists of any kind will seek to minimize composition effort—by operating within existing artistic modes and traditions, by recombining available models, by adopting readymade subjects—as much as is compatible with maximizing the attention and status a work can earn.

Like artists, audiences too seek a favorable cost-benefit ratio. Many seek to reduce comprehension effort for a quick reward: the latest sitcom or Hollywood blockbuster will be easy to follow even if often also easy to forget. Others may prefer more demanding fare that keeps on unfolding slowly in the mind.

Problems emerged with life, with the challenge of maintaining a structure complex enough to maintain and reproduce itself. Biologists accordingly see physiological features and behavioral choices as attempts to solve particular strategic problems. A biocultural approach to art will analyze an individual artist's situation in the process of composition as a series of particular problems and solutions.

Following on from the discussion above, I suggest that we redefine artists' primary problem not as expressing themselves or their times, or as trying to convey meanings, but as creating works to maximize audience attention and response—and hence their own status—within the current economy of attention, given *their* position within *this* art mode. As in biology, solutions may often be compromises, trade-offs between the benefits of one move and its costs for other parts of the emerging work. And just as in biology old elements prove a ready base for many new designs, as fish fins could become amphibian legs and then bird or bat wings, ape arms or seal flippers, so artists can combine ready-made solutions in new ways to answer new problems. Current practices will incorporate more or less successful solutions to prior problems that can then be improved, recombined or redirected to new problems. And individual artists will each have their own personal problem of capitalizing on the attention their previous work has earned, without boring audiences by merely repeating past successes.

The *problem–solution model* applies not only to artists but also to audiences. Selfish gene theory shows that we cannot expect organisms to work routinely for the benefit of others. Audiences will tend to seek the engagement that matters most to them, not necessarily to the artist, and even to appropriate the work in ways the artist did not intend. Those may be personal, like a couple selecting a Shakespeare sonnet for their wedding ceremony; or political, like Aimé Césaire's postcolonial rewriting of *The Tempest*; or artistic, like Picasso's appropriation of African masks. But if audiences engage with art to serve their own purposes, not those of the artists, then artists, intuiting this, will often try to make their interests appear to coincide with those of their audience, especially by promoting prosocial or group values, since we all benefit from associating with altruists or from living within thriving groups.

The problem–solution model emphasizes individual choice within a specific context. Artists make choices within the unique landscape of their individual preferences and capacities. Biology stresses the depth of individual difference at every level and phase: stable polymorphisms within a species (persistent alternative forms like free or attached earlobes, or bold or cautious personalities); genetic differences (the odds against two human parents producing identical offspring in separate conceptions are seventy trillion to one); sensitivity to initial conditions and developmental accident (even identical twins, who begin from the same conception, have different neural folds in the brain by the time of their birth, and hence different synaptic trajectories); neural plasticity, which ensures that the microarchitecture of the brain reflects different individual experience; and the individual choice of niches of difference and specialization that magnify slight initial differences.

Not only is each problem situation subtly different, but the individual mental landscape within which each of us searches for answers will have unique contours and preferences. Criticism has lately tended to underplay individual difference in favor of group-level differences like periods, cultures, classes. But not only do individuals differ more within groups than between them, but we are also finely attuned to perceive and respond to individual difference. Personality differences have been found even in invertebrates, and animals as neurologically simple as guppies can detect and act on differences in personality. In assessing others we respond to specific intentions—to the solutions others have reached within particular problem situations—but also to the individuality that shapes which problems and solutions present themselves to which individuals in the first place. We respond not only to the problems and solutions that arise for a Caravaggio, a Rembrandt or a Vermeer, but to the individual differences, the unique inclinations and associations, that mean each explores a unique landscape of choice.

The consequences of a biocultural approach to art that strike me as most important for considering individual works, then, are (1) multilevel analysis (global, local, individual and particular), applied to both the creation and reception of art and to whatever art represents; (2) a cost-benefit analysis for composition and reception; (3) a problem–solution model for both artists and audiences, stressing especially the artist's generic problem, the need to maximize audience attention and response through inviting rich play with pattern; and (4) a sensitivity to individuality as shaping an artist's problem–solution landscape even before conscious intentions emerge.

How would these consequences affect the way we analyze *The Narrative Corpse*, and especially Spiegelman's contribution? Let me stress two points here. First, I do not suppose that what follows demonstrates the validity of the claims I make for art as adaptive; but I do hope it demonstrates that an evolutionary stance can offer a broader perspective and finer detail than others, that it is expansive rather than reductive. Second, what follows often coincides with other ways of reading art. An evolutionary analysis that suggested the ways we have been reading art until now have been radically wrong would undermine its own claims about art as an evolved and natural human activity.

IV

Comics reflect many human universals. They focus on patterns particularly salient to humans. Sight is our dominant sense, and for good evolutionary reasons visual processing in the brain has specialized subsystems responsive to outlines (which help demarcate objects), others responsive to animate outlines, and still others to faces and eyes. Since we have long had to be alert to the dangers and opportunities other creatures offer, animate forms arouse intense attention and emotion. Some cells in the brain, for instance, fire only when eyes stare straight at us, and amplify the effect emotionally through the amygdala, the brain's main emotional router. In visual art, drawing outlines of animate forms has from the first—phylogenetically, in cave painting, and ontogenetically, in children's stick figures—offered a high benefit-cost ratio, in terms of composition and comprehension effort and intensity of response.

Comics draw on these immemorial features of visual art. Evolutionary anthropologists have also shown that entities that cross intuitive ontological boundaries arrest our attention and persist in our memory—and are often the basis of religion: beings that can be invisible, or immortal, inanimate objects that can speak or move. Creatures with human bodies and animal faces pervade Paleolithic, ancient Egyptian and Mayan visual art; in the art of storytelling, animals that speak have traipsed immemorially through mythology and fable. In comics, talking dogs, cats, mice and ducks have dominated from mass-market work like Disney's or *Garfield* to the sophisticated subversion of George Herriman's *Krazy Kat* or Spiegelman's own *Maus*.

Of course comics include more than visual art. Art offers, when it can, multiple kinds of patterns. Just as opera combines story, drama, music and song, comics conflate the most salient patterns of the visual world and the most salient patterns of narrative: character, event, and speech, often with the telescoped concentration and swift transitions that the frozen poses and the quick saccades to new frames allow. And art is *play* with pattern. That applies even to serious work like epic and tragedy, but all the more to comics, which in newspaper strips and comic books have so often aimed at laughter.

Although comics reflect key human universals, they also have a specific history. Comics emerged in the 1890s in the fierce market for attention between rival newspapers in New York. They quickly spread, primarily because of the high benefit-cost ratio they offered readers. They began and continue as mass art (Boyd, "On the Origin"). Because they can have such low comprehension costs—a newspaper comic strip can offer a familiar character in a new story and a comic resolution at no more expense than a few seconds of time—they readily appeal to the lowest common denominator, including especially to children. But although comics' cost-benefit ratio makes it natural for them to maximize attention by seeking a mass audience, art can aim at different kinds of attention. Indeed this criterion of the kind of attention sought determines the most important distinctions of register within any mode: the attention of all, in a small-scale society; or in a larger and more stratified society, the attention of the widest audience (popular art), or of the most powerful (pharaohs, popes, aristocratic or plutocratic patrons), or of the most discerning, those prepared for greater benefits at greater comprehension costs (highbrow art), or of those most saturated in the art, and therefore prey to the habituation, the dulling of response, induced by repetition (avant-garde art).

Although comics have low comprehension costs, they had high production costs until the development of web-offset printing in the 1960s made small print runs feasible. With this one technological change, underground comics rapidly took off, earning an audience, especially among a student-age population, by their subversion of the expectations and proprieties of mass comic art, as in Robert Crumb's engorged sexual fantasies and fears. As comics proliferated, the competition for attention led to further experiment and to a comics avant-garde, of which Spiegelman had become the most radical figure by the late 1970s.

Deeply committed to comics as a medium, Spiegelman began to feel ambivalent that the further he took the formal and emotional challenges of some of his material the smaller his audience became. Partly in order to reach a wider audience, partly in order to demonstrate the power of comics to deal with serious issues, and partly just to tell a story that meant so much to him and others, he began writing *Maus*. *Maus* showed a wider world that comics, and even the comic convention of talking animals, could be used for weighty matters—nothing less than the Holocaust—and that comics could earn wide attention among a discerning audience, if they aimed at universal tastes and reduced comprehension costs. At the same time, Spiegelman also sought to make the comics avant-garde more visible through *Raw*, his magazine for innovative comics artists—"High Art for Lowbrows," as he jokingly subtitled one issue.

In devising *The Narrative Corpse* Spiegelman adopted the Surrealists' *Corps exquis* to showcase the variety of serious comic art in a way that both subverted the control of the individual artist (since no participant could foresee the direction of the whole) and reasserted it (since the artists wrenched the story to their own ends for their three panels). Despite its avant-garde elements—the individual artists' modes, the disruptive shifts of styles and story—*The Narrative Corpse* also appeals to some human universals even as it subverts others. Not only does it use standard comic features like distinct and sequential frames, outline drawing and speech bubbles, but it also offers a continuous central character, albeit in swiftly changing predicaments—a unifying device for stories at least as far back as the *Epic of Gilgamesh*. In this case the character is a kind of ironic iconic Lowest Common Denominator, a stick figure with a round black head, half child's drawing, half comics minimalism, that nevertheless allows artists to use posture, gesture, and a basic expressiveness in the figure's eyes, mouth, and head position, and through speech, action and the succession of frames to tell a compressed narrative.

The story or anti-story of *The Narrative Corpse* plays brief continuities of character and event against local discontinuities of style and global discontinuities of plot. Unsurprisingly *The Narrative Corpse* has not earned widespread attention: the shocks of discontinuity become repetitive and wearisome, and no meaningful narrative development repays the high costs of repeated reorientation.

As a showcase of artistic individuality, though, *The Narrative Corpse* succeeds triumphantly. Just as comics generally impose low comprehension costs because their schematic visual impact can be so immediate, so the apprehension of artistic difference here has what psychologists call a pop-out effect. Knowing the stylistic diversity of his contributors and their assertiveness about their individuality, and alert to the effects of panel-to-panel transitions, Spiegelman presumably anticipated the strident visual, verbal and narrative cacophonies that would result.

Indeed despite the homogeneity that the lure of a mass market has often imposed on comics styles, stylistic diversity may be more immediately striking in comics than in literature, which uses the common currency of words, even if in different combinations, and in film, whose makers mostly incorporate common real-world objects. Comics can range from detailed realism to the starkly iconic and from loose roughness of finish to painstaking control.

Earlier I stressed individuality at the biological or global level. This does not mean that other levels are irrelevant, even when considering individuality. The expression of individuality can vary markedly from society to society, as the expression of emotions can vary according to different display rules. In all remaining hunter-gatherer and most other small-scale societies, an egalitarian ethos keeps the display of individual differences, and the status distinctions they might earn, firmly in check. But where tribally cohesive Māori culture traditionally stressed tribally sanctioned design, the Big Man culture of Sepik carvers in Papua New Guinea encourages flamboyant displays of individual difference. Western European and especially American cultures have in other ways emphasized individual talent, especially in the post-Romantic era, and especially in the arts. Nevertheless the pressure of lowering comprehension costs for the largest mass-market audience reduced individuality in American mainstream comics from the 1920s onward, even if there had been and would continue to be strikingly individual artists like Winsor McCay and George Herriman.

We can readily account for the flamboyant stylistic differences among independent comic artists of the 1970s and after in terms of the wide possibilities of the medium, the prevailing ethos of Western individualism, and the previous suppression of individualism precisely because the medium could appeal so readily to the lowest common denominator. Here the problem–solution model again proves relevant. Independent and avant-garde artists maximized their individuality in order to earn a different kind of attention from that of mass-market comics, then maintained their personal styles in order to reduce invention costs and to capitalize on the attention they had already earned. Since he expected that his contributors would do nothing to mute their idiosyncrasies to accommodate a collective effort, Spiegelman by juxtaposing their efforts could display both the medium's range of resources and its diversity of practices.

Spiegelman himself, though, stands apart as a genuinely exploratory artist. Where the styles of contributors to *The Narrative Corpse* like Savage Pencil or Robert Crumb could not be more instantly recognizable, Spiegelman avails himself of many styles. His individuality operates at a deeper level, in his posing deeper problems and more searching solutions, in his sharper awareness of the extent to which comics can be rethought, and in his turning technical possibilities into edgy social reflections. Since the early 1970s he has tested the language of comics and stretched it in new ways, exploring and questioning the role of every element of form and content in sometimes disturbing and often hilarious fashion.

Spiegelman faced a particular problem as overseer of the *Narrative Corpse* project: to link, in his allotted three panels, the close of one strand of narrative, ending with the contribution of Sacco (Joe Sacco, the award-winning artist of *Palestine*), and the start of the other, beginning with the contribution of Savage Pencil (Edwin Pouncey) (figures 35.1–35.3).

Spiegelman solves the problem thus. Sacco has added a political twist to an opportunity the previous contribution allowed, but does not know how to develop it within his three frames. Spiegelman establishes narrative continuity with the Sacco panels, intensifying the theatricality of Sacco's central panel, but allows for a narrative direction, as the central character's affirmations turn into boasts that rile his auditors.

He turns the fatuous grins of the audience in Sacco's final panel into marks of disgust (using evolved human expressions that are cross-culturally recognizable, as the work of Paul Ekman has shown, and that he can therefore evoke with the slightest twists to a few short lines). Famous for *Maus*'s excoriation of anti-Semitism, Spiegelman adapts the most famous speech by a fictional Jew, Shylock's "I am a Jew. . . . If you prick us, do we not bleed?" (*Merchant of Venice* 3.1.58–64). But he turns the mark of commonality in Sacco's drawing into an assertion of difference—of a full humanity that he denies the others—in a way that also offers a flash of fictive self-consciousness: "But, YOU? You are mere splats of INK!"

Spiegelman's second panel shows how much he can rethink his solutions from frame to frame. He makes use of the conventions of reading in Western languages from top to bottom and left to right to incorporate three successive times within the same frame: (1) the central character carried away even more by his own grandiloquence, in the speech bubble at the top of the panel; (2) this thereby provokes the attack from the audience, in the motion line and the "SPLAT!" that knocks him from his perch; and (3) the star-crossed eyes and toppling position that indicate his fall. He combines, in other words, cause (the speech) and effect (the object hurled at him) which in turn becomes the cause of the final visual effect in the panel, the character's fall. I have characterized art as cognitive play with pattern. Of all the patterns that human minds attend to in an open-ended fashion, those of cause and effect are the most important and the most central to narrative. Spiegelman exploits them effortlessly.

Notice how he also refreshes visual attention by flipping the black-on-white image of the speaker to white-on-black in the second panel. He has prepared for this by shading in the sky in the first panel, which makes way for a perhaps symbolic night or blackout in the second. Despite the shift, he reduces our comprehension costs through maintaining the orientation and through the little cloud-squiggle on the right that allows us to keep our bearings despite the changed angle, distance and illumination.

Spiegelman operates at a verbal level beyond other comics artists. He transforms the limp repetitions of Sacco's "I am a man" into the dynamic anaphora of "Yes! I am a MAN! If you prick me . . . ," "I am the jewel in the crown," and the final truncated "I am the—" before the missile hits. The speaker's hubris becomes all human hubris, in "I am the jewel in the crown of creation," but jumps from divine creation to the alternative explanation for the origin of life, evolution, and from the cliché "the jewel in the crown" to the ironically contrasting "fly in the soup," often itself the subject of a substantial cartoon subgenre (variations on the theme of "Waiter, there's a fly in my soup") but here punningly incorporating the primordial soup of life's origin. The patterns include syntactic parallelisms, each embracing a clichéd metonymic metaphor with antithetical values (jewel in the crown, fly in the soup) and contrary complements (creation, evolution).

Comics aficionados probably recognize the visual comics allusion in this panel, and certainly do by the third panel. In Herriman's *Krazy Kat* (1913–1944), regarded by most comics fans as the greatest of all comic strips, each strip repeats the same basic characters and schematic story but with endless visual and verbal variations: the mouse Ignatz throws a brick at Krazy, who crazily responds to this with a new rush of love for Ignatz. Spiegelman's third panel shows Ignatz at the right and, walking behind, his own self-image as Artie-Spiegelman-as-mouse from *Maus*. Spiegelman has been preparing for this from the first panel. *Krazy Kat* takes place in a surreal Arizona, and in his first panel Spiegelman quietly flattens the hillock of Sacco's panel 3 down to a mesa, a recurrent backdrop in Herriman's strip. The flat mesa becomes a stylized pair of jagged hills in Spiegelman's panel 2, an even more unmistakable visual echo of Herriman's landscapes, which gleefully violate the artificial–natural boundary. Panel 3 again echoes Herriman, in the vista, the sketchy remote hills, Ignatz fuming off.

FIGURE 35.1 Joe Sacco's three panels in *The Narrative Corpse*, the end of the first of the two story strands. (Panels by Joe Sacco from *The Narrative Corpse: A Chain Story by 69 Artists!* Copyright © 1995 by Raw Books, reprinted with permission of The Wylie Agency LLC on behalf of Raw Books)

FIGURE 35.2 Art Spiegelman's three panels in *The Narrative Corpse*, linking the Sacco and Savage Pencil frames and the two story strands. (Panels by Art Spiegelman from *The Narrative Corpse: A Chain Story by 69 Artists!* Copyright © 1995 by Raw Books, reprinted with permission of The Wylie Agency LLC on behalf of Raw Books)

FIGURE 35.3 Savage Pencil's three panels in *The Narrative Corpse*, the start of the second of the two story strands. (Panels by Savage Pencil from *The Narrative Corpse: A Chain Story by 69 Artists!* Copyright © 1995 by Raw Books, reprinted with permission of The Wylie Agency LLC on behalf of Raw Books)

But what would be a brick in *Krazy Kat* is here a gift-wrapped present "For Inkboy," in pointed refutation of the full humanity "Inkboy" has claimed in contrast to the "mere splats of ink" listening to him in panel 1: he is ink, not flesh and blood, and boy, not man. The gift-wrapped parcel prepares for the gift of a sacrificial mask in Savage Pencil's first panel in the next story strand, and although "Inkboy" remains schematic rather than morphing halfway to Savage Pencil's psychedelic image, Spiegelman prepares visually for his successor's first frame in orienting his character the same way within the frame.

If all art plays with pattern, the best art plays best with pattern. Spiegelman plays with multiple patterns, dramatic, visual and verbal: with verbal allusion, and with repetition, anaphora, syntactic parallelism, metaphor, cliché, antithesis and pun; with the iconic visual language of comics (seeing stars, crossed eyes, the question mark of puzzlement in panel 3, the fume lines expressing irritation) and its spoken and soundtrack language ("splats of ink" in the first speech bubble prefiguring and provoking the comic onomatopoeia of the "SPLAT!" of the missile's trajectory), and the individual alternation between black–white–black representations of his "hero." He mixes high art and low, Shakespearean drama and demotic comic, eloquence and "SPLAT!," the here and now and the high traditions of literature and comics. He makes the most of verbal and visual resources and literary and comics conventions and traditions.

I suggested earlier that an evolutionary approach to art could offer a multilevel analysis not only of artistic composition but also of artistic representation. Three panels with a few identical stick figures and two anthropomorphized mice represent only a thin slice of a possible world, but even here an evolutionary analysis can offer insights that might otherwise be less evident. In the animal world, size matters, and greater height usually corresponds to higher status. "Inkboy" here begins by claiming first a normal then a superior position, silhouetted above the rest, then topples in the second panel, and sits dumped on the ground in the third. As mentioned earlier, humans in small-scale societies enforce an egalitarian ethos, and not because no one seeks status, but because everyone prefers not to have others higher in status than them. Collectively, through counter-dominance measures, they can ensure equality. Here, the stick figures identical with "Inkboy" accept his claim to be a man but protest at his claiming to be more than them and accordingly topple him from his perch.

Evolutionary biologist and anthropologist David Sloan Wilson asks how humans could have crossed the cooperation divide from other apes, from gorillas, where the silverback tyrannically controls the harem, or from chimpanzees, where males successively challenge for the alpha position and dominate others so long as they can (*Evolution* 162–72). He offers Paul Bingham's solution, that because our ancestors with their bipedal gait evolved to throw with a control far beyond that of chimpanzees—an ability that required complex changes in anatomy and apparently in neurology—those of lesser size could together easily bring down one of larger size who would have been able to dominate in a world of hand-to-hand combat alone. Here Spiegelman wryly redeploys the brick Ignatz throws in *Krazy Kat*, but with the intuition of the artist he gains our attention by appealing, in a way truer than he knew, to the combination of the urge to dominate, the urge to resist domination, and the means we have to do so in our species—even if we are represented here by a posse of abstract inkboys and two sulking mice.

V

Some may agree with my analysis of Spiegelman's contribution to *The Narrative Corpse* and concede its unreductiveness, but still ask: Could we not have reached it without the evolutionary framework? Does it not merely reflect ordinary skills of close reading?

Much of the analysis could indeed have been reached without an explicit evolutionary framework, but it would be self-contradictory for evolutionary analysis to claim otherwise. Our ancestors engaged in art compulsively for tens, perhaps hundreds, of thousands of years before they could conceive of evolutionary theory. We *evolved* a predisposition for art, and with this core disposition and an evolved capacity for social learning we have been able to learn the art skills of our own cultures.

A biocultural analysis *accounts* for these human traits rather than merely taking them for granted. But it also corrects old assumptions about art and new mistakes accumulated over the last forty years through the would-be corrections of Theory. We had long supposed our engagement in art was simply human nature, but Theory has called this into question. A biocultural approach to literature can restore much of the earlier common-sense notion of art while offering *valid* versions of the critiques advanced within Theory.

HUMAN NATURE

Theory has execrated as "essentialism" the very notion of a human nature. A biocultural perspective can reinstate human nature, although not as understood only in terms of current Western values and practices, or as a reflection of God's creation or of human dominion over nature, or as uniquely rational, moral, and cultured, in opposition to the rest of the animal world, or as explicable in top-down terms, with conscious reason as both *explicans* and *explicandum*. Instead a biocultural view of the human will restore the notion of a human nature but understand it in bottom-up terms, as we understand the rest of the physical world, the complex emerging from the simpler, in terms of our many continuities with other animals—even in intelligence, morality and culture—as well as our striking discontinuities.

A biocultural perspective makes it possible to retrieve some of the older common-sense approach to literature while still accepting that we should challenge unquestioned assumptions about the naturalness of "our" behavior. But where constructivists have denied human nature to affirm only social construction, bioculturalists can show common human features underlying local cultural differences. In the example before us our common predispositions for metarepresentation, for overattributing agency, especially human-like agency, for responding intensely to visual outlines and for abstracting common visual patterns, for instance, mean that we automatically see frames in *The Narrative Corpse* as representational and interpret its schematic protagonist as quasi-human—as we do stick figures in San rock art or in children's drawings—and not as arbitrary local conventions that need to be culturally learned.

MIND AND SELF

A biocultural view of the human mind will, like the psychoanalytic elements of Theory, question the assumption that the mind and self are transparent, open to introspection, and fundamentally rational. But it will do so on the basis of evolutionary, psychological and neuroscientific evidence and without the fanciful, unsupported and biologically impossible pseudo-explanations of the mind's depths supposedly offered by Freudian and other psychoanalytic traditions. We will not assume Spiegelman to be in complete conscious control of every decision he makes, but we will nevertheless understand that he can modify moves through the inhibitory power of the human executive brain and the inspection facilities afforded within working memory by its two slave systems, the phonological loop and the visuo-spatial sketchpad.

POWER

All large-scale human societies are hierarchical. Much literary criticism of the 1920s to 1960s affirmed hierarchy, authority and order. Theory has rightly questioned authority but has often done so with a suspicion bordering on paranoia yet with a curious deference to the authority of supposedly anti-authoritarian figures like Derrida and Foucault. A biocultural perspective on human nature can appreciate in a much more nuanced way the complex tension in human social life between competition and cooperation and between the urge for dominance and the urge to resist domination. It can therefore analyze, for instance, the artistic tension between the collective project of *The Narrative Corpse* and the individual ambitions of the artists involved, or Spiegelman's shrewd grasp of human nature in focusing his allotted three frames on this central human tension: "Inkboy" versus in-frame audience.

IDEOLOGY

Much of Theory has claimed that what can be thought within a particular time or place reflects an episteme (Foucault, *Archaeology, Order*) or ideology

(Althusser, *Lenin*) that derives from and maintains existing power structures, incorporating even apparent dissent. A biocultural view of social politics will not reify local culture or ascribe it agent-like powers acting irresistibly on individuals. Instead bioculturalists will be more likely to see the power and spread of cultural representations as dependent on species-wide susceptibilities and biases, modified by individual and group variations and interests. They will not assume the political neutrality or disinterestedness of art's production and reception but will see individual self-interest as a natural strategic response to sociality and as compatible nevertheless with cooperative impulses in support of the flexible and permeable groups to which individuals belong.

Spiegelman serves the interests of comics, and especially avant-garde comics artists, and of his own role within comics, as a champion of High Art for Lowbrows, as editor of *Raw*, as the initiator of *The Narrative Corpse*, and as editorial overseer or rescue-worker trying to splice together the two strands of the story. He also serves his own interests as a comics artist of distinction in aligning himself with Herriman and Shakespeare. Nevertheless he avoids anything like the obviously self-interested self-promotion of "Inkboy," which leads only to the audience resisting "Inkboy"'s boasts.

SEX AND GENDER

Feminism has questioned old assumptions of human nature that presupposed masculinity as norm and as natural authority. A biocultural view of the human can reject these masculinist biases but can show that sex is not a mere social construct and that there is no universal (or pan-Western) patriarchal conspiracy. Instead it can offer a nuanced sense of male–female difference and similarity, and of the biological, cultural and ecological reasons for both. It can also explain the greater male drive for social dominance and the consequences this has had in different ecological conditions and cultural traditions—including the ratio of male to female artists (58:11) in *The Narrative Corpse*.

INDIVIDUALITY

Theory has critiqued the notion of the individual, especially the post-Romantic autonomous individual, as defining the core of human nature, particularly in the arts. But this has often led to a denial or diminution of individual authorial agency (frequently by accrediting a pseudo-agency to "the text" [Levin, *Looking* 55–81]) in favor of local conditions such as ideology, episteme, the "circulation of social energy," or simply the ideas or mood of the time.

A biocultural view of human nature also corrects traditional post-Romantic views of autonomous individualism when it stresses human ultrasociality and the importance of group-level differences and of conformity and imitation. Nevertheless it also allows for a reconsidered and amplified notion of individuality, biological (through genetic recombination, epigenetic development, and neural plasticity) and strategic (through our assessing and favoring our own relative strengths and our discriminating response to the individuality of others). It also recognizes that just as different "display rules" determine how much universal human facial expressions will be intensified or damped down in different societies, different cultural traditions and ecological conditions favor different degrees of expression of individuality. A biological notion of organisms as strategic agents allows us to assess individual artists' responses to their own self-assessment within the local economy of artistic attention, and individual audience members' responses as modified by their artistic predispositions, environment, conformist bias and follow-the-best inclinations.

INTENTION AND INTERPRETATION

In keeping with its anti-authoritarian stance, Theory sought to reduce the role of authorial intention: it rejected the mid-twentieth-century notion of the artist in complete command of the work, shaping even large works with a God-like overview of the whole, and asserted instead the importance of social causation and of readers' responses and interpretive freedom. A biocultural criticism complicates and naturalizes artistic intention—without dispensing with it—and audience interpretation. How?

VI

Both evolutionary considerations and naturalistic studies of decision-making reveal that decisions may often be much less conscious and "rational" than we have thought, and involve intuitive, fast-and-frugal—although often surprisingly effective—heuristics. And like other biological processes, decisions tend to involve both recognized trade-offs and unforeseen consequences that compromise any idealized authorial control. Moreover, to lower composition costs authors will often recombine problems posed and solutions offered by others within the same art or elsewhere. They will also tend to reconsider their own partial solutions as they perceive new problems or richer solutions, in what evolutionary analysts of creativity have called a Darwin-machine process—a series of cycles of idea generation, selection and regeneration. Final artistic decisions may often therefore reflect the sedimentation and compacting of earlier intentions rather than a simultaneous God's-eye view of a work.

Like Theory, biocultural criticism understands meaning as quite unlike the transfer of intention from God-like sender to meek receiver that Barthes challenged and that Derrida sought to replace with infinite deferral. But meaning depends not uniquely on the arbitrary conventions of language in structuralist thought or the supposedly endless aporias dear to deconstructionism but on shared processes that often allow the rapid and substantial recuperation of meaning. We can understand so much so fast not only through local conventions of linguistic expression but also through evolved human capacities: for learning a human language, for interpreting other human expressions (facial, postural, gestural, kinetic, prosodic, technological and artistic), for understanding others' intentions through Theory of Mind, and for understanding events and objects through pattern recognition and natural inference systems like cause-and-effect reasoning.

But understanding intended meaning forms only part of our response. A naturalistic criticism assumes continuity between human action and that of other animals and between human activity in art and elsewhere. Animals have learned to assess the intentions of others, especially within their own species. Assessing the intentions of others does not necessarily mean satisfying them—we will tend to serve our own interests wherever these conflict with others'—but being *able* to assess the intentions of others allows us to make appropriate choices for ourselves. We cannot even *object* to a book unless we can make some guess about its author's intentions. Once we have begun to assess intentions, we may appropriate the book to our own ends: we can, for instance, discard, complete, recommend, deprecate, extol, denounce, review, analyze, excerpt or imitate it.

Spiegelman deliberately chooses to entangle and interrogate intention when he denies the contributors to *The Narrative Corpse* access to any shared narrative aim. Nevertheless we can swiftly distinguish not only the different styles but also the differing intentions of each contributor. In Spiegelman's own case we can effortlessly read the actions of "Inkboy" and the reactions, the comic comeuppance, they produce. We can also see Spiegelman's aim, to maximize in his three panels (1) their continuity with those before and after, (2) their internal continuity and development, (3) their individual impact and surprise, and (4) their collective capacity to raise questions about human similarities and differences and our objections to claims of superiority. If we happen to recognize his allusions, we may also see his aim to place himself within the comics tradition and comics within the narrative tradition his whole project simultaneously subverts.

Spiegelman's contribution appeals to the comic aficionado through the allusion to Herriman interlaced with another to Shakespeare. Common cultural referents can create social cohesion, as those who could afford advanced education in early modern Britain could use classical allusion to signal their membership of the genteel classes. But at the same time as his allusions appeal to comics sophisticates, Spiegelman's panels remain accessible through simple human universals, through our sensitivity to patterns of posture, gesture, and expression; of cause and effect; of height and abasement as indices of status. He addresses the cognoscenti by way of culturally local allusions, but he also employs human universals to address anyone who attends. Like a true classic, he plays intensely enough with obvious and arcane patterns to speak to the masses and the elite, the first-time reader and the re-reader, his own time and the future.

Common sense typically accepts the locally familiar as sufficient to define both the natural and the artful. Challenging this, Theory has rejected both the naturalness of human nature and the artfulness of art. An evolutionary criticism, by looking at and far beyond the local but not neglecting the individual and particular, can show both the naturalness and the artfulness of art—even in the avant-garde.

Part V

Literature as Laboratory

36

Literature, Science, and a New Humanities

JONATHAN GOTTSCHALL

I'm sufficiently liberal, I should assume, toward the claims of science, but with a man like Gottlieb—I'm prepared to believe that he knows all about material forces, but what astounds me is that such a man can be blind to the vital force that creates all the others. He says that knowledge is worthless unless it is proven by rows of figures. Well, when one of you scientific sharks can take the genius of a Ben Jonson and measure it with a yardstick, then I'll admit we literary chaps, with our doubtless absurd belief in beauty, loyalty and the world o' dreams, are off on the wrong track!

ENGLISH PROFESSOR DR. BRUMFIT DESCRIBING THE BACTERIOLOGIST,
DR. GOTTLIEB, IN SINCLAIR LEWIS'S *ARROWSMITH*

People agree that the academic field of literary studies is in trouble. Writers in the venerable "crisis in the humanities" genre—of which I am now one— fret over some of the following indicators. Decades of downward trends in undergraduate humanities enrollees and majors mean that the humanities generally, and literary study specifically, "have become a less and less significant part of higher education" (Kernan, *What's Happened* 5). Funding trends among private and public subsidizers show that "people, including Congress, think of the humanities as increasingly marginal contributors to the sum of knowledge and the well-being of society" (5). In fact, systematic "disinvestment in the humanities" means that higher education is going through a decades-long phase of "dehumanization" (Engell and Dangerfield 111). As humanities enrollments have fallen and the professorial job pool has evaporated, the production of new PhDs has not fallen apace, producing yearning masses of adjunct working poor (Stanton et al.). At the same time, the general esteem in which the humanities are held is in the process

---

of "momentous decline" (Hunt 17). In literary study things may be worst of all. It seems that literary scholars are to be the laughingstocks of the academic world (DelBanco; Oakley 67; Patai and Corrall 18). We are savagely parodied in academic novels, humiliated by hoaxers, and held up to ridicule by satirical journalists, who richly feast themselves at the discipline's main conferences. This is all revenge for our perceived pretentiousness, for the impenetrability of our verbiage, for our unearned moral vanity, and for our apparent contempt for reality.

While fewer students hear us in the classroom, far fewer still can be bothered to read what we write. If you visit your local mega-bookstore you will probably find, as I have, shelves groaning under the weight of serious yet lively books about biology, political theory, physics, history, economics (economics!), religion, mathematics, and so on that are pitched to the interests of intelligent lay readers. Yet, Harold Bloom's prodigious output aside, successful trade books in literary study are few and far between. The whole Literary Studies section at my local megastore consists of one eight-foot length of shelving, and most of that consists of reference work and the sprawling corpus of Harold Bloom.

This is obviously not because people have lost interest in literature. My local megastore stocks an impressive array of contemporary literary fiction, the whole range of canonical classics, and tons of genre stuff. People still take great satisfaction in reading; they just don't need academics mediating between them and their texts of choice. While this failure to connect to the public at large is worrying, an even more alarming and poignant indicator of the current troubles is our increasing inability to communicate with each other. Life in the profession of letters is complicated by soaring publishing demands (one harvest of the runaway buyer's market for professorial labor) and a concomitant publishing "crisis" in which academic presses are slashing production of books and journals. There are several important reasons for the cutbacks, but they all come down to the bottom line: books about literature are very likely to lose money because consumers, including academic consumers, are not interested in reading them (Greenblatt, "Stay"; Harpham, "End" 388; Stanton et al.). This situation—"a piling up of books [and articles] that hardly anybody reads" (Kermode, "Changing Epochs" 613)—is palpably absurd. The field has reached the point where even literary scholars are increasingly indifferent to one another's production.

Given these and other indicators, there is remarkably strong agreement among all the squabbling tribes of literary critics and theorists that the field is floundering, aimless, and increasingly irrelevant to the live concerns not only of the "outside world," but also to the world *inside* the ivory tower (for data-based summaries of all the gloomy trends, see Engell and Dangerfield;

Kernan, *What's Happened*; Stanton et al.). Many have gone so far as to say, in hope or in dread, that the discipline is in the midst of an extinction event—that it is nearing its twilight, its death, or its abolition. Jeremiads, obituaries, threnodies, and eschatological meditations abound. Not everyone thinks we are living at the end of days (and these writers are more likely to describe the "crisis" as a "malaise"), but almost everyone seems to agree that the field has "lost a sense of purpose" (Appiah, "Battle" 447), has "run out of steam" (Latour 225), "has lost its authority" (Menand, "Demise" 201), is "uncertain of its role" (Macey v), and is "desperate for a rationale" (Cunningham 24; see also Menand, "Marketplace"); almost everyone seems to agree that all of the trends are bad and that deep change is urgently required. But there is far less agreement on root causes and the most hopeful prescriptions for change.

<hr />

I locate the source of the current troubles not in the surface phenomena that have milked pure vitriol from polemicists, like jargon-clotted language and extremes of political correctness. Rather, I identify deep, elementary weaknesses in the theories that guide literary investigation, in the methods used to explore and validate hypotheses, and in certain prominent attitudinal constellations. In other words, the prognosis is bad: the primary theoretical, methodological, and attitudinal struts that support the field are suffering pervasive rot. If the discipline of literary study ultimately collapses, it will be because of the rot in these primary struts.

I cannot prescribe a panacea for these ills, but my ambitions are high. I seek to sketch rough outlines of a new paradigm in the study of literature. I seek to show how literary analysis can be founded on theoretical, methodological, and attitudinal struts that are strong, and driven in bedrock. The message is simple: literary studies should move closer to the sciences in theory, method, and governing ethos. In the long view, this scientific turn represents the only responsible and attractive correction of course—the only correction with the potential to lift the field from its morass. What I am finally proposing is a bold experiment which, like any experiment, is worth doing only because it might fail.

The Liberationist Paradigm

For the somewhat more than one hundred years that English has existed as a formal academic discipline, it has suffered painful anxieties about its raison d'être. In a world of viciousness and lack, of epochal scientific discovery, how can the literary scholar *possibly* justify (excuse?) a (usually) impressionistic study of the landscape of make-believe? Is it by initiating students into the priceless expansions of attentive and sensitive reading? Is it by providing a criticism of the culture as a whole, partly as a means of producing raw material for creative artists? Is it by making valuable and durable contributions to human knowledge for its own sake? Is it by transforming drawing room natter about stories and poems into an autonomous and rigorous science of the forms, themes, and deep structures of literature? Is it by using literature as a vehicle for advancing political and social goals that the critic holds dear? Is it through good stewardship of culture's sublimest monuments?

While the list of possibilities could be extended, the point is clear: the conscience of the field has been hounded by hard questions about utility; scholars have expended a lot of time, apprehension, and ink trying to produce a satisfying and self-justifying response.

What use are literary scholars? Beginning in the late 1960s, and especially since around 1980, many literary scholars have envisioned themselves striding in the vanguard of noble movements of social liberation and transformation. Drawing energy and impetus from the great emancipation movements of the 1960s and 1970s, from the radical epistemology of poststructuralism, and from immediate catalysts like the Vietnam War and the student uprisings of 1968, literary scholars embarked upon a great project of *denaturalization*. They set out to show that almost everything people considered to be "natural"—gender roles, sexual orientations, suites of attitudes, ideologies, and norms—were actually the local, contingent, and endlessly malleable outgrowths of specific historical and social forces. In Roland Barthes' sense, they were all "myths," designed to "transform history into nature," to give "a historical intention a natural justification," and therefore to make "contingency appear eternal" (29).

The denaturalization process reached its apogee in concerted efforts to identify—like explorers searching out undiscovered countries, chemists seeking new elements, or biologists hunting up the last undescribed megafauna—the specific historical provenances of core aspects of human psychological and social life: romantic and parental love, concepts of homosexuality and heterosexuality, the idea of "man" and of childhood, the modern sense of selfhood, not to mention our very ability to *think* about a multitude of specific concepts and ideas (for a large sample of putatively unthinkable thoughts, see Levin, *Looking*). In this respect, I think Fredric Jameson's terse definition of the postmodern is very apt: "Postmodern is what you have when the modernization process is complete and nature is gone for good" (ix).

The liberationists sought to show—sometimes convincingly—that "the natural" was simply what a certain group of people, mainly privileged men of Western European descent, benefited from naming their artificial and inherently oppressive socio-political innovations. The victims of the doctrine of naturalism (and of scion doctrines of essentialism and humanism) were all the "others": the non-male, the non-white, the non-hetero, the disabled, and the colonized. There was no reference point from which you could judge the "truth" of anything: objectivity was just a synonym for white male subjectivity (Gitlin 404).

Liberationist praxis represented an audacious response to the problem of utility. It was a rebuke to the ivory tower fuddy-duddyism of the historical, philological, and formalist scholars who had dominated academic criticism for most of its history, and it was a repudiation of the psychoanalytic and structuralist dreams of establishing sciences of the literary. In the wake of the great social upheavals of the 1960s and 1970s, sheltering in the cool, echoing halls of academe to ponder, to scrawl, and to shrewdly declaim on the themes, structures, meanings, or historical backgrounds of stories seemed positively indecent—it was to compound key problems instead of working toward their solution. Literary scholars would therefore march with the revolutionaries; they would shout truth to power in the classrooms and in the journals until the hegemons toppled and the masses were free.

The liberationist paradigm has changed over time, mellowing with age. But buzzing rumors of the demise of Theory (which roughly corresponds to what I am calling the liberationist paradigm) are clearly premature. Simply skimming the contents of prominent humanities journals, or the abstracts for the annual convention of the Modern Language Association, is enough to show that this is not the case. On the contrary, variants and offshoots of the liberationist paradigm remain the strongest forces on the contemporary theoretical and critical scene (for similar arguments, see Slingerland).

Surely, this paradigm is the furthest thing from an undifferentiated monolith. Not only are there real and deep differences *between* different liberationist schools, but there are intense debates *within* predominantly Marxist, postcolonial, poststructuralist, new historical, queer, feminist, and psychoanalytic approaches. Indeed, viewed from up close, each school is a zone of pure and often bitter fracas—a zone of faction and fission where aggressive thinkers vie tirelessly for dominance of message. But, for the purposes of this discussion, what is important is not all of the things that divide these tribes (which have been subjects of countless commentaries, in any case), but the things that bind them in confederation. Almost every significant contemporary approach shares some key components (though in different degrees) that typify the liberationist paradigm:

1. Active commitment to achieving radical or progressive political ends *through* scholarly means
2. A "nurturist" commitment to theories of strong socio-cultural constructivism and a rejection of biological "essentialisms"
3. An epistemology strongly influenced by—if not directly based upon—poststructuralist antifoundationalism

Yet for all of its staying power there remains a broad feeling that this movement in libratory scholarship has finally exhausted its force. There is a nervous sense that prime tenets of poststructuralism—which once seemed startlingly radical—amount to the endlessly rococo embellishments of a great banality: we can't be completely sure of anything. There is a feeling that scholars have gone much too far in reducing literary works to the power plays of the weak and strong, to "reading until you find the victim" or "reading for evil." As Robert Scholes, a recent president of the Modern Language Association, ruefully quips, the urgent question now seems to be "wither, or whither?" ("Whither"). That is, Scholes suggests, the academic study of literature will continue to wither away—perhaps deservedly so—unless practitioners eschew the smooth comforts of old grooves and move out in substantially new directions.

We have, in short, an emerging consensus that the dominant paradigm is spent, and that we are urgently in need of massive intellectual overseeding, if not a total break with the old modes (Boyd, "Getting"; C. Butler; Carroll, *Literary Darwinism*; Davis; Delbanco; Eagleton, *After*; Ellis, *Literature*; Goodheart, *Does*; Kernan, *Death, What's Happened*; Lentricchia; López and Potter; McQuillan et al.; Mitchell; Patai and Corral; Scholes, *Rise and Fall*; Weisbuch; Woodring). But while those who are dissatisfied with contemporary critical theory tend to agree on its inadequacies, they have not espoused a new vision that matches the force of their critiques or the attractions of the status quo. For the liberationist paradigm cast literary scholars as academia's questing intellectual and moral heroes. For all of its distortions and exaggerations, it provides many scholars with a self-image that would be absolutely painful to relinquish.

Shrinking Possibility Space

Our troubles are not new. We are simply in a particularly acute stage of a crisis that is as old as the professional study of literature itself. In his 1983 introduction to literary theory, Terry Eagleton notes that the familiar anxieties were already strongly present in the 1920s when "it was desperately unclear why English was worth studying at all" (*Literary Theory* 27). Traveling back even further—to the last quarter of the nineteenth century, when English as an academic discipline was effectively invented (Graff)—many of the modern concerns are already strongly represented. As Graff and Warner write in their collection of founding documents in the field of literary studies, "attempts to diagnose and cure humanities maladies are about as old as the discipline itself" (2).

In my view, the origins of the crisis, and its dogged permanence, are principally traceable to one basic cause: literary scholars only rarely succeed in accumulating more reliable and durable knowledge. As Goodheart writes, "Quarrels among critics have rarely, if ever, been adjudicated. Interpretations and evaluations abound and are often different from or in conflict with one another. The reputations of writers, determined by criticism, fluctuate, sometimes as wildly as the stock market in crisis" ("Casualties" 509). In other words literary studies is not a discipline where we reliably succeed in producing firmer and surer understandings of the things we study. In stark contrast to the sciences, we have only argument and counter-argument. Often these arguments center on "permanent questions" that have been in place, more or less, from the earliest Greek beginnings. And the debates seem to revolve in continuous circles, bending with fashions and the pronouncements of our charismatic leaders.

These are not really controversial observations, and they are not offered in a spirit of meanness. Few literary scholars would leap to refute this assessment of knowledge production in their field. They would be more likely to challenge the assumption latent in the assessment: that literary scholars actually *try*, as one of their primary goals, to generate more reliable and durable understanding (my own view is that most do try, and others should). Or they might dispute my implicit sketch of knowledge progress in other fields, pointing out what they consider to be the main difference between humanities fields and "progressive" fields: we recognize that progress of knowledge is an illusion while researchers in most other fields do not.

Our failures to accumulate knowledge have not been total. There are things we know about different literary questions and phenomena that are as well established as good history. Yet, there is much more that we do not know, and both foundationalists (those who are persuaded that there is a real world that we can know in a reasonably objective way) and anti-foundationalists (those who are not) typically agree that there is no decisive way to address our unanswered questions or to firmly resolve conflicts between two plausible arguments. In short, for many of the most vital questions in literary inquiry the space of possible explanation is absolutely vast, and we have devised no reliable way of shrinking it down.

Science can be understood as the most successful method humans have devised for shrinking the space of possible explanation. The work is carried forward by research communities whose members typically focus on little parts of big problems. Through their competitive, cooperative, and cumulative efforts scientific research communities seek to narrow the range of plausible response to given questions. Sometimes this process is spectacularly successful and possibility space is reduced to a speck. This doesn't mean that we are certain about such issues. But science can achieve understandings that are so theoretically and empirically robust that all reasonable people must provisionally operate on the assumption that the explanation is correct. This spectacular narrowing of the space of possible explanation is what science has achieved in the theories of evolution, of a heliocentric solar system, of the circulation of the blood, of the germ theory of disease, of thermodynamics, of plate tectonics, of the cell theory in biology, of aspects of quantum mechanics, and of countless other phenomena. In other areas, for instance the attempt to derive a unifying theory in physics (the so-called Theory of Everything), the narrowing of possibility space has been more modest and the final result—given that adequately testing an idea like String Theory may be forever beyond human capacities (Smolin; Woit; but see Greene, *Elegant Universe, Fabric*)—is in doubt.

Scientists all close off the space of possible explanation in much the same way. While the instruments, methods, and subject matter differ profoundly, they all proceed by systematically testing and ruling out competing hypotheses until (ideally) only the most robust possibility remains. The result is the triumph of the scientific method: a steady, usually slow process that often backslides or blunders into cul-de-sacs but, in the long view, results in an undeniable narrowing of the space of possible explanation.

These contractions of the space of possible *explanation* are not, of course, tantamount to contractions of *general possibility*. If anything, I would argue that genuine contractions of the space of possible *explanation* are more typically associated with rapid expansions of human possibilities, as the great imperium of ignorance, superstition, prejudice, and simple falsehood is robbed of reach and power (for similar arguments, see Sagan).

This ability to systematically and decisively narrow our allotted portion of possibility space—to zoom in toward truth in the immense, multidimensional hyperspace of error and vacuity—is precisely what literary studies, and other humanities disciplines, have always lacked. We have not developed ways of putting our ideas to rigorous tests. These tests—the core of scientific methodology—seek to limit the scope for various forms of bias (subjectivity, selection, confirmation, and so on) to distort human perception (Kahneman and Tversky).

Again, this is basically an uncontroversial position; actually, it is a self-evident one. I am saying that literary study is not a scientific field. Literary scholars may bemoan or celebrate this fact, but almost none dispute it. Most literary scholars proceed as though a universal and inexorable law of epistemology forever sunders the sorts of questions they ask from those that scientists ask. Literary questions simply *do not* submit to falsifying tests: they are "unquantifiable" (Fludernik 64), "they cannot be captured by scientific reason" (Harpham, "Response" 105); and they cultivate mental activity that shares more in common with artistic or "religious consciousness" than with scientific consciousness (Fludernik 59; see also Delbanco). For whatever intercourse is possible and valuable between the two cultures—for all the enriching cross-fertilization that can and should occur—the scientific method cannot be dragged across the divide. This is because, as Woodrow Wilson announced when he was still a young professor at Princeton University, the literary "expression of the spirit . . . escapes all scientific categories. It is not pervious to research" (82).

This reasoning about methodological incommensurability is wrong (Gottschall, *Literature*; Mantzavinos; Martindale, "Empirical Questions"; Van Peer, Hakemulder, and Zyngier). It has helped mire the discipline in a deep rut where progress comes slowly if at all. But to claim that the humanities and sciences are not divided by an unbridgeable methodological chasm is not to claim that all our questions are within, or will come within, the practical reach of scientific methodology. I suspect that there will always be vital humanities questions that deflect every tool and device in science's organon. Moreover, I do not mean to hint that sound qualitative studies cannot help us generate more reliable knowledge. I have personally devoted a great deal of energy to this kind of work (for instance, *The Rape of Troy*), and I am not repudiating that work here. I therefore reject the adamantine positivism attributed to the bacteriologist Max Gottlieb, and other "scientific sharks," by the English professor Dr. Brumfit in Sinclair Lewis's *Arrowsmith* (see epigraph): "He says that knowledge is worthless unless it is proven by rows of figures." But I equally condemn the smug and slothful apriorism of Brumfit's mere fiat that phenomena like literary genius and beauty are the products of a mysterious "vital force" that will forever frustrate scientific exploration. If I am more critical of Brumfits than of Gottliebs it is because Brumfitian apriorism is a chief obstacle to knowledge generation in the humanities and Gottliebian positivism is not.

The soul of the paradigm I seek to inaugurate, is that we have been much too pessimistic about the capacities of humans—including even literary scholars—to produce durable and reliable knowledge. For decades, many literary scholars have expressed infinite pessimism about the possibility of knowledge generation, about the "decidability" of anything save "undecidablity." Literary scholarship in the poststructuralist era has been aptly characterized as entailing "a kind of despair about the Enlightenment-derived public functions of reason" (C. Butler 11). At some point, literary academics began seeing themselves not as knowledge generators, but as uncompromising knowledge dissolvers whose acid was perfect skepticism. The liberationists distinguished themselves from their predecessors through their special knowledge that none of the questions they were asking had real answers (it has often been asked, How, then, did *they know* the questions had no answers? How did they achieve certainty about uncertainty?). Starting in the late 1960s, claims of interest, insight, and political usefulness have been applied to distinguished literary investigations, not claims of reliability, validity, or accuracy. Among the many flailing and contradictory attempts to define "Theory," one of the most concise and appropriate is Leitch's: "It's all too easy to think of Theory as a body of knowledge rather than as *an approach to insoluble problems*" (2318; italics added). Another major theorist, Jonathan Culler, concurs. In the concluding paragraph of his *Literary Theory: A Very Short Introduction*, he writes, "Theory, then, *offers not a set of solutions but the prospect of further thought. It calls for commitment to the work of reading, of challenging presuppositions, of questioning the assumptions on which you proceed*" (120; italics added).

Thus praxis in the liberationist era has amounted to endlessly asking questions while despairing of more valid answers; to deconstruction without a clear sense of how to reconstruct once things are made to fall apart; to smothering the flame of Enlightenment without a clear vision of how to light a world which—deprived of reason—is "demon haunted" (Sagan). The quintessence of the liberationist paradigm is, then, constitutional and reflexive pessimism about the ability of humans to really know *anything.* Poststructuralism, which has so profoundly influenced every significant contemporary approach, has introduced a literally abysmal vision of the human capacity to understand—a vision of the "ultimate signified" skidding just beyond the fingertips and into the abyss. In general, literary scholars are not so naïve as to hope for progress of knowledge because, as Lyotard said, the idea of progress has disappeared (1612–1613).

But long before Derrida's neon declarations, the scientists were comfortable with the idea that it is impossible for humans to know the truth of something in the sense of its ultimate reality (Boyd, "Getting"). Popper's concept of falsifiability, which has been a guiding philosophical principle of scientific investigation for more than a half-century, is an attempt to grapple with the fact that it is not logically possible to *prove* any scientific claim by experiment. But science's response to this realization was more reasonable and productive than that of the "great generation" of liberationist theorists (Menand, "Dangers"). We can't know for certain what is true. Science makes no ultimate claims. But through a gradual process of rational thinking and falsifying tests communities of scientists can show where the preponderance of evidence lies. This is the best that humans can do, and this is no small thing.

I argue that rescuing the field of literary study requires moving closer to the sciences. Of course, I am far from the first to feel dissatisfied with the soft modes of literary analysis. Nor am I first to propose that the study of literature should move in a scientific direction. Thus, in one sense, my message sits snugly within a "tradition" of rebellion. Since the very beginning of institutionalized literary study, prominent thinkers have advanced specific plans for how the study of literature should become more serious, more systematic, more rigorous, more scientific. Louis Menand goes so far as to describe the whole history of university literary study as a long sequence of "mood swings of the field—from attraction to controlled scientific methods to distaste for them and back again" ("Marketplace" 100). The philologists, the structuralists, the semiologists, the psychoanalysts, the narratologists, the Marxists, and some others have bemoaned the failures of their discipline, derided the efforts of dilettante amateurs engaged in "chatter about Shelley," and sought to move toward a science of literary analysis. All along the way these systemizers have been opposed by belletrists who—like the New Critic Cleanth Brooks—have branded attempts "to be objective and scientific" as "quixotic" (235).

But Menand's thumbnail history of a field whiplashing back and forth between attraction to scientific methods and distaste for them gets it wrong in one profound way. While literary scholars may have been, at various times, *attracted* to science's methods they have almost never *employed* them. The main attraction, in fact, has been to science's theories, vocabularies, taxonomies, and aura of logical rigor, not its methods. While different literary schools have attempted to apply strict systems to literature, and while they have imported concepts and impressive vocabularies from "more scientific" fields, each one has lacked utterly the *sine qua non* of a true science: the method. Science *is* the method. No matter how systematic the underlying conceptual structure, questions were ultimately explored with the old methods of careful reading and close argument. What is more, some of the most prominent "scientific" approaches (branches of psychoanalysis and Marxism stand out) have, through their determined efforts to inoculate themselves against the inconveniences of negative evidence, devolved into authentic anti-sciences—or, as Kernan puts it—"more numerology than numbers" (*Death* 41).

The methodologically unscientific nature of previous "scientific" schools has convinced the great majority of contemporary scholars that the entire concept of "literary science" is risibly oxymoronic. The frustrated scientific ambitions of previous schools are said to illustrate the fatuity of attempting to cram a non-scientific subject into science's alluring, but ill-fitting mold. As Roland Barthes wrote of his own transition away from structuralism, it may be said that most literary intellectuals are now wide awake from the "euphoric dream of scientificity" (97)—and they are grumpily disillusioned.

But I would argue that three important elements distinguish my proposal from previous attempts to establish a scientific study of literature. The first departure: I argue for an approach that is based in scientific theory and grounded, ultimately, in the bedrock of evolution by natural selection. This approach will therefore be based on theoretical foundations that blend seamlessly into scientific foundations, and that are, as a result, more robust, more accurate and—most vitally (if also most surprisingly)—more *supple and holistic* than those it displaces. The second departure: I argue for a much more vigorous branch of literary research based in the scientific method. I do not merely suggest that literary studies should be "more scientific" (whatever that means), or that scholars should know more about science (as C. P. Snow averred in *The Two Cultures*), but that literary scholars should actually *do* science; where possible, we can and should make use of science's powerful methodology. In addition to these major theoretical and methodological shifts, I propose important adjustments in governing attitudes which will be necessary for the theoretical and methodological innovations to take hold.

In summary, I argue that by emulating aspects of scientific theory, method, and ethos literary scholars can gather much more reliable knowledge about many of the questions we ask, and the discipline can be one where—along with all of the words—real understanding accumulates.

37

Slash Fiction and Human Mating Psychology

CATHERINE SALMON AND DONALD SYMONS

Marriage is a romance in which the hero dies in the first chapter.

LAURENCE J. PETER

"Slash fiction" or "slash" is a kind of romance fiction, usually but not always very sexually graphic, in which both of the lovers are male. To be considered "true" slash the lovers must be an expropriated media pairing, such as Captain Kirk/Mr. Spock (K/S) from the original *Star Trek* series or Sherlock Holmes/Dr. Watson (H/W). The term "slash" arose from the convention of using the slash punctuation mark to unite the lovers' names or initials.

Like mainstream genre romance novels, slash is written almost exclusively by and for women. It originated in the mid-1970s when female *Star Trek* fans began to write and disseminate narratives in which Kirk and Spock fall in love and become lovers. As time went by, virtually every cop, spy, adventure, and science fiction television series featuring two male partners was "slashed" (i.e., slash stories focusing on the main characters were written and disseminated) by some of its female fans.

In the early years slash was disseminated primarily via fan magazines ("fanzines" or "zines"), which were sold by mail order and at fan conventions.

"Slash Fiction and Human Mating Psychology." *The Journal of Sex Research* 41, no. 1 (2004): 94–100, reprinted by permission of the publisher (Taylor & Francis, http://www.informaworld.com).

Today, slash is disseminated primarily via the Internet. Enter *slash* into any search engine and you will find hundreds of sites, most of which are devoted to only one or a few male pairings, though some archives contain thousands of stories featuring many pairings and TV shows.

Slash is not distinctively American. It seems to have arisen spontaneously at about the same time in the U.S., the U.K., Germany, Australia, and Canada, and many of the most frequently slashed TV shows are British. A similar but professionally produced genre, with sales in the millions, arose in Japan: comic books for girls called *shounen ai* (boy's love) (Buckley; Thorn).

When most people, including the second author of this article, first learn of the existence of slash they are deeply puzzled. Why, they wonder, would any woman want to write or read such fiction? Our goal, when we began collaborating in 1994, was to answer this question. We also hoped that if we could discover why slash appeals to so many women we would thereby discover something new about human female mating psychology; in other words, we hoped that slash could be used as an unobtrusive measure of human female mating psychology.

An *unobtrusive measure* is any research method that does not require the cooperation of subjects (Webb et al.). For example, if one were interested in studying the human walking gait one could employ obtrusive measures, such as using electrodes to record the timing and strength of muscle contractions as subjects walked back and forth in the laboratory, or unobtrusive measures, such as analyzing the wear patterns on old shoes. If one were interested in studying human mating psychology, one could employ standard obtrusive measures, such as surveys, questionnaires, and laboratory measurements of genital blood flow, or one could use less common but perhaps no less useful unobtrusive measures, such as analyses of commercially successful erotica (Ellis and Symons; Symons, Salmon, and Ellis). We reasoned that slash exists because a sizable international community of women derives pleasure from writing and reading it; hence, the essential features of this genre must contain information about human female mating psychology.

Our research on slash has not, to date, led to the discovery of a heretofore-undreamed-of psychological mechanism or even to a novel hypothesis about such a mechanism. Instead, slash has turned out to be an exception that proves (tests) the rules, and the rules remain essentially intact. That is to say, it was more the case that our previously held views of female mating psychology led to a deeper understanding of slash than the other way around. We did, however, develop testable hypotheses to account for the appeal of slash, which we describe later in this article.

Before discussing slash and its fans, however, we first consider the general question of why human beings enjoy fiction at all. Our discussion is animated by the premises that mental phenomena, such as enjoyment, are the products of brain states and that the human brain, like every organ in every species, is the product of evolution by natural selection.

Why Do Humans Enjoy Fiction?

In their article "Does Beauty Build Adapted Minds?" Tooby and Cosmides noted that involvement in fictional, imagined worlds appears to be a human universal. They define *fiction* broadly to include "any representation intended to be understood as nonveridical, whether story, drama, film, painting, sculpture, and so on" (7). To the evolutionist, there are two possible explanations for the human tendency to create and consume fictional representations.

The first explanation is that human engagement in fictional experience is a functionless byproduct of psychological (brain) adaptations that were designed by natural selection to serve other functions. In this view, engagement in fictional experience is not something that we are *designed* to do but, rather, something that we are *susceptible* to, as we are susceptible to becoming addicted to drugs. This hypothesis is elaborated and championed by Pinker (*How*), who argues that many of the arts are best understood as evolutionarily novel technologies that effectively "pick the locks" of our brain's pleasure circuits.

The second explanation is that human engagement in fictional experience is itself an adaptation; that is, it is something we are designed to do. Tooby and Cosmides champion this hypothesis ("Does Beauty"). They begin by noting that some psychological adaptations may be designed to operate in two different modes: a functional mode and an organizational mode. When an adaptation is operating in the latter mode, it becomes better organized to carry out its function (the first mode). For example, rhesus monkey fighting is the behavioral outcome of the underlying psychological adaptations operating in their functional mode, whereas rhesus play-fighting is the behavioral outcome of these adaptations operating in their organizational mode (Symons, *Play*). In other words, the fighting mode and the play-fighting mode are both functional, in the sense of being the designed products of evolution by natural selection, but their functions are different: the function of fighting is to harm one's opponent, while the function of play-fighting is to safely practice and thereby improve (organize) fighting skills (without harming one's play partner).

Tooby and Cosmides argue that human engagement in fictional experience may have been favored by natural selection over the course of human evolutionary history because it produced adaptive benefits:

Fictional information input as a form of simulated or imagined experience presents various constellations of situation-cues, unlocking [emotional] responses, and making this value information available to systems that produce foresight, planning, and empathy. With fiction unleashing our reactions to potential lives and realities, we feel more richly and adaptively about what we have not actually experienced. This allows us not only to understand others' choices and inner lives better, but to feel our way more foresightfully to adaptively better choices ourselves. ("Does Beauty" 23)

Tooby and Cosmides add, however, that the psychological adaptations underpinning our enjoyment of fiction may not detect what experiences are actually adaptively organizing in the evolutionarily novel environments in which we live today, only what experiences manifest cues that would have made them adaptively organizing in the circumstances and conditions prevailing throughout the vast majority of human evolutionary history (in the human environment of evolutionary adaptedness, or EEA).

Not only do current human environments differ in many respects from the EEA, but, more specifically, most current modes of disseminating fictional experiences also differ profoundly from the oral storytelling of our Pleistocene ancestors. Although it is often alleged that the human EEA is unknowable or merely conjectural because the past cannot be observed directly, in fact it is a dead certainty that writing, photography, motion pictures, videos, and the Internet are evolutionarily novel technologies that played no role in shaping human psychological adaptations. It would be surprising indeed if these new technologies did *not* provide a broad scope for "picking the locks" of our brains' pleasure circuits, even if, as Tooby and Cosmides convincingly argue, human engagement in fictional experience is an adaptation.

Written fiction probably contains elements both of engagement of organizing adaptations and of pleasure circuit lock-picking, and different kinds of fiction may contain different proportions. Perhaps "great" works of fiction are those that most fully engage organizing adaptations, which is why they have survived the tests of time and translation, while "lesser" fiction, including genre romance novels, may primarily pick the locks of the brain's pleasure circuits.

The Nature of the Genre Romance Novel

Romance novels have been called, with some justification, "women's pornography." While there are profound differences, which we outline below, between male-oriented porn and female-oriented romances, we would like to begin by considering pornography in light of Tooby and Cosmides's analysis of the nature of fiction and the psychology that underpins it ("Does Beauty"). If we can persuade the reader that porn consists almost entirely of lock-picking rather than engagement of organizing adaptations, our subsequent argument that the same is true of genre romances may be more persuasive.

To begin with, Tooby and Cosmides characterize fiction as consisting of representations understood to be nonveridical. Whether porn meets even this most basic test is questionable. The sex presented in film and video porn is, in an important sense, veridical—a fact that pornographers consciously emphasize by requiring ejaculation to occur externally—and the more porn sex looks like "real" sex, the more effectively it achieves its goal of sexually arousing the male viewer (e.g., porn videos are more sexually arousing when shot in color than in black and white). We suspect that surreptitiously filmed "real life" sex would be even more popular and more sexually arousing than ordinary porn (all else being equal). This may help to explain the current popularity of amateur home video porn. Another of Tooby and Cosmides's fiction criteria is that fictional worlds engage emotion systems while disengaging action systems. In at least one sense porn does not disengage action systems: its main purpose is to facilitate action, in the form of masturbation.

More importantly, it seems unlikely that by viewing porn men come to better understand other people's choices and inner lives, to feel their way to adaptively better choices, or to learn generally applicable truths about female sexual psychology or male–female mating relations. On the contrary, the portrayal of female sexual psychology and male–female relations in porn are systematically and deliberately falsified to conform to men's wishful fantasies and thereby to enhance sexual arousal. Porn is set in a male fantasy realm, dubbed *Pornotopia* by historian Steven Marcus, in which sex is sheer desire and physical gratification with an endless succession of lustful, physically attractive women who are always eager to have impersonal sex with strangers and who are always orgasmic.

In sum, porn probably is not very useful for organizing our understanding of human relations because it systematically and deliberately falsifies those relations (although it may have other uses, such as teaching young people about the nature of human genitals and the mechanics of sex). To use Pinker's analogy, porn, like drugs, seems to be something we are susceptible to (*How*).

That porn is probably best understood as almost pure lock-picking, however, in no way diminishes its usefulness for the study of male sexual psychology. On the contrary, the nature of a successful lock picker constitutes potent evidence about the nature of the lock. Commercial pornography is a multibillion-dollar world-wide industry the essence of which remains fundamentally constant through time and space and the nature of which has been shaped primarily by consumer preferences. Analyzing the essential features of porn thus has great potential to illuminate the pleasure circuits of the human male brain (Ellis and Symons; Symons, Salmon, and Ellis). For example, the fact that one of the most popular kinds of Internet porn is the subgenre called "barely legal" (featuring actresses who are at least 18 years old but who strive to, and often do, appear younger) is an unobtrusive measure of a fundamental aspect of human male mating psychology.

In addition, experimental exposure to pornography does not make the vast majority of men more likely to commit acts of violence against women (Allen, D'Alessio, and Emmers-Sommer; Malamuth, Addision, and Koss), implying that violence is not a component of most men's sexual psychology. The small minority of men who are more likely to commit acts of violence (in an experimental setting) after viewing porn have a preexisting hostility toward women or a violence-prone personality, which better predicts their behavior than does the exposure to porn.

We believe that genre romances are analogous to male-oriented porn in the sense that they are wish-fulfilling fantasies, well designed to pick the locks of the pleasure circuits in female brains, and largely worthless as guides to male mating psychology or male–female relations in the real world. Romances probably are not very useful for organizing female psychological adaptations because they deliberately falsify reality, not just in creating fictional characters, settings, and plots, as all fiction does, but in systematically falsifying male mating psychology and human mating relations to conform with female wishes (just as porn systematically falsifies female mating psychology and human mating relations to conform with male wishes). But precisely because genre romances have been shaped in free markets by the cumulative choices of tens or hundreds of millions of women to effectively pick the locks of the pleasure circuits of women's brains, they have great potential to illuminate female mating psychology.

The extremely constrained (and hence extremely psychologically informative) nature of genre romances springs into focus most sharply when existing romances are viewed against the literally infinite background of possible kinds of erotic fiction that women could be writing and reading but aren't. The argument that women's erotica necessarily contains information about women's mating psychology implies the corollary that the *absence* of certain kinds of erotica contains such information as well. Consider that any of the following kinds of erotic fiction could be disseminated via the Internet, at virtually no cost, if even a handful of women were disposed to write and read them:

1. Narratives in which heroines have sexual relations with trees, ferrets, isosceles triangles, or any other random creature or object
2. Narratives with little development of character, plot, or setting in which heroines have brief, impersonal sexual encounters with attractive male strangers, with no obstacles, no falling in love, no strings attached, and no happily-ever-after endings (i.e., narratives that directly mimic male-oriented porn)
3. Narratives in which heroines have sex with their husbands or long-time lovers
4. Narratives in which heroines meet, win the hearts of, and ultimately marry gentle, sensitive, mild-mannered, hard-working, non-threatening heroes with slightly feminine facial features who are anxious to shoulder half the housework and childcare
5. Narratives in which heroines marry such men as described above, and then have impersonal, short-term extramarital flings during ovulation with Mr. Good Genes macho studs

Any of these erotic genres could exist but, to our knowledge, none of them does. Just as the dog that didn't bark in the night was, to Sherlock Holmes, a clue to a murderer's identity, the genres of erotic fiction that women don't write and post on the Internet is a clue to the nature of human female mating psychology.

What, then, actually does characterize women's erotic fiction? The genre romance novel has the following features. The goal of the heroine is never sex for its own sake, much less sex with strangers. The core of a romance novel's plot is a love story in which the heroine overcomes obstacles to identify, win the heart of, and marry the one man in the world who is right for her. (The romance novel's obligatory happily-ever-after ending precludes the possibility of serial romances featuring the same heroine with different heroes.)

Romances vary dramatically in how much explicit sex is portrayed—from none at all to a lot—because romance readers vary dramatically in how much explicit sex they enjoy reading about. In a romance novel, sex serves the plot rather than the other way around, as in porn. The hero finds in the heroine a focus for his passion that binds him to her and ensures his future fidelity. Sex scenes depict the heroine's control of the hero, not her sexual submission. The emotional focus of a romance is not on sex but, rather, on love, domesticity, and mutual nurturing.

Janice Radway found that the members of an American romance novel readers' club were acutely aware of men's tastes for impersonal sex and sexual variety, and they didn't like it. They did not want to adopt men's standards, either in real life or in their erotica; they wanted men to adopt their standards. In short, the genre romance novel is set in a female wish-fulfilling fantasy realm that one might call *Romantopia*.

Anthropologist April Gorry analyzed the characteristics of the heroes of 45 popular romance novels, each of which had been independently nominated for its excellence by a minimum of three romance readers or writers. She found that, overwhelmingly, the heroes of these popular romances shared the following traits.

In terms of physical characteristics, heroes were uniformly described as tall (in the novels in which height was stated exactly, the heroes were 6 feet, 6 feet 2 inches, and, in one case, 6 feet 3 inches). In addition to height, the adjectives most frequently used to describe the hero's physical appearance were muscular, handsome, strong, large, tanned, masculine, and energetic. (Not every one of these traits was explicitly mentioned in every novel, but no hero was described as lacking any of them.)

For physical and social competence, heroes were described as sexually bold, calm, confident, and intelligent. No hero was described as lacking any of these traits, and no hero was described as being a gentle, sensitive fellow (except with respect to his feelings for and actions toward the heroine).

The most consistently described traits had to do with the heroes' feelings for the heroines. The hero wanted the heroine more than he had ever wanted a woman; he had never been so deeply in love; he had intrusive thoughts about the heroine; he was gentle with the heroine; he considered the heroine to be unique; he wanted to protect the heroine; he was possessive of the heroine; he was sexually jealous of the heroine. As Helen Harris documented in her meticulous survey of the psychological literature and the ethnographic, historical, and literary records, this is essentially a textbook list of the syndrome of traits that universally constitutes the experience of romantic love ("Human Nature").

Being rich and having high socioeconomic status (which many researchers, especially evolutionary psychologists, emphasize as important female mating criteria), while undoubtedly much more common among romance heroes than among the average man in the street, were by no means necessary characteristics of a romance hero. In Gorry's sample, the hero had high socioeconomic status (SES) or occupation in 23 of the 45 novels, but low SES or occupation in 5, while the hero was rich in 19 novels, but poor in 10. From an evolutionary perspective, this finding perhaps should not be too surprising. Social classes, money, and formal education did not exist for the overwhelming majority of human evolutionary history. Heroes in Romantopia may or may not be aristocratic, rich, or well educated, but they consistently possess the characteristics that would have reliably indexed high mate quality in the human EEA: they are tall, strong, handsome, intelligent, confident, competent, "dangerous" men whose overwhelming passion for the heroine ensures that she and her children will reap the benefits of these sterling qualities. Romance novel heroes are "warriors," not necessarily literally but in the sense that they possess the physical, intellectual, and temperamental characteristics of successful warriors.

Slash Fiction and Its Fans

One hypothesis that might account for the existence of slash is simply that the women who write and read this erotic subgenre are, in some yet-to-be-determined way, psychosexually unusual—analogous, for example, to male paraphiliacs. We will discuss two lines of evidence that we believe suggest otherwise: First, the results of our empirical research with a group of mainstream romance novel readers suggest that most ordinary romance fans can enjoy reading a male–male romance. Second, slash fiction is so similar to mainstream genre romances that it could reasonably be classified as a species of that genus.

OUR RESEARCH ON ROMANCE NOVEL READERS

Our participants (22 women, members of a Canadian romance readers club), none of whom had previously read a male–male romance, agreed to read such a romance (*The Catch Trap* by Marion Zimmer Bradley, in which the two lovers are trapeze artists, a flyer and a catcher). After reading the novel, participants completed (anonymously) a long questionnaire, which included questions about their reactions to *The Catch Trap*, their views on romances in general, and various personal and demographic information. For our purposes the key question was "Compared to other romances you have read, how much did you enjoy *The Catch Trap*?" Seventy-eight percent of the participants reported enjoying it at least as much as they enjoy other romances, and significantly more reported enjoying it more than average than reported enjoying it less than average. Although this was admittedly only one small study, we think this result is dramatic enough to tell against if not to outright disconfirm the "psychosexual quirk" explanation for the existence of slash; that is, our research implies that romance readers in general, not just slash fans, can enjoy reading a romance in which both lovers are men.

We also tested which participant characteristics did and did not correlate with degree of enjoyment of *The Catch Trap*. Some correlations were unsurprising and not very interesting (e.g., enjoyment and homophobia were negatively correlated). But other correlations were less obvious and more interesting. For example, participants who enjoyed *The Catch Trap* more than the average romance were significantly more likely to report that they had been considered tomboys as girls. Also, these participants were significantly more likely to report that they enjoy buddy, action, science fiction, and horror movies. Finally, these participants were especially attracted to the protagonists' working partnership.

SLASH AS A SUBGENRE OF ROMANCE FICTION

Slash is much more similar to mainstream romance novels than most academic students of slash have realized (e.g., Jenkins; Lamb and Veitch; Penley; Russ). For example, a slash story is in essence a love story in which two long-term male partners, usually depicted as heterosexual (however unlikely this may seem), suddenly realize that they have come to love one another. Slash stories typically have a happily-ever-after ending, namely the establishment of a permanent, monogamous romantic and sexual union.

In addition, while the average slash story probably contains more graphic sex than the average romance novel does, as many students of slash have noted, graphic sex is not a necessary component of either genre; there are PG, R, and X versions of both, and in both the emphasis is always on the emotional rather than the purely physical aspects of sex. In slash and mainstream romances alike, sex occurs within a committed relationship as part of an emotionally meaningful exchange, and it serves rather than dominates the plot.

While slash fans produce a great deal of slash-related artwork, it is unabashedly romantic, very much in the vein of romance novel cover art. It may depict nudity, but it almost never depicts penetration. A further similarity is that the heroine of a romance novel is always the main point-of-view (POV) character, but it is common for the POV to shift between heroine and hero, because many romance readers enjoy having a direct pipeline to the hero's thoughts and feelings. In slash, one of the lovers is always the main POV character, but the POV commonly shifts between the two.

In describing their characters, slash writers are to some extent constrained by the physical traits of the actors who play the roles in the series being slashed, but poetic license regularly enables the main POV character to be "feminized": that is, to be portrayed as the smaller of the two, physically weaker, lighter in coloring, more seductive, more in touch with his emotions, and quicker to perceive the development of mutual love. During an episode of anal sex in slash fiction, each protagonist may play each role, but the main POV character is much more often depicted as the bottom and his partner as the top. In the early days of slash, writers often made technical mistakes in their descriptions of male–male anal sex (e.g., easy anal intercourse without lubricant), which suggests the possibility that they were not literally imagining anal sex at all. In light of this hypothesis, it may be significant to note that in Internet slash discussion groups it sometimes emerges that female fans who enjoy reading and writing about anal sex in slash do not actually enjoy having anal sex. In fact, it has been argued that slash is not really about male homosexuality at all; rather, it is about a female fantasy of heterosexual sex acted out via ostensibly male bodies (Russ).

A frequent romance novel theme is the heroine's giving of her virginity to the hero. Although slash protagonists usually are depicted as having had a great deal of sex with women, they are also usually depicted as "anal virgins." The loss of their anal virginity affirms a bond they share with no one else, and "first time" slash stories are extremely popular. Although male-oriented and female-oriented erotica differ profoundly and in many respects, they are alike in that each typically depicts sex between new rather than long-established partners. Romances, of course, depict sex primarily as a vehicle for establishing a permanent relationship, whereas porn depicts sex primarily as an end in itself. Nevertheless, this similarity between these otherwise so different eroticas suggests the hypothesis that men and women alike experience sex with new partners as especially exciting.

A final similarity between slash and mainstream romance novels is that themes of sexual exclusivity and sexual jealousy are prominent features of both.

In sum, perhaps the main lesson to be learned from analyzing slash is the rather banal one that the more things seem to change in the domain of human mating psychology, the more they actually remain the same. Romances—mainstream novels and slash stories alike—are in essence female fantasies about overcoming obstacles to achieve the perfect mateship.

Then Why Does Slash Exist?

We propose two kinds of answers to this question, which should be regarded as complementary rather than as competing and as hypotheses to be tested rather than as established conclusions.

First, although the heroes of mainstream romance novels are "warriors," the heroines are not warriors, no matter how intelligent, well educated, fiercely independent, professionally successful, and spunky they may be. In slash, however, both lovers are warriors. Slash is based on shared adventure, and its protagonists slay each other's dragons. This, we believe, is the most significant difference between slash and mainstream romances.

The typical slash fan may be a woman who is psychosexually unexceptional but who, for whatever reason, prefers the fantasy of being a co-warrior to the fantasy of being Mrs. Warrior, the fantasy of being a hero who triumphs over the forces of evil to the fantasy of being a heroine who triumphs over an alpha male.

Who might such women be? Our research suggests at least one hypothesis: they might be, disproportionately, former tomboys. Research on tomboys suggests that most do not reject traditionally female activities but rather include traditionally male ones (e.g., they may play with both dolls and trucks). As adults, they typically score high on tests of assertiveness, competitiveness, and willingness to take risks. Slash may have a special appeal to such women because it uniquely fuses traditionally female romance with traditionally male camaraderie, adventure, and risk taking.

The second reason for the existence of slash may be that it solves some of the problems inherent in the genre romance formula better than genre romances themselves do. Here is one example (see Salmon and Symons, *Warrior Lovers*, for others). For the happily-ever-after ending to be credible, the reader of a genre romance must suspend disbelief regarding the way male mating psychology and male–female mating relations are portrayed. (Of course, slash fans must suspend disbelief that two heterosexual men could fall in love with each other and sexually desire each other because of that love. Women who cannot do this do not become slash fans.) In the real world, intense sexual passion and romantic love are evanescent, but in Romantopia they are not: the hero's sexual and romantic passions bind him permanently to the heroine. To find the happily-ever-after ending credible and satisfying, the reader of a genre romance must believe that this bond is so durable that the hero will never be tempted by the opportunities that are bound to come the way of a warrior who possesses every trait that young women seek in a prospective mate.

However, the essence of slash is that a deep, abiding, and most importantly *tested* friendship is firmly in place long before the scales fall from the protagonists' eyes and they realize that they love each other. The partners have put their hands in the fire for each other in the past and they will do so again in the future. They have fully earned each other's trust. In short, before they fell in love, before they had sex, the partners were united by a bond that is plausibly more durable and secure than sexual or romantic passions.

We hypothesize that slash writers and readers derive pleasure from imagining romantic or sexual relationships built on the foundation of an established friendship. If we are right, it seems unlikely that male–female slash will ever replace male–male slash in the hearts of fans, even if many more TV series come to feature male–female partnerships. With a male–female pairing, such as Mulder and Scully from the TV series *The X-Files*, sexual tension—or at least the mutual recognition of sexual and romantic possibilities—exists from the very inception of the partners' relationship and ineluctably muddies the motivational waters. Even if the partners are portrayed as being "just" friends, it is impossible to know—in fiction or in real life—whether a male–female friendship is partly or wholly motivated by sexual or romantic attraction.

Female/Female Slash: A Test Case

Homosexuals constitute the acid test for hypotheses about the nature of psychological domains that are (or may be) sexually dimorphic (Symons, *Evolution*). For example, mainstream male-oriented porn is sometimes claimed to constitute evidence of men's contempt for, disrespect for, or objectification of women. If this claim were correct, one would expect gay male porn either not to exist at all, or, if it did exist, to differ in significant ways from straight male porn. But gay male porn does exist, and it is essentially identical to straight male porn except with respect to the sex of the actors. The existence and nature of gay male porn thus supports our alternative hypothesis, that the essential features of porn illuminate human male mating psychology, not men's attitudes toward or feelings about women (Salmon and Symons, *Warrior Lovers*).

If, as we have argued, slash fiction illuminates human female mating psychology, then slash that is written and read primarily by lesbians should be essentially identical to male–male slash except with respect to the sex of the protagonists. This prediction appears to be confirmed. Some lesbian (and bisexual) fans of TV shows featuring female comrades who face danger and adventure together (e.g., Xena and Gabrielle from *Xena: Warrior Princess*) slash these shows, and their stories have all the earmarks of male–male slash: they are based on friendship and shared adventure, their protagonists are "warriors" who battle the forces of evil, and their plots are resolved when two soul-mates recognize their mutual love and consummate that love sexually.

Coda

Without question, there are now and always have been happy marriages. That said, we recommend the following armchair experiment in unobtrusive measurement: open any book of quotations and read the entries on marriage (this article's epigraph is a typical, relatively mild example). After reading a few dozen quotations on this topic you may conclude that the core fantasy that animates slash fiction—erecting a "marriage" on the foundation of an established, trusted, and tested friendship—might not be such a bad idea.

38

Cultural Variation Is Part of Human Nature

LITERARY UNIVERSALS, CONTEXT-SENSITIVITY,
AND "SHAKESPEARE IN THE BUSH"

MICHELLE SCALISE SUGIYAMA

More than three decades ago, anthropologist Laura Bohannan wrote a now classic (Fung), widely anthologized (e.g., Angeioni; Podolefsky) essay that many have perceived as a challenge to the supposition that "great literature speaks deeply to all and expresses universally held beliefs about the human condition" (Bazerman 44). Prior to her fieldwork among the Tiv of West Africa, Bohannan had believed that "human nature is pretty much the same the whole world over," and that "the general plot and motivation of the greater tragedies would always be clear" to people everywhere (44). When she got to the field, however, this belief was thrown into doubt when the Tiv asked her to tell them about the book she was reading. As she proceeded to relate the story of *Hamlet,* she discovered—to her great surprise and frustration—that their interpretation of the story was "different" from hers.

This is an odd reaction, especially since she begins her essay by explaining that she was reading *Hamlet* in the first place because of a dispute

Michelle Scalise Sugiyama. "Cultural Variation Is Part of Human Nature: Literary Universals, Context-Sensitivity, and 'Shakespeare in the Bush.'" *Human Nature* 14 (2003): 383–396. With kind permission of Springer Science and Business Media.

with a friend at Oxford over the proper interpretation of the play: "You Americans," this friend told her, "often have difficulty with Shakespeare. He was, after all, a very English poet, and one can easily misinterpret the universal by misunderstanding the particular" (Bohannan 44). If people with a common language and culture can disagree over the interpretation of a work, it should come as no surprise—especially to an anthropologist—that people living on a different continent, speaking a different language, and practicing a completely different way of life bring different experiences and opinions to bear on their judgments of story events and characters.

Although she does not explicitly say so, Bohannan implies that, because the Tiv interpret *Hamlet* differently from Westerners, there is no such thing as a universal human nature. Certainly, her essay has been received in this spirit. One anthropologist, for example, describes Bohannan's experience as a "failure to get Tiv elders to see *Hamlet* as a story about incest, revenge and justice" (Fung). With its suggestion that cultural variation (i.e., cross-cultural differences in literary interpretation) precludes psychic unity, the essay implies that cultural variation and cognitive universality are mutually exclusive phenomena.

An adaptationist perspective suggests otherwise. Throughout their evolutionary history, humans have faced a recurrent set of tasks requisite to survival and reproduction, such as food acquisition, face recognition, communication, coalition formation, mate acquisition and retention, cheater detection, childrearing, defense against pathogens, and so on. Collectively, these adaptive problems correspond to what humanists often refer to as "the human condition." The psychological software that evolved to address these tasks, in turn, corresponds to what in lay terms is referred to as "human nature." Adaptive problems and their evolved cognitive solutions are constant across cultures. What varies between cultures is habitat and historical happenstance. Because humans are able to occupy a wide range of habitats, and because the same set and sequence of events does not occur in every locale, a solution that works in one place may not be available in another. Indeed, local solutions may vary *within* a culture group if it is dispersed over a sufficiently large area (see, e.g., Birket-Smith 9, 35; Rasmussen 15, 38–39, 50, 92, 102–3, 104; Wilbert and Simoneau). Complex adaptations are expected to be sensitive to environmental variation—that is, to calibrate themselves to local conditions. Specifically, cognitive algorithms are expected to consist of context-sensitive rules that generate different psychological and behavioral outputs in response to different inputs (Tooby and Cosmides, "Cognitive Adaptations"). We experience the effects of this context-sensitive design as cultural variation.

We are compensated for our mistakes whenever we can profit by them, and "Shakespeare in the Bush" is a case in point. Bohannan's examples of cultural difference demonstrate how the concept of context-sensitivity can be applied to the study of world literature. This, in turn, can help lay the foundations for an adaptationist model of the complex, species-typical, universal behavior whereby humans use words to create a facsimile of their environment.

The principle of context-sensitivity is an important stepping stone on the path to understanding cross-cultural differences in literary interpretation. Literature can be seen as a verbal facsimile of human experience composed largely of representations of human beings (e.g., "characters") and human habitats (e.g., "setting") (Scalise Sugiyama, "Narrative Theory," "On the Origins"). Because it is impossible to depict a human character without also depicting the human psyche, we would expect any given literary character to consist of at least a partial representation of our evolved psychology ("human nature"). And because all habitats present the same basic set of obstacles to survival and reproduction, we would expect any given literary setting to represent one or more adaptive problems ("the human condition"). However, because literary settings are representations of human habitats, and human habitats vary, we would expect local solutions to adaptive problems to vary in literature as they do in real life. On this view, then, we would expect literary art to express both human universals (e.g., adaptive problems, cognitive adaptations) and cultural variation (e.g., local solutions to adaptive problems).

Human habitats vary geographically, economically, technologically, and demographically. These differences affect local expression and solution of adaptive problems and, hence, narrative content. For example, survival as a forager requires detailed knowledge of local terrain and resident flora and fauna. Accordingly, forager oral traditions contain a wealth of information regarding orientation and geography, environmental hazards, plant and animal characteristics and uses, and hunting, gathering, and processing techniques (Scalise Sugiyama, "Food," "Lions," "Narrative Theory"). In contrast, survival in industrialized, hierarchical, state societies depends on access to capital and influence; we would thus expect the literature of such peoples to prominently feature strategies for socioeconomic advancement. Consider the classic British novel: Fielding, Richardson, Austen, Dickens, Thackeray, and their peers do not tell us how to track a tapir or what wood to use for a bow; rather, they recount the mores and manners, the *bon mots* and *faux pas* of a wide spectrum of social climbers.

Context-sensitivity affects literary interpretation in the same way it affects literary production. It stands to reason that, when an individual is presented with an adaptive problem, the solution that comes to mind is the local one. When I'm hungry, I don't go out hunting or gathering; rather, I go to the refrigerator, the cupboard, or, if those avenues fail, the grocery store. When we experience a literary work, we are often presented with a habitat that is unfamiliar to us—ancient Greece, Saxon England, 1920s Paris, etc. This situation is analogous to visiting a foreign country and being unacquainted with local customs and history. "Misinterpretation" may be said to occur when we apply our local solution to a given problem instead of the solution local to the setting of the work. In literary terms, this is known as the error of not situating a work in its historical context.

This is precisely what happens in "Shakespeare in the Bush." Being unacquainted with the political, economic, and intellectual terrain of Hamlet's (or, perhaps more accurately, Shakespeare's) world, the Tiv respond to Hamlet's difficulties by consulting their own cultural map. For example, they object to the ghost's choice of Hamlet as his avenger because, according to Tiv custom, important matters such as revenge and punishment are handled by chiefs and elders. As one man explains, "If your father's brother has killed your father, you must appeal to your father's age-mates; *they* may avenge him. No man may use violence against his senior relatives" (Bohannan 51). The interpretive difference that occurs here does not stem from cognitive differences between the Tiv and Westerners. The Tiv do not lack the concept of revenge; rather, they have somewhat different rules for exacting it. I say "somewhat" because these rules are fairly consonant with Western practice: our judges and advisors tend to be people with considerable life experience—that is, "elders" rather than youths—and guilt and punishment are typically determined not by the plaintiff but by a neutral third party (i.e., a judge or jury).

Although the Tiv do not believe Hamlet should be the one to avenge his father's murder—"For a man to raise his hand against his father's brother and the one who has become his father—that is a terrible thing" (Bohannan 51)—they nevertheless believe that Claudius deserves punishment if he is the cause of Hamlet's "madness." As they explain, "if his father's brother had indeed been wicked enough to bewitch Hamlet and make him mad that would be a good story indeed, for it would be his fault that Hamlet, being mad, no longer had any sense and thus was ready to kill his father's brother" (51). The notion of fair play expressed in this passage is quite consonant with Western ideas of justice. Because the Tiv believe that madness can only be caused by witchcraft, and that one can only be bewitched by one's male relatives, they conclude that "it had to be Claudius who was attempting to harm him [Hamlet]. And, of course, it was" (49). Both cultures agree, then, that Claudius is responsible for his nephew's "madness" and that he should be punished for his actions. Two different cultures, same conclusion.

Another interpretive difference identified in the essay is the Tiv reaction to Gertrude and Claudius' marriage. The Tiv are a levirate-practicing society, so when Bohannan tells them that the dead chief's brother has become the new chief and married the dead chief's widow, they commend Claudius' actions: he has performed his duty to his brother, his brother's widow, and his brother's orphan, and has boosted his own status in the process. Bohannan is "upset" and thrown "off balance" by this reaction (46). But, by her own admission, the Tiv lack a crucial piece of information regarding Hamlet's uncle—she has not told them that Claudius murdered his brother: "I decided to skip the soliloquy. Even if Claudius was here thought quite right to marry his brother's widow, there remained the poison motif, and I knew they would disapprove of fratricide" (47). When she finally does tell the Tiv Claudius's dark secret, they do not disagree that he has committed a crime and, indeed, specify the proper means of redress: "If your father's brother has killed your father, you must appeal to your father's age-mates; *they* may avenge him" (51). The Tiv clearly understand the concepts of kinship, familial obligation, homicide, and justice. Indeed, their attitude toward murder accords with Western practice: as Bohannan herself admits, the Tiv "disapprove of fratricide." Again, the interpretive difference identified here is attributable to environmental differences rather than psychological ones: both cultures agree on the nature of the crime and the necessity of punishing it; they merely disagree on administrative details.

Bohannan is also frustrated by the Tiv's apparent failure to appreciate the heartlessness of Gertrude's actions and tries to explain to them that Hamlet is upset because his mother has remarried so soon. To persons unfamiliar with the political and economic workings of feudal society, however, the implications of Gertrude's hasty remarriage (i.e., adulterous relationship with Claudius, complicity in her husband's murder) are not at all evident. The demands of life in a subsistence economy are not compatible with long periods of mourning. As one Tiv wife explains, "Two years is too long. Who will hoe your farms for you while you have no husband?" (Bohannan 47). With her vassals and revenues, a widowed queen in an agrarian kingdom need not worry where her next meal is coming from and can thus afford to mourn her husband in a leisurely fashion. The Tiv, however, respond to Gertrude's actions in the context of their own political and economic realities, which make prompt remarriage a necessity.

Another interpretive difference concerns the ghost of Hamlet's father. The Tiv do not believe in ghosts (Bohannan 48), so they claim that the apparition must be an omen sent by a witch. However, this alleged "difference" is undermined (1) by the long Western tradition of belief in witchcraft and (2) by the fact that, like the Tiv, many Westerners do not believe in ghosts either.

Which brings us to a serious problem with Bohannan's analysis: it is not clear which Western culture her essay makes reference to—1960s America, Elizabethan England, the world of the play, or something else. What is overwhelmingly clear is that both Tiv and Western peoples (past and present) have the cognitive ability to imagine and believe in the existence of supernatural beings. Moreover, each culture attributes the same motivation to these beings: ghosts and witches alike seek to influence the behavior of the living and are often malevolent.

Indeed, it is fair to say that Bohannan's essay is bursting with cultural convergences. For example, one elder comments, "Polonius sounds like a fool to me," to which Bohannan concedes, "Many people think he was" (49). As with foolishness, the Tiv understanding of madness—"no longer ha[ving] any sense" (51)—is for all practical purposes the same as the Western one. Similarly, in terms of their qualities and duties, Tiv leaders bear a striking resemblance to their Western counterparts. Ideally, a Tiv leader is a man skilled in discussion, a man who acts and speaks quietly but firmly, and a man of sincerity and integrity (Bohannan and Bohannan 32–34). However, it is also the case that, as in Western society, some men achieve positions of power through "intimidation, a talent for intrigue, and bribery" (Bohannan and Bohannan 32). The duties of a Tiv compound head include admitting new members to and/or expelling troublesome members from the compound, arbitrating disputes, halting punishment deemed too harsh or unwarranted, provisioning compound members in times of hunger, collecting food for rituals and feasts, and serving as the spiritual head of the compound. At higher levels of influence, Tiv leaders furnish safe-conduct to passers-through, arbitrate quarrels, and act as spokesmen and representatives in both external and internal affairs. Many of these duties and powers echo those of our own government to grant and revoke citizenship, negotiate with foreign powers, collect and dispense revenues, resolve civil conflict, and punish infractions.

Perhaps the most striking convergence is found in the Tiv reaction to the relationship between Hamlet and Ophelia. Bohannan introduces Polonius as an important advisor to the chief and member of his household, whose daughter Hamlet has been courting but is unlikely to marry. When a Tiv wife asks why, her husband retorts, "They lived in the same household" (Bohannan 48). Bohannan quickly explains, "That was not the reason. . . . Polonius was a stranger who lived in the household because he helped the chief, not because he was a relative" (48). The "reason" to which both parties allude here is incest, yet neither has made an explicit reference to it. There is no need to: both parties understand that members of the same household tend to be related and that sexual intercourse between relatives is undesirable.

In sum, when we look closely at the Tiv response to *Hamlet*, we see that the same issues that resonate with Western audiences also resonate with Bohannan's audience: kinship, fratricide, revenge, justice, madness. The few genuine interpretive differences identified in the essay can be attributed to differences in local custom and economy rather than differences in cognitive design. Thus, rather than contradicting the claim that "human nature is pretty much the same the whole world over," Bohannan's observations support the hypothesis that cultural variation and universal cognitive design are complementary phenomena.

In its preoccupation with interpretive differences, Bohannan's essay overlooks the most impressive literary universal of all: despite the linguistic, political, economic, and other cultural obstacles between Bohannan and the Tiv, she effectively communicates the story of *Hamlet* to her audience. In other words, her essay overlooks the astounding psychological fact that all people everywhere are capable of telling and processing stories, and that this ability develops reliably without any special education. This is the real lesson of Bohannan's experience: storytelling would be impossible if human minds were not fundamentally the same in design and function. We must therefore take care, in the words of Bohannan's colleague, not to "misinterpret the universal by misunderstanding the particular"—that is, we must not mistake local solutions to adaptive problems for differences in human cognitive design.

39

Paleolithic Politics in British Novels of the Longer Nineteenth Century

JOSEPH CARROLL, JONATHAN GOTTSCHALL,
JOHN JOHNSON, AND DANIEL KRUGER

Moving Past the Two Cultures

Scientists typically operate by formulating testable hypotheses and producing data to test the hypotheses. Students of literature, in contrast, typically proceed by way of argument and rhetoric. In their most scholarly guise, they aim at producing objective textual and historical information, but all such information must ultimately be interpreted within some larger order of ideas. During roughly the first two-thirds of the twentieth century, the most prominent theoretical systems taken from outside the humanities and used for literary study were Marxist social theory, Freudian and Jungian psychology, and structuralist linguistics. Even in their own fields, these systems were only quasi-scientific, more speculative than empirical, and in literary study, speculative ideas served chiefly as sources for imaginative stimulation. Most critics operated as eclectic free agents, spontaneously gleaning materials from every region of knowledge—from philosophy, the sciences, history, the arts, and especially literature itself. Although using selected bits of information from the sciences, students of literature commonly regarded their own kind of knowledge—imaginative, subjective, qualitative—as an autonomous order of discourse incommensurate with the quantitative reductions of science. Over the past three decades or so, these older forms of literary criticism have been superseded by a new theoretical paradigm designated variously as poststructuralism or postmodernism. It incorporates psychoanalysis and Marxism in their Lacanian and Althusserian forms, but poststructuralists explicitly reject the idea that scientific methods secure the highest standard of epistemic validity. Instead, they include science itself within the rhetorical domain formerly set aside as the province of the humanities.

As literary culture has been moving steadily farther away from the canons of empirical inquiry, the sciences have been approaching ever closer to a commanding and detailed knowledge of the subjects most germane to literary culture: human motives, human feelings, and the operations of the human mind. Evolutionary psychology and affective neuroscience have been penetrating the inner sanctum of the "qualitative" and making it accessible to precise empirical knowledge. Since the early 1990s, a few literary scholars have been assimilating the insights of evolutionary social science and envisioning radical changes in the conceptual foundations of literary study. These "literary Darwinists" have produced numerous theoretical and interpretive essays. Until recently, though, most literary Darwinists have remained within the methodological boundaries of traditional humanistic scholarship. Their work has been speculative, discursive, and rhetorical. They have drawn on empirical research but have not, for the most part, adopted empirical methods. Instead, they have used Darwinian theory as a source of theoretical and interpretive concepts. In respect to method, then, their work is similar to that of old-fashioned Freudians and Marxists.

In the project we describe here, we aimed to move past the barrier that separates the methods of the humanities from the methods of the social sciences. Building on research in evolutionary social science, we aimed to (1) construct a model of human nature consisting of motives, emotions, features of personality, and preferences in marital partners; (2) use that model to analyze a specific body of literary texts and the responses of readers to those texts; and (3) produce data—information that could be quantified and used to test specific hypotheses about those texts.

Project Design

We created an online questionnaire, listed about 2,000 characters from 201 canonical British novels of the nineteenth and early twentieth centuries, and asked respondents to select individual characters and answer questions about each character selected. We identified potential research participants by scanning lists of faculty in hundreds of English departments worldwide and selecting specialists in nineteenth-century British literature, especially scholars of the novel. We also sent invitations to multiple Listservs dedicated to the discussion of Victorian literature or specific authors or groups of authors in our study. Judging from demographic information provided by the respondents, approximately 519 respondents completed a total of 1,470 protocols on 435 characters.*

The questionnaire contains three sets of categories. One set consists of elements of personal identity: age, attractiveness, motives, criteria of mate selection, and personality. (The sex of the characters was a given.) The second set of categories consists of readers' emotional responses to characters. We listed ten possible emotional responses and asked the respondents to rate the intensity of their response on each of the ten emotions. The third set consists of four possible "agonistic" role assignments: (1) protagonists, (2) friends and associates of protagonists, (3) antagonists, and (4) friends and associates of antagonists. Dividing the four agonistic character sets into male and female sets produced a total of eight character sets. We conducted statistical tests to determine which scores on various categories differed significantly among the character sets. The patterned contrast between protagonists and antagonists is a contrast between desirable and undesirable traits in characters— a contrast we designate "agonistic structure." We also calculated degrees of correlation among the various categories of analysis—motives, criteria for selecting mates, personality factors, and the emotional responses of readers (for more technical statistical details on the project, see Johnson et al.).

TESTING A HYPOTHESIS

The questionnaire we used to collect data is couched in the common language and pitched at the level of common understanding, but it is also formulated within the framework of an evolutionary model of human nature. The questions we posed are thus situated at the point at which the evolutionary model converges with the common understanding. The questions register the common understanding, quantify it, and locate it within the context of empirical social science. Quantification enabled us to conduct an objective, formal analysis of the common understanding and to assess statistically the structural relations among its conceptual elements. One chief purpose of our study was simply to demonstrate that major features of literary meaning

---

*A copy of the questionnaire used in the study can be accessed at http://www-personal .umich.edu/~kruger/carroll-survey.html. The form is no longer active and will not be used to collect data.

can be effectively reduced to simple categories grounded in an evolutionary understanding of human nature.

Generating empirical knowledge in this way has an intrinsic value, but empirical findings clearly gain in value when they are brought to bear as evidence for specific hypotheses about important problems. Perhaps the most important problem in evolutionary literary study concerns the adaptive functions of literature and other arts—whether there are any adaptive functions and, if there are, what they might be. Our central hypothesis was that protagonists and their associates would form communities of cooperative endeavor and that antagonists would exemplify dominance behavior. If this hypothesis proved correct, the ethos reflected in the agonistic structure of the novels would replicate the egalitarian ethos of hunter-gatherers, who stigmatize and suppress status seeking in potentially dominant individuals (Boehm). If suppressing dominance in hunter-gatherers fulfills an adaptive social function, and if agonistic structure in the novels engages the same social dispositions that animate hunter-gatherers, our study would lend support to the hypothesis that this specific body of literary texts fulfills at least one adaptive social function.

Getting Motivated All species have a "life history," a species-typical pattern for birth, growth, reproduction, social relations (if the species is social), and death. For each species, the pattern of life history forms a reproductive cycle. In humans, that cycle centers on parents, children, and the social group. Successful parental care produces children capable, when grown, of forming adult pair bonds, becoming functioning members of a community, and caring for children of their own. "Human nature" is the set of species-typical characteristics that form adaptively functional parts of the human reproductive cycle.

For the purposes of this study, we reduced human life history to a set of twelve basic motives—that is, goal-oriented behaviors regulated by the reproductive cycle. For survival, we included two motives: survival itself (fending off immediate threats to life) and performing routine work to earn a living. We also asked questions about the importance of acquiring wealth, power, and prestige. We asked respondents to rate characters on how important acquiring a mate was to them in both the short term and the long term. In the context of these novels, short-term would mean flirtation or illicit sexual activity; long-term would mean seeking a marital partner. Taking account of "reproduction" in its wider significance of replicating genes that one shares with kin ("inclusive fitness"), we asked about the importance of helping offspring and other kin. For motives oriented to positive social relations beyond one's own kin, we included a question on "acquiring friends and making alliances" and another on "helping non-kin." And finally, to capture the uniquely human dispositions for acquiring culture, we included "seeking education or culture" and "building, creating, or discovering something."

Factor analysis is a statistical process in which terms that correlate with one another are grouped together to form a smaller number of terms designated "factors." When we submitted scores on the twelve separate motives to factor analysis, five main factors emerged: we refer to these as Social Dominance, Constructive Effort, Romance, Subsistence, and Nurture. Seeking wealth, power, and prestige all have strong positive loadings on Social Dominance, and helping non-kin has a moderate negative loading. (That is, helping non-kin correlates negatively with seeking wealth, power, and prestige.) Constructive Effort loads most strongly on the two cultural motives— "seeking education or culture" and "building, creating, or discovering something"—and has substantial loadings on two pro-social or affiliative motives: making friends and forging alliances, and helping non-kin. Romance is a mating motive, chiefly loading on short-term and long-term mating. Subsistence combines two motives: survival and performing routine tasks to earn a livelihood. Nurture loads most heavily on helping offspring or other kin, which correlates negatively with short-term mating. Helping non-kin also loads moderately on this factor, bringing affiliative kin-related behavior into association with generally affiliative social behavior.

Both male and female antagonists display a pronounced and exclusive emphasis on Social Dominance (figure 39.1). Male protagonists score higher than any other character set on Constructive Effort and Subsistence. Female protagonists score higher than any other character set on Romance, but their positive motives are fairly evenly balanced among Constructive Effort, Romance, and Nurture. In the novels in this study, female protagonists are largely restricted to the nubile age range. That restriction corresponds with a pronounced emphasis on Romance as a motive.

The opposition between dominance and affiliation in the novels is clearly linked with a robust and often replicated finding in psychological studies of motives and personality. Summarizing research into basic motives, David Buss observes that in cross-cultural studies, the two most important dimensions of interpersonal behavior are "power and love" (*Evolutionary Psychology* 21). Surveying the same field and citing still other antecedents, Delroy Paulhus and Oliver John observe that in debates about "the number of important human values," there are two, above all, that are "never overlooked" (1039). They designate these values "agency and communion" and associate them with contrasting needs: the need for "power and status," on the one side, and for "approval," on the other (1045).

Paulhus and John link the contrasting needs for power and approval with contrasting forms of bias in self-perception. "Egoistic" bias attributes

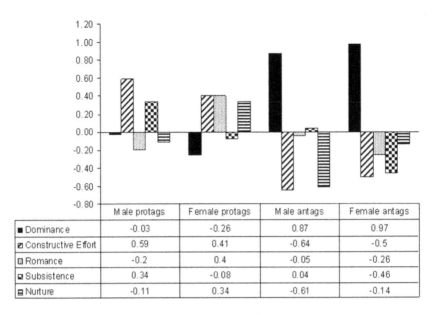

	M ale protags	F emale protags	M ale antags	F emale antags
■ Dominance	-0.03	-0.26	0.87	0.97
☑ Constructive Effort	0.59	0.41	-0.64	-0.5
▣ Romance	-0.2	0.4	-0.05	-0.26
▢ Subsistence	0.34	-0.08	0.04	-0.46
▤ Nurture	-0.11	0.34	-0.61	-0.14

FIGURE 39.1 Motive factors in protagonists and antagonists.

exaggerated "prominence and status" to oneself, and "moralistic" bias gives an exaggerated picture of oneself as a "nice person" and "a good citizen" (1045, 1046). Adopting these terms, we can say decisively that the novels in this study, taken collectively, have a moralistic bias. In protagonists, striving for personal predominance is strongly subordinated to communitarian values. Protagonists and their friends typically form communities of affiliative and cooperative behavior, and antagonists are typically envisioned as a force of social domination that threatens the very principle of community.

Most of the novels included in this study are "classics." One chief reason novels become classics is that they gain access to the deepest levels of human nature—not necessarily because they produce mimetically accurate representations of human nature, but because they evoke elemental human passions and deploy elemental forms of imaginative organization. The novels contain a vast fund of realistic social depiction and profound psychological analysis. In their larger imaginative structures, though, the novels evidently do not just represent human nature; they embody the impulses of human nature. Those impulses include a need to derogate dominance in others and to affirm one's identity as a member of a social group. Our evidence strongly suggests that those needs provide the emotional and imaginative force that shapes agonistic structure in the novels.

The novels create a virtual imaginative world designed to give concentrated emotional force to the opposition between dominance and affiliation. This virtual imaginative world provides a medium in which readers participate in a shared social ethos. The social ethos shapes agonistic structure, and agonistic structure in turn feeds back into the social ethos—affirming it, reinforcing it, integrating it with the changing circumstances of material and social life, and illuminating it with the aesthetic, intellectual, and moral powers of individual artists

Choosing a Partner Most of the novels in this study are "love stories." Plots usually include choosing a marital partner. Along with questions about motives, we asked questions about the criteria that characters use to select mates. Evolutionary psychologists have identified mating preferences that males and females share and those in which they differ (Buss, *Evolution*; Gangestad; Geary, *Male, Female*; Gottschall, "Greater Emphasis"; Kruger, Fisher, and Jobling; Schmitt; Symons, *Evolution*). Both males and females value kindness, intelligence, and reliability in mates. Males preferentially value physical attractiveness, and females preferentially value wealth, prestige, and power. These sex-specific preferences are rooted in the logic of reproduction. Physical attractiveness in females correlates with youth and health in a woman—hence with reproductive potential—and wealth, power, and prestige enable a male to provide for a mate and her offspring. We anticipated that scores for mate selection would correspond to the differences between males and females found in studies of mate selection in the real world. We also anticipated that protagonists would exhibit a stronger preference for intelligence, kindness, and reliability than would antagonists (figure 39.2).

In the results of the factor analyses for mate selection, the loadings divide with the sharpest possible clarity into three distinct factors: Extrinsic Attributes (a desire for wealth, power, and prestige in a mate), Intrinsic Qualities (a desire for kindness, reliability, and intelligence in a mate), and Physical Attractiveness (that one criterion by itself).

Female protagonists and antagonists show a stronger preference for Extrinsic Attributes than do male protagonists and antagonists, but female antagonists exaggerate the female tendency toward preferring Extrinsic Attributes. The emphasis that female antagonists give to Extrinsic Attributes parallels their single-minded pursuit of Social Dominance. Female protagonists display a more marked preference for Intrinsic Qualities than do male protagonists.

We did not anticipate, though, that male protagonists would be so strongly preoccupied with Physical Attractiveness relative to other qualities,

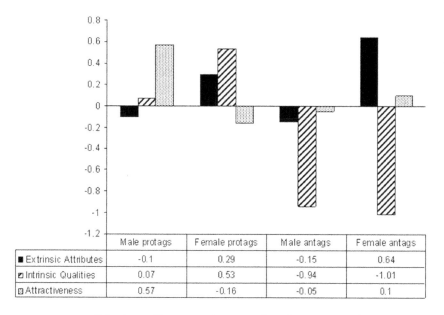

	Male protags	Female protags	Male antags	Female antags
■ Extrinsic Attributes	-0.1	0.29	-0.15	0.64
◪ Intrinsic Qualities	0.07	0.53	-0.94	-1.01
▨ Attractiveness	0.57	-0.16	-0.05	0.1

FIGURE 39.2 Criteria used by major characters for selecting marital partners.

nor did we foresee that male antagonists would be so relatively indifferent to Physical Attractiveness. The inference we draw from these findings is that the typically male desire for physical beauty in mates is part of the ethos of the novels. It is part of the charm and romance of the novels, part of the glamour. The relative indifference of male antagonists to Physical Attractiveness seems to reflect their general indifference to the quality of their personal relations.

If one were to look at only the motive factors, one might speculate that male antagonists correspond more closely to their gender norms than do female antagonists. Male antagonists could be conceived of as personified reductions to male dominance striving. The relative indifference of male antagonists to any differentiating features in mates might then look like an exaggeration of the male tendency toward interpersonal insensitivity. Thought of in this way, male antagonists would appear to be ultra-male, and female antagonists, in contrast, would seem to cross the gender divide. Their reduction to dominance striving would be symptomatic of a certain masculinization of motive and temperament. They would be, in an important sense, de-sexed. Plausible as this line of interpretation might seem, it does not bear up under the weight of the evidence about male antagonists' relative indifference to Physical Attractiveness in a mate. Like female antagonists' dominance striving, male antagonists' indifference to physical beauty is a form of de-sexing. Dominance striving devoid of affiliative disposition constitutes a reduction to sex-neutral egoism. The essential character of male and female antagonists is thus not a sex- or gender-specific tendency toward masculinization, but a tendency toward sexual neutralization in the isolation of an ego disconnected from all social bonds.

Developing a Personality When we speak of "human nature," we focus first of all on "human universals"—cognitive and behavioral features that everyone shares and that thus merge individuals into the common mass of humanity. We typically use personality, in contrast, to distinguish one person from another—for example, a friendly, careless extravert in contrast to a cold, conscientious introvert. The factors of personality can nonetheless themselves be conceived as stable, shared components of human nature. Each factor has a common substratum; individuals differ only in degree on each factor.

Current research into personality commonly distinguishes five broad factors (Buss, "Social Adaptation"; Costa and McCrae; MacDonald; Smits and Boeck). Extraversion signals assertive, exuberant activity in the social world, versus a tendency to be quiet, withdrawn, and disengaged. Agreeableness defines a pleasant, friendly disposition and a tendency to cooperate and compromise, versus a tendency to be self-centered and inconsiderate. Conscientiousness refers to an inclination toward purposeful planning, organization, persistence, and reliability, versus impulsivity, aimlessness, laziness, and undependability. Emotional Stability reflects a temperament that is calm and relatively free from negative feelings, versus a temperament marked by extreme emotional reactivity and persistent anxiety, anger, or depression. Openness to Experience describes a dimension of personality that distinguishes open (imaginative, intellectual, creative, complex) people from closed (down-to-earth, uncouth, conventional, simple) people.

We predicted (1) that protagonists and their associates would on average score higher than antagonists and their associates on the personality factor Agreeableness, a measure of warmth and affiliation, and (2) that protagonists would score higher than antagonists and minor characters on the personality factor Openness to Experience, a measure of intellectual vivacity.

Female protagonists score higher than any other set on Agreeableness, Conscientiousness, and Openness, and they score in the positive range on Stability (figure 39.3). Male protagonists look like muted or moderated versions of female protagonists. The personality profiles of male and female antagonists are very similar to each other—both somewhat extraverted, highly disagreeable, and low in Stability and Openness. Female antagonists are somewhat more conscientious than male antagonists.

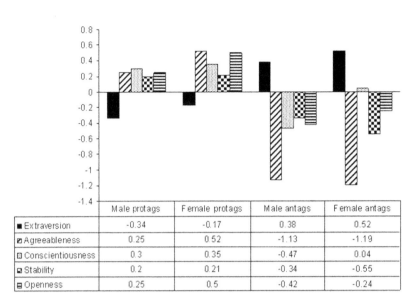

	Male protags	Female protags	Male antags	Female antags
■ Extraversion	-0.34	-0.17	0.38	0.52
◪ Agreeableness	0.25	0.52	-1.13	-1.19
▣ Conscientiousness	0.3	0.35	-0.47	0.04
▨ Stability	0.2	0.21	-0.34	-0.55
▤ Openness	0.25	0.5	-0.42	-0.24

FIGURE 39.3 Personality factors in protagonists and antagonists.

In the value structures implicit in the organization of characters in agonistic structure, Introversion, Agreeableness, Conscientiousness, Stability, and Openness are all positively valenced features. Agreeableness is the most strongly marked component of this normative array. Being agreeable is a trait that distinguishes good characters generally, but being conscientious and open to experience are more specifically characteristic of protagonists. With respect to personality, female protagonists are clearly the normative character set.

The value system embodied in agonistic structure links a volatile temperament with relatively weak self-discipline and a bad temper. Openness would be associated with the desire to seek education or culture and to build, discover, or create, and that whole complex of cognitive features is one of the two basic elements in Constructive Effort. As one would anticipate, then, Openness correlates moderately with Constructive Effort (r = 0.41). The total profile for protagonists is that of quiet, steady people—curious and alert but not aggressive, friendly but not particularly outgoing. The antagonists, in contrast, are assertive, volatile, and unreliable, but also intellectually or imaginatively dull and conventional. The main antagonistic motive factor is Social Dominance, which correlates strongly and negatively with Agreeableness (r = −0.54). Antagonists score in the extreme range on both Agreeableness (negatively) and Social Dominance (positively).

Repudiating a conventional distinction between "the novel of character" and "the novel of incident," Henry James poses a set of rhetorical questions: "What is character but the determination of incident? What is incident but the illustration of character? What is either a picture or a novel that is not of character? What else do we seek in it and find in it?" (55). In *The Mayor of Casterbridge*, more succinctly but in an equally emphatic way, Thomas Hardy (following Novalis) declares, "Character is Fate" (chap. 17). In what, then, does character consist? Strip away the now standard triad of race, class, and sex, and what is left? More than has been taken away. Beneath ethnic and class identity, beneath even the two basic human morphs of male and female, there are elemental features of human nature, the bedrock of personal identity. The composition of that bedrock can be assessed with the five factors of personality: the biologically elemental interaction between an organism and its environment; the capacity of all higher organisms to feel pain and react against it; the disposition of all mammals for affiliative bonding; and the specifically human capacities for organizing behavior over time, carrying out plans, and generating imaginative culture.

Becoming Emotionally Involved One of our chief working hypotheses was that when readers respond to characters in novels, they respond in much the same way, emotionally, as they do to people in everyday life. They like or dislike them, admire or despise them, fear them, feel sorry for them, or are amused by them. In writing fabricated accounts of human behavior, novelists select and organize their material in order to generate such responses, and readers willingly cooperate with this purpose. They participate vicariously in the experiences depicted and form personal opinions about the qualities of the characters. Authors and readers thus collaborate in producing a simulated experience of emotionally responsive evaluative judgment.

In building emotional responses into our research design, we sought to identify emotions that are universal and that are thus likely to be grounded in evolved features of human psychology. Emotions at that conceptual level would be on the same level as the basic motives extrapolated from human life history. Over the past forty years or so, adaptationist psychologists have made substantial progress in identifying basic emotions. Much of this work was pioneered by Paul Ekman. By isolating emotions that can be universally or almost universally recognized from facial expressions, Ekman and other researchers ultimately produced a core set of seven "basic" emotions: anger, fear, disgust, contempt, joy, sadness, and surprise. Different researchers sometimes use slightly different terms, register different degrees of intensity in emotions (for instance, anxiety, fear, terror, panic), organize the emotions in various patterns and combinations, or link them with self-awareness or social awareness to produce complex social emotions such as embarrassment, shame, guilt, and envy (M. Lewis; Panksepp 143). The core group of seven emotions nonetheless has widespread support as a usable taxonomy of basic emotions (Lewis and Haviland-Jones; Plutchik, *Emotions*).

Our questionnaire includes a list of ten emotional responses. To produce this list, we started with seven basic emotions defined by Ekman and adapted them for registering graded responses specifically to persons or characters. We used four of the seven terms unaltered: "anger," "disgust," "contempt," and "sadness." We also retained "fear" but divided it into two distinct items: fear *of* a character, and fear *for* a character. Ekman observes that the positive emotions have been less carefully observed and differentiated than the emotions that reflect emotional upset. The simple terms "joy" and "enjoyment" cover a wide spectrum of possible pleasurable or positive emotions, ranging from "amusement" to "schadenfreude" to "bliss" (Ekman, *Emotions* 191). In adapting the term "joy" or "enjoyment," we sought to register some qualitative differences and also devise terms appropriate to responses to a person. We chose three: "liking," "admiration," and "amusement." "Liking" is an emotionally positive response to a person, but it does not contain a specific element of approval or disapproval. "Admiration" combines positive emotionality with a measure of approval or respect (Darwin, *Expression* 289; Dutton 190–92; Plutchik, "Nature" 349). By itself, "surprise," like "joy," seems more appropriate as a descriptor for a response to a situation than to a person. Consequently, we did not use the word "surprise" by itself. Instead, along with "admiration," we used "amusement," which combines the idea of surprise with that of positively valenced emotionality. "Amusement" extends emotional response to take in responses to comedy. (Fear and sadness take in responses to tragedy; and anger and contempt, mingled with amusement, take in responses to satire.)

We included one further term in our list of possible emotional responses: "indifference." A number of researchers have included a term such as "interest" to indicate general attentiveness, the otherwise undifferentiated sense that something matters, that it is important and worthy of attention. "Indifference" can be regarded as the inverse of "interest." "Indifference" provides a qualitatively neutral measure of emotional reaction to a character.

We predicted (1) that protagonists would receive high scores on the positive emotional responses "liking" and "admiration"; (2) that antagonists would receive high scores on the negative emotional responses "anger," "disgust," "contempt," and "fear *of*" the character; (3) that protagonists would score higher than antagonists on "sadness" and "fear *for*" the character; and

(4) that major characters (protagonists and antagonists) would score lower than minor characters on "indifference."

Factor analysis produced three clearly defined emotional response factors: (1) Dislike, which includes anger, disgust, contempt, and fear *of* the character as well as negative correlations with liking and admiration; (2) Sorrow, which includes sadness and fear *for* the character as well as a negative correlation with amusement; and (3) Interest, which consists chiefly of a negative correlation with indifference.

Male and female protagonists score relatively low on Dislike and relatively high on Sorrow (figure 39.4). Male and female antagonists score very high on Dislike—higher than any other set—low on Sorrow, and somewhat above average on Interest. Female protagonists score high on Interest, whereas male protagonists, contrary to our expectations, score below average on Interest. They score lower even than good minor males, although not lower than the other minor characters.

Once one has isolated the components of agonistic structure and deployed a model of reading that includes basic emotions as a register of evaluatively polarized response, most of the scores on emotional response factors are predictable. There is, however, one surprising and seemingly anomalous finding that emerges from the scores on emotional responses—the relatively low score received by male protagonists on Interest. It ran contrary to our

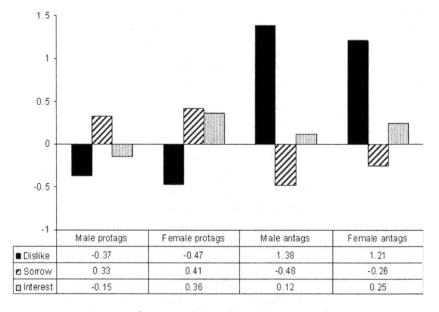

	Male protags	Female protags	Male antags	Female antags
■ Dislike	-0.37	-0.47	1.38	1.21
▨ Sorrow	0.33	0.41	-0.48	-0.26
▨ Interest	-0.15	0.36	0.12	0.25

FIGURE 39.4 Emotional response factors for protagonists and antagonists.

expectation that protagonists, both male and female, would score lower on indifference than would any other character set. We think this result can be explained by the way agonistic polarization feeds into the psychology of cooperation. Male protagonists in our data set are relatively moderate, mild characters. They are introverted and agreeable, and they do not seek to dominate others socially. They are pleasant and conscientious, and they are also curious and alert. They are attractive characters, but they are not very assertive or aggressive. They excite very little Dislike at least in part because they do not excite much competitive antagonism. They are not intent on acquiring wealth and power, and they are thoroughly domesticated within the forms of conventional propriety. They serve admirably to exemplify normative values of cooperative behavior, but in serving this function they seem to be diminished in some vital component of fascination, some element of charisma. They lack power, and they thus also seem to lack some quality that excites intensity of interest in emotional response.

In the novels in this study, the aggressive pursuit of Social Dominance—wealth, prestige, and power—is morally demonized. The desire for Social Dominance is overwhelmingly the single most distinctive motivational trait of both male and female antagonists. That motivational trait correlates with low scores on the affiliative personality factor Agreeableness and high scores on the emotional response factor Dislike. Despite this strongly valenced cluster of correlations, male and female antagonists score higher on Interest (lower on indifference) than do male protagonists. Readers dislike antagonists, but antagonists are sometimes more exciting than protagonists, especially male protagonists.

What Do We Make of It All?

Agonistic structure in these novels displays a systematic contrast between desirable and undesirable traits in characters. Protagonists exemplify traits that evoke admiration and liking in readers, and antagonists exemplify traits that evoke anger, fear, disgust, and contempt. Antagonists virtually personify Social Dominance—the self-interested pursuit of wealth, prestige, and power. In these novels, those ambitions are sharply segregated from prosocial and culturally acquisitive dispositions. Antagonists are not only selfish and unfriendly but also undisciplined, emotionally unstable, and intellectually dull. Protagonists, in contrast, display motive dispositions and personality traits that exemplify strong personal development and healthy social adjustment. They are agreeable, conscientious, emotionally stable, and open to experience. Protagonists clearly represent the apex of the positive values implicit in agonistic structure. Both male and female protagonists score high on the motive factor Constructive Effort, which combines prosocial and culturally acquisitive dispositions. Their introversion, in this context, seems part of their mildness. The extraversion of antagonists, in contrast, seen in the context of their other scores, seems to indicate aggressive self-assertion.

In the past thirty years or so, more criticism on the novel has been devoted to the issue of gender identity than to any other topic. The data in our study indicate that gender can be invested with a significance out of proportion to its true place in the structure of interpersonal relations in novels and that it can be conceived in agonistically polarized ways out of keeping with the forms of social affiliation depicted in novels. In this data set, differences between males and females are less prominent than differences between protagonists and antagonists. If polarized emotional responses were absent from the novels, or if those polarized responses co-varied with differences between males and females, the differences between male and female characters would have to be conceived agonistically, as a conflict (as it is, for instance, in Gilbert and Gubar). The differences between male and female characters in motives and personality could be conceived as competing value structures. From a Marxist perspective, that competition would be interpreted as essentially political and economic in character (as it is, for instance, in Armstrong). From the deeper Darwinian perspective, it would ultimately be attributed to competing reproductive interests. The subordination of sex to agonistic role assignment, though, suggests that in these novels conflict between the sexes is subordinated to their shared and complementary interests. In the agonistic structure of plot and theme, male and female protagonists are allies. They cooperate in resisting the predatory threats of antagonists, and they join together to exemplify the values that elicit readers' admiration and sympathy. Both male and female antagonists are massively preoccupied with material gain and social rank. That preoccupation stands in stark contrast to the more balanced and developed world of the protagonists—a world that includes sexual interest, romance, the care of family, friends, and the life of the mind. By isolating and stigmatizing dominance behavior, the novels affirm the shared values that bind its members into a community.

In *Hierarchy in the Forest: The Evolution of Egalitarian Behavior*, Christopher Boehm offers a cogent explanation for the way interacting impulses of dominance and affiliation have shaped the evolution of human political behavior. In an earlier phase of evolutionary social science, sociobiological theorists repudiated the idea of "altruistic" behavior and restricted prosocial dispositions to nepotism and to the exchange of reciprocal benefits. In contrast, Boehm argues that at some point in their evolutionary history—at the latest, 100,000 years ago—humans developed a special capacity, dependent on their symbolic and cultural capabilities, for enforcing altruistic or group-oriented norms. By enforcing these norms, humans succeeded in controlling "free riders" or "cheaters," and they thus made it possible for genuinely altruistic genes to survive within a social group. Such altruistic dispositions, enforced by punishing defectors, would have enabled some social groups to compete more successfully against other groups and would thus have made "group selection" or "multilevel selection" an effective force in subsequent human evolution. The selection for altruistic dispositions—and dispositions for enforcing altruistic cultural norms—would have involved a process of gene–culture co-evolution that would snowball in its effect of altering human nature itself (Darwin, *Descent*; Richerson and Boyd; Wade, *Before*; Wilson, *Evolution*).

Taking into account not just the representation of characters but the emotional responses of readers, we can identify agonistic structure as a simulated experience of emotionally responsive social interaction. That experience has a clearly defined moral dimension. Agonistic structure precisely mirrors the kind of egalitarian social dynamic documented by Boehm in hunter-gatherers—our closest contemporary proxy to ancestral humans. As Boehm and others have argued, the dispositions that produce an egalitarian social dynamic are deeply embedded in the evolved and adapted character of human nature. Humans have an innate desire for power and an innate dislike of being dominated. Egalitarianism as a political strategy arises as a compromise between the desire to dominate and the dislike of being dominated. By pooling their power in order to exercise collective social coercion, individuals in groups can repress dominance behavior in other individuals. The result is autonomy for individuals. No one gets all the power he or she would like, but then, no one has to accept submission to other dominant individuals. Boehm describes in detail the pervasive collective tactics for repressing dominance within social groups organized at the levels of bands and tribes (see also Salter 63–70).

An egalitarian social dynamic is the most important basic structural feature that distinguishes the social organization of humans from that of chimpanzees. In chimpanzee society, social organization is regulated exclusively by dominance—that is, power. In human society, social organization is regulated by interactions between impulses of dominance and impulses for suppressing dominance. State societies with elaborate systems of hierarchy emerged only very recently in the evolutionary past, about 10,000 years ago, after the agricultural revolution made possible concentration of resources and, therefore, power. Before the advent of despotism, the egalitarian disposition for suppressing dominance had, at a minimum, 100,000 years in which to become entrenched in human nature—more than sufficient time for significant adaptive change to take place (Wade, *Before*). In highly stratified societies, dominance assumes a new ascendancy, but no human society dispenses with the need for communitarian association. It seems likely, then, that agonistic structure in fictional narratives emerged in tandem with specifically human adaptations for cooperation and specifically human adaptations for creating imaginative constructs that embody the ethos of the tribe.

In preliterate cultures, social dynamics take place in face-to-face interactions, through the perpetual hubbub of dialogue, gossip, and the telling of tales. That kind of interaction is necessarily restricted to relatively small populations, to bands or tribes. In literate cultures, in contrast, social dynamics can take place vicariously through the shared imaginative experience of literature. In responding to literary characters, readers join the community of all readers responding in similar ways to the social dynamics depicted in novels. Authors and readers thus collaborate in producing a virtual imaginative world. In this world, readers affirm and reinforce cooperative dispositions on a large social scale. Agonistic structure thus functions as a cultural technology that extends an adaptive social process across social groups larger than the band or tribe. It is a medium both for gene–culture co-evolution and for natural selection at the level of social groups. It is, in other words, an adaptively functional feature of human nature.

Bibliography

Abbott, H. Porter. "The Evolutionary Origins of the Storied Mind: Modelling the Prehistory of Narrative Consciousness and Its Discontents." *Narrative* 8 (2000): 247–56.

Abrams, M. H. "The Correspondent Breeze: A Romantic Metaphor." In *English Romantic Poets: Modern Essays in Criticism*, edited by M. H. Abrams, 37–54. New York: Oxford University Press, 1960.

———. *Natural Supernaturalism: Tradition and Revolution in Romantic Literature*. New York: Norton, 1971.

Adamson, Jane. *"Othello" as Tragedy: Some Problems of Judgment and Feeling*. Cambridge: Cambridge University Press, 1980.

Aiello, Leslie C. "Thumbs Up for Our Early Ancestors." *Science* 265 (1994): 1540–41.

Aiello, Leslie C., and Peter Wheeler. "The Expensive Tissue Hypothesis." *Current Anthropology* 36 (1995): 199–211.

Aiken, Nancy E. *The Biological Origins of Art*. Westport, Conn.: Praeger, 1998.

Alcock, John. *Animal Behavior: An Evolutionary Approach*. 4th ed. Sunderland, Mass.: Sinauer, 1989.

Alexander, Richard D. *The Biology of Moral Systems*. Hawthorne, N.Y.: Aldine de Gruyter, 1987.

Allen, Michael, David D'Alessio, and Tara Emmers-Sommer. "Reactions of Criminal Sexual Offenders to Pornography: A Meta-Analytic Summary." In *Communication Yearbook* 22, edited by Michael E. Roloff, 139–70. Thousand Oaks, Calif.: Sage, 2000.

Allen, Woody. *Without Feathers*. New York: Ballantine, 1983.

Allott, Miriam, ed. Introduction to *Emily Brontë: Wuthering Heights: A Casebook*. Houndmills: Macmillan, 1970.

———. "The Rejection of Heathcliff?" *Essays in Criticism* 8 (1958): 27–47.

Althusser, Louis. *The Future Lasts Forever: A Memoir*. Translated by Richard Veasey. New York: New Press, 1992.

———. *Lenin and Philosophy and Other Essays*. Translated by Ben Brewster. New York: Monthly Review Press, 1971.

Ambady, Nalini, and Robert Rosenthal. "Thin Slices of Expressive Behavior as Predictors of Interpersonal Consequences: A Meta-Analysis." *Psychological Bulletin* 111 (1992): 256–74.

Amussen, Susan Dwyer. *An Ordered Society: Gender and Class in Early Modern England*. Oxford: Blackwell, 1988.

Anderson, Joseph D. *The Reality of Illusion: An Ecological Approach to Cognitive Film Theory*. Carbondale: Southern Illinois University Press, 1996.

Anderson, Joseph D., and Barbara Fisher Anderson, eds. *Moving Image Theory: Ecological Considerations*. Carbondale: Southern Illinois University Press, 2005.

Andersson, Malte. *Sexual Selection*. Princeton, N.J.: Princeton University Press, 1994.

Angeioni, Elvio. *Annual Editions: Anthropology 01101*. 24th ed. New York: McGraw-Hill, 2000.

Appiah, Kwame Anthony. "The Battle of the Bien-Pensant." In *Theory's Empire: An Anthology of Dissent*, edited by Daphne Patai and Will H. Corral, 441–48. New York: Columbia University Press, 2005.

———. *Cosmopolitanism: Ethics in a World of Strangers*. New York: Norton, 2006.

———. *Experiments in Ethics*. Cambridge, Mass.: Harvard University Press, 2008.

———. *In My Father's House: Africa in the Philosophy of Culture*. London: Methuen, 1992.

———. "Whose Culture Is It?" *New York Review of Books*, February 9, 2006, 38–41.

Argyros, Alexander J. *A Blessed Rage for Order: Deconstruction, Evolution, and Chaos*. Ann Arbor: University of Michigan Press, 1991.

Aristotle. *The Poetics*. Translated by Richard Janko. Indianapolis: Hackett, 1965.

Armstrong, Nancy. *Desire and Domestic Fiction: A Political History of the Novel*. New York: Oxford University Press, 1987.

———. "Emily Brontë in and out of Her Time." *Genre* 15 (1982): 243–64.

Atran, Scott. *In Gods We Trust: The Evolutionary Landscape of Religion*. Oxford: Oxford University Press, 2002.

Balázs, Béla. *Theory of the Film*. New York: Arno Press, 1972.

Barkow, Jerome H. "Beneath New Culture Is Old Psychology: Gossip and Social Stratification." In *The Adapted Mind: Evolutionary Psychology and the Generation of Culture*, edited by Jerome H. Barkow, Leda Cosmides, and John Tooby, 627–37. New York: Oxford University Press, 1992.

Barkow, Jerome H., Leda Cosmides, and John Tooby, eds. *The Adapted Mind: Evolutionary Psychology and the Generation of Culture*. New York: Oxford University Press, 1992.

Baron-Cohen, Simon. "The Empathizing System: A Revision of the 1994 Model of the Mindreading System." In *Origins of the Social Mind: Evolutionary Psychology and Child Development*, edited by Bruce J. Ellis and David F. Bjorklund, 468–92. New York: Guilford Press, 2005.

——. *The Essential Difference: The Truth About the Male and Female Brain*. New York: Basic Books, 2003.

——. *Mindblindness: An Essay on Autism and Theory of Mind*. Cambridge, Mass.: MIT Press, 1995.

Barrett, Louise, Robin Dunbar, and John Lycett. *Human Evolutionary Psychology*. Princeton, N.J.: Princeton University Press, 2002.

Barrow, John D. *The Artful Universe*. Oxford: Clarendon Press, 1995.

Barry, Peter. *Beginning Theory: An Introduction to Literary and Cultural Theory*. Manchester: Manchester University Press, 1995.

Barthelemy, Anthony Gerard, ed. *Critical Essays on Shakespeare's "Othello."* New York: Hall, 1994.

——. "Ethiops Washed White: Moors of the Nonvillainous Type." In *Critical Essays on Shakespeare's "Othello,"* edited by Anthony Gerard Barthelemy, 91–103. New York: Hall, 1994.

Barthes, Roland. *Mythologies*. Translated by Annette Lavers. New York: Hill and Wang, 1972.

Bate, Jonathan. *The Genius of Shakespeare*. London: Macmillan, 1997.

Batten, Mary. *Sexual Strategies: How Females Choose Their Mates*. New York: Putnam, 1992.

Battersby, Christine. *Gender and Genius: Towards a Feminist Aesthetics*. London: Women's Press, 1989.

Baum, Rosalie Murphy. "The Shape of Hurston's Fiction." In *Zora in Florida*, edited by Steve Glassman and Kathryn Lee Seidel, 94–109. Orlando: University of Central Florida Press, 1991.

Baumeister, Roy F. *The Cultural Animal: Human Nature, Meaning, and Social Life*. Oxford: Oxford University Press, 2005.

Bazerman, Charles, ed. *The Informed Reader: Contemporary Issues in the Disciplines*. Boston: Houghton Mifflin, 1989.

Beauchamp, Gorman. "Of Man's Last Disobedience: Zamiatin's *We* and Orwell's *1984*." *Comparative Literature Studies* 10 (1973): 285–301.

Beebe, Beatrice. "Mother–Infant Mutual Influence and Precursors of Self- and Object-Representations." In *Empirical Studies of Psychoanalytic Theories*, edited by Joseph Masling, 2:27–48. Hillsdale, N.J.: Analytic Press, 1986.

Beebe, Beatrice, and Louis Gerstman. "A Method of Defining 'Packages' of Maternal Stimulation and Their Functional Significance for the Infant with Mother and Stranger." *International Journal of Behavioral Development* 7 (1984): 423–40.

Bell, Quentin. *On Human Finery*. 1976. Reprint, London: Allison and Busby, 1992.

Belsey, Catherine. "Biology and Imagination: The Role of Culture." Paper presented at the Institute of Contemporary Arts, London, May 8, 2004.

——. *Critical Practice*. London: Methuen, 1980.

——. *Desire: Love Stories in Western Culture*. Oxford: Blackwell, 1994.

——. "The Name of the Rose in *Romeo and Juliet*." *Yearbook of Shakespeare Studies* 23 (1993): 125–42.

——. *The Subject of Tragedy*. London: Methuen, 1985.

Berman, Jeffrey. *Narcissism and the Novel*. New York: New York University Press, 1990.

Berry, Philippa, and Margaret Trudeau-Clayton. Introduction to *Textures of Renaissance Knowledge*, edited by Phillipa Berry and Margaret Trudeau-Clayton, 1–14. Manchester: Manchester University Press, 2003.

Bersani, Leo. *A Future for Astyanax: Character and Desire in Literature*. Boston: Little, Brown, 1969.

Betzig, Laura. *Despotism and Differential Reproduction: A Darwinian View of History*. Hawthorne, N.Y.: Aldine de Gruyter, 1986.

——, ed. *Human Nature: A Critical Reader*. New York: Oxford University Press, 1997.

Bevington, David. "*Othello:* Portrait of a Marriage." In *"Othello": New Critical Essays*, edited by Philip Kolin, 221–29. New York: Routledge, 2002.

Bingham, Paul. "Human Uniqueness: A General Theory." *Quarterly Review of Biology* 74 (1999): 133–69.

Birket-Smith, Kaj. *The Caribou Eskimos: Material and Social Life and Their Cultural Position*. Translated by W. E. Calvert. Vol. 5 of *Report of the Fifth Thule Expedition, 1921–1924*. Copenhagen: Gyldendalske Boghandel, Nordisk Forlag, 1929.

Bischof, Norbert. "Comparative Ethology of Incest Avoidance." In *Biosocial Anthropology*, edited by Robin Fox, 37–67. London: Malaby, 1975.

Bjorklund, David F., and Anthony D. Pellegrini. *The Origins of Human Nature: Evolutionary Developmental Psychology*. Washington, D.C.: American Psychological Association, 2002.

Blank, G. Kim. *Wordsworth and Feeling: The Poetry of an Adult Child*. Madison, N.J.: Fairleigh Dickinson University Press, 1995.

Bloom, Allan. *Shakespeare on Love and Friendship*. Chicago: University of Chicago Press, 2000.

Bloom, Harold. *Shakespeare: The Invention of the Human*. New York: Riverhead Books, 1998.

——. *The Western Canon: The Books and School of the Ages*. Orlando, Fla.: Harcourt, Brace, 1994.

Boas, Franz. *Primitive Art*. 1927. Reprint, New York: Dover, 1995.

Boehm, Christopher. *Hierarchy in the Forest: The Evolution of Egalitarian Behavior*. Cambridge, Mass.: Harvard University Press, 1999.

Bohannan, Laura. "Shakespeare in the Bush." In *The Informed Reader: Contemporary Issues in the Disciplines*, edited by Charles Bazerman, 43–53. Boston: Houghton Mifflin, 1966.

Bohannan, Laura, and Paul Bohannan. *The Tiv of Central Nigeria*. London: International African Institute, 1953.

Boker, Pamela A. *The Grief Taboo in American Literature: Loss and Prolonged Adolescence in Twain, Melville, and Hemingway*. New York: New York University Press, 1996.

Bordwell, David. "Contemporary Film Studies and the Vicissitudes of Grand Theory." In *Post-Theory: Reconstructing Film Studies*, edited by David Bordwell and Noël Carroll, 1–36. Madison: University of Wisconsin Press, 1996.

——. *Figures Traced in Light: On Cinematic Staging*. Berkeley: University of California Press, 2005.

——. Foreword to *Moving Image Theory: Ecological Considerations*, edited by Joseph D. Anderson and Barbara Fisher Anderson, ix–xii. Carbondale: Southern Illinois University Press, 2005.

——. *Making Meaning: Inference and Rhetoric in the Interpretation of Cinema*. Cambridge, Mass.: Harvard University Press, 1991.

——. *Narration in the Fiction Film*. Madison: University of Wisconsin Press, 1985.

——. *Planet Hong Kong: Popular Cinema and the Art of Entertainment*. Cambridge, Mass.: Harvard University Press, 2000.

——. *Poetics of Cinema*. New York: Routledge, 2008.

Bordwell, David, and Kristin Thompson. *Film Art: An Introduction*. 7th ed. New York: McGraw-Hill, 2004.

Borgia, Gerald. "Complex Male Display and Female Choice in the Spotted Bowerbird: Specialized Functions for Different Bower Decorations." *Animal Behaviour* 49 (1995): 1291–1301.

——. "Sexual Selection in Bowerbirds." *Scientific American* 254, no. 6 (1986): 92–100.

Bourdieu, Pierre. *Distinction: A Social Critique of the Judgement of Taste*. Translated by Richard Nice. Cambridge, Mass.: Harvard University Press, 1984.

——. *The Field of Cultural Production: Essays on Art and Literature*. Edited, with an introduction, by Randal Johnson. New York: Columbia University Press, 1993.

——. *The Logic of Practice*. Translated by Richard Nice. Stanford, Calif.: Stanford University Press, 1990.

——. *Outline of a Theory of Practice*. Translated by Richard Nice. Cambridge: Cambridge University Press, 1977.

Bower, Gordon H., and Daniel G. Morrow. "Mental Models in Narrative Comprehension." *Science*, n.s., 247 (1990): 44–48.

Bowlby, John. *Attachment*. Vol. 1 of *Attachment and Loss*. New York: Basic Books, 1969.

Boyd, Brian. "Art and Evolution: The Avant-Garde as Test Case: Spiegelman in *The Narrative Corpse*." *Philosophy and Literature* 32 (2008): 31–57. [For a fuller text and detailed notes, see http://www.arts.auckland.ac.nz/staff/index.cfm?P=3566]

——. "Getting It All Wrong." *American Scholar* 75, no. 4 (2006): 18–30.

——. "Literature and Evolution: A Bio-Cultural Approach." *Philosophy and Literature* 29 (2005): 1–23.

——. "On the Origin of Comics." *Evolutionary Review* 1 (2010).

——. *On the Origin of Stories: Evolution, Cognition, and Fiction*. Cambridge, Mass.: Harvard University Press, 2009.

Boyer, Pascal. *The Naturalness of Religious Ideas: A Cognitive Theory of Religion*. Berkeley: University of California Press, 1994.

——. "Specialised Inference Engines as Precursors of Creative Imagination?" *Proceedings of the British Academy* 147 (2007): 239–58.

Bradshaw, Graham. *Misrepresentations: Shakespeare and the Materialists*. Ithaca, N.Y.: Cornell University Press, 1993.

Brauner, Sigrid. *Fearless Wives and Frightened Shrews: The Construction of the Witch in Early Modern Germany*. Amherst: University of Massachusetts Press, 2001.

Brazelton, T. Berry. "Implications of Infant Development Among the Mayan Indians of Mexico." In *Culture and Infancy: Variations in the Human Experience*, edited by P. Herbert Leiderman, Steven R. Tulkin, and Anne Rosenfeld, 151–87. New York: Academic Press, 1977.

Breitenberg, Mark. *Anxious Masculinity in Early Modern England*. Cambridge: Cambridge University Press, 1996.

Brockman, John. *The Third Culture: Beyond the Scientific Revolution*. New York: Simon and Schuster, 1994.

Brontë, Emily. *Wuthering Heights: The 1847 Text, Backgrounds and Criticism*. Edited by Richard J. Dunn. 4th ed. New York: Norton, 2003.

Brooks, Cleanth. *The Well Wrought Urn: Studies in the Structure of Poetry*. 1942. Reprint, New York: Macmillan, 1975.

Broude, Gwen J., and Sarah J. Greene. "Cross-Cultural Codes on Husband–Wife Relationships." *Ethnology* 22 (1983): 263–80.

Brown, Donald E. *Hierarchy, History, and Human Nature: The Social Origins of Historical Consciousness*. Tucson: University of Arizona Press, 1988.

——. *Human Universals*. Philadelphia: Temple University Press, 1991.

Bruce, Vicki, Tim Valentine, and Alan Baddeley. "The Basis of the 3/4 View Advantage in Face Recognition." *Applied Cognitive Psychology* 1 (1987): 109–20.

Buckley, Sandra. "'Penguin in Bondage': A Graphic Tale of Japanese Comic Books." In *Technoculture*, edited by Constance Penley and Andrew Ross, 163–95. Minneapolis: University of Minnesota Press, 1991.

Burford, Bronwen. "Action Cycles: Rhythmic Actions for Engagement with Children and Young Adults with Profound Mental Handicap." *European Journal of Special Needs Education* 3, no. 4 (1988): 189–206.

Burkert, Walter. *Creation of the Sacred: Tracks of Biology in Early Religions*. Cambridge Mass.: Harvard University Press, 1996.

Burton, Robert. *The Anatomy of Melancholy*. 1621. Edited by Thomas C. Faulkner, Nicolas K. Kiessling, and Rhonda L. Blair. 6 vols. Oxford: Clarendon Press, 1989–2000.

Buss, David M. *The Dangerous Passion: Why Jealousy Is as Necessary as Love and Sex*. New York: Free Press, 2000.

——. *The Evolution of Desire: Strategies of Human Mating*. Rev. ed. New York: Basic Books, 2003.

——. *Evolutionary Psychology: The New Science of the Mind*. Boston: Allyn and Bacon, 1999.

——. ed. *The Handbook of Evolutionary Psychology*. Hoboken, N.J.: Wiley, 2005.

——. "Sex Differences in Human Mate Preferences.: Evolutionary Hypothesis Testing in 37 Cultures." *Behavioral and Brain Sciences* 12 (1989): 1–49.

——. "Sexual Conflict: Evolutionary Insights into Feminism and the 'Battle of the Sexes.'" In *Sex, Power, Conflict: Evolutionary and Feminist Perspectives*, edited by David M. Buss and Neil M. Malamuth, 296–315. New York: Oxford University Press, 1996.

———. "Social Adaptation and Five Major Factors of Personality." In *The Five-Factor Model of Personality: Theoretical Perspectives*, edited by Jerry S. Wiggins, 180–207. New York: Guilford Press, 1996.

Buss, David M., and Neil M. Malamuth, eds. *Sex, Power, Conflict: Evolutionary and Feminist Perspectives*. New York: Oxford University Press, 1996.

Butler, Christopher. *Postmodernism: A Very Short Introduction*. Oxford: Oxford University Press, 2002.

Butler, Samuel. *The Authoress of the Odyssey*. 1922. Reprint, London: Cape, 1987.

Byron, George Gordon Byron. *Don Juan*. Edited by T. G. Steffan, E. Steffan, and W. W. Pratt. Introduction by Susan J. Wolfson and Peter J. Manning. London: Penguin, 2004.

Calderwood, James L. "Appalling Property in *Othello*." *University of Toronto Quarterly* 57, no. 3 (1988): 353–75.

Callaghan, John W. "A Comparison of Anglo, Hopi, and Navajo Mothers and Infants." In *Culture and Early Interactions*, edited by Tiffany M. Field, Anita M. Sostek, Peter Vietze, and P. Herbert Leiderman, 115–31. Hillsdale, N.J.: Erlbaum, 1981.

Campbell, Anne. *A Mind of Her Own: The Evolutionary Psychology of Women*. Oxford: Oxford University Press, 2002.

Campbell, Donald T. "Blind Variation and Selective Retention in Creative Thought and Other Knowledge Processes." *Psychological Review* 67 (1960): 380–400.

———. "Evolutionary Epistemology." In *The Philosophy of Karl Popper*, edited by Paul A. Schilpp, 413–63. LaSalle, Ill.: Open Court, 1974.

Carroll, Joseph. "The Cuckoo's History: Human Nature in *Wuthering Heights*." *Philosophy and Literature* 32, no. 2 (2008): 241–57.

———. "The Deep Structure of Literary Representations." *Evolution and Human Behavior* 20 (1999): 159–73.

———. *Evolution and Literary Theory*. Columbia: University of Missouri Press, 1995.

———. "An Evolutionary Paradigm for Literary Study." *Style* 42 (2008): 103–35.

———. *Literary Darwinism: Evolution, Human Nature, and Literature*. New York: Routledge, 2004.

———. "Rejoinder to the Responses." *Style* 42 (2008): 308–79.

Carroll, Joseph, Jonathan Gottschall, John A. Johnson, and Daniel J. Kruger. *Graphing Jane Austen: Paleolithic Politics in British Novels of the Nineteenth Century*. Under submission.

———. "Paleolithic Politics in British Novels of the Nineteenth Century." In *Integrating Science and the Humanities: Interdisciplinary Approaches*, edited by Edward Slingerland and Mark Collard. New York: Oxford University Press, forthcoming.

Carroll, Noël. *Mystifying Movies: Fads and Fallacies in Contemporary Film Theory*. New York: Columbia University Press, 1988.

———. "The Power of Movies." In *Theorizing the Moving Image*, edited by Noël Carroll, 78–93. Cambridge: Cambridge University Press, 1996.

———. "Toward a Theory of Point-of-View Editing: Communication, Emotion, and the Movies." *Poetics Today* 14, no. 1 (1993): 127–31.

Cecil, David. *Early Victorian Novelists: Essays in Revaluation*. Indianapolis: Bobbs-Merrill, 1935.

Chagnon, Napoleon A. *Yanomamö.* 5th ed. Fort Worth, Tex.: Harcourt Brace, 1997.

Charney, Maurice. *Shakespeare on Love and Lust.* New York: Columbia University Press, 2000.

Charron, Pierre. *Of Wisdome.* Translated by Samson Lennard. London, 1606.

Chinn, Nancy, and Elizabeth E. Dunn. "'The Ring of Singing Metal on Wood': Zora Neale Hurston's Artistry in 'The Gilded Six-Bits.'" *Mississippi Quarterly: The Journal of Southern Cultures* 49, no. 4 (1996): 2–10.

Chitham, Edward. *A Life of Emily Brontë.* Oxford: Blackwell, 1987.

Chodorow, Nancy. *The Reproduction of Mothering: Psychoanalysis and the Sociology of Gender.* Berkeley: University of California Press, 1978.

Chomsky, Noam. "Language and Freedom." *TriQuarterly* 23–24 (1972): 29–30.

——. *Language and Thought.* Wakefield, R.I.: Moyer Bell, 1993.

Clay, Felix. *The Origin of the Sense of Beauty.* London: John Murray, 1908.

Coleman, Janet. "Machiavelli's *Via Moderna*: Medieval and Renaissance Attitudes to History." In *Niccolò Machiavelli's "The Prince": New Interdisciplinary Essays,* edited by Martin Coyle, 40–64. Manchester: Manchester University Press, 1995.

Commission on the Humanities. *The Humanities in American Life: Report of the Commission on the Humanities.* Berkeley: University of California Press, 1980.

Connors, James. "Zamyatin's *We* and the Genesis of *1984.*" *Modern Fiction Studies* 21 (1975): 107–24.

Constable, John. "Verse Form: A Pilot Study in the Epidemiology of Representations." *Human Nature* 8 (1997): 171–203.

Cooke, Brett. *Human Nature in Utopia: Zamyatin's "We."* Evanston, Ill.: Northwestern University Press, 2002.

Cosmides, Leda, and John Tooby. "Consider the Source: The Evolution of Adaptations for Decoupling and Metarepresentation." In *Metarepresentations: A Multidisciplinary Perspective,* edited by Dan Sperber, 53–115. New York: Oxford University Press, 2000.

——. "Evolutionary Psychology: A Primer." CEP Home Page at UCSB. 1997. Available at http://www.psych.ucsb.edu/research/cep (accessed September 8, 2009).

——. "Neurocognitive Adaptations Designed for Social Exchange." In *The Handbook of Evolutionary Psychology,* edited by David M. Buss, 584–627. Hoboken, N.J.: Wiley, 2005.

Costa, Paul T., Jr., and Robert R. McCrae. "Personality Trait Structure as a Human Universal." *American Psychologist* 52 (1997): 509–16.

Count, Earl W. "The Biological Basis of Human Sociality." *American Anthropologist* 60 (1958): 1049–85.

Cousins, Norman. *Anatomy of an Illness as Perceived by the Patient.* New York: Norton, 1979.

——. *The Healing Heart: Antidotes to Panic and Helplessness.* New York: Norton, 1983.

Cowhig, Ruth. "The Importance of Othello's Race." *Journal of Commonwealth Literature* 12 (1977): 153–61.

Crews, Frederick C. *Skeptical Engagements.* New York: Oxford University Press, 1986.

Cronin, Helena. "Adaptation: A Critique of Some Current Evolutionary Thought." *Quarterly Review of Biology* 80 (2005): 19–27.

——. *The Ant and the Peacock: Altruism and Sexual Selection from Darwin to Today*. New York: Cambridge University Press, 1992.

Csermely, D., and D. Mainardi. "Infant Signals." In *The Behavior of Human Infants*, edited by Alberto Oliverio and Michelle Zappella, 1–19. New York: Plenum Press, 1983.

Culler, Jonathan. *Literary Theory: A Very Short Introduction*. Oxford: Oxford University Press, 1997.

Cummins, Denise. "Dominance, Status, and Social Hierarchies." In *The Handbook of Evolutionary Psychology*, edited by David M. Buss, 676–97. Hoboken, N.J.: Wiley, 2005.

Cunningham, Valentine. "Theory, What Theory?" In *Theory's Empire: An Anthology of Dissent*, edited by Daphne Patai and Will H. Corral, 24–41. New York: Columbia University Press, 2005.

Daly, Martin, and Margo Wilson. "Evolutionary Psychology and Marital Conflict." In *Sex, Power, Conflict: Evolutionary and Feminist Perspectives*, edited by David M. Buss and Neil M. Malamuth, 9–28. New York: Oxford University Press, 1996.

——. *Homicide*. New York: Aldin de Gruyter, 1988.

——. *Sex, Evolution, and Behavior*. Belmont, Calif.: Wadsworth, 1983.

——. *The Truth About Cinderella: A Darwinian View of Parental Love*. New Haven, Conn.: Yale University Press, 1999.

Danto, Arthur. "Description and the Phenomenology of Perception." In *Visual Theory: Painting and Interpretation*, edited by Norman Bryson, Michael Ann Holly, and Keith Moxey, 209–11. New York: HarperCollins, 1991.

Darwin, Charles. *The Descent of Man, and Selection in Relation to Sex*. 1871. Edited by John Tyler Bonner and Robert M. May. 2 vols. Princeton, N.J.: Princeton University Press, 1981.

——. *The Expression of the Emotions in Man and Animals*. 1872. Edited by Paul Ekman. 3rd ed. Oxford: Oxford University Press, 1998.

——. *On the Origin of Species*. 1859. Reprint, Cambridge, Mass.: Harvard University Press, 1964.

Davis, Colin. *After Poststructuralism: Reading, Stories, and Theory*. New York: Routledge, 2003.

Dawkins, Richard. *The Ancestor's Tale: A Pilgrimage to the Dawn of Evolution*. Boston: Houghton Mifflin, 2004.

——. *The Blind Watchmaker: Why the Evidence of Evolution Reveals a Universe Without Design*. New York: Norton, 1987.

——. *The Extended Phenotype: The Gene as the Unit of Selection*. Oxford: Freeman, 1982.

——. *The God Delusion*. Boston: Houghton Mifflin, 2006.

——. *River Out of Eden: A Darwinian View of Life*. New York: Basic Books, 1995.

——. *The Selfish Gene*. 1976. Reprint, New York: Oxford University Press, 1989.

Degler, Carl N. *In Search of Human Nature: The Decline and Revival of Darwinism in American Social Thought*. New York: Oxford University Press, 1991.

Delbanco, Andrew. "The Decline and Fall of Literature." *New York Review of Books*, November 4, 1999, 36.

Deresiewicz, William. "Professing Literature in 2008." Review of *Professing Literature: An Institutional History*, by Gerald Graff. Twentieth Anniversary ed. *The Nation*, March 11, 2008. Available at http//www.thenation.com/doc/20080324/deresiewicz (accessed April 21, 2008).

Derrida, Jacques. "Structure, Sign, and Play in the Discourse of the Human Sciences." In *Writing and Difference*, translated by Alan Bass, 278–94. Chicago: University of Chicago Press, 1978.

DeStano, David, and Peter Salovey. "Evolutionary Origins of Sex Differences in Jealousy? Questioning the 'Fitness' of the Model." In *Emotions in Social Psychology: Essential Readings*, edited by W. Gerrod Parrot, 150–56. Philadelphia: Psychology Press, 2001.

Devaney, M. J. *"Since at Least Plato" and Other Postmodernist Myths*. New York: St. Martin's Press, 1997.

Devereux, George. "A Typological Study of Abortion in 359 Primitive, Ancient, and Pre-Industrial Societies." In *Abortion in America: Medical, Psychiatric, Legal, Anthropological, and Religious Considerations*, edited by Harold Rosen, 97–152. Boston: Beacon Press, 1967.

de Waal, Frans. *The Ape and the Sushi Master: Cultural Reflections of a Primatologist*. New York: Basic Books, 2001.

——. *Chimpanzee Politics: Power and Sex Among Apes*. London: Cape, 1982.

——. *Good Natured: The Origins of Right and Wrong in Humans and Other Animals*. Cambridge, Mass.: Harvard University Press, 1996.

Diamond, Jared. *The Third Chimpanzee: The Evolution and Future of the Human Animal*. New York: HarperCollins, 1992.

Dickemann, Mildred. "Paternal Confidence and Dowry Competition: A Biocultural Analysis of Purdah." In *Human Nature: A Critical Reader*, edited by Laura Betzig, 311–28. Oxford: Oxford University Press, 1997.

Dickens, Charles. *Oliver Twist*. 1838. New York: Signet, 1961.

Dijkstra, Bram. *Idols of Perversity: Fantasies of Feminine Evil in Fin-de-Siècle Culture*. New York: Oxford University Press, 1986.

Dissanayake, Ellen. *Art and Intimacy: How the Arts Began*. Seattle: University of Washington Press, 2000.

——. *Homo Aestheticus: Where Art Comes From and Why*. New York: Free Press, 1992.

——. *What Is Art For?* Seattle: University of Washington Press, 1988.

Dobzhansky, Theodosius. *Genetics and the Origin of Species*. New York: Columbia University Press, 1937.

Dollimore, Jonathan. *Radical Tragedy: Religion, Ideology and Power in the Drama of Shakespeare and His Contemporaries*. 3rd ed. Basingstoke: Palgrave Macmillan, 2004.

Drakakis, John. "The Engendering of Toads: Patriarchy and the Problem of Subjectivity in Shakespeare's 'Othello.'" *Shakespeare Jahrbuch* 124 (1988): 62–80.

Dryden, John. *"Of Dramatic Poesy" and Other Critical Essays*. Edited, with an introduction, by George Watson. 2 vols. London: Dent, 1962.

Dunbar, Robin. "Coevolution of Neocortical Size, Group Size and Language in Humans." *Behavioural and Brain Sciences* 16 (1993): 681–735.

——. *Grooming, Gossip, and the Evolution of Language*. London: Faber and Faber, 1996.

——. *Primate Social Systems*. London: Chapman and Hall, 1988.

——. "Why Are Good Writers So Rare? An Evolutionary Perspective on Literature." *Journal of Evolutionary and Cultural Psychology* 3 (2005): 7–22.

Dunbar, Robin, and Louise Barrett. "Evolutionary Psychology in the Round." In *The Oxford Handbook of Evolutionary Psychology*, edited by Robin Dunbar and Louise Barrett, 3–9. Oxford: Oxford University Press, 2007.

——, eds. *The Oxford Handbook of Evolutionary Psychology*. Oxford: Oxford University Press, 2007.

Dunbar, Robin, Anna Marriott, and N. D. C. Duncan. "Human Conversational Behavior." *Human Nature* 8 (1997): 231–46.

Dunn, Judy, and Carol Kendrick. *Siblings: Love, Envy, and Understanding*. Cambridge, Mass.: Harvard University Press, 1982.

Dutton, Denis. *The Art Instinct: Beauty, Pleasure, and Human Evolution*. New York: Bloomsbury, 2009.

Eagleton, Terry. *After Theory*. New York: Basic Books, 2003.

——. *The Illusions of Postmodernism*. Oxford: Blackwell, 1996.

——. *Literary Theory: An Introduction*. Oxford: Blackwell, 1983.

——. *Myths of Power: A Marxist Study of the Brontës*. Houndmills: Palgrave, 2005.

Eagly, Alice H., and Wendy Wood. "The Origins of Sex Differences in Human Behavior: Evolved Dispositions versus Social Roles." *American Psychologist* 54 (1999): 408–23.

Easterlin, Nancy. "Making Knowledge: Bioepistemology and the Foundations of Literary Theory." *Mosaic* 32 (1999): 131–47.

——. "Psychoanalysis and the 'Discipline of Love.'" *Philosophy and Literature* 24 (2000): 261–79.

——. "Voyages in the Verbal Universe: The Role of Speculation in Darwinian Literary Criticism." *Interdisciplinary Literary Studies* 2, no. 2 (2001): 59–73.

——. *Wordsworth and the Question of "Romantic Religion."* Lewisburg, Pa.: Bucknell University Press, 1996.

Easterlin, Nancy, and Barbara Riebling, eds. *After Poststructuralism: Interdisciplinarity and Literary Theory*. Evanston, Ill.: Northwestern University Press, 1993.

Eibl-Eibesfeldt, Irenäus. *Human Ethology*. Hawthorne, N.Y.: Aldine de Gruyter, 1989.

——. *Love and Hate: The Natural History of Behavior Patterns*. Translated by Geoffrey Strachan. New York: Schocken Books, 1974.

——. "Patterns of Parent–Child Interaction in a Cross-Cultural Perspective." In *The Behavior of Human Infants*, edited by Alberto Oliverio and Michelle Zappella, 177–217. New York: Plenum Press, 1983.

Eimas, Peter D. "Infant Competence and the Acquisition of Language." In *Biological Perspectives on Language*, edited by David Caplan, André Roch Lecours, and Allan Smith, 109–29. Cambridge, Mass.: MIT Press, 1984.

Einstein, Albert. "What Life Means to Einstein: An Interview by George Sylvester Viereck." *Saturday Evening Post*, October 26, 1929.

Ekman, Paul, ed. *Emotion in the Human Face*. 2nd ed. Cambridge: Cambridge University Press, 1982.

——. *Emotions Revealed: Recognizing Faces and Feelings to Improve Communication and Emotional Life*. New York: Holt, 2003.

——. "Facial Expressions of Emotion: An Old Controversy and New Findings." *Philosophical Transactions of the Royal Society of London* 335 (1992): 63–70.

Ekman, Paul, Richard J. Davidson, and Wallace V. Friesen. "Emotional Expression and Brain Physiology: II. The Duchenne Smile." *Journal of Personality and Social Psychology* 58 (1990): 342–53.

Ekman, Paul, Robert W. Levenson, and Wallace V. Friesen. "Autonomic Nervous System Activity Distinguishes Among Emotions." *Science* 218 (1983): 1208–10.

Elfenbein, Hillary Anger, and Nalini Ambady. "When Familiarity Breeds Accuracy: Cultural Exposure and Facial Emotion Recognition." *Journal of Personality and Social Psychology* 85 (2003): 276–90.

Eliot, George. *Middlemarch: An Authoritative Text, Backgrounds, Reviews, and Criticism*. Edited by Bert G. Hornback. 2nd ed. New York: Norton, 2000.

Ellis, Bruce J., and David F. Bjorklund, eds. *Origins of the Social Mind: Evolutionary Psychology and Child Development*. New York: Guilford Press, 2005.

Ellis, Bruce J., and Doug Symons. "Sex Differences in Sexual Fantasy: An Evolutionary Psychological Approach." *Journal of Sex Research* 7 (1990): 527–55.

Ellis, John M. *Against Deconstruction*. Princeton, N.J.: Princeton University Press, 1989.

——. *Literature Lost: Social Agendas and the Corruption of the Humanities*. New Haven, Conn.: Yale University Press, 1997.

Ember, Melvin. "On the Origins and Extension of the Incest Taboo." *Behavior Science Research* 10 (1975): 250–77.

Engell, James, and Anthony Dangerfield. "The Market-Model University: Humanities in the Age of Money." *Harvard Magazine*, May–June 1998, 48–55, 111.

Erickson, Mark. "Rethinking Oedipus: An Evolutionary Perspective of Incest Avoidance." *American Journal of Psychiatry* 150 (1993): 411–16.

Etlin, Richard A. *In Defense of Humanism: Value in the Arts and Letters*. Cambridge: Cambridge University Press, 1996.

Evans, Dylan. *Emotion: The Science of Sentiment*. Oxford: Oxford University Press, 2001.

Eysenck, Hans. *Decline and Fall of the Freudian Empire*. New York: Viking, 1985.

Fajado, Barbara F., and Daniel G. Freedman. "Maternal Rhythmicity in Three American Cultures." In *Culture and Early Interactions*, edited by Tiffany M. Field, Anita M. Sostek, Peter Vietze, and P. Herbert Leiderman, 133–47. Hillsdale, N.J.: Erlbaum, 1981.

Feiring, Candice, and Michael Lewis. "Middle-Class Differences in the Mother–Child Interaction and the Child's Cognitive Development." In *Culture and Early Interactions*, edited by Tiffany M. Field, Anita M. Sostek, Peter Vietze, and P. Herbert Leiderman, 63–91. Hillsdale, N.J.: Erlbaum, 1981.

Fernald, Anne. "Human Maternal Vocalizations to Infants as Biologically Relevant Signals: An Evolutionary Perspective." In *The Adapted Mind: Evolutionary Psychology and the Generation of Culture*, edited by Jerome H. Barkow, Leda Cosmides, and John Tooby, 391–428. New York: Oxford University Press, 1992.

Field, Tiffany M., and Susan M. Widmayer. "Mother–Infant Interaction Among Lower SES Black, Cuban, Puerto Rican, and South American Immigrants." In *Culture and Early Interactions*, edited by Tiffany M. Field, Anita M. Sostek, Peter Vietze, and P. Herbert Leiderman, 41–62. Hillsdale, N.J.: Erlbaum, 1981.

Figueredo, Aurelio José, Geneva Vásquez, Barbara H. Brumbach, and Stephanie Schneider. "The K-Factor, Covitality, and Personality: A Psychometric Test of Life History Theory." *Human Nature* 18 (2007): 47–73.

Fisher, Helen E. *Anatomy of Love: The Natural History of Monogramy, Adultery, and Divorce*. New York: Norton, 1992.

Fisher, Philip. *The Vehement Passions*. Princeton, N.J.: Princeton University Press, 2002.

Flesch, William. *Comeuppance: Costly Signaling, Altruistic Punishment, and Other Biological Components of Fiction*. Cambridge, Mass.: Harvard University Press, 2007.

Flinn, Mark V., and Carol V. Ward. "Ontogeny and Evolution of the Social Child." In *Origins of the Social Mind: Evolutionary Psychology and Child Development*, edited by Bruce J. Ellis and David F. Bjorklund, 19–44. New York: Guilford Press, 2005.

Fludernik, Monika. "Threatening the University: The Liberal Arts and the Economization of Culture." *New Literary History* 36 (2005): 57–70.

Forster, E. M. *A Passage to India*. Melbourne: Penguin, 1936.

Foucault, Michel. *The Archaeology of Knowledge*. Translated by A. M. Sheridan Smith. New York: Pantheon, 1972.

——. *The History of Sexuality: An Introduction*. Translated by Robert Hurley. London: Penguin, 1979.

——. *The Order of Things: An Archaeology of the Human Sciences*. London: Tavistock, 1977.

Fowler, Alastair. *Kinds of Literature: An Introduction to the Theory of Genres*. Cambridge, Mass.: Harvard University Press, 1982.

Fox, Robin. *The Red Lamp of Incest*. New York: Dutton, 1980.

Frank, Robert H. *Passions Within Reason: The Strategic Role of the Emotions*. New York: Norton, 1988.

Freadman, Richard, and Seumas Miller. *Re-thinking Theory: A Critique and an Alternative Account*. Cambridge: Cambridge University Press, 1992.

French, Marilyn. *Shakespeare's Division of Experience*. London: Cape, 1981.

Freud, Sigmund. *The Complete Introductory Lectures on Psychoanalysis*. 1916–1917. Edited and translated by James Strachey. Reprint, New York: Norton, 1966.

——. "Leonardo da Vinci and a Memory of His Childhood." In *The Standard Edition of the Complete Psychological Works of Sigmund Freud*, edited by James Strachey, 10:153–318. London: Hogarth Press, 1930.

——. *Totem and Taboo: Some Points of Agreement Between the Mental Lives of Savages and Neurotics*. 1912–1913. Translated by James Strachey. Reprint, New York: Norton, 1950.

Frith, Gillian. "Decoding *Wuthering Heights*." In *Critical Essays on "Wuthering Heights*," edited by Thomas John Winnifrith, 243–61. New York: Hall–Simon and Schuster, 1997.

Frith, Uta. "Autism: Beyond 'Theory of Mind.'" *Cognition* 50 (1995): 13–30.

——. *Autism: Explaining the Enigma*. Oxford: Blackwell, 1989.

Fromm, Harold. *Academic Capitalism and Literary Value*. Athens: University of Georgia Press, 1991.

——. *The Nature of Being Human: From Ecology to Consciousness*. Baltimore: Johns Hopkins University Press, 2009.

Fromm, Harold, and Robert Scholes. Exchange in "Forum." *PMLA* 121 (2006): 297–98.

Frye, Northrop. *Anatomy of Criticism: Four Essays*. Princeton, N.J.: Princeton University Press, 1957.

Fulkerson, Laurel. "Epic Ways of Killing a Woman: Gender and Transgression in the *Odyssey* 22.465–472." *Classical Journal* 97 (2002): 335–50.

Fung, Christopher. Review of *Applying Cultural Anthropology: An Introductory Reader*, edited by Aaron Podolefsky and Peter J. Brown. Available at http://www.amazon.com/Applying-Cultural-Anthropology-Introductory-Reader/dp/0073530921 (accessed May 3, 2009).

Gadamer, Hans-Georg. *Truth and Method*. Translated by Joel Weinsheimer and Donald G. Marshall. 2nd rev. ed. New York: Continuum, 1989.

Gangestad, Steven W. "Reproductive Strategies and Tactics." In *The Oxford Handbook of Evolutionary Psychology*, edited by Robin Dunbar and Louise Barrett, 321–32. Oxford: Oxford University Press, 2007.

Gangestad, Steven W., and Jeffry A. Simpson, eds. *The Evolution of Mind: Fundamental Questions and Controversies*. New York: Guilford Press, 2007.

Gates, Henry Louis, Jr., and Sieglinde Lenke. Introduction to *The Complete Stories*, by Zora Neale Hurston. Edited by Henry Louis Gates Jr. and Sieglinde Lenke. New York: HarperCollins, 1995.

Gaulin, Steven J. C., and Donald H. McBurney. *Psychology: An Evolutionary Approach*. Saddle River, N.J.: Prentice Hall, 2001.

Gavish, Leah, Joyce E. Hofmann, and Lowell L. Getz. "Sibling Recognition in the Prairie Vole, *Microtus ochrogaster*." *Animal Behaviour* 23 (1984): 362–66.

Geary, David C. "Evolution of Paternal Investment." In *The Handbook of Evolutionary Psychology*, edited by David M. Buss, 483–505. Hoboken, N.J.: Wiley, 2005.

——. *Male, Female: The Evolution of Human Sex Differences*. Washington, D.C.: American Psychological Association, 1998.

——. *The Origin of Mind: Evolution of Brain, Cognition, and General Intelligence*. Washington, D.C.: American Psychological Association, 2005.

Geary, David C., and Mark V. Flinn. "Evolution of Human Parental Behavior and the Human Family." *Parenting: Science and Practice* 1 (2001): 5–61.

Geerken, Ingrid. "'The Dead Are Not Annihilated': Mortal Regret in *Wuthering Heights*." *Journal of Narrative Theory* 34 (2004): 374–76, 385–86.

Geertz, Clifford. *The Interpretation of Cultures*. New York: Basic Books, 1973.

Gellner, Ernest. *Relativism and the Social Sciences*. Cambridge: Cambridge University Press, 1985.

German, Norman. "Counterfeiting and a Two-bit Error in Zora Neale Hurston's 'The Gilded Six-Bits.'" *Xavier Review* 19, no. 2 (1999): 5–15.

Gibson, James J. *The Ecological Approach to Visual Perception.* Boston: Houghton Mifflin, 1979.

Gigerenzer, Gerd, Peter M. Todd, and the ABC Research Group. *Simple Heuristics That Make Us Smart.* Oxford: Oxford University Press, 1999.

Gilbert, Sandra M., and Susan Gubar. *The Madwoman in the Attic: The Woman Writer and the Nineteenth-Century Literary Imagination.* New Haven, Conn.: Yale University Press, 1979.

Gilligan, Carol. *In a Different Voice: Psychological Theory and Women's Development.* Cambridge, Mass.: Harvard University Press, 1982.

Gitlin, Todd. "The Cant of Identity." In *Theory's Empire: An Anthology of Dissent*, edited by Daphne Patai and Will H. Corral, 400–410. New York: Columbia University Press, 2005.

Gladwell, Malcolm. *Blink: The Power of Thinking Without Thinking.* New York: Little, Brown, 2005.

Glass, Bentley, Owsei Temekin, and William L. Straus Jr., eds. *Forerunners of Darwin, 1745–1859.* Baltimore: Johns Hopkins University Press, 1959.

Gombrich, Ernst H. *Art and Illusion: A Study in the Psychology of Pictorial Representation.* 11th ed. Princeton, N.J.: Princeton University Press, 2000.

——. "Illusion and Art." In *Illusion in Nature and Art*, edited by Richard L. Gregory and Ernst H. Gombrich, 199–213. New York: Scribner, 1973.

——. "Image and Code: Scope and Limits of Conventionalism in Pictorial Representation." In *The Image and the Eye: Further Studies in the Psychology of Pictorial Representation*, 278–97. Ithaca, N.Y.: Cornell University Press, 1982.

——. *The Sense of Order: A Study in the Psychology of Decorative Art.* 2nd ed. London: Phaidon Press, 1984.

Good, Kenneth, with David Chanoff. *Into the Heart: One Man's Pursuit of Love and Knowledge Among the Yanomami.* New York: Simon and Schuster, 1991.

Goodheart, Eugene. "Casualties of the Culture Wars." In *Theory's Empire: An Anthology of Dissent*, edited by Daphne Patai and Will H. Corral, 508–21. New York: Columbia University Press, 2005.

——. *Does Literary Studies Have a Future?* Madison: University of Wisconsin Press, 1999.

Gorry, April. "Leaving Home for Romance: Tourist Women's Adventures Abroad." Ph.D. diss., University of California, Santa Barbara, 1999.

Gottschall, Jonathan. "Greater Emphasis on Female Attractiveness in *Homo sapiens*: A Revised Solution to an Old Evolutionary Riddle." *Evolutionary Psychology* 5 (2007): 347–58.

——. *Literature, Science, and a New Humanities.* London: Palgrave, 2008.

——. *The Rape of Troy: Evolution, Violence, and the World of Homer.* Cambridge: Cambridge University Press, 2008.

Gottschall, Jonathan, Johanna Martin, Hadley Quish, and John Rea. "Sex Differences in Mate Choice Criteria Are Reflected in Folktales from Around the World and in Historical European Literature." *Evolution and Human Behavior* 25 (2004): 102–12.

Gottschall, Jonathan, and Marcus Nordlund. "Romantic Love: A Literary Universal?" *Philosophy and Literature* 30 (2006): 432–52.

Gottschall, Jonathan, and David Sloan Wilson, eds. *Literature and the Human Animal: Evolution and the Nature of Narrative*. Evanston, Ill.: Northwestern University Press, 2005.

Gould, Stephen Jay. "Biological Potentiality versus Biological Determinism." In *Ever Since Darwin: Reflections in Natural History*, 251–59. New York: Norton, 1977:

——. *The Flamingo's Smile: Reflections in Natural History*. New York: Norton, 1985.

——. *Wonderful Life: The Burgess Shale and the Nature of History*. New York: Norton, 1989.

Gould, Stephen Jay, and Niles Eldredge. "Punctuated Equilibria: The Tempo and Mode of Evolution Reconsidered." *Paleobiology* 3 (1977): 115–51.

Gould, Stephen Jay, and Richard C. Lewontin. "The Spandrels of San Marco and the Panglossian Program: A Critique of the Adaptationist Programme." *Proceedings of the Royal Society of London* 205 (1979): 281–88.

Graff, Gerald. *Professing Literature: An Institutional History*. Chicago: University of Chicago Press, 1989.

Graff, Gerald, and Michael Warner, eds. *The Origins of Literary Study in America: A Documentary Anthology*. New York: Routledge, 1989.

Grandin, Temple, and Margaret M. Scariano. *Emergence: Labeled Autistic*. Tunbridge Wells: Costello, 1986.

Grayling, A. C. *What Is Good? The Search for the Best Way to Live*. London: Weidenfeld & Nicolson, 2003.

Greenblatt, Stephen. "Resonance and Wonder." In *Learning to Curse: Essays in Early Modern Culture*, 161–83. New York: Routledge, 1990.

——. "'Stay, Illusion'—On Receiving Messages from the Dead." *PMLA* 118 (2003): 417–26.

Greene, Brian. *The Elegant Universe: Superstrings, Hidden Dimensions, and the Quest for the Ultimate Theory*. New York: Vintage, 2003.

——. *The Fabric of the Cosmos: Space, Time, and the Texture of Reality*. New York: Vintage, 2005.

Greenspan, Stanley I. *The Growth of the Mind and the Endangered Origins of Intelligence*. Reading, Mass.: Addison-Wesley, 1997.

Griffiths, Paul W. *What Emotions Really Are: The Problem of Psychological Categories*. Chicago: University of Chicago Press, 1997.

Griswold, Jerry. *Audacious Kids: Coming of Age in America's Classic Children's Books*. New York: Oxford University Press, 1992.

Grodal, Torben. *Moving Pictures: A New Theory of Film Genres, Feelings, and Cognition*. New York: Oxford University Press, 1997.

Grosse, Ernst. *The Beginnings of Art*. New York: Appleton, 1897.

Grunbaum, Adolph. *The Foundations of Psychoanalysis: A Philosophical Critique*. Berkeley: University of California Press, 1984.

Guss, David M. *To Weave and Sing: Art, Symbol, and Narrative in the South American Rain Forest*. Berkeley: University of California Press, 1989.

Hadfield, Andrew. *Shakespeare and Renaissance Politics*. London: Arden Shakespeare, 2004.

Hagen, Margaret A. *Varieties of Realism: Geometries of Representational Art*. Cambridge: Cambridge University Press, 1986.

Hagstrum, Jean H. *The Romantic Body: Love and Sexuality in Keats, Wordsworth, and Blake*. Knoxville: University of Tennessee Press, 1985.

Hamilton, W. D. "The Evolution of Altruistic Behavior." *American Naturalist* 97 (1963): 354–56.

———. "The Genetical Evolution of Social Behaviour, I and II." *Journal of Theoretical Biology* 7 (1964): 1–52.

———. *Narrow Roads of Gene Land: The Collected Papers of W. D. Hamilton*. Vol. 1, *Evolution of Social Behaviour*. New York: Freeman, 1996.

Haney, David P. *Wordsworth and the Hermeneutics of Incarnation*. University Park: Penn State University Press, 1993.

Hardy, Thomas. *The Mayor of Casterbridge: An Authoritative Text, Backgrounds and Contexts, Criticism*. Edited by Phillip Mallett. 2nd ed. New York: Norton, 2001.

Harpending, Henry. "Gene Frequencies, DNA Sequences, and Human Origins." *Perspectives in Biology and Medicine* 37 (1994): 384–95.

Harpham, Geoffrey Galt. "The End of Theory, the Rise of the Profession: A Rant in Search of Responses." In *Theory's Empire: An Anthology of Dissent*, edited by Daphne Patai and Will H. Corral, 381–94. New York: Columbia University Press, 2005.

———. "Response." *New Literary History* 36 (2005): 105–9.

Harris, C. Leon. *Concepts in Zoology*. New York: HarperCollins, 1992.

Harris, Helen. "Human Nature and the Nature of Romantic Love." Ph.D. diss., University of California, Santa Barbara, 1995.

———. "Rethinking Polynesian Heterosexual Relationships: A Case Study on Mangaia, Cook Islands." In *Romantic Passion: A Universal Experience?* edited by William Jankowiak, 95–127. New York: Columbia University Press, 1995.

Harris, Roy. *Reading Saussure: A Critical Commentary on the "Cours de linguistique générale."* La Salle, Ill.: Open Court, 1987.

Hassin, Ran R., James S. Uleman, and John A. Bargh. *The New Unconscious*. New York: Oxford University Press, 2005.

Hatfield, Elaine, John T. Cacioppo, and Richard L. Rapson. *Emotional Contagion: Studies in Emotional and Social Interaction*. Cambridge: Cambridge University Press, 1994.

Hauge, Michael. *Writing Screenplays That Sell*. New York: McGraw-Hill, 1988.

Hausfater, Glenn, and Sarah Blaffer Hrdy, eds. *Infanticide: Comparative and Evolutionary Perspectives*. New York: Aldine de Gruyter, 1984.

Haviland, William A. *Cultural Anthropology*. 8th ed. New York: Harcourt Brace Jovanovich, 1996.

Hawkes, Terence. *That Shakespeherian Rag: Essays on a Critical Process*. London: Methuen, 1986.

———, ed. *Alternative Shakespeares*. Vol. 2. London: Routledge, 1996.

Heffernan, James A. W. "The Presence of the Absent Mother in Wordsworth's *Prelude*." *Studies in Romanticism* 27 (1988): 253–72.

Heider, Fritz. *The Psychology of Interpersonal Relations*. New York: Wiley, 1958.

Hemenway, Robert E. *Zora Neale Hurston: A Literary Biography*. Urbana: University of Illinois Press, 1980.

Hernadi, Paul. "Literature and Evolution." *SubStance* 30 (2001): 55–71.

Hersey, George L. *The Evolution of Allure: Art and Sexual Selection from Aphrodite to the Incredible Hulk*. Cambridge, Mass.: MIT Press, 1996.

Hesiod. *The Homeric Hymns and Homerica*. Translated by Hugh G. Evelyn-White. Loeb Classical Library, no. 57. 1914. Reprint, Cambridge, Mass.: Harvard University Press, 1982.

Hilfer, Tony. *The New Hegemony in Literary Studies: Contradictions in Theory*. Evanston, Ill.: Northwestern University Press, 2003.

Hill, Kim. "Evolutionary Biology, Cognitive Adaptations, and Human Culture." In *The Evolution of Mind: Fundamental Questions and Controversies*, edited by Steven W. Gangestad and Jeffry A. Simpson, 348–56. New York: Guilford Press, 2007.

Hill, Russell, and Robin Dunbar. "Social Network Size in Humans." *Human Nature* 14 (2003): 53–72.

Hillner, Jennifer. "J. J. Abrams, Spymaster." *Wired*, May 2006, 160.

Hjort, Mette, ed., *Rules and Conventions: Literature, Philosophy, Social Theory*. Baltimore: Johns Hopkins University Press, 1992.

Hobbs, Jerry R. *Literature and Cognition*. Stanford, Calif.: Center for the Study of Language and Information, 1990.

——. "Will Robots Ever Have Cognition?" Paper presented at the Workshop on Literary Cognition, Information, and Communication, Carleton University, Ottawa, Canada, June 1993. Available at http://www.isi.edu/%7Ehobbs/ottawa.pdf (accessed September 1, 2009).

Hochberg, Julian. "Representation of Motion and Space in Video and Cinematic Displays." In *Sensory Processes and Perception*, vol. 1 of *Handbook of Perception and Human Performance*, edited by Kenneth R. Boff, Lloyd Kaufman, and James P. Thomas, 31–40. New York: Wiley, 1986.

——. "The Representation of Things and People." In *Art, Perception, and Reality*, edited by Ernst H. Gombrich, Julian Hochberg, and Max Black, 67–73. Baltimore: Johns Hopkins University Press, 1972.

Hockett, Charles F. *Man's Place in Nature*. New York: McGraw-Hill, 1973.

Hoeller, Hildegard. "Racial Currency: Zora Neale Hurston's 'The Gilded Six-Bits' and the Gold-Standard Debate." *American Literature* 77, no. 4 (2005): 761–85.

Hogan, Patrick Colm. "Literary Universals." *Poetics Today* 18 (1997): 223–49.

——. *Mind and Its Stories: Narrative Universals and Human Emotion*. New York: Cambridge University Press, 2003.

——. "*Othello*, Racism, and Despair." *CLA Journal* 41 (1998): 431–51.

Homans, Margaret. *Bearing the Word: Language and Female Experience in Nineteenth-Century Women's Writing*. Chicago: University of Chicago Press, 1986.

——. "The Name of the Mother in *Wuthering Heights*." In Emily Brontë, *Wuthering Heights: Complete, Authoritative Text with Biographical and Historical Contexts, Critical History, and Essays from Five Contemporary Critical Perspectives*, edited by Linda H. Peterson, 341–58. Boston: Bedford–St. Martin's Press, 1992.

Homer. *Iliad*. Books 1–12. Translated by A. T. Murray. Revised by William F. Wyatt. Loeb Classical Library, no. 170. Cambridge, Mass.: Harvard University Press, 1924.

——. *Iliad*. Books 13–24. Translated by A. T. Murray. Revised by William F. Wyatt. Loeb Classical Library, no. 171. Cambridge, Mass.: Harvard University Press, 1925.

——. *Odyssey*. Books 1–12. Translated by A. T. Murray. Revised by George E. Dimock. Loeb Classical Library, no. 104. Cambridge, Mass.: Harvard University Press, 1919.

——. *Odyssey*. Books 13–24. Translated by A. T. Murray. Revised by George E. Dimock. Loeb Classical Library, no. 105. Cambridge, Mass.: Harvard University Press, 1919.

Horace. *Satires, Epistles, and Ars Poetica, with English Notes*. Edited by W. Brownrigg Smith. New ed. London: Crosby Lockwood, 1878.

Horton, Robin. "Tradition and Modernity Revisited." In *Rationality and Relativism*, edited by Martin Hollis and Steven Lukes, 201–60. Cambridge, Mass.: MIT Press, 1982.

Howard, David. *How to Build a Great Screenplay: A Master Class in Storytelling for Film*. New York: St. Martin's Press, 2004.

Howard, Jean. "The New Historicism in Renaissance Studies." *English Literary Renaissance* 16 (1986): 13–43.

Howard, Lillie P. "Marriage: Zora Neale Hurston's System of Values." *CLA Journal* 21 (1977): 256–68.

Hrdy, Sarah Blaffer. *Mother Nature: A History of Mothers, Infants, and Natural Selection*. New York: Pantheon, 1999.

——. *The Woman That Never Evolved*. Cambridge, Mass.: Harvard University Press, 1981.

Humphrey, Nicholas K. *Consciousness Regained*. Oxford: Oxford University Press, 1983.

Hunt, Lynn. "Democratization and Decline: The Consequences of Demographic Change in the Humanities?" In *What's Happened to the Humanities?* edited by Alvin Kernan, 17–31. Princeton, N.J.: Princeton University Press, 1997.

Hurston, Zora Neale. *Dust Tracks on a Road*. 1942. Reprint, New York: Arno Press, 1969.

——. "The Gilded Six-Bits." In *The Complete Stories*. Edited by Henry Louis Gates Jr. and Sieglinde Lemke, 86–98. New York: HarperCollins, 1995.

Huxley, Julian S. *Evolution: The Modern Synthesis*. London: Allen and Unwin, 1942.

Izard, Carroll E. *The Face of Emotion*. New York: Appleton-Century-Crofts, 1971.

Jackson, Leonard. *The Dematerialisation of Karl Marx: Literature and Marxist Theory*. London: Longman, 1994.

——. *The Poverty of Structuralism: Literature and Structuralist Theory*. London: Longman, 1991.

Jackson, Tony. "Questioning Interdisciplinarity: Cognitive Science, Evolutionary Psychology, and Literary Criticism." *Poetics Today* 21 (2000): 319–47.

Jacobs, Carol. *Uncontainable Romanticism: Shelley, Brontë, Kleist*. Baltimore: Johns Hopkins University Press, 1989.

Jacobus, Mary. *Romanticism, Writing, and Sexual Difference: Essays on "The Prelude."* Oxford: Clarendon Press, 1989.

James, Henry. *Literary Criticism: Essays on Literature, American Writers, English Writers*. Edited by Leon Edel and Mark Wilson. New York: Library of America, 1984.

Jameson, Fredric. *Postmodernism, or, the Cultural Logic of Late Capitalism*. Durham, N.C.: Duke University Press, 1991.

Jankowiak, William, and Ted Fischer. "A Cross-Cultural Perspective on Romantic Love." *Ethnology* 31 (1992): 149–55.

Jenkins, Henry. *Textual Poachers: Television Fans and Participatory Culture*. New York: Routledge, 1992.

Jenkins, Jennifer M., Keith Oatley, and Nancy L. Stein, eds. *Human Emotions: A Reader*. Oxford: Blackwell, 1998.

Johnson, Allen, and Douglass Price-Williams. *Oedipus Ubiquitous: The Family Complex in World Folk Literature*. Stanford, Calif.: Stanford University Press, 1996.

Johnson, John A., Joseph Carroll, Jonathan Gottschall, and Daniel Kruger. "Hierarchy in the Library: Egalitarian Dynamics in Victorian Novels." *Evolutionary Psychology* 6 (2008): 715–38.

Johnson, Samuel. *The Rambler*. Vol. 4 of *The Yale Edition of the Works of Samuel Johnson*, edited by W. J. Bate and Albrecht B. Strauss. New Haven, Conn.: Yale University Press, 1969.

Johnston, Kenneth R. *Wordsworth and "The Recluse."* New Haven, Conn.: Yale University Press, 1984.

Jones, Evora W. "The Pastoral and Picaresque in Zora Neale Hurston's 'The Gilded Six-Bits.'" *CLA Journal* 35 (1992): 316–24.

Jones, Gayl. "Breaking Out of the Conventions of Dialect: Dunbar and Hurston." *Présence Africaine: Revue Culturelle du Monde Noir / Cultural Review of the Negro World* 144 (1987): 32–46.

Jones, Stephen, Robert D. Martin, and David R. Pilbeam, eds. *The Cambridge Encyclopedia of Human Evolution*. New York: Cambridge University Press, 1992.

Kahneman, Daniel, Paul Slovic, and Amos Tversky, eds. *Judgment Under Uncertainty: Heuristics and Biases*. New York: Cambridge University Press, 1982.

Kahneman, Daniel, and Amos Tversky. "On the Reality of Cognitive Illusions: A Reply to Gigerenzer's Critique." *Psychological Review* 103 (1996): 582–91.

Kaplan, Hillard S., and Steven W. Gangestad. "Optimality Approaches and Evolutionary Psychology: A Call for Synthesis." In *The Evolution of Mind: Fundamental Questions and Controversies*, edited by Steven W. Gangestad and Jeffry A. Simpson, 121–29. New York: Guilford Press, 2007.

Keating, Patrick. "Emotional Curves and Linear Narratives." *Velvet Light Trap* 58 (2006): 4–15.

Kellert, Stephen R., and Edward O. Wilson, eds. *The Biophilia Hypothesis*. Washington, D.C.: Island Press, 1993.

Kelly, Robert L. *The Foraging Spectrum: Diversity in Hunter-Gatherer Lifeways.* Washington, D.C.: Smithsonian Institution Press, 1995.

Kenrick, Douglas T., Melanie R. Trost, and Virgil L. Sheets. "Power, Harassment, and Trophy Mates: The Feminist Advantages of an Evolutionary Perspective." In *Sex, Power, Conflict: Evolutionary and Feminist Perspectives,* edited by David M. Buss and Neil M. Malamuth, 29–53. New York: Oxford University Press, 1996.

Kermode, Frank. "Changing Epochs." In *Theory's Empire: An Anthology of Dissent,* edited by Daphne Patai and Will H. Corral, 605–20. New York: Columbia University Press, 2005.

——. Introduction to *The Shakespearean International Yearbook.* Vol. 3, *Where Are We Now in Shakespearean Studies?* edited by Graham Bradshaw, John M. Mucciolo, Angus Fletcher, and Tom Bishop. Farnham: Ashgate, 2003.

Kernan, Alvin. *The Death of Literature.* New Haven, Conn.: Yale University Press, 1992.

——, ed. *What's Happened to the Humanities?* Princeton, N.J.: Princeton University Press, 1997.

Kernan, Alvin, Peter Brooks, and Michael Holquist. *Man and His Fictions: An Introduction to Fiction-Making, Its Forums, and Uses.* New York: Harcourt Brace Jovanovich, 1973.

Keverne, Eric B., Nicholas D. Martensz, and Bernadette Tuite. "Beta-Endorphin Concentrations in Cerebrospinal Fluid of Monkeys Are Influenced by Grooming Relationships." *Psychoneuroendocrinology* 14 (1989): 155–61.

Kevles, Bettyann. *Females of the Species: Sex and Survival in the Animal Kingdom.* Cambridge, Mass.: Harvard University Press, 1986.

Kirsch, Arthur. *Shakespeare and the Experience of Love.* Cambridge: Cambridge University Press, 1981.

Knight, Chris. *Blood Relations: Menstruation and the Origins of Culture.* New Haven, Conn.: Yale University Press, 1995.

Knight, Chris, Camilla Power, and Ian Watts. "The Human Symbolic Revolution: A Darwinian Account." *Cambridge Archaelogical Journal* 5 (1995): 75–114.

Kolin, Philip C., ed. *"Othello": New Critical Essays.* New York: Routledge, 2002.

Kott, Jan. *Shakespeare Our Contemporary.* Translated by Boleslaw Taborski. New York: Norton, 1974.

Krieger, Murray. "The Ambiguities of Representation and Illusion: An E. H. Gombrich Retrospective." *Critical Inquiry* 11 (1984): 195–201.

Kruger, Daniel J., Maryanne Fisher, and Ian Jobling. "Proper and Dark Heroes as Dads and Cads: Alternative Mating Strategies in British Romantic Literature." *Human Nature* 14 (2003): 305–17.

Kudo, H., and Robin Dunbar. "Neocortex Size and Social Network Size in Primates." *Animal Behaviour* 62 (2001): 711–22.

Kuleshov, Lev. *Kuleshov on Film: Writings by Lev Kuleshov.* Translated and edited, with an introduction, by Ronald Levaco. Berkeley: University of California Press, 1974.

Kurland, Jeffrey A., and Steven J. C. Gaulin. "Cooperation and Con Among Kin." In *The Handbook of Evolutionary Psychology,* edited by David M. Buss, 447–82. Hoboken, N.J.: Wiley, 2005.

Laland, Kevin N., and Gillian R. Brown. *Sense and Nonsense: Evolutionary Perspectives on Human Behaviour*. Oxford: Oxford University Press, 2002.

Lamarque, Peter, and Stein Haugom Olsen. *Truth, Fiction, and Literature: A Philosophical Perspective*. Oxford: Clarendon Press, 1994.

Lamb, Patricia Frazer, and Diana L. Veitch. "Romantic Myth, Transcendence, and *Star Trek* Zines." In *Erotic Universe: Sexuality and Fantastic Literature*, edited by Donald Palumbo, 235–56. Westport, Conn.: Greenwood Press, 1986.

Landsburg, Steven E. *The Armchair Economist: Economics and Everyday Life*. New York: Free Press, 1993.

Latour, Bruno. "Why Has Critique Run Out of Steam? From Matters of Fact to Matters of Concern." *Critical Inquiry* 30 (2004): 225–48.

Layton, Robert. *The Anthropology of Art*. 2nd ed. Cambridge: Cambridge University Press, 1991.

Lazarus, Richard. "The Past and Present in Emotion." In *The Nature of Emotion: Fundamental Questions*, edited by Paul Ekman and Richard J. Davidson, 306–10. New York: Oxford University Press, 1994.

Leavis, Q. D. "A Fresh Approach to *Wuthering Heights*." In *Lectures in America*, by F. R. Leavis and Q. D. Leavis, 85–138. New York: Pantheon, 1969.

Lederer, Richard, and Michael Gilleland. *Literary Trivia: Fun and Games for Book Lovers*. New York: Vintage, 1994.

LeDoux, Joseph. *Synaptic Self: How Our Brains Become Who We Are*. London: Macmillan, 2002.

Lee, John. *Shakespeare's "Hamlet" and the Controversies of the Self*. Oxford: Clarendon Press, 2000.

Leitch, Vincent B., William E. Cain, Laurie A. Finke, Barbara E. Johnson, John P. McGowan, and Jeffrey L. Williams, eds. *The Norton Anthology of Theory and Criticism*. New York: Norton, 2001.

Lentricchia, Frank. "Last Will and Testament of an Ex-Literary Critic." *Lingua Franca*, September–October 1996, 59–67.

Leslie, Alan M. "Pretense and Representation: The Origins of Theory of Mind." *Psychological Review* 94 (1987): 412–26.

——. "Spatiotemporal Continuity and the Perception of Causality in Infants." *Perception* 13 (1984): 287–305.

Lessa, William A. "Oedipus-Type Tales in Oceania." In *Oedipus: A Folklore Casebook*, edited by Lowell Edmunds and Alan Dundes, 56–75. New York: Garland, 1983.

Levin, Richard. "Feminist Thematics and Shakespearean Tragedy." *PMLA* 103 (1988): 125–38.

——. *Looking for an Argument: Critical Encounters with the New Approaches to the Criticisms of Shakespare and His Contemporaries*. Madison, N.J.: Farleigh Dickinson University Press, 2003.

——. *New Readings vs. Old Plays: Recent Trends in the Reinterpretation of English Renaissance Drama*. Chicago: University of Chicago Press, 1979.

Levine, Laura. *Men in Women's Clothing: Anti-theatricality and Effeminization, 1579–1642*. Cambridge: Cambridge University Press, 1994.

LeVine, Robert. "Intergenerational Tensions and Extended Family Structures in Africa." In *Social Structure and the Family: Generational Relations*, edited by Ethel Shanas and Gordon F. Streib, 188–204. Englewood Cliffs, N.J.: Prentice Hall, 1965.

Levins, Richard, and Richard C. Lewontin. *The Dialectical Biologist*. Cambridge, Mass.: Harvard University Press, 1985.

Lewis, David. *Convention: A Philosophical Study*. Cambridge, Mass.: Harvard University Press, 1969.

Lewis, Michael. "Self-Conscious Emotions: Embarrassment, Pride, Shame, and Guilt." In *Handbook of Emotions*, edited by Michael Lewis and Jeannette M. Haviland-Jones, 623–36. 2nd ed. New York: Guilford Press, 2000.

Lewis, Michael, and Peggy Ban. "Variance and Invariance in the Mother–Infant Interaction: A Cross-Cultural Study." In *Culture and Early Interactions*, edited by Tiffany M. Field, Anita M. Sostek, Peter Vietze, and P. Herbert Leiderman, 329–55. Hillsdale, N.J.: Erlbaum, 1981.

Lewis, Michael, and Jeannette M. Haviland-Jones, eds. *Handbook of Emotions*. 2nd ed. New York: Guilford Press, 2000.

Lewis, Sinclair. *Arrowsmith*. New York: Harcourt, Brace, 1925.

Lewkowicz, David J., and Gerald Turkewitz. "Cross-Modal Equivalence in Early Infancy: Auditory-Visual Intensity Matching." *Developmental Psychology* 16 (1980): 597–607.

Lewontin, Richard C., Steven Rose, and Leon J. Kamin. *Not in Our Genes*. New York: Pantheon, 1984.

Lindenberger, Herbert. *On Wordsworth's "Prelude."* Princeton, N.J.: Princeton University Press, 1963.

Livingston, Paisley. *Literature and Rationality: Ideas of Agency in Theory and Fiction*. Cambridge: Cambridge University Press, 1991.

Locke, John. *An Essay Concerning Human Understanding*. 1690. Edited, with an introduction, by Peter H. Nidditch. Oxford: Clarendon Press, 1975.

——. *Two Treatises of Government*. 1689. Cambridge: Cambridge University Press, 1960.

Locke, John L. *The Child's Path to Spoken Language*. Cambridge, Mass.: Harvard University Press, 1993.

Long, Diane Hoeveler. *Romantic Androgyny: The Woman Within*. University Park: Penn State University Press, 1990.

Loomba, Ania. "Sexuality and Racial Difference." In *Critical Essays on Shakespeare's "Othello,"* edited by Anthony Gerard Barthelemy, 162–86. New York: Hall, 1994.

López, José, and Garry Potter, eds. *After Postmodernism: An Introduction to Critical Realism*. New York: Continuum, 2005.

López-Morillas, Juan. "Utopia and Anti-Utopia: From 'Dreams of Reason' to 'Dreams of Unreason.'" *Survey* 18 (1972): 47–62.

Lorenz, Konrad "Vergleichende Bewegungsstudien an Anatiden." *Journal of Ornithology* 89 (1941): 194–294.

——. *Evolution and Modification of Behavior*. Chicago: University of Chicago Press, 1965.

Low, Bobbi S. "Sexual Selection and Human Ornamentation." In *Evolutionary Biology and Human Social Behavior: An Anthropological Perspective*, edited by Napoleon A. Chagnon and William Irons, 462–87. Boston: Duxbury Press, 1979.

——. *Why Sex Matters: A Darwinian Look at Human Behavior*. Princeton, N.J.: Princeton University Press, 2000.

Lowe, David G. "The Viewpoint Consistency Constraint." *International Journal of Computer Vision* 1 (1987): 57–72.

Lowe, John. *"Jump at the Sun": Zora Neale Hurston's Cosmic Comedy*. Excerpted in *Sweat / Zora Neale Hurston*, edited by Cheryl A. Wall, 183–92. New Brunswick, N.J.: Rutgers University Press, 1997.

Ludvico, Lisa R., and J. A. Kurland. "Symbolic or Not-So-Symbolic Wounds: The Behavioral Ecology of Human Scarification." *Ethology and Sociobiology* 16 (1995): 155–72.

Lumsden, Charles J., and Edward O. Wilson. *Promethean Fire: Reflections on the Origin of Mind*. Cambridge, Mass.: Harvard University Press, 1983.

Lupton, Julia Reinhard, and Kenneth Reinhard. *After Oedipus: Shakespeare in Psychoanalysis*. Ithaca, N.Y.: Cornell University Press, 1993.

Luria, Aleksandr R. *Higher Cortical Functions in Man*. Translated by Basil Haigh. London: Tavistock, 1966.

Luthi, Max. *The European Folktale: Form and Nature*. Translated by John D. Niles. Philadelphia: Institute for the Study of Human Issues, 1982.

Lydenberg, Robin. "Freud's Uncanny Narratives." *PMLA* 112 (1997): 1072–86.

Lynch, Michael, D. K. Oller, M. L. Steffens, and E. H. Buder. "Phrasing in Prelinguistic Vocalizations." *Developmental Psychobiology* 28, no. 1 (1995): 3–25.

Lyotard, Jean-François. "Defining the Postmodern." In *The Norton Anthology of Theory and Criticism*, edited by Vincent B. Leitch, William E. Cain, Laurie A. Finke, Barbara E. Johnson, John P. McGowan, and Jeffrey L. Williams, 1612–15. New York: Norton, 2001.

MacArthur, Robert H., and Edward O. Wilson. *The Theory of Island Biogeography*. Princeton, N.J.: Princeton University Press, 1967.

MacDonald, Kevin B. "Evolution, Culture, and the Five-Factor Model." *Journal of Cross-Cultural Psychology* 29 (1998): 119–49.

Macey, David. *The Penguin Dictionary of Critical Theory*. New York: Penguin, 2000.

Macfarlane, Alan. *Marriage and Love in England: Modes of Reproduction, 1300–1840*. Oxford: Blackwell, 1986.

Maddock, Kenneth. *The Australian Aborigines: A Portrait of Their Society*. London: Allen Lane, 1973.

Malamuth, Neil M., Tamara Addison, and Mary Koss. "Pornography and Sexual Aggression: Are There Reliable Effects and Can We Understand Them?" *Annual Review of Sex Research* 11 (2000): 26–91.

Malik, Kenan. *Man, Beast and Zombie: What Science Can and Cannot Tell Us About Human Nature*. London: Weidenfeld & Nicolson, 2000.

Malthus, Thomas. *An Essay on the Principle of Population, As It Affects the Future Improvement of Society: With Remarks on the Speculations of Mr. Godwin, M. Condorcet and Other Writers*. London: J. Johnson, 1798.

Mantzavinos, C. *Naturalistic Hermeneutics*. Translated by Darrell Arnold and C. Mantzavinos. Cambridge: Cambridge University Press, 2005.

Marcus, Steven. *The Other Victorians: A Study of Sexuality and Pornography in Mid-Nineteenth-Century England*. New York: Basic Books, 1966.

Marks, Lawrence E., Robin J. Hammeal, and Marc H. Bornstein. "Perceiving Similarity and Comprehending Metaphor." *Monographs of the Society for Research in Child Development* 52 (1987): 1–92.

Marr, David. *Vision: A Computational Investigation into the Human Representation and Processing of Visual Information*. San Francisco: Freeman, 1982.

Martindale, Colin. *The Clockwork Muse: The Predictability of Artistic Change*. New York: Basic Books, 1990.

———. "Empirical Questions Deserve Empirical Answers." *Philosophy and Literature* 29 (1996): 347–61.

Maryanski, Alexandra, and Jonathan H. Turner. *The Social Cage: Human Nature and the Evolution of Society*. Stanford, Calif.: Stanford University Press, 1992.

Massé, Michelle A. "'He's More Myself Than I Am': Narcissism and Gender in *Wuthering Heights*." In *Psychoanalyses / Feminisms*, edited by Peter L. Rudnytsky and Andrew M. Gordon, 135–53. Albany: State University of New York Press, 2000.

Mathison, John K. "Nelly Dean and the Power of *Wuthering Heights*." *Nineteenth-Century Fiction* 11 (1956): 106–29.

Maynard, John. "The Brontës and Religion." In *The Cambridge Companion to the Brontës*, edited by Heather Glen, 192–213. Cambridge: Cambridge University Press, 2002.

Mayr, Ernst. *The Growth of Biological Thought: Diversity, Evolution, and Inheritance*. Cambridge, Mass.: Harvard University Press, 1982.

———. "How to Carry Out the Adaptationist Program." *American Naturalist* 121 (1983): 324–34.

———. *Systematics and the Origin of Species from the Viewpoint of a Zoologist*. New York: Columbia University Press, 1942.

McCabe, Justine. "FBD Marriage: Further Support for the Westermarck Hypothesis of the Incest Taboo?" *American Anthropologist* 85 (1983): 50–69.

McEwan, Ian. "The Great *Odyssey*." *Guardian Saturday Review*, June 9, 2001, 1–3.

McLain, D. Kelly, Deanna Setters, Michael P. Moulton, and Ann E. Pratt. "Ascription of Resemblance of Newborns by Parents and Nonrelatives." *Evolution and Human Behavior* 21 (2000): 11–23.

McQuillan, Martin, Graeme Macdonald, Robin Purves, and Stephen Thomson, eds. *Post-Theory: New Directions in Criticism*. Edinburgh: Edinburgh University Press, 2000.

Mead, Margaret. *Blackberry Winter: My Earlier Years*. New York: Simon and Schuster, 1972.

Mellor, Anne K. *Romanticism and Gender*. London: Routledge, 1993.

Meltzer, Françoise. "Future? What Future?" *Critical Inquiry* 30 (2004): 468–71.

Menand, Louis. "Dangers Within and Without." *Profession* (2005): 10–17.

———. "The Demise of Disciplinary Authority." In *What's Happened to the Humanities?* edited by Alvin Kernan, 201–19. Princeton, N.J.: Princeton University Press, 1997.

———. "The Marketplace of Ideas." American Council of Learned Societies Occasional Paper, no. 49. Paper presented at the "Crisis in the Humanities?" series, Fanny and Alan Leslie Center for the Humanities, Dartmouth College, Hanover, N.H., October 2, 2001. Available at http//www.acls.org/op-49.htm (accessed April 21, 2008).

Mendelson, Edward. *The Things That Matter: What Seven Classic Novels Have to Say About the Stages of Life*. New York: Pantheon, 2006.

Messaris, Paul. *Visual Literacy: Image, Mind, and Reality*. Boulder, Colo.: Westview Press, 1994.

Miall, David S. "The Alps Deferred: Wordsworth at the Simplon Pass." *European Romantic Review* 9 (1998): 87–102.

——. "Wordsworth and 'The Prelude': The Problematics of Feeling." *Studies in Romanticism* 31 (1992): 233–53.

Midgley, Mary. *Beast and Man: The Roots of Human Nature*. 2nd rev. ed. London: Routledge, 1995.

Miller, Geoffrey. "Looking to Be Entertained: Three Strange Things That Evolution Did to Our Minds." *Times Literary Supplement*, October 16, 1998, 14–15.

——. *The Mating Mind: How Sexual Choice Shaped the Evolution of Human Nature*. New York: Doubleday, 2000.

——. "How Mate Choice Shaped Human Nature: A Review of Sexual Selection and Human Evolution." In *Handbook of Evolutionary Psychology: Ideas, Issues, and Applications*, edited by Charles Crawford and Dennis L. Krebs, 87–130. Mahwah, N.J.: Erlbaum, 1998.

Miller, J. Hillis. *Fiction and Repetition: Seven English Novels*. Cambridge, Mass.: Harvard University Press, 1982.

Mineka, Susan, and Michael Cook. "Mechanisms Involved in the Observational Conditioning of Fear." *Journal of Experimental Psychology: General* 122 (1993): 23–38.

Mitchell, W. J. T., ed. "The Future of Criticism: A Critical Inquiry Symposium." Special issue, *Critical Inquiry* 30 (2004): 324–483.

Mithen, Steven. *The Prehistory of the Mind: The Cognitive Origins of Art, Religion, and Science*. London: Thames and Hudson, 1996.

Moglen, Helene. "The Double Vision of *Wuthering Heights*: A Clarifying View of Female Development." *Centennial Review* 15 (1971): 391–405.

Montaigne, Michel de. *The Essays*. 1603. Translated by John Florio. Edited by W. E. Henley. 3 vols. New York: AMS Press, 1967.

Moon, Michael. "A Small Boy and Others: Sexual Disorientation in Henry James, Kenneth Anger, and David Lynch." In *Comparative American Identities: Race, Sex, and Nationality in the Modern Text*, edited by Hortense J. Spillers, 141–56. New York: Routledge, 1991.

Munsterberg, Hugo. *The Photoplay: A Psychological Study*. New York: Appleton, 1916.

——. "Why We Go to the 'Movies.'" *Cosmopolitan*, December 1915, 31.

Murdock, George P. "The Common Denominator of Cultures." In *The Science of Man in the World Crisis*, edited by Ralph Linton, 123–42. New York: Columbia Univerity Press, 1945.

Myers, David G. *Intuition: Its Power and Perils*. New Haven, Conn.: Yale University Press 2002.

Nakano, Shigeru. "Heart-to-Heart (Inter-*jo*) Resonance: A Concept of Intersubjectivity in Japanese Everyday Life." In *Annual Report* 19. Research and Clinical Center for Child Development, Faculty of Education, Hokkaido University, Sapporo, Japan, 1996.

Neely, Carol Thomas. "Women and Men in *Othello*." In *Critical Essays on Shakespeare's "Othello*," edited by Anthony Gerard Barthelemy, 68–90. New York: Hall, 1994.

Nesse, Margaret H. "Guinevere's Choice." *Human Nature* 6, no. 2 (1995): 145–63.

Nettle, Daniel. *Strong Imagination: Madness, Creativity, and Human Nature.* Oxford: Oxford University Press, 2001.

——. *Personality: What Makes You the Way You Are.* Oxford: Oxford University Press, 2007.

——. "What Happens in *Hamlet*? Exploring the Psychological Foundations of Drama." In *Literature and the Human Animal: Evolution and the Nature of Narrative*, edited by Jonathan Gottschall and David Sloan Wilson, 56–75. Evanston, Ill.: Northwestern University Press, 2005.

——. "The Wheel of Fire and the Mating Game: Explaining the Origins of Tragedy and Comedy." *Journal of Cultural and Evolutionary Psychology* 3, no. 1 (2005): 39–56.

Newman, Karen. "'And Wash the Ethiop White': Femininity and the Monstrous in *Othello*." In *Shakespeare Reproduced: The Text in History and Ideology*, edited by Jean E. Howard and Marion F. O'Connor, 143–62. New York: Methuen, 1987.

Nichols, Ashton. *The Poetics of Epiphany: Nineteenth-Century Origins of the Modern Literary Movement.* Tuscaloosa: University of Alabama Press, 1987.

Nordau, Max Simon. *Paradoxes.* Chicago: Laird and Lee, ca. 1895.

Nordlund, Marcus. "Consilient Literary Interpretation." *Philosophy and Literature* 26, no. 2 (2002): 312–33.

——. *Shakespeare and the Nature of Love: Literature, Culture, Evolution.* Evanston, Ill.: Northwestern University Press, 2007.

Nussbaum, Martha C. "The Oedipus Rex and the Ancient Unconscious." In *Freud and Forbidden Knowledge*, edited by Peter L. Rudnytsky and Ellen Handler Spitz, 42–71. New York: New York University Press, 1994.

——. "*Wuthering Heights*: The Romantic Ascent." *Philosophy and Literature* 20 (1996): 362–82.

Oakley, Francis. "Ignorant Armies and Nighttime Clashes." In *What's Happened to the Humanities?* edited by Alvin Kernan, 63–83. Princeton, N.J.: Princeton University Press, 1997.

Oatley, Keith. "Why Fiction May Be Twice as True as Fact: Fiction as Cognitive and Emotional Simulation." *Review of General Psychology* 3 (1999): 101–17.

Ochs, Elinor, and Bambi B. Schieffelin. "Language Acquisition and Socialization: Three Developmental Stories and Their Implications." In *Culture Theory: Essays on Mind, Self, and Emotion*, edited by Richard A. Shweder and Robert A. Levine, 276–320. Cambridge: Cambridge University Press, 1984.

Odgen, Daniel. *Greek Bastardy in the Classical and Hellenic Periods.* Oxford: Oxford Univerity Press, 1996.

Ogden, John T. "The Structure of Imaginative Experience in Wordsworth's *Prelude*." *Wordsworth Circle* 6 (1975): 290–98.

Oliverio, Alberto, and Michelle Zappella, eds. *The Behavior of Human Infants.* New York: Plenum Press, 1983.

Olson, Janet L. "Encouraging Visual Storytelling." In *Creating Meaning Through Art: Teacher as Choice Maker*, edited by Judith W. Simpson, Jean M. Delaney, Karen Lee Carroll, Cheryl M. Hamilton, Sandra L. Kay, Marianne S. Kerlavage, and Janet L. Olson, 163–205. Columbus, Ohio: Prentice Hall, 1997.

Orians, Gordon H. "Habitat Selection: General Theory and Applications to Human Behavior." In *The Evolution of Human Social Behavior*, edited by Joan S. Lockard, 49–66. New York: Elsevier North Holland, 1980.

Otterbein, Keith F. *The Ultimate Coercive Sanction: A Cross-Cultural Study of Capital Punishment*. New Haven, Conn.: HRAF Press, 1987.

Ovid. *"The Art of Love" and Other Poems*. Translated by J. H. Mozley. Loeb Classical Library 97. London: Heinemann, 1947.

Page, Judith W. *Wordsworth and the Cultivation of Women*. Berkeley: University of California Press, 1994.

Panksepp, Jaak. "Emotions as Natural Kinds Within the Mammalian Brain." In *Handbook of Emotions*, edited by Michael Lewis and Jeannette M. Haviland-Jones, 137–56. 2nd ed. New York: Guilford Press, 2000.

Panksepp, Jaak, and Jules B. Panksepp. "The Seven Sins of Evolutionary Psychology." *Evolution and Cognition* 6 (2000): 108–31.

Parker, Seymour. "The Precultural Basis of the Incest Taboo: Toward a Biosocial Theory." *American Anthropologist* 78 (1976): 285–305.

Parmigiani, Stefano, and Frederick S. vom Saal, eds. *Infanticide and Parental Care*. Langhorne, Pa.: Harwood Academic, 1994.

Parrott, W. Gerrod. "Volume Overview." In *Emotions in Social Psychology: Essential Readings*, edited by W. Gerrod Parrot, 1–20. Philadelphia: Psychology Press, 2001.

Patai, Daphne, and Will H. Corral, eds. *Theory's Empire: An Anthology of Dissent*. New York: Columbia University Press, 2005.

Paulhus, Delroy L., and Oliver P. John. "Egoistic and Moralistic Biases in Self-Perception: The Interplay of Self-Deceptive Styles with Basic Traits and Motives." *Journal of Personality* 66 (1998): 1025–60.

Penley, Constance. "Brownian Motion: Women, Tactics, and Technology." In *Technoculture*, edited by Constance Penley and Andrew Ross, 135–61. Minneapolis: University of Minnesota Press, 1991.

Perner, Josef. *Understanding the Representational Mind*. Cambridge, Mass.: MIT Press, 1991.

Persson, Per. *Understanding Cinema: A Psychological Theory of Moving Images*. Cambridge: Cambridge University Press, 2003.

Peters, Pearlie Mae Fisher. *The Assertive Woman in Zora Neale Hurston's Fiction, Folklore, and Drama*. New York: Garland, 1998.

Pfeiffer, John E. *The Creative Explosion: An Inquiry into the Origins of Art and Religion*. New York: Harper & Row, 1982.

Pfitzer, Gregory M. "Thoreau and Mother Nature: 'Ktaadn' as an Oedipal Tale." *American Transcendental Quarterly* 2 (1988): 301–11.

Pilbeam, David R. *The Ascent of Man: An Introduction to Human Evolution*. New York: Macmillan, 1972.

——. "What Makes Us Human?" In *The Cambridge Encyclopedia of Human Evolution*, edited by Stephen Jones, Robert D. Martin, and David R. Pilbeam, 1–5. New York: Cambridge University Press, 1992.

Pinker, Steven. *The Blank Slate: The Modern Denial of Human Nature*. New York: Viking, 2002.

——. Foreword to *The Handbook of Evolutionary Psychology*, edited by David M. Buss, xi–xvi. Hoboken, N.J.: Wiley, 2005.

——. *How the Mind Works*. New York: Norton, 1997.

——. *The Language Instinct*. New York: HarperCollins, 1994.

Plantinga, Carl, and Greg Smith, eds., *Passionate Views: Thinking About Film and Emotion*. Baltimore: Johns Hopkins University Press, 1999.

Plotkin, Henry. *Darwin Machines and the Nature of Knowledge*. Cambridge, Mass.: Harvard University Press, 1994.

Plutchik, Robert. *Emotions and Life: Perspectives from Psychology, Biology, and Evolution*. Washington, D.C.: American Psychological Association, 2003.

——. "The Nature of Emotions." *American Scientist* 89 (2001): 344–50.

Podolefsky, Aaron, and Peter J. Brown, eds. *Applying Cultural Anthropology: An Introductory Reader*. 5th ed. New York: WCB/McGraw, 2000.

Polti, Georges. *Thirty-six Dramatic Situations*. Translated by Lucille Ray. 1917. Reprint, Whitefish, Mont.: Kessinger, 2003.

Pomeroy, Sarah. *Goddesses, Whores, Wives, and Slaves: Women in Classical Antiquity*. New York: Schocken Books, 1975.

Popper, Karl. *Conjectures and Refutations: The Growth of Scientific Knowledge*. London: Routledge, 1963.

——. *Logik der Forschung*. Vienna: Springer, 1935. Translated by Karl Popper as *The Logic of Scientific Discovery*. New York: Basic Books, 1959.

——. *Objective Knowledge: An Evolutionary Approach*. Oxford: Clarendon Press, 1972.

——. "Truth, Rationality, and the Growth of Scientific Knowledge." In *Philosophical Problems of Science and Technology*, edited by Alexandros C. Michalos, 76–117. Boston: Allyn & Bacon, 1974.

Power, Camilla. "'Beauty' Magic: The Origins of Art." In *The Evolution of Culture: A Historical and Scientific Overview*, edited by Robin Dunbar, Chris Knight, and Camilla Power, 92–112. New Brunswick, N.J.: Rutgers University Press, 1999.

Premack, David. "Cause/Induced Motion: Intention/Spontaneous Motion." In *Origins of the Human Brain*, edited by Jean-Pierre Changeaux and Jean Chavaillon, 286–308. Oxford: Clarendon Press, 1995.

Premack, David, and Ann James Premack. "Origins of Human Social Competence." In *The Cognitive Neurosciences*, edited by Michael S. Gazzaniga, 205–18. Cambridge, Mass.: MIT Presss, 1995.

Pudovkin, V. I. *Film Technique*. 1929. Translated by Ivor Montagu. Reprint, New York: Grove Press, 1970.

——. *Film Technique; and Film Acting: The Cinema Writings of V. I. Pudovkin*. Translated by Ivor Montagu. London: Vision Press, 1954.

Pusey, Anne E. "Inbreeding Avoidance in Chimpanzees." *Animal Behaviour* 28 (1980): 543–52.

Pylyshyn, Zenon W. *Seeing and Visualizing: It's Not What You Think*. Cambridge, Mass.: MIT Press, 2006.

Radway, Janice A. *Reading the Romance: Women, Patriarchy, and Popular Literature*. Chapel Hill: University of North Carolina Press, 1984.

Ramachandran, V. S. "Mirror Neurons and Imitation Learning as the Driving Force Behind 'The Great Leap Forward' in Human Evolution." Available at http://www.edge.org/3rd_culture/ramachandran/ramachandran_p1.html (accessed May 2, 2009).

Rasmussen, Knud. *Intellectual Culture of the Copper Eskimos*. Translated by W. E. Calvert. Vol. 9 of *Report of the Fifth Thule Expedition, 1921–1924*. Copenhagen: Gyldendalske Boghandel, Nordisk Forlag, 1932.

Reisz, Karel, and Gavin Millar. *The Technique of Film Editing*. 2nd ed. London: Focal Press, 1968.

Reynolds, Vernon. *The Biology of Human Action*. San Francisco: Freeman, 1976.

Richardson, Alan. "Rethinking Romantic Incest: Human Universals, Literary Representation, and the Biology of Mind." *New Literary History* 31, no. 3 (2000): 553–72.

——. "Romanticism and the Colonization of the Feminine." In *Romanticism and Feminism*, edited by Anne K. Mellor, 13–25. Bloomington: Indiana University Press, 1988.

Richerson, Peter J., and Robert Boyd. *Not by Genes Alone: How Culture Transformed Human Evolution*. Chicago: University of Chicago Press, 2005.

Ricks, Christopher. "The Pursuit of Metaphor." In *What's Happened to the Humanities?* edited by Alvin Kernan, 179–200. Princeton, N.J.: Princeton University Press, 1997.

Ridley, Mark. *The Cooperative Gene: How Mendel's Demon Explains the Evolution of Complex Beings*. New York: Free Press, 2001.

Ridley, Matt. *Nature via Nurture: Genes, Experience, and What Makes Us Human*. New York: HarperCollins, 2003.

——. *The Origins of Virtue: Human Instincts and the Evolution of Cooperation*. London: Viking, 1996.

Rizzolatti, Giacomo, and Leonardo Fogassi. "Mirror Neurons and Social Cognition." In *The Oxford Handbook of Evolutionary Psychology*, edited by Robin Dunbar and Louise Barrett, 179–95. Oxford: Oxford University Press, 2007.

Root-Bernstein, Robert. "The Sciences and Arts Share a Common Creative Aesthetic." In *The Elusive Synthesis: Aesthetics and Science*, edited by Alfred I. Tauber, 49–82. Dordrecht: Kluwer, 1996.

Ross, Marlon. *The Contours of Masculine Desire: Romanticism and the Rise of Women's Poetry*. Oxford: Oxford University Press, 1989.

Rossiter, A. P. *Angel with Horns: Fifteen Lectures on Shakespeare*. Edited by Graham Storey. London: Longman, 1961.

Rostand, Edmond. *Cyrano de Bergerac*. Translated by Anthony Burgess. London: Nick Hern Books, 1993.

Rozin, Paul. "Towards a Psychology of Food and Eating: From Motivation to Module to Model to Marker, Morality, Meaning, and Metaphor." *Current Directions in Psychological Science* 5 (1996): 18–24.

Rubinstein, Annette T. "Bourgeois Equality in Shakespeare." *Science and Society* 41 (1977): 25–35.

Rudnytsky, Peter L. "Freud and Augustine." In *Freud and Forbidden Knowledge*, edited by Peter L. Rudnytsky and Ellen Handler Spitz, 128–52. New York: New York University Press, 1994.

Russ, Joanna. *Magic Mommas, Trembling Sisters, Puritans and Perverts: Feminist Essays.* Trumansburg, N.Y.: Crossing Press, 1985.

Sacks, Oliver. *An Anthropologist on Mars: Seven Paradoxical Tales.* London: Picador, 1995.

——. *The Man Who Mistook His Wife for a Hat and Other Clinical Tales.* London: Duckworth, 1985.

Sade, Donald Stone. "Inhibition of Son–Mother Mating Among Free-Ranging Rhesus Monkeys." *Science and Psychoanalysis* 12 (1968): 18–38.

Sagan, Carl. *The Demon-Haunted World: Science as a Candle in the Dark.* New York: Ballantine Books, 1997.

Salmon, Catherine, and Todd K. Shackelford, eds. *Family Relationships: An Evolutionary Perspective.* Oxford: Oxford University Press, 2008.

Salmon, Catherine, and Donald Symons. "Slash Fiction and Human Mating Psychology." *Journal of Sex Research* 41 (2004): 94–100.

——. *Warrior Lovers: Erotic Fiction, Evolution and Human Sexuality.* London: Weidenfeld & Nicolson, 2001.

Salt, Barry. *Film Style and Technology: History and Analysis.* London: Starword, 1983.

Salter, Frank K. *Emotions in Command: Biology, Bureaucracy, and Cultural Evolution.* New Brunswick, N.J.: Transaction, 2008.

Sandars, N. K. *Prehistoric Art in Europe.* London: Penguin, 1985.

Saunders, Judith. *Edith Wharton Through a Darwinian Lens: Evolutionary Biological Issues in Her Fiction.* Jefferson, N.C.: MacFarland, 2009.

Scalise Sugiyama, Michelle. "Cultural Variation Is Part of Human Nature: Literary Universals, Context-Sensitivity, and 'Shakespeare in the Bush.'" *Human Nature* 14 (2003): 383–96.

——. "Food, Foragers, and Folklore: The Role of Narrative in Human Subsistence." *Evolution and Human Behavior* 22 (2001): 221–40.

——. "Lions and Tigers and Bears: Predators as a Folklore Universal." In *Heuristiken Der Literaturwissenschaft: Disziplinexterne Perspectiven auf Literatur*, edited by Uta Klein, Katja Mellmann, and Stephanie Metzger, 319–31. Paderborn: Mentis, 2006.

——. "Narrative Theory and Function: Why Evolution Matters." *Philosophy and Literature* 25 (2001): 233–50.

——.. "New Science, Old Myth: An Evolutionary Critique of the Oedipal Paradigm." *Mosaic* 34, no. 1 (2001): 121–36.

——. "On the Origins of Narrative: Storyteller Bias as a Fitness-Enhancing Strategy." *Human Nature* 7 (1996): 403–25.

——. "Reverse-Engineering Narrative: Evidence of Special Design." In *The Literary Animal: Evolution and the Nature of Narrative*, edited by Jonathan Gottschall and David Sloan Wilson, 177–96. Evanston, Ill.: Northwestern University Press, 2005.

——. "What's Love Got to Do with It? An Evolutionary Analysis of 'The Short Happy Life of Francis Macomber.'" *Hemingway Review* 15, no. 2 (1996): 15–32.

Schank, Roger C. *Dynamic Memory: A Theory of Reminding and Learning in Computers and People*. New York: Cambridge University Press, 1982.

Schapiro, Barbara Ann. *Literature and the Relational Self*. New York: New York University Press, 1994.

——. *The Romantic Mother: Narcissistic Patterns in Romantic Poetry*. Baltimore: Johns Hopkins University Press, 1983.

Schlaug, Gottfried, Lutz Jäncke, Yanxiong Huang, Jochen F. Staiger, and Helmuth Steinmetz. "Increased Corpus Callosum Size in Musicians." *Neuropsychologia* 33 (1995): 1047–55.

Schlaug, Gottfried, Lutz Jäncke, Yanxiong Huang, and Helmuth Steinmetz. "In Vivo Evidence of Structural Brain Asymmetry in Musicians." *Science* 267 (1995): 699–71.

Schmitt, David P. "Fundamentals of Human Mating Strategies." In *The Handbook of Evolutionary Psychology*, edited by David M. Buss, 258–91. Hoboken, N.J.: Wiley, 2005.

Scholes, Robert. *The Rise and Fall of English: Reconstructing English as a Discipline*. New Haven, Conn.: Yale University Press, 1998.

——. "Whither, or Wither, the Humanities?" *Profession* (2005): 7–9.

Schore, Allan N. *Affect Regulation and the Origin of the Self: The Neurobiology of Emotional Development*. Hillsdale N.J.: Erlbaum, 1994.

Selby, Henry A. *Zapotec Deviance: The Convergence of Folk and Modern Sociology*. Austin: University of Texas Press, 1974.

Shakespeare, William. *The Arden Shakespeare: Complete Works*. Edited by Richard Proudfoot, Ann Thompson, and David Scott Kastan. Walton-on-Thames: Thomas Nelson, 1998.

Shapiro, James A. "Adaptive Mutation: Who's Really in the Garden?" *Science* 268 (1995): 373–74.

Sharpe, J. A. *Early Modern England: A Social History, 1550–1760*. London: Edward Arnold, 1987.

Shaw, George Bernard. Preface to *On the Rocks*. London: Constable, 1934.

Shepher, Joseph. "Mate Selection Among Second Generation Kibbutz Adolescents and Adults: Incest Avoidance and Negative Imprinting." *Archives of Sexual Behavior* 1 (1971): 293–307.

Shostak, Marjorie. *Nisa: The Life and Words of a !Kung Woman*. New York: Vintage, 1983.

Sidanius, Jim, and Felicia Pratto. *Social Dominance: An Intergroup Theory of Social Hierarchy and Oppression*. Cambridge: Cambridge University Press, 1999.

Simpson, George Gaylord. *Tempo and Mode in Evolution*. New York: Columbia University Press, 1944.

Sinfield, Alan. *Faultlines: Cultural Materialism and the Politics of Dissident Reading*. Oxford: Clarendon Press, 1992.

Singer, Ben. "A Taxonomy of Pathos." Paper presented at the conference "Narration, Imagination, and Emotion in the Moving Image Media," Center for the Cognitive Study of the Moving Image, Calvin College, Grand Rapids, Mich., July 22, 2004.

Singh, Devandra, and P. Matthew Bronstad. "Sex Differences in the Anatomical Locations of Human Body Scarification and Tattooing as a Function of Pathogen Prevalence." *Evolution and Human Behavior* 18 (1997): 403–16.

Singh, Devandra, Peter Renn, and Adrian Singh. "Did the Perils of Abdominal Obesity Affect Depiction of Feminine Beauty in the Sixteenth to Eighteenth Century British Literature? Exploring the Health and Beauty Link." *Proceedings of the Royal Society: Biological Sciences* 274 (2007): 891–94.

Slingerland, Edward. *What Science Offers the Humanities: Integrating Body and Culture*. Cambridge: Cambridge University Press, 2008.

Small, Meredith F. "The Evolution of Female Sexuality and Mate Selection in Humans." *Human Nature* 3 (1992): 133–56.

Smith, Adam. *The Theory of Moral Sentiments*. 1759. Reprint, Kila, Mont.: Kessinger, 2004.

Smith, Bruce R. *Homosexual Desire in Shakespeare's England: A Cultural Poetics*. Chicago: University of Chicago Press, 1991.

——. *Shakespeare and Masculinity*. Oxford: Oxford University Press, 2000.

Smith, Greg M. *Film Structure and the Emotion System*. Cambridge: Cambridge University Press, 2003.

Smith, Murray. "Darwin and the Directors." *Times Literary Supplement*, February 7, 2003, 13–15.

——. *Engaging Characters: Fiction, Emotion, and the Cinema*. Oxford: Oxford University Press, 1995.

Smits, Dirk J. M., and P. D. Boeck. "From BIS/BAS to the Big Five." *European Journal of Personality* 20 (2006): 255–70.

Smolin, Lee. *The Trouble with Physics: The Rise of String Theory, the Fall of a Science, and What Comes Next*. New York: Houghton Mifflin, 2006.

Smuts, Barbara B. "Apes of Wrath." *Discover*, August 1995, 103–5.

——. "The Evolutionary Origins of Patriarchy." *Human Nature* 6 (1995): 1–32.

——. "Male Aggression Against Women: An Evolutionary Perspective." In *Sex, Power, Conflict: Evolutionary and Feminist Perspectives*, edited by David M. Buss and Neil M. Malamuth, 231–68. New York: Oxford University Press, 1996.

——. *Sex and Friendship in Baboons*. New York: Aldine de Gruyter, 1985.

Smuts, Barbara B., Dorothy L. Cheney, Robert M. Seyfarth, Richard W. Wrangham, and Thomas T. Struhsaker, eds. *Primate Societies*. Chicago: University of Chicago Press, 1987.

Snow, C. P. *The Two Cultures*. Introduction by Stefan Collini. 1959. Reprint, Cambridge: Cambridge University Press, 1993.

Sober, Elliott, and David Sloan Wilson. *Unto Others: The Evolution and Psychology of Unselfish Behavior*. Cambridge, Mass.: Harvard University Press, 1998.

Sokal, Alan D. "Transgressing the Boundaries: Toward a Transformative Hermeneutics of Quantum Gravity." *Social Text* 14 (1996): 217–52.

Sokal, Alan D., and Jean Bricmont. *Fashionable Nonsense: Postmodern Intellectuals' Abuse of Science*. New York: Picador, 1998.

——. *Intellectual Impostures*. London: Profile Books, 1998.

Sompayrac, Lauren M. *How the Immune System Works*. Malden, Mass.: Blackwell Science, 1999.

Spacks, Patricia Meyer. *The Female Imagination*. New York: Knopf, 1975.

Spelke, Elizabeth S. "Perceptual Knowledge of Objects in Infancy." In *Perspectives on Mental Representation: Experimental and Theoretical Studies of Cognitive Processes and Capacities*, edited by Jacques Mehler, Edward C. T. Walker, and Merrill Garrett, 409–30. Hillsdale, N.J.: Erlbaum, 1982.

Sperber, Dan. *Explaining Culture: A Naturalistic Approach*. Oxford: Blackwell, 1996.

Sperber, Dan, and Deirdre Wilson. *Relevance: Communication and Cognition*. Cambridge, Mass.: Harvard University Press, 1986.

Spiegelman, Art. *Maus: A Survivor's Tale*. New York: Pantheon, 1986.

——. *Maus II: A Survivor's Tale: And Here My Troubles Began*. New York: Pantheon, 1991.

Spiegelman, Art, and R. Sikoryak, eds. *The Narrative Corpse: A Chain Story by 69 Artists!* New York: Gates of Heck, 1995.

Staiger, Janet. *Perverse Spectators: The Practices of Film Reception*. New York: New York University Press, 2000.

Stanton, Domna C., Michael Bérubé, Leonard Cassuto, Morris Eaves, John Guillory, Donald E. Hall, and Sean Latham. "Report of the MLA Task Force on Evaluating Scholarship for Tenure and Promotion." 2007. Available at http://www.mla.org/tenure_promotion (accessed April 21, 2008).

Steen, Francis F., and Stephanie A. Owens. "Evolution's Pedagogy: An Adaptationist Model of Pretense and Entertainment." *Journal of Cognition and Culture* 1 (2001): 289–321.

Steiner, George. "The High Noon of the Arts, Music, and Possibly Literature Lies Behind Us in the West." *Chronicle of Higher Education*, June 21, 1996, B6.

Sterelny, Kim. *Thought in a Hostile World: The Evolution of Human Cognition*. Oxford: Oxford University Press, 2003.

Stern, Daniel N. *The First Relationship: Infant and Mother*. Cambridge, Mass.: Harvard University Press, 1977.

——. *The Interpersonal World of the Infant: A View from Psychoanalysis and Developmental Psychology*. New York: Basic Books, 1985.

Sternberg, Robert J. *Cupid's Arrow: The Course of Love Through Time*. Cambridge: Cambridge University Press, 1998.

——. "Triangulating Love." In *The Psychology of Love*, edited by Robert J. Sternberg and Michael L. Barnes, 119–38. New Haven, Conn.: Yale University Press, 1988.

Stiller, James, Daniel Nettle, and Robin Dunbar. "The Small World of Shakespeare's Plays." *Human Nature* 14 (2003): 397–408.

Stoneman, Patsy. "The Brontë Myth." In *The Cambridge Companion to the Brontës*, edited by Heather Glen, 214–41. Cambridge: Cambridge University Press, 2002.

——, ed. *Emily Brontë: "Wuthering Heights."* Cambridge: Icon, 2000.

Storey, Robert. "'I Am Because My Little Dog Knows Me': Prolegomenon to a Theory of Mimesis." In *After Poststructuralism: Interdisciplinarity and Literary Theory*, edited by Nancy Easterlin and Barbara Riebling, 45–70. Evanston, Ill.: Northwestern University Press, 1993.

———. *Mimesis and the Human Animal: On the Biogenetic Foundations of Literary Representation*. Evanston, Ill.: Northwestern University Press, 1996.

Sugiyama, Lawrence. "Physical Attractiveness in Adaptationist Perspective." In *The Handbook of Evolutionary Psychology*, edited by David M. Buss, 292–343. Hoboken, N.J.: Wiley, 2005.

Sulloway, Frank J. *Born to Rebel: Family Conflict and Radical Genius*. New York: Pantheon, 1996.

Symons, Donald. *The Evolution of Human Sexuality*. New York: Oxford University Press, 1979.

———. *Play and Aggression: A Study of Rhesus Monkeys*. New York: Columbia University Press, 1978.

Symons, Donald, Catherine Salmon, and Bruce Ellis. "Unobtrusive Measures of Human Sexuality." In *Human Nature: A Critical Reader*, edited by Laura Betzig, 209–12. New York: Oxford University Press, 1997.

Tallis, Raymond. "A Dark Mirror: Reflections on Dementia." *News from the Republic of Letters* 2 (1997): 12–16.

———. *Enemies of Hope: A Critique of Contemporary Pessimism*. Basingstoke: Macmillan, 1997.

———. *Not Saussure: A Critique of Post-Saussurean Literary Theory*. 2nd ed. London: Macmillan, 1995.

Talmon, Yonina. "Mate Selection in/on Collective Settlements." *American Sociological Review* 29 (1964): 491–508.

Tan, Ed S. *Emotion and the Structure of Narrative Film: Film as an Emotion Machine*. Translated by Barbara Fasting. Mahwah, N.J.: Erlbaum, 1996.

Tavinor, Grant. "Video Games and Interactive Fiction." *Philosophy and Literature* 29 (2005): 24–40.

Taylor, Timothy L. *The Prehistory of Sex: Four Million Years of Human Sexual Culture*. London: Fourth Estate, 1996.

Thompson, Kristin. "The Continuity System." In *The Classical Hollywood Cinema: Film Style and Mode of Production to 1960*, edited by David Bordwell, Janet Staiger, and Kristin Thompson, 208–10. New York: Columbia Unversity Press, 1985.

Thompson, Stith. *Motif-Index of Folk Literature*. 6 vols. Bloomington: Indiana University Press, 1932.

Thorn, Matt. "What Japanese Girls Do with Manga and Why." Paper presented at the Japan Anthropology Workshop, University of Melbourne, Melbourne, Australia, July 10, 1997.

Tiger, Lionel. *The Pursuit of Pleasure*. Boston: Little, Brown, 1992.

Tiger, Lionel, and Robin Fox. *The Imperial Animal*. New York: Holt, Rinehart and Winston, 1971.

Tinbergen, Nikolaas. *The Study of Instinct*. Oxford: Clarendon Press, 1951.

Tomasello, Michael, Malinda Carpenter, Josep Call, Tanya Behne, and Henrike Moll. "Understanding and Sharing Intentions: The Origins of Cultural Cognition." *Behavioral and Brain Sciences* 28 (2005): 675–735.

Tomasulo, Frank P. "Narrate and Describe? Point of View and Narrative Voice in *Citizen Kane*'s Thatcher Sequence." *Wide Angle* 8 (1986): 45–62.

Tooby, John, and Leda Cosmides. "Cognitive Adaptations for Social Exchange." In *The Adapted Mind: Evolutionary Psychology and the Generation of Culture*, edited by Jerome H. Barkow, Leda Cosmides, and John Tooby, 163–228. New York: Oxford University Press, 1992.

——. "Conceptual Foundations of Evolutionary Psychology." In *The Handbook of Evolutionary Psychology*, edited by David M. Buss, 5–67. Hoboken, N.J.: Wiley, 2005.

——. "Does Beauty Build Adapted Minds? Toward an Evolutionary Theory of Aesthetics, Fiction, and the Arts." *SubStance* 30, nos. 1–2 (2001): 6–27.

——. "The Past Explains the Present: Emotional Adaptations and the Structure of Ancestral Environments." *Ethology and Sociobiology* 11 (1990): 375–424.

——. "The Psychological Foundations of Culture." In *The Adapted Mind: Evolutionary Psychology and the Generation of Culture*, edited by Jerome H. Barkow, Leda Cosmides, and John Tooby, 19–136. New York: Oxford University Press, 1992.

Traversi, Derek. "*Wuthering Heights* After a Hundred Years." *Dublin Review* 202 (1949): 154–68.

Trevarthen, Colwyn. "Communication and Cooperation in Early Infancy: A Description of Primary Intersubjectivity." In *Before Speech: The Beginning of Human Communication*, edited by Margaret Bullowa, 321–47. Cambridge: Cambridge University Press, 1979.

——. "The Concept and Foundations of Infant Intersubjectivity." In *Intersubjective Communication and Emotion in Early Ontogeny*, edited by Stein Bråten, 15–46. Cambridge: Cambridge University Press, 1998.

——. "Interpersonal Abilities of Infants as Generators for Transmission of Language and Culture." In *The Behavior of Human Infants*, edited by Alberto Liverio and Michele Zappella, 145–76. New York: Plenum Press, 1983.

Trigg, Roger. *Reality at Risk: A Defence of Realism in Philosophy and the Sciences.* 2nd ed. New York: Harvester Wheatsheaf, 1989.

Trivers, Robert. "The Evolution of Reciprocal Altruism." *Quarterly Review of Biology* 46 (1971): 35–57.

——. "Parent–Offspring Conflict." *American Zoologist* 14 (1974): 249–64.

——. "Parental Investment and Reproductive Success." In *Natural Selection and Social Theory: Selected Papers of Robert Trivers*, 56–106. New York: Oxford University Press, 2002.

——. "Parental Investment and Sexual Selection." In *Sexual Selection and the Descent of Man*, edited by Bernard Campbell, 136–79. Chicago: Aldine de Gruyter, 1972.

——. "Self-Deception in Service of Deceit." In *Natural Selection and Social Theory: Selected Papers of Robert Trivers*, 255–93. New York: Oxford University Press, 2002.

Turner, Frederick. *Beauty: The Value of Values.* Charlottesville: University Press of Virginia, 1991.

——. "The Neural Lyre: Poetic Meter, the Brain, and Time." In *Natural Classicism: Essays on Literature and Science*. New York: Paragon House, 1985.

Turner, Mark. *The Literary Mind: The Origins of Thought and Language*. New York: Oxford University Press, 1996.

——. *Reading Minds: The Study of English in the Age of Cognitive Science*. Princeton, N.J.: Princeton University Press, 1991.

Valero, Helena. *Yanoáma: The Story of a Woman Abducted by Brazilian Indians, as Told to Ettore Biocca*. Translated by Dennis Rhodes. London: George Allen & Unwin, 1969.

Van den Berghe, Pierre L. "Human Inbreeding Avoidance: Culture in Nature." *Behavioral and Brain Sciences* 6 (1983): 91–123.

Vandermassen, Griet. *Who's Afraid of Charles Darwin? Debating Feminism and Evolutionary Theory*. Lanham, Md.: Rowman & Littlefield, 2005.

Van Ghent, Dorothy. *The English Novel: Form and Function*. 1953. Reprint, New York: Harper Torchbooks, 1961.

Van Peer, Willie, Jan Hakemulder, and Sonia Zyngier. *Muses and Measures: Empirical Research Methods for the Humanities*. Cambridge: Cambridge Scholars, 2007.

Vaughan, Virginia Mason. *"Othello": A Contextual History*. 1994. Reprint, Cambridge: Cambridge University Press, 2002.

Veblen, Thorstein. *The Theory of the Leisure Class*. 1899. Reprint, New York: Penguin, 1994.

Vernant, Jean-Pierre. "Oedipus Without the Complex." In *Myth and Tragedy in Ancient Greece*, edited by Jean-Pierre Vernant and Pierre Vidal-Naquet; translated by Janet Lloyd, 85–111. New York: Zone, 1988.

Vickers, Brian. *Appropriating Shakespeare: Contemporary Critical Quarrels*. New Haven, Conn.: Yale University Press, 1993.

Wachterhauser, Brice R., ed. *Hermeneutics and Truth*. Evanston, Ill.: Northwestern University Press, 1994.

Wade, Nicholas. *Before the Dawn: Recovering the Lost History of Our Ancestors*. New York: Penguin, 2006.

——. "Dainty Worm Tells Secrets of the Human Genetic Code." *New York Times*, June 24, 1997, B9.

Wagner, Sheldon, Ellen Winner, Dante Cicchetti, and Howard Gardner. "'Metaphorical' Mapping in Human Infants." *Child Development* 52 (1981): 728–31.

Wall, Cheryl A. Introduction to *Sweat / Zora Neale Hurston*, 3–19. New Brunswick, N.J.: Rutgers University Press, 1997.

Webb, Eugene J., Donald T. Campbell, Richard D. Schwartz, and Lee Sechrest. *Unobtrusive Measures: Nonreactive Research in the Social Sciences*. Chicago: Rand McNally, 1966.

Webster, Richard. *Why Freud Was Wrong: Sin, Science and Psychoanalysis*. London: HarperCollins, 1996.

Weiner, Jonathan. *The Beak of the Finch: A Story of Evolution in Our Time*. New York: Vintage, 1994.

Weisbuch, Robert. "Six Proposals to Revive the Humanities." *Chronicle of Higher Education*, March 26, 1999, 64–65.

Wellman, Henry W., and Susan A. Gelman. "Knowledge Acquisition in Foundational Domains." In *Handbook of Child Psychology*. Vol. 2, *Cognition, Perception, and Language*, edited by Deanna Kuhn and Robert S. Siegler, 523–73. New York: Wiley, 1998.

Wells, Robin Headlam. *Shakespeare's Humanism*. Cambridge: Cambridge University Press, 2005.

West-Eberhard, Mary Jane. *Developmental Plasticity and Evolution*. Oxford: Oxford University Press, 2003.

Westermarck, Edward. *The History of Human Marriage*. 3 vols. London: Macmillan, 1891–1926.

Wheatcroft, Geoffrey. "A Prophet Without Honour." *Times Literary Supplement*, October 24, 2003, 13–16.

Whissell, Cynthia. "Mate Selection in Popular Women's Fiction." *Human Nature* 7 (1996): 427–47.

Wiederman, Michael W., and Erica Kendall. "Evolution, Sex, and Jealousy: Investigation with a Sample from Sweden." *Evolution and Human Behavior* 20 (1999): 121–28.

Wilbert, Johannes, and Karin Simoneau, eds. *Folk Literature of the Yanomami Indians*. Los Angeles: UCLA Latin American Center Publications, 1990.

Williams, George C. *Adaptation and Natural Selection: A Critique of Some Current Evolutionary Thought*. Princeton, N.J.: Princeton University Press, 1966.

Williams, Linda Ruth. *Critical Desire: Psychoanalysis and the Literary Subject*. London: Edward Arnold, 1995.

Wilson, David Sloan. *Evolution for Everyone: How Darwin's Theory Can Change the Way We Think About Our Lives*. New York: Delacorte, 2007.

——. "Evolutionary Social Constructionism." In *Literature and the Human Animal: Evolution and the Nature of Narrative*, edited by Jonathan Gottschall and David Sloan Wilson, 20–37. Evanston, Ill.: Northwestern University Press, 2005.

——. Foreword to *Evil Genes: Why Rome Fell, Hitler Rose, Enron Failed, and My Sister Stole My Mother's Boyfriend*, by Barbara Oakley. Amherst, Mass.: Prometheus Books, 2007.

Wilson, David Sloan, Elliott Sober, and commentators. "Reintroducing Group Selection to the Human Behavioral Sciences." *Behavioral and Brain Sciences* 17 (1994): 585–608.

Wilson, David Sloan, and Edward O. Wilson. "Rethinking the Theoretical Foundation of Sociobiology." *Quarterly Review of Biology* 82 (2007): 327–48.

Wilson, Edward O. *Biophilia*. Cambridge, Mass.: Harvard University Press, 1984.

——. *Consilience: The Unity of Knowledge*. New York: Knopf, 1998.

——. *The Insect Societies*. Cambridge, Mass.: Harvard University Press, 1971.

——. *Naturalist*. Washington, D.C.: Island Press, 1994.

——. *On Human Nature*. Cambridge, Mass.: Harvard University Press, 1978.

——. *Sociobiology: The New Synthesis*. Cambridge, Mass.: Harvard University Press, 1975.

——. *Sociobiology: The New Synthesis*. Twenty-fifth Anniversary ed. Cambridge, Mass.: Harvard University Press, 2000.

Wilson, Margo, and Martin Daly. "Life Expectancy, Economic Inequality, Homicide, and Reproductive Timing in Chicago Neighborhoods." *British Medical Journal* 314 (1997): 1271–74.

——. "The Man Who Mistook His Wife for a Chattel." In *The Adapted Mind: Evolutionary Psychology and the Generation of Culture*, edited by Jerome H. Barkow, Leda Cosmides, and John Tooby, 289–322. New York: Oxford University Press, 1992.

Wilson, Timothy D. *Strangers to Ourselves: Discovering the Adaptive Unconscious.* Cambridge, Mass.: Harvard University Press, 2002.

Wilson, Woodrow. "Mere Literature." In *The Origins of Literary Studies in America: A Documentary Anthology*, edited by Gerald Graff and Michael Warner, 82–89. New York: Routledge, 1989.

Winslow, James T., Nick Hastings, C. Sue Carter, Carroll R. Harbaugh, and Thomas R. Insel. "A Role for Central Vasopressin in Pair Bonding in Monogamous Prairie Voles." *Nature* 365 (1993): 544–48.

Wion, Philip K. "The Absent Mother in Emily Brontë's *Wuthering Heights*." *American Imago* 42 (1985): 143–64.

Woit, Peter. *Not Even Wrong: The Failure of String Theory and the Search for Unity in Physical Law.* New York: Basic Books, 2006.

Wolf, Arthur P. "Childhood Association and Sexual Attraction: A Further Test of the Westermarck Hypothesis." *American Anthropologist* 72 (1970): 503–15.

Wolf, Arthur P., and Chieh-shan Huang. *Marriage and Adoption in China, 1845–1945.* Stanford, Calif.: Stanford University Press, 1980.

Wolfe, Tom. *The Painted Word.* New York: Bantam Books, 1975.

——. *The Right Stuff.* New York: Farrar, Straus and Giroux, 1979.

Wolfson, Susan J. "Lyrical Ballads and the Language of (Men) Feeling: Wordsworth Writing Women's Voices." In *Men Writing the Feminine: Literature, Theory, and the Question of Genders*, edited by Thais E. Morgan, 29–58. Albany: State University of New York Press, 1994.

Wolpert, Lewis. *The Triumph of the Embryo.* New York: Oxford University Press, 1992.

Wood, Wendy, and Alice H. Eagly. "A Cross-Cultural Analysis of the Behavior of Women and Men: Implications for the Origins of Sex Differences." *Psychological Bulletin* 128 (2002): 699–727.

Woodring, Carl. *Literature: An Embattled Profession.* New York: Columbia University Press, 1999.

Woolf, Virginia. "Character in Fiction." In *The Essays of Virginia Woolf.* Vol. 3, *1919–1924.* New York: Harvest, 1991.

——. *Mr Bennett and Mrs Brown.* London: Hogarth Press, 1924.

Wordsworth, William. *The Thirteen-Book "Prelude."* Vol. 1. Edited by Mark L. Reed. Ithaca, N.Y.: Cornell University Press, 1991.

Wrangham, Richard, and Dale Peterson. *Demonic Males: Apes and the Origins of Human Violence.* Boston: Houghton Mifflin, 1996.

Wright, Chauncey. "On the Limits of Natural Selection." *North American Review.* October 1870, 295.

Wynne-Edwards, V. C. *Animal Dispersion in Relation to Social Behavior.* Edinburgh: Oliver & Boyd, 1962.

Zahavi, Amotz. "Decorative Patterns and the Evolution of Art." *New Scientist* 19 (1978): 182–84.

Zamyatin, Yevgeny. *A Soviet Heretic.* Translated by Mirra Ginsburg. Chicago: University of Chicago Press, 1970.

——. *We.* Translated by Mirra Ginsburg. New York: Avon, 1972.

Contributors

JOSEPH ANDERSON is Distinguished Professor Emeritus at the University of Central Arkansas and founding director of the Society for Cognitive Studies of the Moving Image. He is the author of *The Reality of Illusion: An Ecological Approach to Cognitive Film Theory* (1996, 1998) and coeditor (with Barbara Fisher Anderson) of *Moving Image Theory: Ecological Considerations* (2005, 2006) and *Narration and Spectatorship in Moving Images* (2007).

DAVID BORDWELL is Jacques Ledoux Professor Emeritus of Film Studies at the University of Wisconsin–Madison. He has written several books on the aesthetics and history of cinema, including *Narration in the Fiction Film* (1985), *On the History of Film Style* (1997), and, most recently, *Poetics of Cinema* (2007).

BRIAN BOYD is University Distinguished Professor of English at the University of Auckland, New Zealand. His work on Vladimir Nabokov (biography, criticism, editions) has been translated into twelve languages. His book *On the Origin of Stories: Evolution, Cognition and Fiction* (2009) proposes art and storytelling as adaptations, explains the evolved cognitive mechanisms underpinning fiction, and shows the implications for reading classic fictions. He has also published many evolutionary literary essays, theoretical and interpretive.

DONALD E. BROWN is professor emeritus of anthropology at the University of California, Santa Barbara. His research focuses on the peoples and cultures of Southeast

Asia, social structure, ethnohistory, human universals, and ethnicity/ethnocentrism. His principal publications include *Brunei: The Structure and History of a Bornean Malay Sultanate* (1970), *Principles of Social Structure: Southeast Asia* (1976), *Hierarchy, History, and Human Nature: The Social Origins of Historical Consciousness* (1988), and *Human Universals* (1991).

DAVID M. BUSS is professor of psychology at the University of Texas, and past president of the Human Behavior and Evolution Society. He is the author of more than 200 scientific publications, as well as a number of books, including *Evolutionary Psychology: The New Science of the Mind* (1999), *The Dangerous Passion: Why Jealousy Is as Necessary as Love and Sex* (2000), and *The Evolution of Desire: Strategies of Human Mating* (2003).

JOSEPH CARROLL is Curators' Professor of English at the University of Missouri, St. Louis. His *Evolution and Literary Theory* (1995) integrates traditional humanist theory with evolutionary psychology and sets both into sharp opposition with poststructuralist theory. The essays collected in *Literary Darwinism: Evolution, Human Nature, and Literature* (2004) takes in new developments in the field and works toward a comprehensive theory of human nature and literature, and he has written many subsequent essays. His annotated edition of Charles Darwin's *On the Origin of Species* (2003) is widely used in classes on the history of science.

BRETT COOKE is professor of Russian at Texas A&M University. In addition to Darwinist studies of opera, Western science fiction, Irish art, ballet, and, naturally, Russian literature, he is the author of *Human Nature in Utopia: Zamyatin's "We"* (2002). He is also coeditor (with Frederick Turner) of *Biopoetics: Evolutionary Explorations in the Arts* (1999) and (with Jan Baptist Bedaux) of *Sociobiology and the Arts* (1999).

CHARLES DARWIN—independent scholar, biologist, and discoverer of the theory of evolution by natural selection—was the author of *On the Origin of Species by Means of Natural Selection* (1859), *The Descent of Man and Selection in Relation to Sex* (1871), *The Expression of Emotions in Animals and Man* (1872), and many other works.

RICHARD DAWKINS retired in 2009 as the first Charles Simonyi Professor of the Public Understanding of Science at Oxford University. He is the author of many books, including *The Selfish Gene* (1976), *The Blind Watchmaker* (1987), *The Ancestor's Tale* (2004), *The God Delusion* (2006), and, most recently, *The Greatest Show on Earth: The Evidence for Evolution* (2009). He is a Fellow of the Royal Society and the Royal Society of Literature. His prizes include the International Cosmos Prize, the Kistler Prize, the Lewis Thomas Prize, the Shakespeare Prize, and the Nierenberg Prize.

ELLEN DISSANAYAKE is an independent scholar, a writer, and a lecturer, whose writings about the arts over thirty-five years synthesize many disciplines and draw on fifteen years of living and working in non-Western countries. The author of three books and numerous scholarly and general articles, she is currently Affiliate Professor in the School of Music at the University of Washington. Her book *Homo Aestheticus: Where Art Comes From and Why* (1992) has been translated into Chinese and Korean.

DENIS DUTTON is professor of philosophy at the University of Canterbury, New Zealand. He edits the Web site "Arts & Letters Daily" and the journal *Philosophy and Literature*, published by Johns Hopkins University Press, and is the author of *The Art Instinct: Beauty, Pleasure, and Human Evolution* (2009).

NANCY EASTERLIN is Research Professor of English at the University of New Orleans. She is coeditor (with Barbara Riebling) of *After Poststructuralism: Interdisciplinarity and Literary Theory* (1993) and author of *Wordsworth and the Question of "Romantic Religion"* (1996) as well as numerous essays on biocultural approaches to literature. She is currently completing a book titled *What Is Literature For? Biocultural Theory and Interpretation*.

WILLIAM FLESCH teaches English, film, and sometimes philosophy at Brandeis University. In addition to *Comeuppance* (2007), he is the author of *Generosity and the Limits of Authority: Shakespeare, Herbert, Milton* (1992), and *The Facts on File Companion to British Poetry: Nineteenth Century* (2009).

JONATHAN GOTTSCHALL is an adjunct professor in the Department of English at Washington & Jefferson College. He is the author of *The Rape of Troy: Evolution, Violence, and the World of Homer* (2008) and *Literature, Science, and a New Humanities* (2008), and coeditor (with David Sloan Wilson) of *The Literary Animal: Evolution and the Nature of Narrative* (2005).

JOHN JOHNSON is professor of psychology at Pennsylvania State University, DuBois. He has published many articles on the personality and evolutionary psychology of moral and educational development, career choice, and work performance. He is currently coediting a book, *Advanced Methods for Conducting Online Behavioral Research*.

DANIEL KRUGER is an assistant research professor at the University of Michigan, where he is affiliated with the Prevention Research Center in the School of Public Health and the Life Course: Evolutionary and Ontogenetic Dynamics program at the Institute for Social Research. His evolutionary research interests include altruism, cooperation, competition, risk taking, mortality patterns, mating strategies, and applications for social and ecological sustainability. Many of his research projects are grounded in evolutionary life history theory.

GEOFFREY MILLER teaches evolutionary psychology and human sexuality, and does research on mate choice, sexual selection, intelligence, creativity, art, music, personality, psychopathology, consumer behavior, and behavior genetics. His books include *The Mating Mind: How Sexual Choice Shaped the Evolution of Human Nature* (2000), *Mating Intelligence: Sex, Relationships, and the Mind's Reproductive System* (co-edited with Glen Geher, 2008), and *Spent: Sex, Evolution, and Consumer Behavior* (2009). He is associate professor of psychology at University of New Mexico.

DANIEL NETTLE is a reader in the Centre for Behaviour and Evolution at Newcastle University, where he studies the application of evolutionary theory to human behavior in many domains. He is the author of several books, including, most recently, *Personality: What Makes You the Way You Are* (2007). He also spent several years as a professional actor and theater director, and still retains an interest in theater in practice as well as theory.

MARCUS NORDLUND is associate professor of English at the University of Gothenburg, Sweden, and specializes in English Renaissance literature and biocultural literary theory. He has published two books—*The Dark Lantern: A Historical Study of Sight in Shakespeare, Webster, and Middleton* (1999) and *Shakespeare and the Nature of Love: Literature, Culture, Evolution* (2007)—and articles on a variety of literary subjects from Shakespeare to Toni Morrison.

STEVEN PINKER is Harvard College Professor and Johnstone Family Professor of Psychology at Harvard University, and the author of seven books on language, cognition, and human nature, including, most recently, *The Stuff of Thought: Language as a Window into Human Nature* (2007).

CATHERINE SALMON is associate professor of psychology at the University of Redlands. She is coauthor (with Donald Symons) of *Warrior Lovers: Erotic Fiction, Evolution and Human Sexuality* (2001) and coeditor (with Charles Crawford) of *Evolutionary Psychology, Public Policy and Personal Decisions* (2004) and (with Todd Shackelford) of *Family Relationships: An Evolutionary Perspective* (2008). Her research interests include female sexuality and pornography, eating disorders, and birth order and family relationships.

JUDITH P. SAUNDERS is professor of English at Marist College. Her articles take in a wide range of literary figures, including Edgar Allan Poe, Henry David Thoreau, Stephen Crane, Sherwood Anderson, Gertrude Stein, Virginia Woolf, Gwendolyn Brooks, and Elizabeth Bishop. She is the author of a book-length study of the British poet Charles Tomlinson. She has undertaken Darwinian analysis of a variety of literary works, most recently in *Edith Wharton and Evolutionary Biology: Reading Her Fiction Through a Darwinian Lens* (2009).

MICHELLE SCALISE SUGIYAMA is research associate in the Institute of Cognitive and Decision Sciences at the University of Oregon, and founder and director of the Cognitive Cultural Studies Division at the Center for Evolutionary Psychology, University of California, Santa Barbara. Her research focuses on cognitive adaptations for cultural transmission, with an emphasis on narrative and art behavior in foraging societies. She has written numerous articles on the origins of storytelling, the role of folklore in foraging societies, and the cognitive foundations of narrative.

EDWARD SLINGERLAND is associate professor of Asian studies and Canada Research Chair in Chinese Thought and Embodied Cognition at the University of British Columbia. His first book, *Effortless Action: Wu-wei as Conceptual Metaphor and Spiritual Ideal in Early China* (2003), won the American Academy of Religion's award for the Best First Book in the History of Religions. His most recent monograph, *What Science Offers the Humanities: Integrating Body and Culture* (2008), argues for the relevance of the natural sciences to the humanities.

MURRAY SMITH is professor of film studies at the University of Kent, Canterbury, England. He is the author of *Engaging Characters: Fiction, Emotion, and the Cinema* (1995) and *Trainspotting* (2002), and coeditor (with Richard Allen) of *Film Theory and Philosophy* (1997), (with Steve Neale) of *Contemporary Hollywood Cinema* (1998), and (with Thomas Warentberg), of *Thinking Through Cinema: Film as Philosophy* (2006). His research interests include the psychology of film viewing and the place of emotion in film reception, as well as the philosophy of film, music, and art more generally. He is currently working on the implications of evolutionary theory for film culture.

DONALD SYMONS is professor emeritus in the Department of Anthropology at the University of California, Santa Barbara. He studied social play among free-ranging rhesus monkeys (*Play and Aggression*, 1978) and was an early and important contributor to the field of evolutionary psychology (*The Evolution of Human Sexuality*, 1979).

JOHN TOOBY AND LEDA COSMIDES are best known for their work in pioneering the field of evolutionary psychology. They are professors of anthropology (Tooby) and psychology (Cosmides) at the University of California, Santa Barbara, where they codirect the Center for Evolutionary Psychology. Awards for their research include the NIH Director's Pioneer Award, the American Association for the Advancement of Science Prize for Behavioral Science Research, the American Psychological Association's Early Career Award, a National Science Foundation Presidential Young Investigator Award, and J. S. Guggenheim Fellowships. John Tooby has been president of the Human Behavior and Evolution Society.

ROBIN HEADLAM WELLS is Professor Emeritus of English Literature at Roehampton University, London. His most recent book is *Shakespeare's Humanism* (2005). He is currently writing *A Short History of Human Nature*.

DAVID SLOAN WILSON is SUNY Distinguished Professor of Biology and Anthropology at Binghamton University. He applies evolutionary theory to all aspects of humanity in addition to the rest of life, both in his own research and as director of EvoS, a unique campuswide evolutionary studies program. He is known for championing the theory of multilevel selection, which has implications ranging from the origin of life to the nature of religion. His books include *Darwin's Cathedral: Evolution, Religion, and the Nature of Society* (2002) and *Evolution for Everyone: How Darwin's Theory Can Change the Way We Think About Our Lives* (2007). His next book is titled *Evolving the City: An Evolutionist Contemplates Changing the World— One City at a Time*.

EDWARD O. WILSON is University Research Professor, Emeritus, at Harvard University. His twenty-five books include *On Human Nature* (1978) and (with Bert Hölldobler) *The Ants* (1990), both of which won the Pulitzer Prize for General Nonfiction; *Consilience: The Unity of Knowledge* (1998), and, most recently, *The Future of Life* (2002), and (again with Bert Hölldobler) *The Superorganism* (2009).

Index

Numbers in italics refer to pages on which illustrations appear.

Cuvier, Baron Georges, 22

Cypria (ancient epic), 293

Cyrano de Bergerac (Rostand), 170–73

Daly, Martin, 116, 313–14, 336

Dante Alighieri, 136

Darwin, Charles, 10–11, 21, 22–30, 32, 36, 41–54, 56, 75–78, 97, 241, 244, 315, 317; on sexual selection, 158, 166–67, 171, 200, 202, 205

Darwin machine, 453

Dawkins, Richard, 4, 10, 28–29, 55–71, 166, 241, 400

deception, 242, 264, 399; self-deception, 397

deconstruction, 2, 6, 111, 139, 212, 214, 240, 371, 453, 466

demonic males, in Homer, 302–5

Deresiewicz, William, 1

Derrida, Jacques, 139, 199–200, 202, 203, 204, 206, 450, 453, 466

Descartes, René, 259

Descent of Man, The (Darwin), 10, 75–78; on sexual selection, 166–67, 171

despotism, 505

determinism: biological, 4, 112, 268–69, 284, 430–31; genetic, 4, 98, 115–17; linguistic, 220

developmental psychology, 348–59, 379, 435–36

dialectics, 249

Dial M for Murder (film; Hitchcock), 365–66

Diamond, Jared, 242

Dickens, Charles, 218, 361–66, 485

Dieckemann, Mildred, 336

Die Hard (film; McTiernan), 277–78, 331

différance/difference, 199, 204–10, 222, 453

disgust, as basic emotion, 262, 384, 500

Disney, Walt, 441

Dissanayake, Ellen, 12, 144–55, 157; on art as "making special," 168–69

diversity, human, 4, 99, 407

DNA, 10, 28, 57–71, 237–38

"Does Beauty Build Adapted Minds?" (Tooby and Cosmides), 174–83, 471–73

Dollimore, Jonathan, 233–34

domain specificity, 212, 213. *See also* intelligence: general; modularity

dominance, 6, 110, 160, 213, 327, 378, 451, 493–94, 497, 501, 505–6; counter-dominance and, 5, 14, 16, 448, 450, 493, 495, 501, 504–6. *See also* hierarchy; status

Don Juan (Byron), 322

Dostoevsky, Fyodor, 133, 153

drama, 192; appeal of, to evolved mind, 322–23; group size in, 323

dream, 387

Dryden, John, 130, 211, 236

Duchess of Malfi, The (J. Webster), 240

Dunbar, Robin, 160, 318

Dunn, Elizabeth E., 399

Dürer, Albrecht, 209

Durkheim, Emile, 159

Dutton, Denis, 12, 184–93

dystopia, 381

Eagleton, Terry, 234, 236, 244, 379, 462–63

Easterlin, Nancy, 14, 15, 348–59

Eastwood, Clint, 263

ecocriticism, 1

ecological perception, 13, 252–53, 256, 285, 411, 415

ecology, culture and, 14

economics, evolutionary, 4–5

editing, film, 249, 261–62, 276, 418. *See also* montage; shot/reverse-shot editing

EEA. *See* environment of evolutionary adaptedness

egalitarianism, 388, 443, 448, 493, 505–6

Eibl-Eibesfeldt, Irenäus, 389

Einstein, Albert, 6, 10, 122, 202

Eisenstein, Sergei, 249, 261

Eisner, Will, 437

Ekman, Paul, 262–64, 444, 500

Eliot, T. S., 127

embryology, 22, 47–48, 52, 69–71, 75–76

emotion(s), 2, 110, 221; in artistic response, 436, 440; attachment theory and, 355–57; basic, 2, 262, 500; commitment and, 280; emotional contagion and, 264–65; emotional intelligence and, 281; evolutionary interest in, 259–60; as evolved, 198;

emotion(s) (*continued*)

facial expression and, 148–49, 260–69; fiction and, 176, 181–82, 191, 491; in film, 13, 260–69, 280–82; genre and, 216; in human nature, 491; memory and, 282, 436; motivation and, 260, 385; in nineteenth-century British fiction, 500–503, *502*; perception and, 275; psychology of, 108; recognition of, 273; in response to characters, 500–502; social and moral, 2, 77, 108, 501

empathy, 2, 94–95, 109, 265

English Patient, The (film; Mingella), 260

Enlightenment, 466

environment: cultural variation and, 484; genes and, 4

environment of evolutionary adaptedness (EEA), 472, 476–77

Ephron, Nora, 278

Epic of Gilgamesh, 192, 442

epigenetics, 101–2

episteme, 450, 452

epistemology, evolutionary, 200–202, 203

Erickson, Mark, 311

erotica, 469–70, 472–82

essentialism, 226, 231, 232–34, 334, 449, 461, 462

ethnicity, in criticism, 272

ethology, 30–32

Eumenides (Aeschylus), 133

Evans, Dylan, 260

event, 500

evidence: confirming, 15, ignoring counterevidence and, 339; negative, 468

evolution, 256–57, 472; art and, 125–93, 437–40; brain size and, 80–81, *81*; change and, 5, 21–22, 111–22; early theories of, 21–22, 205; film and, 253–56; founder effects and, 26–27; genetic bottlenecks and, 27; genetic drift and, 26; literature and, 307, 367–68, 437–40, 449, 468; major transitions in, 5; misapprehensions about, 4–5, 7–9; mutation and, 26; theory of, by natural selection; 10, 22–29, 41–54, 75–78; top-down/bottom-up distinction in, 449

evolutionary psychology, 4–5, 9, 10, 29, 32, 107–10, 111, 112–13, 114, 115, 121, 122, 212,

241, 259–60, 384; behavior depicted in literature and, 393–94; critique of, 119; Darwin on, 53, 76–77, 244; literary criticism and, 237, 368; sociobiology and, 99, 212

explanation, 105–10; of character, 337–39; evolution and, 368; over-explanation, 337; science and, 220, 463–64; versus interpretation, 284

expression, facial. *See* facial expression

Expression of the Emotions in Man and Animals, The (Darwin), 98, 261, 262

Exquisite Corpse. See *Le Corps exquis*

Extended Phenotype, The (Dawkins), 166

eye, direction of: in film, 427; salience of, 440; in shared attention, 435–36

face: recognition of, 484; visual salience of, 440

face-to-face interaction, and film, 425–26

facial expression, 260–69; cultural display rules for, 263–64, 443, 452; emotion and, 500; gender display rules for, 263

fairness, 2, 5, 242, 281, 486

false belief, 210. *See also* Theory of Mind

falsifiability, 467

family, 385

Fatal Attraction (film; Adrian Lyne), 134

father–son conflict, Oedipus complex and, 313–14

fear, 260, 385; as basic emotion, 262; display rules for, in film, 263

females: choices of, in marriage, 297–300; conflict of, with males, 372; as strategic, 289–90

female sexual strategies: in Homer, 294–95; infidelity and, 398; resource extraction and, 396

feminism, 214, 296–97, 451; evolutionary psychology and, 396

feminist criticism, 348–50, 371, 461

fiction: as adaptive, 175–83, 184–93, 259, 470–72, 493, 506; appeal of, to evolved mind, 316–32; definition of, 175; effects of, 385–86; embedded fact in, 188, 485; scenario simulation theory of, 133–34,

Printed in the USA
CPSIA information can be obtained
at www.ICGtesting.com
JSHW011518221024
72172JS00008B/58